Thomas Bates and the Kirklevington Short Horn Cattle
A Contribution to the History of the Pure Durham Cattle

by Cadwaller John Bates

with an introduction by Jackson Chambers

This work contains material that was originally published in 1897.

This publication is within the Public Domain.

*This edition is reprinted for educational purposes
and in accordance with all applicable Federal Laws.*

Introduction Copyright 2017 by Jackson Chambers

Self Reliance Books

Get more historic titles on animal and stock breeding, gardening and old fashioned skills by visiting us at:

http://selfreliancebooks.blogspot.com/

Introduction

I am pleased to present another title in the "Cattle" series.

The work is in the Public Domain and is re-printed here in accordance with Federal Laws.

As with all reprinted books of this age that are intended to perfectly reproduce the original edition, considerable pains and effort had to be undertaken to correct fading and sometimes outright damage to existing proofs of this title. At times, this task is quite monumental, requiring an almost total "rebuilding" of some pages from digital proofs of multiple copies. Despite this, imperfections still sometimes exist in the final proof and may detract from the visual appearance of the text.

I hope you enjoy reading this book as much as I enjoyed making it available to readers again.

Jackson Chambers

THOMAS BATES OF KIRLEVINGTON.

PREFACE

AT the suggestion of two eminent French Shorthorn-breeders, I have undertaken the work of editing the papers and correspondence of my great-uncle, Thomas Bates of Kirklevington.

'Modern History,' says 'The Druid' (H. H. Dixon), 'has been much too sparing in its prose pictures of pastoral life. A great general or statesman has never lacked the love of a biographer; but the thoughts and labours of men who lived remote from cities, and silently built up an improved race of sheep or cattle, whose influence was to be felt in every market, have had no adequate record.' The contempt with which the average *fin-de-siècle* Englishman regards practical country-life is emphasised by the fact that the name of Thomas Bates has been accorded no niche in the much-lauded *Dictionary of National Biography*.

Bates had himself intended to write a History of Shorthorns, and to publish it in America. A fear that this would be interpreted by his enemies as a wish to blazon the merits of his own cattle ultimately deterred him from so doing. The rambling notes which his tenant Mr. Thomas Bell wrote down at his dictation in 1846 represent all the progress he actually made. Revelations, based on irrefutable evidence, which would, he anticipated, startle the whole shorthorn world, and explain much that was enigmatical in his criticisms, have thus been suffered to remain dormant for yet another half century.

Mr. Thomas Bell did, it is true, publish, in 1871, *The History of Improved Shorthorn Cattle from the Notes of the late Thomas Bates*. But from a variety of conflicting causes, for which Mr. Bell was not wholly responsible, this book, though containing an immense mass of valuable material, was so peculiarly arranged as, in the words of the American historian, Mr. Lewis F. Allen, to be neither 'a lucid history of shorthorns nor even of Mr. Bates.'

At first, I thought of attempting nothing more than a new edition of Mr. Bell's work (of which I had acquired the copyright), so convinced was I that a simple re-casting of the contents would turn a sealed book into an interesting narrative. I soon found, however, that much that was most important, and most of what was most piquant in the Kirklevington Papers, had been withheld from publication. Personal reasons for this censorship which were in operation twenty-five years ago have lost much of their force to-day. I have determined to make use of all the materials left at my disposal by the capricious selection of rats and housemaids.

I have been compelled to enter at some length into details of family history. No man was ever more predestined for his life's work than was Thomas Bates, from the circumstances of his birth and the disappointments of his early manhood. His was not a case of a farmer picking up a fat cow and breeding another fat cow from her. He was born with a natural instinct for adapting the methods of genealogy to the improvement of live stock. Although he never forgot the maxims on the value of really pure shorthorns to which he listened as a child, he was for years the most bitter adversary of shorthorns in general, and the zealous advocate of a West Highland cross. Quite contrary to the popular impression, everything with him was made the subject of rigid experiments. Even the block-test was insufficient to decide on the merits of an animal: a final appeal lay to the knife and fork. That he ultimately fell back on the purest lines of shorthorns he could find was the result, not of fancy and prejudice, but of prosaic weights and measures. People ignorant of the scientific methods pursued by a man who at the age of five-and-thirty left his farm to spend the better part of two years in agricultural studies at the University of Edinburgh, could not understand that while he was ever ready to prove by experiments any position he advanced, he enforced the results of his experiments when once obtained as though they were gospel-truth itself. Weak minds, whose only idea was to attain to wealth or celebrity by following fashion blindfold, refused to submit the comparative merits of their cattle to what he regarded as the only rational tests, and writhed in consequence under his unsparing logic.

His disregarded Address to the Board of Agriculture in 1807, on the subject of cattle tests, is still of the greatest practical value. Milk, butter, and even cheese were, in his opinion, of almost equal importance with beef in the selection of a breed of cattle. The outward appearance of an animal was only of consequence as indicating its fattening and milking properties—cows were cows, and not 'Veiled Prophetesses' nor 'Visions of Fair Women.' Their function was not to have cigars smoked over them on Sunday afternoons, but to yield milk in the first place, and then a handsome residue for the butcher.

PREFACE

The hundred guineas Bates paid to the brothers Colling for a cow was the first thing that brought their herds into public prominence. His offer to prepare and print, at his own expense, the pedigrees of the cattle sold at Ketton in 1810, misunderstood and neglected though it was, proved the germ of the British Herd Book. Had his views prevailed, this would not now have as little visible connection with shorthorns as have the opening chapters of the Hebrew 'Chronicles.' Portraits of the best animals, on a fixed scale of measurement, would be inserted, and the entries be supplemented with obligatory milk and block-test records. On the other hand, cattle fourth in descent from yaks and bisons would not be admitted; nor any shorthorns, however well bred, whose merits fell short of an impartial standard. Had attention been paid to his suggestions, practical lessons in breeding would be afforded by the exhibition of animals in family groups, and the honours of the show-ring would be awarded, not to the chance obesity of the individual, pampered at a ruinous expense, but to collective merit, attained by scientific means and at a remunerative cost. The national improvement of shorthorns as paying farm-stock, and not the conversion of his own particular herd into the most money possible, was the unswerving aim of Bates's later life.

I have traced the history of shorthorns from the earliest times, showing that whatever truth there may be in the subsequent use of bulls from Holland, there were shorthorns in the North Riding in the fourteenth century. I have given full credit to the brothers Colling for the preservation of the race. As regards their best tribes, however, I have not claimed for them an improvement which they themselves denied they had effected. I fail to see why the buyers and collectors should be the great men, while Mr. Watson of Manfield, from whom the Red Roses came, is almost forgotten, and Mr. Appleby of Stanwick, the breeder of the first Duchess, is never so much as named. The Collings' premature disposal of Hubback (319), and their immoderate use of Favourite (252), become all the more reprehensible when it turns out that Favourite's family were not originally shorthorns at all. That Bates concealed his knowledge of this, though he always deplored the harm which the crosses by Favourite and Comet (155) had done to the Duchess tribe, was an undoubted weakness. So, too, was the mystery in which he enveloped the origin of his Oxfords and his Waterloos. A breeder is, however, not bound to disclose the secrets of his success. Not to give a pedigree is a very different matter from giving a false one. Bates must have laughed in his sleeve while his calumniators were charging him, most baselessly, with striking Lawn-sleeves (365) out of Belvedere's (1706) ancestry, whereas, had they known it, the really vulnerable point was the possible descent of the Oxfords from Anna-by-Lawn-sleeves.

With regard to the vexed question of the 'alloy,' we must remember that Charles Colling's introduction of the red Galloway blood into his herd was perfectly fortuitous, and that it was the ridiculous puffs of this vaunted improvement that led Bates and others to vindicate so strenuously the merits of pure blood. It is for admirers of alloy cattle to explain why they have never again had recourse to a Galloway cross in their leading herds. The fact that they have not done so has reduced the admixture to so infinitesimal a fraction as practically to terminate the controversy in a purist victory. No shorthorns can, of course, be called 'pure,' except in a relative sense. But to deny that epithet to those that have been

PREFACE

purposely bred as close as possible to Hubback and Favourite, merely on account of some chance strain of mixed blood that could not be avoided, is as childish as it would be to decline calling those boasting a descent from Grandson of Bolingbroke by any other name than that of Galloways. Neither glamour nor stigma now attaches to the word 'alloy' in its technical application. Breeders of pure can admire the good points of alloyed Durhams equally with those of Herefords or Aberdeen-Angus. The history of the other shorthorn hemisphere is most interesting and most instructive, only it does not fall within the scope of these pages.

However strongly Bates objected to the Galloway alloy as then entailing hard handling and lack of milk, constitutional disease was the principal reason for his condemning certain fashionable herds. With his unfortunate experience of the blood of Ben (70) and St. John (572), it is hard to see how he could avoid warning his friends on the subject.

I have told my story as I believe my great-uncle would have wished it to be told. I have told the whole truth unreservedly, and nothing besides. I have abstained from much comment of my own, having the ambitions rather of a chronicler than of a historian. A certain vein of pessimism may seem to run through portions of the tale. Those who, imagining all other persons are like themselves, begin life with too high an estimate of human nature are prone, with the experience of middle age, to lapse into the contrary extreme. It was not Bates's fault but his misfortune, if so many of those who crossed his path were actuated by sordid sentiments and mercenary instincts. As has been well remarked, a considerable sportive element was mingled in his criticisms of man and beast.

PREFACE

I have added a summary of the history of the Kirklevington cattle after their dispersion. The fabulous prices they fetched, thanks to the good hands into which most of them fell, at the very time of their disappearance from the show-ring, adds, whatever we may think of it, a most curious chapter to the annals of agriculture. A sharp reaction necessarily followed; no one could afford to retain stock of such value for the purpose of producing mere beef and milk; every animal, however indifferent, was kept to breed from; and pedigree, instead of being regarded as an explanation of merit, came to be looked upon as merit itself.

In addition to the kind assistance on special points that I have received from various quarters, as is acknowledged in the footnotes, my best thanks are due to Messrs. John Thornton and Co. for permission to make somewhat lengthy extracts from their *Shorthorn Circular*. It is Mr. Thornton's rigid impartiality, ready wit, and honourable dealing that have contributed so much to maintain the popularity of the shorthorn breed. An intimate acquaintance with the past volumes of the *Circular* is the first essential for all practical breeding, while the current numbers need only a little more general support in order fully to meet all up-to-date requirements.

I am indebted to the kindness of Mr. Ernest Clarke for many particulars connected with the early history of the Royal Agricultural Society of England. On some of these, as is perhaps natural, we place, I know, different interpretations. I may have rendered myself liable to the charge of out-heroding Herod on the subject of the exhibition of breeding stock, but in other respects I yield to no one in my admiration of the Society, as continuing loyal to that device of 'Practice with Science,' that formed so long the sheet-anchor of our

PREFACE

industrial supremacy. The Marquis de Chauvelin and Mr. R. H. Allen of New Jersey have saved me from several inaccuracies; Mr. John Wood of Darlington, Mr. J. W. P. Page of Norton, the Rev. F. D. Brock, and Mr. T. Gisborne Fawcett have been of considerable local help; and my old friend, the Rev. A. C. Jennings, and Mr. J. Crawford Hodgson, with his unique knowledge of northern family history, have once again tendered much valued advice and recondite information.

The state of British agriculture, desperate though it be, is probably better than it was during the gloomy period that Bates lived through without ever losing faith in its ultimate recovery. Whatever be the future in store for pure shorthorns, it is hardly yet time to haul down the tricolour of the red, white, and roan, and put all our cattle into deep mourning. Should the shorthorn ultimately go the way of the longhorn, it will be because Bates's views, as expressed in the following pages, have been systematically set at naught.

C. J. B.

LANGLEY CASTLE, NORTHUMBERLAND,
    3rd December 1896.

CONTENTS

CHAPTER I
EARLY YEARS, 1775–1800

Aydon Castle—Descent by the distaff—Bates's father at Buttermere—Thomas Bates of Prudhoe—Aydon White House—John Cook—Dutch cattle from Witton Castle—An agricultural library—The Moores of the Moore—Arthur Blayney—The Stamfordham milch cow—Visit of William Wastell—Haydon Bridge School—Witton-le-Wear—Schoolboy politics—A Blackwell cow by Hubback—Mensuration of cattle—At Aydon White House—Black-fleshed shorthorns—Joyous Moore—Gregynog left to a stranger in blood—Arthur Blayney's character—Wark Eals—Alliance with the Culleys—Darlington Great Markets—Charge's two steers—Lady Maynard—The Ketton ox and the Duchess heifer—Lease of Halton Castle—Arrowsmith's steers and Wastell's heifer . Pages 1-22

CHAPTER II
PRE-COLLINGITE SHORTHORNS, 1400–1800

The Improved Shorthorn, a *Bos compositus*—Mediaeval shorthorns at Haram—Black cattle—Septimius Severus—The Lindisfarne ox—The dun cow at Durham—Black Yorkshires, pied Lincolns, and red Somersets—Welsh and Scotch runts—The red English cow—Holderness dairy cattle—White Dutch cows—Teeswaters—Stanwick, Newby, and Studley Royal—Dutch bulls—Modern cattle in Holland—Michael Dobinson—The Clarewood herd—Sir William St. Quintin—Sir James Pennyman—Milbank of Barningham—The Studley Bull—Horace and Virgil—The Princess family—Stephenson's cow—Snowdon's bull—Hubback—The brothers Colling—Robert Bakewell—Dishley longhorns—Arthur Young's criticism—'Fawcett's little bull'—Appleby's Duchess—The Stanwick herd—Little Hollon—The Hills of Blackwell—Old Sockburn—The new Dun Cow—Daisy—Lady Maynard—Eryholme Necklace—Jolly's Bull—Sale of Hubback—Foljambe—Phœnix—Lord Bolingbroke—Lame Bull—Johanna—Broken Horn, Punch, and Ben—The Collings in-breed accidentally—Favourite—Galloway Alloy—Lady—Lord Bolingbroke in Northumberland—Transformation of the Dutch breed—In-and-in 23-57

CONTENTS

CHAPTER III

HALTON CASTLE, 1800–1809

Halton Castle—The Collings' first hundred-guinea heifer—'Direct from Hubback to Favourite'—Barmpton steers—West Highlanders—Cross-breds—Broadpool Common—Excused being a soldier—The Hexham cavalry—Bonaparte, a puppet—Alarm of French landing—Death of Joyous Moore—Daisy Bull and Styford—Duchess-by-Daisy-Bull—Robert Colling's milch cow—The Grey Bull's handling—Leicester ewe drowned—Use of Duke (224) refused—The best cow in England—Charles Colling's apology—Highland tour—Births of Ketton and Laird—Recourse to St. John—Fortune and Lily—The Agricultural Society of Tindale Ward—Duchess-by-Daisy-Bull as a milk-and-butter cow—Sale of cattle by auction—A Bath show—Address to the Board of Agriculture—The Brindled Ox—Lord Somerville's dinner—Laird—Portrait of the Brindled Ox—The beauties of the Tyne—Newcastle Race Week—Stagshawbank Fair—Madam Diana—Address ignored—Cheese to Cadiz—Curwen of Workington—Fire—Prejudice against mixed breeds—Selection of kyloes—Visit to Baron Hepburn at Smeaton—Workington Meeting—Peat composts—Motive in going to Edinburgh . . . . Pages 58-93

CHAPTER IV

EDINBURGH AND HALTON, 1810–1812

Dr. Coventry's lectures—Temptations of freehold—Tyneside gossip—Newton Hall sale—Walter Blackett Trevelyan—Visit to Lincolnshire—On storing turnips—Purchase of Kirklevington—Rotations in Cleveland—Application of lime—Duchess-by-Daisy-Bull—Donkin of Sandhoe—A blind bargain—Ketton sale—Young Duchess—Donkin's cows wintered—*Survey of Durham*—Churning experiments—Offer to print Ketton pedigrees—Milking properties hereditary—Bailey's want of logic—Duchess's pedigree—Correspondence with Curwen—Gratitude in politics—The White Ox—*Ipse dixit* judgments of shorthorns—Curwen's *Report*—Maynard's use of Laird—A declaration of war—Letters to Campbell of Ardnave—Cross-bred *versus* Highland beef—A fore-chine to Edinburgh—Cross-bred *versus* shorthorn steers—Mackenzie of Applecross—Mason and the butter-pats—Death of Baroness—St. John's blood—Northumberland backbiting 94-134

CHAPTER V

HALTON: AGRICULTURAL DISTRESS, 1813–1820

Evil course of the Tyneside Society—Suggested Scotch matrimonial cross—Prosperity of East Lothian—William Nelson—A last look at Comet—'The Emperor of all bulls'—Fall of prices—Abyssinian cattle—Repeal

CONTENTS

of the duty on cart-horses—Wages—Correspondence with Sir John Graham—Parliament should not interfere—Mackenzie's bailiff—Ross-shire election—No canvass and no expenses—Mr. Fawcett's reminiscences—Families with numbers—Lord Althorp—A new hat from Old Daisy—Acklam Red Rose—Spot and Sparkles purchased—Young Star—Fatuity of breeders—Spot and Sparkles at Halton—Ketton 2nd—Marske—Jonas Whitaker—Elemore—A ruined stock—'The Americans know nothing about Hermit'—A cantankerous uncle-by-marriage—The *Herd Book*—George Coates—His Grace—Visit to Wiseton—'Might become Prime Minister if he would not talk so much'—'The only hope of the shorthorns'—Fracas over Firby—Lord Althorp on a Ministry of All Parties—Hope against hope—Radical—Greenholme Intelligence—Alleged defeats of Ketton 3rd—'Is it pounds or guineas?'—Removal to Ridley Hall—The vicar of Haltwhistle—*The Romish Goliath slain with His Own Weapon*—'Wanted, a schoolmistress possessing genuine piety'—'The modus covers everything'—A judge of mankind   Pages 135-173

CHAPTER VI

RIDLEY HALL, 1821–1830

A second paradise—Imprudent treatment of The Earl—Rev. James Armitage Rhodes—Furioso, Orlando, and Norman Willy dance through the wilderness of fancy and romance—First attempts at shorthorn celebrity—A shorthorn poet—*Origin and Pedigrees of the Sockburn Short-horns*—Libel on Sir Leoline—Kate's black-nosed calf—'Old Ketton's stock, the up-making of me'—*The Herd Book to be all read to Charles Colling, two persons being present, or disputes will be endless*—The black-nosed calf again—Constitutional ardour—Quest of a prize-winner for Whitaker—Enchanter—Disappointments attendant on breeding—Whitaker's tour in Herefordshire—'Disregarding all consequences to myself, I follow wherever Truth leads'—Agricultural politicians—Lord Althorp proposes an import duty and an export bounty concurrently—The apathy of great men—Flight from Westminster—*Herd Book* illustrations—Everything flat and gloomy—Jolly's sale at Worsall—Coates's red-ink criticisms—*The Herd Book merely a copy of pedigrees furnished by subscribers*—No guarantee of purity of blood—Fabrication of pedigrees—A calf with a correct nose—Second Hubback—Successful opposition in the House of Lords—Philip Skipworth in search of fashionable blood—A *History of Shorthorns* needed—Rev. Henry Berry—Death of The Earl—White Rose at Jervaux Abbey—Laird's descendants at Marton-le-Moor—Bearl sale—Lord Althorp's self-sufficiency—An underhand offer—Hollingsworth wants Ridley Hall—The conditional mood—Arbitration without arbitrators—A court of conciliation—Dr. Collingwood Bruce—The Marquis of Bowmont delves at Gilsland—*Fleurs: a Poem*—Christopher North on *The Leg of Mutton School of Poetry*—Spite and prayer—Scandals of episcopal patronage—Chilton sale—St. Alban's—£20,000 lost in breeding shorthorns—Retired life—Intended emigration—The Poor Law—Old Age Pensions—The Yarm Bench—'Not to America but to Kirklevington'   174-217

CONTENTS

CHAPTER VII

KIRKLEVINGTON, 1830–1837

Second Hubback's stock—Philippa 2nd—'Hoodle'—John Brown—Howitt's rebuff at Seaton Delaval—Nunstainton sale—The Matchem cow—John Porritt of Claxton—The Oxford mystery—One out of seventeen—*Omne ignotum pro turpissimo*—Atkinson Greenwell—Second Hubback and Rosanne—Bertram—Gambier—Belvedere—The Wynyard Herd—Sir Henry Vane Tempest's sale—A lady's cow—Young Wynyard—Anna-by-Lawnsleeves—The Countess of Antrim's sale—John Stephenson—Waterloo—An error rectified—The Waterloo Cow—'Cush, Waterloo!' —Middlesbrough sale—Wild Eyes—Hutchinson's calf-house—Blanche—Alleged double cross by Belvedere—Grassy Nook sale—The wrong brother—Lady Barrington, Fletcher and Shorthorns—Nonpareil—Her real origin—Ocular demonstration—Belvedere's steers *versus* Gambier's—Felix Renick—Maynard's *dictum* on shorthorns—A surrender of judgment—Red Rose 11th and Teeswater—Whitaker's invective against Duchess 33rd—The Skipton Bridge Bull—Whitaker's Duke of York—16 guineas turned into £175—A retractation—Rose of Sharon and Duchess 19th—The Matchem cow in Darlington market—Welham sale—The Skipton Bridge Bull again—Comet Halley—Timothy Metcalf—Lady Colling—Norfolk's stock—Duke of Northumberland—Chillicothe sale—'My object has never been to make money of breeding, but to improve the breed of shorthorns'—Decadence of cattle on Tyneside—Wetherell's bull-buying—At home—Dislike of ostentation   Pages 218-259

CHAPTER VIII

KIRKLEVINGTON: THE SHOW-YARDS, 1838–1844

Bates resolves to let the world see what pure shorthorns really are—Golden calves in high places—The York Show, 1838—Shorthorns 4th and Young Matchem—The champion bull Hecatomb—This picture and that—John Grey's judgment—A family group—Lord Spencer pooh-poohs reaping machines—Opportunism of judges—Form nothing without quality—Foggathorpe—Journey of the Duke and his companions—The Oxford Show—Daniel Webster's oration—Oxford Premium Cow—American impressions—The *saga* of John Grey of Dilston—Print of Duke of Northumberland—The Duchess Tribe—Duke of Wellington—The Cambridge Show, 1840—Judges should be blindfolded—Locomotive—2nd Duke of Northumberland—The Alnwick Show, 1840—Old Mr. Jolly dines at Kirklevington—George offered to Mr. Stratton—Cleveland Lad at Liverpool, 1841—Mr. Pusey's toast—Elizabethan cattle—Buffoonery at Berwick—The Hull Show, 1841—A rod in pickle—'The world-renowned twins'—Necklace-by-Priam—Old Brokenleg—When Killerby met Kirklevington—Fatal effects of *bonnes bouches*—Henry

CONTENTS

Colman—Bates intends to publish a Shorthorn History in America—Cleveland Lad—'A hundred men may be found to make a Prime Minister for one fit to judge of the real merits of shorthorns'—Improvements achieved—Ten tribes selected out of two hundred—The six best tribes—Richmond, Durham, and Stockton Shows—Banks Stanhope—Young Parkinson's challenge—Demerits of Cramer's blood—Whitaker's blue-black cow—Forced milking trials—Challenges declined . . . . Pages 260-289

CHAPTER IX

KIRKLEVINGTON: LAST YEARS, 1844-1849

Land and Money—Twenty-five years of agricultural depression—Ready to do his duty—Address to Lords, M.P.'s, and Agricultural Protection Societies—A Government Paper Currency—The establishment of a Sinking Fund—Aggrandisement of Cotton Lords—Egyptian bondage of agriculturists—Distrust of Sir Robert Peel—Agricultural prosperity the only road to universal prosperity—Loans to carry on war, a combination of avarice and bloodthirstiness—Are English farmers to be ruined?—English settlers on the Elbe—Walton at Dumfries, 1845—Predetermination of prizes—Impossibility to obtain solid facts as to milking and feeding properties—Belleville, very fat and very quiet—An unwilling critic—2nd Cleveland Lad at Duncombe Park—Chillingham sale, 1846—3rd Duke of Cambridge with Sir Anthony Buller—'No fault, except his temper'—Beauford's charge of taurine incapacity—The umpire holds that silence gives consent—A very good calf at Keal Hall—Euclid and Lord Hardinge at Yanwath—A sure calf-getter and valuable sire after all—Mr. Glass's last bull—The *Herd Book* deceptive—A true forecast: 'the day when my stock will be duly appreciated'—A family group at Scarborough, 1847—Vail of Troy—The Kirklevington blood wins no less fame in America—Hilpa and Meteor—Ambrose Stevens—The Royal Agricultural Society ignores pedigrees—An ultimatum—The York Show, 1848—The four generations—Arrival of Prince Albert—Dissatisfaction in the bull classes—'Not shorthorns at all'—Ellison of Sizergh protests—'How can pigs unable to stand or walk promote *the breeding and improvement of live stock?*'—Energy and popularity of Prince Albert—An intellectual feast—'The Council of the R.A.S.E. determined to resist any proper investigation into the merits of shorthorns'—'In all probability the best will go abroad and never return till the present generation die out'—American sense and intelligence—The Shipman heifers—Handling has to give way to fat—Torr's admission—Conspiracy to defame Belvedere—Stevens's investigation of his origin—Cock-and-bull stories—Vail and Stevens—American confidence in the superiority of the Duchess tribe—Grant Duff agrees that the falsification of pedigrees should be prevented and prizes be given for blood, not beef—Cowper Hincks appreciates The Duke—Friendly relations with neighbours of position—With hoof and horn to the last—Monument and memorial window—Criticisms, half-sportive, half-sarcastic—Character—Gentle and yielding: staunch and unflinching—A Christian, when religion not so fashionable . 290-328

CONTENTS

CHAPTER X

THE DISPERSION, 1850–1895

Edward Bates at Cloeden-on-the-Elbe—Freiherr von Kleist and Herr von Nathusius—Comparative failure of shorthorns in Germany—Kirklevington sale—No stow-aways and no pricking—Bates no line-breeder himself—A rush to out-crosses—Lord Ducie's experience—The Grand Duchesses—More *mésalliances*—Tortworth sale—A Libyan King of Beasts—Oxford 15th—The Duchesses of Airdrie—Mr. Tanqueray—Hendon sale—Mr. Gunter—' No place for shorthorns like the valley of the Wharfe '—Wetherby Grange—Irony of Mr. Brandreth Gibbs's testimonial—Mr. Thorne's misfortunes—' That brand plucked from the fire '—The Dukes and Duchesses of Thorndale—Mr. Harvey Combe—' The candle prophet '—Cobham sale—The first American bull—Thorndale Rose and The Beauty—Mr. Atherton and Mr. Hegan—Willis's Rooms—' No such test of value as that obtained without shows and judges and prizes in the auction room '—Weak places corrected at the cost of fecundity and constitution—Rheumatic Grand Dukes—' Burghley nods '—Preston Hall sale—Grand Duchess 17th—Captain Oliver's success—Grand Dukes 4th, 16th, and 17th—The ten descendants of Cambridge Rose 6th—Havering sale, 1867—A Scottish reporter dumfounded—Mr. McIntosh succeeded where Sir Charles Knightley failed—Sheldon of Geneva, Illinois—Windsor sale—Candles and champagne—Joe Culshaw's humble pie—' Another bit of Bates '—' The Druid ' at Wetherby and Holker—Era of long prices—Winterfold sale—' Bates and no surrender '—Tea and coffee—Clear Star—Nervous high-bred appearance of the young stock—8th Duke of Geneva ' came out as only Bates bulls can '—' Who says a thousand?'—*Parallel and Parallax*—' John Storer, bold, of Hellidon '—A clerical effusion—' Those *parvenus* the Oxfords '—*Omne ignotum pro turpissimo* again—The wolf and the lamb—Lord Dunmore's reply—Show-ring honours not worth the risk—Plenty of room for both the Bates and the Booth strains—The Great Shorthorn Controversy closed—Importations from America in 1871 and 1872—The New York Mills sale—8th Duchess of Geneva sold for 8120 guineas—' Higher than the cow that jumped over the moon '—Increased demand for Princesses and the minor Kirklevington tribes—The climax in England—Duke of Connaught fetches 4500 guineas—Unprecedented and unsurpassed—Comparative depression in 1876—Foundation of the Brailes herd—Millbeckstock sale—The International Show at Paris—French discrimination—Dispersion of the Dunmore and Gaddesby herds—Audley End and Lathom—Revulsion against line-breeding—*A Review of Shorthorn Prices to the Close of 1880*—Sales by Auction, 1881-1883—Sholebroke Lodge, Hindlip, Rowfant, and Audley End—Sales, 1886-1888—Whittlebury—Sales, 1890-1895—Afghan sagacity—Handling and milk—The duplex qualification—Scottish shorthorns—The true indications of taurility—A unique standpoint—The real and only question for the farmer—Hopes for the future . . . . Pages 329-385

CONTENTS

PEDIGREES

	PAGE
MALE DESCENT OF APPLEBY'S DUCHESS AND WATSON'S RED ROSE	34
THE PRINCESS (LATE BRIGHT EYES) FAMILY TO 1806	35
THE FAVOURITE (or LADY MAYNARD) FAMILY	48
BROKEN HORN, PUNCH, BEN, etc.	53
THE DUCHESS TRIBE TO THE TIME OF THE KIRKLEVINGTON SALE, 1850	*To face* 64
THE RED ROSE AND FOGGATHORPE TRIBES TO THE TIME OF THE KIRKLEVINGTON SALE, 1850	,, 150
THE OXFORD, WATERLOO, AND WILD EYES TRIBES TO THE TIME OF THE KIRKLEVINGTON SALE, 1850	,, 226

GENEALOGICAL DIAGRAMS

HUBBACK	37
FOLJAMBE	51
LAME BULL (357)	52
HERMIT, LAWNSLEEVES, etc.	157
BELVEDERE, WATERLOO, etc.	230
DUKE OF NORTHUMBERLAND, DUCHESS 34th, etc.	253
GRAND DUCHESS 17th, etc.	351

APPENDIX

NOTE A. Alexander Hall's Memoranda	386
B. John Chapman's Relation	388
C. Mr. Maynard's Recollections	389
D. Atkinson Greenwell's Narrative	390
E. Mr. Harrison's Use of Waterloo	393
F. Mr. John Wood's Statement	393
G. George Dobling's Testimony	394
H. Robert Waistell's Tales	396
I. Edward Hall's Allegations	398
K. Handy Smith's Story	399
L. Extracts from the Kirklevington Accounts, 1837-1849	400
M. A Selection of Prices paid for Kirklevington Cattle at Public Auctions	406
GENERAL INDEX	479
SHORTHORN INDEX	497

LIST OF ILLUSTRATIONS

PLATES

		PAGE
PORTRAIT OF THOMAS BATES	*Frontispiece.*	
THE DUN COW, DURHAM, 1895	*To face*	46
BELVEDERE	,,	230
DUCHESSES 42nd AND 43rd	,,	266
DUKE OF NORTHUMBERLAND, 1839	,,	272
DUKE OF NORTHUMBERLAND, 1843	,,	284

VIGNETTES

ST. LUKE'S OX, LINDISFARNE GOSPELS	24
THE DUN COW, DURHAM, *circ.* 1300	25
HOLSTEIN OR DUTCH BULL, 1792	30
HOLSTEIN OR DUTCH COW, 1792	31
THE NEW DUN COW, DURHAM, 1784	45
IMPROVED HOLSTEIN OR DUTCH BULL, 1800	56
IMPROVED HOLSTEIN OR DUTCH COW, 1800	56
BRINDLED OX, HALTON, 1808	80
DUCHESS 1st, 1815	108
WHITE OX, HALTON, 1810	119
COMET, 18th JUNE 1815	138
KETTON, 1814	139
DUCHESS-BY-DAISY-BULL, 1815	147
KETTON 2nd	154

LIST OF ILLUSTRATIONS

	PAGE
PHILIPPA 2nd, 1836	219
WATERLOO	239
DUKE OF NORTHUMBERLAND, AUGUST 1836	254
OXFORD PREMIUM COW	269
DUCHESS 34th, 1836	280
WALTON	296
3rd DUKE OF CAMBRIDGE	299
OXFORD 6th	333
GRAND DUCHESS 2nd	338
OXFORD 15th	341
DUCHESS 66th	344
GRAND DUCHESS 17th	350
LADY OXFORD 5th	354
MAID OF OXFORD 7th	356
DUCHESS 84th	358
DUKE OF GENEVA 8th	361
DUCHESS OF AIRDRIE 8th	365
DUKE OF CONNAUGHT	367

MAPS

TYNESIDE SHORTHORNS	3
THE OLD SHORTHORN COUNTRY	*At end*

THOMAS BATES

AND

THE KIRKLEVINGTON SHORTHORNS

CHAPTER I

EARLY LIFE

1775–1800

ABOUT sixteen miles to the west of Newcastle-upon-Tyne and two above the village of Corbridge stands Aydon Castle, celebrated as one of the most perfect examples that have come down to us of a fortified English manor-house in the days of the Edwards. Built by Sir Robert de Reymes in 1305, besieged and taken by David Bruce on his march to Neville's Cross, Aydon Hall, as it was originally called, occupies a situation of singular beauty. Many of the windows open on a deep dene, long the haunt of badgers, the banks so luxuriantly hung with trees and brushwood that the burn rushing to join the Tyne beneath the site of the old Roman city of CORSTOPITUM is heard more readily than seen. From the battlements, the openings of which still retain the pivot-holes of the swing-shutters that protected the archers after the discharge of their bolts, a charming peep of Hexham, with the tower of the priory church, is obtained in the grey distance. The inner courtyard, with the external stair leading to the Great Hall, has a look strangely Venetian; the walls of the outer bailey, which contains the farm-buildings and garden, are carried on the east along overhanging crags, and rise to a great height near the entrance archway, where the ground lends itself less naturally to purposes of defence.

EARLY LIFE

In the last quarter of the eighteenth century, Aydon Castle was occupied by George Bates, his wife, and their two sons, the younger of whom, Thomas, was destined to preserve and improve the valuable short-horned cattle of the North of England in a way to make his name lastingly famous on both sides of the Atlantic.

> THOMAS BATES, born Feb. 16, 1775 ;[1] father, George Bates ;
> mother, Diana, daughter of Thomas Moore ;
> grandmother, Ann, daughter of Henry Blayney ;
> great-grandmother, Mary, daughter of Lawrence Seddon ;[2]
> great-great-grandmother, Ann, daughter of Richard Blunden ;
> great-great-great-grandmother, Jane, daughter of Richard Otley ;
> great-great-great-great-grandmother, Catherine, daughter of John Mackworth ;
> great-great-great-great-great-grandmother, Elizabeth, daughter of Thomas Hosier.

Such would be the entry relating to the founder of the Kirklevington Herd if the genealogies of the landed gentry were registered on the same lines as those of pure bred shorthorns. Dr. Lawrence Seddon was a Royalist divine, who suffered considerably under the Commonwealth ; the Blundens of Bishop's Castle, the Otleys of Pitchford, and the Mackworths of Mackworth, are well-known families ; the memory of the Hosiers is still kept green in the picturesque alms-houses they founded at Ludlow in the fifteenth century. Probably the importance attached to male descents in our own pedigrees, as a guarantee of the same hereditary characteristics, is often as much misplaced as that given to female descents in the case of cattle. The only rational mode of calculating the

---

[1] This date is definitely established by entries in the Family Bible, his father's account-books, and his aunt Joyous Moore's diary.

[2] Sometimes called Sidney, but the evidence points to his being of the Lancashire family of Seddon of Wavertree.

possible results of atavism is to follow up, like horse-breeders, all the ramifications of a genealogical tree.

In the case of Thomas Bates it is easy to determine the special bent he received from each parent; the marriage of George Bates to Diana Moore, celebrated at St. Chad's, Shrewsbury, on 15th September 1769, was as it were, the espousals of Agriculture and Genealogy.

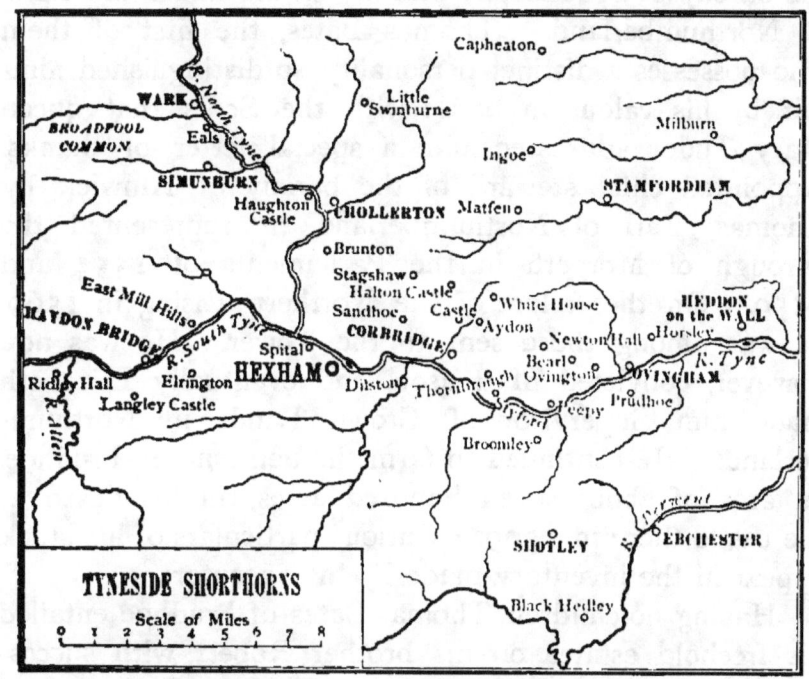

George Bates was devoted heart and soul to farming. His highest ideal of happiness was the Horatian picture of a man owning and occupying a hundred acres, undisturbed by anything passing in the world outside. For those who regarded their estates merely as game preserves or props to their self-importance, and took no active interest in their cultivation, he entertained the bitterest contempt. Originally intended for the legal profession, he had been educated at the Newcastle grammar-school, and while there boarded with the aunts of the great Lord Collingwood. On his refusing absolutely to lead any but a country life, he was sent to a mathematical tutor at

Buttermere. The gibbets on which the Stuart loyalists had been recently executed must have been still standing by the roadside. The boyish surveys which he made in 1750 of the shores of the beautiful Cumbrian lake, as well as of his own paternal acres of Aydon White House, are still in existence.[1]

The Bates family had long held property at Horsley, in the barony of Prudhoe, by military tenure under the Earls of Northumberland. Thomas Bates, the first of them who possesses a distinct personality, so distinguished himself by his valour in battle with the Scots that queen Mary Tudor addressed him a special letter of thanks. Appointed chief steward of the barony of Alnwick by Thomas, Earl of Northumberland, he represented the borough of Morpeth in the Parliaments of 1554 and 1559. On the failure of the Northern Rising in 1569, he was among those sent to the Tower. He was not, however, convicted of treason, and eventually Elizabeth made him Supervisor of Crown Lands in Northumberland. He continued to farm the demesne of Prudhoe, an area of about seven hundred acres, till his death at the castle there in 1599. Curious particulars of his stock appear in the inventory made by his executors.[2]

Having no children, Thomas Bates of Prudhoe entailed his freehold estates on his brother Robert, with successive remainders to Robert's sons, Cuthbert, Thomas, and George; Robert Bates left his own property to his son

---

[1] One of these was of 'a parrock belonging to Mr. Charles Norman of Buttermere, laid down by George Bates 25th May 1750.' On the survey of Aydon White House, laid down at Buttermere on 29th May, are the mottos, only too true:

'One generation passeth away, and another cometh on, but the earth abideth for ever' (*Ecclesiastes* i. 4)

and   Passibus ambiguis Fortuna volubilis errat
                Et manet in nullo certa tenaxque loco.

[2] 'The inventorie of all the goodes, debtes, and cattails of Thomas Baites late of Proudhowe . . . indifferently proved by Roberte Erington, William Erington, and George Baites. . . . In primis eleven oxen priced to £17 : 10s. Item, eighteen kine, £24. Item, ten younge cattall, £6 : 13 : 4. Item, eighte whies, £8. Item, eighteen other younge cattel, £24. Item, one bull, £2, etc. etc.'—Durham Probate Registry.

Thomas, with reversions to Cuthbert and George in turn. By virtue of these singular arrangements, Cuthbert succeeded to Milburn and the other freeholds, while George was so far disinherited, and disappears.

Early in the seventeenth century, on the main stem, George Bates of Horsley married Catherine Surtees, a match of some importance owing to the kinship it ultimately established between the families of Bates and Culley. The marriage of George Bates, grandson of the last, with Catherine Cook in 1699, brought with it the small estate of Aydon White House, situated on rising ground about a mile and a half to the east of Aydon Castle. Thomas Bates of Kirklevington was pleased at the end of his life to relate an anecdote of his ancestor, John Cook the younger, an intimate neighbour of the unfortunate Earl of Derwentwater :—

John Cook, before he went to reside on his estate at Aydon White House, which I understand was built by him when he retired from farming, save only his own property, lived in a house at Aydon, with stone steps going up on the outside. At that time he was in active business and reported to be very rich, and as in those days there were no banks in which to deposit money in safety, it was conceived he had much of it in his house. A gang of housebreakers who infested the district, and had committed many depredations, agreed all to meet on a fixed night to break into it. As they were aware he was a very stout man, and would defend himself, they decided to murder him before taking his property. A very great number in all, they kept their appointment; when just before they began to break into the house, John Cook came out in his shirt and walked about for some time. They could easily have secured him and, the door being unlocked, could have entered and ransacked the place, but (as some of the gang confessed when taken and condemned years afterwards for other acts of villainy) they were all struck with terror and became like dead men. None of them were able to move or speak till John Cook again entered the house and locked the door. Holding a consultation they all admitted their terror on seeing him, and declared: 'This man is protected by God; we will never again attempt to rob or murder him.'[1]

[1] Letter of Thomas Bates to John Reed of Warkworth Barns, Kirklevington, 9th February 1849.—Bell MSS., Alnwick Castle.

The hundred and twenty acres round Aydon White House did not offer sufficient scope for the agricultural energies of the two sons of George and Catherine. John Bates, the elder brother, took a lease of the adjoining farm of Thornbrough, which had formed part of the Derwentwater estates. Happening to be at Yarm Fair in October 1730 (the year his elder son, George, was born), he especially noticed a shorthorn cow, the property of a Mr. Dobinson who lived at the Isle near Sedgefield. Learning that this came of a tribe brought over from Holland by her owner's brother, Mr. Michael Dobinson, he went home with him to Witton Castle and bought six cows and a white bull of the same tribe from which he bred ever after.

Practice is often little in accordance with precept; notwithstanding his love of yeoman independence, George Bates became in his turn the tenant of the excellent farm of Aydon Castle in 1761. Conservative in the extreme in all he deemed good, he is said to have been the last to continue ploughing with oxen on Tyneside; but every new book that appeared on agricultural topics was regularly added to his library—to mention only a few, Ellis's *Modern Husbandman*, Mills's *Husbandry*, the *Museum Rusticum*, the Memoirs of Robert Dossie, the Annals of Arthur Young, the invaluable surveys of the several counties commenced by Marshall, Lord Kames's *Gentleman Farmer*, and the book George Bates prized more than all the others, Swayne's *Gramina Pascua*.[1]

Enough has now been said to explain the ingrained enthusiasm for agriculture that Thomas Bates of Kirklevington inherited from his father. His mother was the youngest daughter of Thomas Moore, the representative of one of the oldest of the old families of Salop. According to Camden, Thomas de la More, the founder of the family, fell at Hastings, while his son married Constance de Umfreville, daughter of Robert With-the-Beard, lord of Redesdale in Northumberland. Be this as it may, an

---

[1] *Gramina Pascua, a Collection of the Common Pasture Grasses*, by G. Swayne, A.M., Vicar of Pucklechurch, Bristol, 1790.

inquisition of the time of King John records that the Moore had then been held by the Moore family for five generations, by the service as dangerous as it was honourable of carrying the royal banner in the van of the army whenever a king of England marched into Wales.[1] The marriage of Sir Robert de la More with the heiress of Adam de Montgomery, the resolute defender of Montgomery Castle against Simon de Montfort, brought an increase of affluence, but with the murder of John Moore, while sheriff of Salop in 1399, the fortunes of the house began to wane. By the time of Elizabeth most of the family property had been alienated, but none of the various alliances with Corbets, Lees, and Needhams, set forth in the heralds' visitations, could have seemed more auspicious than Thomas Moore's marriage with Ann Blayney in 1714. The Blayneys of Gregynog in Montgomeryshire were the direct lineal descendants of Brochwel, prince of Powys, the opponent of Ethelfrid of Northumberland at the battle of Chester in 607. A second son of the family served with distinction in Ireland, where he was raised to the peerage by James I. Sir Arthur, the younger son of the first Lord Blayney, returning to Wales, married the heiress of the elder line, and the wife of Thomas Moore was his grand-daughter.

At the time of the marriage of George Bates with Diana Moore, the Gregynog estates were in the possession of her first cousin, Arthur Blayney. Cadwallader, ninth Lord Blayney, the next known representative in the male line, had offended the morbid susceptibilities of this confirmed bachelor by arriving on a visit at Gregynog with too numerous a retinue.[2] Of Arthur's first cousins on his

[1] As according to Blackstone services by grand serjeantry could not be alienated, and were excepted from the action of the statute 12 Car. II., the hereditary office of Royal Standard-Bearer in Wales would appear to have passed eventually to Thomas Bates's elder brother as heir-general of the Moores of the Moore.

[2] *Ex inf.* Cadwallader twelfth and last Lord Blayney. All efforts to find any male descendants of Sir Arthur Blayney, who would have succeeded to the peerage, have failed. Most of the pretenders put themselves out of court by tracing their pedigrees back beyond the births of Sir Arthur's sons.

father's side, the children of his aunt Mary had behaved so badly to their mother that she utterly disowned them and carried her portrait to Castle Blayney lest it should ever fall into their hands;[1] the only child of his aunt Bridget had married the Rev. Sir Thomas Edwardes, rector of Frodesley, and Arthur Blayney had a holy horror of parsons. He therefore gave Diana Bates, with whom he remained always on good terms, to understand that should she have a son he was to be called Arthur and to be brought up as his heir. George Bates was always incredulous, and refused to lend his first-born to this arrangement. The second son appeared so weak and unlikely to live that he was privately baptized under the name of Thomas, and this once given was not changed, though Arthur Blayney stood godfather at the public christening at Stamfordham, on 12th April 1775.

Thomas Bates was duly taken to Gregynog by his mother, and Arthur Blayney expressed himself 'much pleased with the child.' Of this or a similar visit the only two things which John Moore Bates, the elder brother, recollected were that the pump in the back court-yard had a long handle with which he was delighted to play, and that the Blayney pedigree was painted in the old Welsh fashion on the dining-room panelling.[2]

The first event which fixed itself in Thomas's mind was his father's purchase of a calf in 1778, from a cow belonging to Mr. Dixon of Ingoe, which gave forty quarts of milk a day. This cow's dam, the only one kept by its owner at Stamfordham, had given the prodigious quantity of twenty-four quarts twice a day. The calf purchased by George Bates was eventually 'fed off at seventeen years

---

The name of Blayney was common enough in Montgomeryshire, being borne by the steward and others at Gregynog who, however, disclaimed relationship with the ennobled family.

[1] The picture is now at Langley. A recent application from one of her descendants for permission to have a copy of it was made in such a way that it had ultimately to be refused, and her wishes thus rigidly respected.

[2] The pedigree has disappeared; the splendid panelling, dated 1636, with armorial shields of Blayney, Loftus, Herbert, and many Welsh chieftains, carved in high relief, has been fitted into a smaller room in the renovated house.

of age, and made a very fat and handsome carcase. Her descendants were good milkers and quick grazers when put dry, though none of them gave so much milk as she did. They were not great consumers of hay, and turnips were then never given to milch cows.'[1]

In the spring of 1782, Mr. William Wastell of Great Burdon,[2] near Darlington, paid a visit to George Bates at Aydon Castle. Thomas, now a boy of seven, accompanied them to view the cattle and sheep on his father's farms, as well as those on neighbouring ones in the district, from which Mr. Wastell had had the cast stock of cows and steers for feeding, many years in succession, he furnishing with the shorthorn bulls as needed. This was about the time that Mr. Wastell bought two shorthorn oxen, bred by Sir Henry Grey of Howick; after he had fed them they weighed, at six years old, 130 stones each, their proof being most extraordinary. Another ox, one of six Mr. Wastell bought in Northumberland, yielded 26 stones of tallow. His steer by Masterman's Bull (422) the grandsire of Hubback (319), weighed 110 stones when four years old.[3] Mr. Wastell's remarks and observations on the subject of shorthorns made an indelible impression on Thomas Bates's mind.[4]

A year or two later the two brothers were sent to the grammar-school at Haydon Bridge. This enjoyed a great reputation at the time, and was attended by the sons of the best families in the neighbourhood. The Rev. William Hall, a former fellow of St. John's College, Cambridge, and brother of the bishop of Dromore, was then master. In acknowledging a remittance from Mr. Bates in discharge of his modest quarter's account at Christmas

---

[1] Bates MS. fo. 22.

[2] 'William Wastell died 5th July 1788, in the 80th year of his age' —M.I. Haughton-le-Skerne. His family had long held lands in Burdon: —'Wm. son of Nicholas Wastwell of Great Burdon, bapt. 4th Feb. 1648.'— Surtees, *Durham*, iii. p. 341. The spellings 'Wastell' and 'Waistell' seem convertible.

[3] *Improved Short-horns: their Pretensions stated*, by the Rev. Henry Berry, Liverpool, 1824, pp. 20, 21.

[4] Bates MS. fo. 16.

1785, Mr. Hall enclosed a proof of his own scholarship in the form of a translation into Latin verse of some lines on the tragic fate of the aeronaut, Pilatre de Rozier, who had been killed the June previous in consequence of his balloon catching fire.[1]

Even at this early age, Thomas Bates expressed a decided wish to enter the service of the Church. The social condition of the clergy in the North of England had, however, little to commend it. George Bates would not hear of any such intention on his son's part, and turned a deaf ear to offers of influence for his obtaining an entrance to the Eton foundation, and even to the express wish of Mr. Hall that so promising a pupil should in any case proceed to Cambridge. When his elder brother was sent to Richmond School, Thomas continued under the Rev. John Farrer at Witton-le-Wear. He thus refers to the assiduous nature of his early studies, and to the dawnings of his spiritual life:—

I began at school to rise every morning at four o'clock, and got my Greek grammar off before school hours; nor was one moment at my disposal unoccupied till bed-time. Dr. Watts' *Improvement of the Mind* was the first work that made me think of the importance both of time and eternity.[2]

His schoolfellows remarked that he was not like other boys. He never joined in their games, but would sit for hours in the churchyard with a book.[3] A letter to his father conveys the same impression, and amusingly illustrates his introduction to politics:

[1] The Latin verses are headed, "Obsunt auctoribus artes.—In casum D^{ni} Pilatri de Rozier, 15to Junii, 1785." The English lines commencing—

> The silken Ball now filled with fiery air,
> Th' undaunted Heroes to their car repair

are initialed J. R.

[2] Letter of Thomas Bates to Philander Chase, Bishop of Ohio, Kirklevington, 31st May 1831.

[3] *History of Improved Short-horn or Durham Cattle*, by Thomas Bell, Newcastle, 1871, p. 110.

THOMAS BATES *to* GEORGE BATES

WITTON-LE-WEAR, *6th August* 1789.

HONOURED FATHER—Mr. Burdon came here to-day after Mr. Milbank's agents, accompanied by Lord Strathmore and others. Both parties made much diversion. Mr. Burdon got us play all day and no task, so we are all Burdon's men. We expected Sir John Eden, but he did not come, and it is the opinion of all that he will not stand unless he gets Lord Barnard's interest. When he canvassed Sunderland he gave nothing to drink, neither was there one huzza all the time he was there. There is a report that Lord Barnard and the Earl of Darlington have split their interest between Mr. Milbank and Sir John Eden, but I cannot say for the truth of it.[1]

Yesterday was Hamsterley hopping,[2] but I did not go because Mr. Farrer was not agreeable. He gave some leave, but those that did not go he promises that he will let into his garden, which I thought was much better. I was at Mr. Farrer's hay on Tuesday evening as we had play.

Thomas Bates was soon at the head of Witton School. He contracted lasting friendships with Mr. Farrer and his family, and with the Rev. George Newby, the tutor and subsequent headmaster. The next year, however, when only fifteen, he was called home to assist in the management of his father's farms. The first fat animal that appears to have impressed itself on his memory was a cow by Hubback (319), from the stock of Mr. Hill of Blackwell, which was killed by Mr. Richley, a butcher at Corbridge. The cow weighed 127 stones. Bates went to see her every day during the week before she was slaughtered.[3] He began a series of experiments to ascertain the results obtained from the food given to milch cows and feeding cattle. The latter he measured in order to

[1] The election does not seem to have actually taken place until between the 28th June and 8th July 1790, when Mr. Burdon of Castle Eden polled 2073, Mr. Milbank 1799, and Sir John Eden 1696 votes.

[2] The 5th of August was the ancient festival of St. Oswald, king of Northumberland.

[3] MS. note on p. 20 of Berry's *Improved Short-horns*, 'transcribed by Mr. Ambrose Stevens from Thomas Bates's copy, which he took with him to America.'

determine the improvement they made, as he found measurement at different periods to be a surer method than weighing them alive. It was thus that he came to know the great differences which existed between the several varieties of stock, learning by the external characteristics of each their real merits, since the same data extended to others gave a like result.

Thomas Radcliffe, afterwards for more than twenty years the principal butcher in North Shields, had just commenced business in Newcastle. For some years he came every fortnight to Thomas Bates, and bought two beasts, taking one the same week and the other the week following. He was a well-educated, intelligent young man, and very attentive to business. Bates told him the quantity of food consumed by each animal and the details of its gaining weight; Radcliffe made notes of how it killed and cut up. This intercourse improved the judgment of both, and Radcliffe often expressed his sense of the benefit he derived.

One morning Bates saw six cattle killed, and put down what he considered each would weigh. The next day the machine proved that he was very few pounds wrong in the weight of any one of them, and hardly wrong at all in the weight of the whole six. Still he did not allow his judgment to be fettered by measurements, recognising that differences of form and of degrees of fatness must necessarily cause differences in weight that no mere mensuration can accurately determine. It was his long perseverance during seventeen years (1790-1807) in this plan of weighing the food consumed by his cattle, ascertaining by the tape line the improvement each made, and correcting the result by the block-test, that matured Bates's judgment, and enabled him afterwards to decide with precision on the merits of animals by handling and external appearance, a practice which, as he said, when once learnt is never forgotten, but remains fixed in the memory like the multiplication table.[1]

[1] Bell, pp. 75, 76.

Before he was eighteen, in order to attain more independence, Bates became, he tells us, the tenant of his father's patrimony at Aydon White House:

> When I began the world it was as an occupier, not as owner, on 120 acres of land that barely kept (by its returns) the working cattle and horses and men employed upon it. My own yearly expenditure for three years was under £20; my capital was not £500. I planned and arranged everything according to my means, making my expenditure according to my income, and though I was under twenty years of age, I soon produced annually crops worth more than the land was worth when I began.[1]

> I had not been long engaged in farming before I became thoroughly convinced that the atmosphere contained the great ingredients for the amelioration of the soil. This I discovered by seeing the good effects of the same surface being exposed to the atmosphere, as long as possible before turnip seed was sown. A field ploughed in October after oats, ribbed across in February, and then only worked with a scuffler till drilled for turnips, produced in 1793 the heaviest crop of white turnips I ever saw. Mr. George Culley and Mr. Bailey, who were making the survey of Northumberland for the Board of Agriculture, both said they never before had seen so large a crop of turnips; nor have I since. They were sown in June, and considerably exceeded 50 tons per acre when they had stood till February.[2]

The shorthorns on Tyneside were at the time very inferior to what they subsequently became; many of them bore a strong resemblance to Ayrshire cattle.[3] There were still a great many of the coarse black-fleshed shorthorns in Northumberland, of which George Culley rightly said that no man would buy one of them if he knew anything of the matter, or if he should be once taken in he would remember it ever after.[4] Tempted by the low price of 2s. 6d. per stone (14 lbs.), Bates bought some three-year-old steers of this kind in the spring of 1794, He kept them well on turnips for three months along with

---

[1] Letter of Thomas Bates to Bishop Chase, 31st May 1831.
[2] *Farmer's Magazine*, xxi. p. 3.   [3] Bell, p. 76.
[4] *Observations on Live Stock, containing Hints for Choosing and Improving the Best Breeds of Domestic Animals*, by George Culley, 3rd. ed. 1801, p. 44.

some good grazing steers of the same age, purchased on the same day at double the price per stone. These latter increased to twice the weight the former did though they consumed far less food, and finally left four times as much for their keep. Bates never bought another black-fleshed shorthorn.[1]

It was in June 1794 that Joyous Moore, Diana Bates's surviving sister, came to Aydon Castle.[2] This lady had moved in the best society of Salop, and the correspondence she continued with her friends there throws many side lights on the inner life of that county. She had, it would seem, been engaged to one of the last of the Moores of Millichope, who, on his return from a voyage to the Philippines, presented her with a celebrated diamond. A later aspirant sent her one morning a copy of Thomson's *Seasons* with the assurance—

> Each Nymph he paints some single Charm may boast,
> That forms a Belle, or nominates a Toast;
> Center'd in thee while all the Charms we find
> That grace, or ever grac'd the female kind,
> Yet Beauty's self attacks the Heart in vain,
> Till Wit, like thine secures the captive Swain.

He returned to the charge with a fresh set of verses indited 'past twelve at night':

> Awhile the Heart enthrall'd may wear
>   Soft Beauty's silken Chains;
> But soon will break the tender Snare
>   If nothing else detains.
>
> . . . . .
>
> What Sweets, dear Maid, may flow from hence!
>   Each day how happy prove!
> Since you have Goodness, Charms, and Sense,
>   And I have Youth and Love.

The maladroitness of the last couplet probably proved fatal to the suitor.

---

[1] Bell, p. 77.
[2] The quaint name of Joyous was derived originally from a direct ancestress, Joyous Herbert, aunt of the historian, Lord Herbert of Cherbury.

After Joyous Moore had been some months at Aydon Castle she told her sister that she intended remaining there for the rest of her days. George Bates immediately bought a single-horse chaise, in which Thomas drove his mother and aunt to church and to see their acquaintances, as Joyous Moore was 'always fearful of riding on horseback.' The two sisters 'amused themselves with relating the events of their younger years,'[1] and in so doing confirmed Thomas Bates's hereditary taste for genealogical studies.

Joyous Moore's settlement in Northumberland snapped the last link which connected Arthur Blayney with his father's family. He died on 1st October 1795 at the age of eighty-five leaving £300,000. Much surprise was naturally felt when it appeared that, disregarding his long-expressed intentions, he had made a will shortly before his death, under which the ancient heritage of his house passed to Lord Tracy, a stranger in blood. It was noted by the superstitious in Montgomeryshire that all the witnesses to this will came to untimely ends, and the alienated estates do not appear to have brought much happiness or good fortune to their subsequent possessors. 'The Father of Montgomeryshire,' as Arthur Blayney was called (perhaps with the same *double entente* that was applied to Caesar's title of *Pater Patriae*) did not appear in the light of a worthy to his nearest relatives, though some of his eccentricities merely show that his opinions were a hundred years in advance of the times. Joyous Moore's nephew, who was a solicitor in Hereford, did not disguise his disappointment:

WILLIAM HOLMES *to* JOYOUS MOORE

HEREFORD, 1st *November* 1795.

I did expect that Arthur Blayney would have left me some acknowledgment for about three years' trouble and expense in his affairs with the Dean and Chapter of Hereford, and with Gilbert Jones about the purchase of the Newton estate. I paid even the

[1] Letter of Thomas Bates to William Holmes.

postage of all his letters except two. In 1783 he desired me to come to Gregynog in order to hear from me all the particulars of his affairs more fully than could be related in letters. I was there nearly three weeks. He never offered me, nor did I ever receive one farthing or the value of one farthing from him or his agents. It appears that his friendship lasted no longer than it served his own views. One day I happened to say, I supposed that the farmers were very diligent in getting in their hay as in all situations near hills and woods the weather was apt to be uncertain. He answered everything was uncertain, blasphemously adding that he put no trust in anything human or divine. As to his wines he brought me one day a book to make a preparation to discover if they had not been fined with some poisonous drug. He said he kept as few servants as possible as they were all pickpockets, and that he was obliged to do largely for the poor, lest they should do him a mischief, for they were always gabbling Welsh together, which language he did not understand, and the Welsh did not scruple to say it was no sin to cheat a Saxon, meaning himself for one.

Arthur Blayney's eccentricity had developed in his dotage, until it was difficult to distinguish it from the insanity prevalent in his mother's family. George Bates, however, declined to contest the validity of the will, considering that Sir Thomas Edwardes, the husband of the co-heiress-at-law nearest the spot, would move if there were any chance of success. By some extraordinary crotchet, Sir Thomas Edwardes thought 'it would not be proper to say anything to Lord Tracy till the death year had expired';[1] so to the surprise of all Montgomeryshire, the matter was allowed to slide.

The disappointment of Thomas Bates, who had always been worshipped by his mother and her sister as the heir to Gregynog, must have been very keen. Though an engraving of Arthur Blayney, from a portrait by Beechey, continued to hang in his dining-room till the last, he never alluded to him except as having been his godfather, and as having, by a singular coincidence, died on the same day as that famous improver of farm-stock, Robert Bakewell. The strong family likeness between Thomas Bates

[1] Letter of William Holmes.

and his godfather in the engraving was the subject of general remark at Kirklevington.[1] Nearly every trait, too, in the character of Arthur Blayney, as delineated by his friend Philip Yorke of Erthig, was reduplicated in that of the godson he had disinherited:

He valued himself on his pedigree no otherwise than by taking care that his conduct should not disgrace it. His loyalty did not preclude him from freely censuring, upon proper occasions, both the measures and instruments of Government. Uncorruptable himself he detested venality in others. He was of no party but that of honest men. He was by no means partial to Lords or Placemen. The active part he took in behalf of candidates for Parliament was so pure in its motives, that his support gave a decided superiority over the highest rank and influence. Few gentlemen were better qualified for the magistracy or more sensible of its importance, but he could never be prevailed on to act in the commission, though always ready to applaud and second the just efforts of those who did. Of the established religion he was a steady member; defended its rights and respected its ministers, where they respected themselves. His tenants, from their relation, he considered as friends, and not only allowed them ample profit from his estates, but encouraged and assisted them in every rational attempt to improvement. He was undoubtedly an economist on system which enabled him to do what he did; when the object of expense was a proper one, he never regarded the sum. He could never be persuaded to keep a carriage, and very seldom hired one, performing, till his infirmities disabled him, his longest journeys on horseback. His place was neither elegant nor ornamented, but comfortable in the most extended sense of the word; inasmuch that it would be difficult to find another house, where the visitor was more perfectly at his ease, from the titled tourist to the poor benighted way-worn exciseman, who knew not where else to turn in either for refreshment or lodging. In his conversation he was affable, polite, instructive, and cheerful. Order and regularity pervaded his whole household. He was never married, but was remarkably pleased with, and pleasing to the ladies who visited him, and they were not a few. He carried his notions of independence to a pitch that bordered upon excess; always ready to confer reasonable favours, he reluctantly accepted them. In his temper he was constitutionally warm; what true Welshman is otherwise? His

[1] Bell, pp. 108, 110.

resentments, generally well founded, were consequently strong, and sometimes permanent. He could forgive an injury, but if his confidence was forfeited, it was nearly impossible to retrieve it.[1]

Instead of wasting time in repining at the cruel blow which had deprived him of a princely succession, Bates threw himself with quadrupled energy into an agricultural career, and henceforth applied his genealogical proficiency to the problems of cattle-breeding. In order to attain a position of more independence than had been afforded by his occupation of Aydon White House, he became, on coming of age, the tenant of his father's small estate of Wark Eals on North Tyne. This property had been sold in 1663 (when wheat was allowed to be exported, if not above 48s. a quarter, at a duty of 5s. 4d. a quarter) for £223 : 5s. to a yeoman who kept it in his own hands for forty years. It was then sold for £1000. In about 1772 it was again offered for sale; John Bates, then an old man, went to view it with his son George, and plucking a stalk of clover wound it thrice round his hat. This evidence of the fertility of the soil so fixed itself in George's memory, that after it had been more than fourteen years in the market he bought it for £4000 in 1787, without having ever again been to inspect it. Mr. Thomas Ridley, the owner of the adjoining allotment of Park End, had offered only £2500, and said he would not have made any advance on this at the time of the commencement of Bates's tenancy in 1796.[2] The property consisted partly of a considerable tract of haugh-land, which had once been a series of *eals*, or islets, and was liable to be flooded, and partly of allotments on high-lying commons. During the first three years Bates did not make the farm pay the unavoidable expenses of cultivation and management, and the rent, small as it was, had to be paid out of capital. His skill, judgment, and attention were however indefatigable. He ran an embankment—the first work of

---

[1] *The Royal Tribes of Wales*, Wrexham, 1799, pp. 161-164.
[2] Letter of Thomas Bates to Bishop Chase, 31st May 1831.

kind on the Tyne—along the riverside; tile-drained the fields that required it, stubbed up the old straggling fences, planted the bare slopes, and collected a mill-stream to drive his threshing-machine. During the next seven years he made every pound he had invested at the Eals into fifty.[1]

The old connection subsisting between the families of Bates and Culley had been confirmed in 1783 by the betrothal of Matthew Culley of Akeld to Elizabeth Bates of Halton. It is said that he was staying at Aydon Castle at the time, and sorrowfully confided to George Bates that he had been rejected that morning by his sister Catherine, then living at Aydon White House. 'If that's all,' said Bates, 'we'll go over to Halton this afternoon and see cousin Bessy.' The visit proved eminently soothing, and before long 'Bonny Bessy Bates,' as she was called, transferred her charms from Tyneside to Glendale.

The herd of shorthorns which Matthew Culley and his brother George had taken with them from their old home at Denton, near Darlington, in 1767, to the large farm of Fenton in North Northumberland, was always regarded by Robert Colling as the best lot he ever saw together. The buying and selling of the stock from these extensive farms, for which the brothers paid at one time no less than £10,000 a year in rent, was principally managed by George Culley. Bates paid him several long visits, and profited greatly from his experience. The only point on which they differed was that Culley held that good dairy cows were generally poor grazers, while Bates contended that it was possible to combine a large yield of milk with a propensity to fatten when set dry.

His intimacy with the Culleys introduced Bates to a large circle of agricultural acquaintance on the Tees. He generally attended the Great Markets at Darlington, where the winter-fed cattle were exposed for sale. As

[1] *Ibid.*; Bell, pp. 115, 151.

these were always held on a Monday, he usually stayed over the Sunday with Charles Colling at Ketton, with Robert Colling at Barmpton, or with Christopher Mason at Chilton.

In March 1791 he saw two excellent three-year-old steers exhibited by Mr. Charge of Newton Morrell. Both were by the same shorthorn bull, but one was out of a pure shorthorn cow and the other out of a pure West Highland. Charge put a price on each and would not sell them together. He had many customers, and after some time sold them separately for nearly the same money. 'Can any man,' he said, 'doubt which of the two steers has been most profitable? They have gone together and been treated alike from being calves, and the dam of the half-highlander was not one-third of the weight of the dam of the shorthorn.' This made a great impression upon Bates, for Charge was looked up to as a very superior man in his day, and one of the best judges of cattle. He had repeatedly visited Bakewell even before 1760, and Bakewell had confided to him his opinion that it was from the West Highland heifer that the best breed of cattle might be produced.[1]

Seven years later (5th March 1798) Bates saw the celebrated cow Lady Maynard sold in the same market. There were two young cows sold with her at twenty guineas each, but the old cow, though she was nineteen years old and had bred twenty calves in all, was still lively and fresh looking, certainly a far better cow than any bred from her that grew up to maturity.

Mr. Maynard of Eryholme was accustomed to exhibit eight steers and eight heifers opposite the King's Head Inn, but at the Great Market on the first Monday in March 1799, it was a white ox and a roan heifer from Ketton, both three-year-olds by Favourite (252), that attracted the gaze of the multitude. No animals had ever before been seen so good at that age. The 'Wonderful

---

[1] Letter of Thomas Bates to the editor of the *Mark Lane Express*, Kirklevington, 10th November 1840.

Ox,' out of a daughter, by Hubback (319),[1] of the Stick-a-Bitch cow, was considered the great sight of the year. After being exhibited all over the kingdom by his purchaser, Mr. Day, as the Durham or Ketton ox, he expired at Oxford in 1807; at ten years old he weighed 270 stones. Bates, however, considered the roan heifer (dam by Foljambe (263), grandam Duchess by Hubback (319)) by far the finer beast of the pair, and his opinion was confirmed by thrice finding Mr. Thompson of Stamford, a well-known judge of stock from Northumberland, lost in admiration by her side. She was more weight than the ox then, and should have been kept as the show animal. Unfortunately Charles Colling was advised to keep the ox and sell the heifer. This he did and she scaled either 1 lb. under 100 stones (of 14 lbs.) or 1 lb. over. Of a beautiful roan colour, she was admitted by all who saw her to have the most perfectly shaped and uniformly covered frame of beef, equally good in every point, her breast the nearest the ground of any animal ever seen. Often afterwards Charles Colling said to Bates, 'It was the heifer I ought to have kept for exhibition.' Robert Colling's 'White Heifer that Travelled,' estimated to weigh 130 stones when four years old, was nothing like so fine an animal.[2] Bates resolved to take the first opportunity of becoming possessed of the Duchess breed.

At the close of the century his cousin, Thomas Bates, the brother-in-law of Matthew Culley, determined to give up the important farm of Halton Castle and to retire to his own estate of Brunton near Chollerford. Another farm at Halton that had been occupied by the Messrs. Bell became vacant at the same time. The two together contained about 650 acres, some of them valuable old grass. Much against his father's wishes, who advised him to be content with Wark Eals, the freehold of which he had given him as a reward for his exertions, Bates offered

---

[1] 'As Thomas Bates knows from the testimony of both the Messrs. Colling.'—MS. note on p. 27 of Berry's *Improved Short-horns*, see n. 16. Stick-a-Bitch is the euphonious name of a farm between Darlington and Croft.

[2] Bell, p. 45.

£750 a year (£125 more than any one else) for a twenty-one-years lease of both the Halton farms.

JOYOUS MOORE *to* WILLIAM HOLMES

AYDON CASTLE, 5*th April* 1800.

In this country they let their farms before Christmas, but do not enter on them till the twelfth of May. I do not doubt but Halton will answer Thomas (though it is a large undertaking, several hundreds a year). He is sensible, active and steady, and in the prime of life. He was twenty-five in February last.

At Durham Fair, on 31st March 1800, Bates bought two twin steers from Mr. Arrowsmith of Ferryhill. They were descended from a heifer of the breed of Mr. Wastell of Great Burdon, which cost fifty guineas at the sale at Barmpton after the death of his brother-in-law, Mr. Harrison, in 1782, and at the time of Robert Colling's entry on that farm, prices being then much depressed at the close of the American war. These twin steers were by Favourite (252) and proved extraordinarily good, making a greater improvement for the time Bates had them than any he had then tried.[1] He never saw any better steers either at Barmpton or Ketton while the Colling stock was in its greatest perfection. The descendants of Mr. Wastell's heifer continued to produce superior animals long after both the Colling herds had ceased to produce good ones, with the exception of the Duchesses, Daisies, Cherries, and Princesses.[2]

---

[1] Bell, p. 45.
[2] Bates MS. fo. 17. There are some grounds for supposing the Waterloo tribe to be descended from this heifer, see below p. 238.

CHAPTER II

PRE-COLLINGITE SHORTHORNS

1400—1800

IT seems becoming more and more a matter of general belief that bulls and cows of the Aberdeen-Angus, Devon, Galloway, Guernsey, Hereford and other breeds stepped solemnly two and two out of Noah's Ark. Even shorthorn breeders are not altogether free from this egregious fallacy, although the Improved Shorthorn is confessedly, to a great extent, a *Bos compositus*.

The comparative antiquity of short-horned cattle in the North Riding of Yorkshire is attested by the will of John Percy of Haram, near Helmsley, made in the year 1400: 'To my son John, I bequeath two stots with short horns; to John Webster a small horned stot; to John Belby a cow with a white " leske "; to my son John a heifer with a white head; to Thomas Peke a heifer called Meg, and to Margaret Percy another heifer.'[1] This specific mention of the short-horned stots points to their being something out of the common at that time, and possibly possessed of especial value. The white 'leske' of the cow and the white head of the heifer show too that these cattle were not all of whole colours.

---

[1] 'Johanni, filio meo, ij stottys with schorthornes. Johanni Webster a smal horned stott. Johanni Belby iii s. iiij d. et j vaccam with a whyte leske. Johanni filio meo, j juvencam cum albo capite. . . . Thomæ Peke j juvencam vocatam le Meg. Margaretæ Percy j aliam juvencam.'—*Testamenta Eboracensia*, iii. Surtees Soc. Publ. 1865, p. 188.

Black appears to have been for many ages the general colour of Yorkshire cattle. As far back as the beginning of the third century, the Emperor Septimius Severus, returning from his Caledonian campaigns, was conducted by mistake into the temple of Bellona at York, where black beasts were presented to him for sacrifice. He refused to offer up animals of that ominous colour, and returned to

ST. LUKE'S OX, LINDISFARNE, *circa* A.D. 680.[1]

the palace. The black cattle, let go by the priests, followed him to the very gates, and the incident was regarded as a presage of his death, which occurred soon after.[2] The Romans, indeed, considered black cattle the strongest and hardiest, and white ones the most tender; but they preferred red or dun ones, especially the former, to either black or white.[3]

[1] Brit. Mus. *Cotton* MS. *Nero* D. IV. That this ox or bull-calf, for it is marked *imago vituli*, is a genuine specimen of Anglo-Celtic art representing the contemporary cattle of Northumberland, and not a mere imitation of the conventional Roman type, seems proved by the amusing attempt of the illuminator to evolve the Lion of St. Mark out of his inner consciousness; the result is enough to show that he had never seen a lion himself, nor even the representation of one.

[2] L. Aelii Spartiani, *Vita Severi* (in *Historia Augusta*), cap. xviii.

[3] 'corium attactu non asperum ac durum, colore potissimum nigro, dein rubeo, tertio helvo, quarto albo: mollissimus enim hic, ut durissimus primus. De mediis duobus prior quam posterior melior; utrique pluris, quam nigri et albi.'—M. Terentii Varronis *de Re Rustica*, lib. ii. v.

After the English conquest we find a cream-coloured short-horned ox associated with St. Luke in the splendid Gospels written in the seventh century by Eadfrid, afterwards bishop of Lindisfarne. It was, too, a dun cow that, according to the local tradition, the 'family of St. Cuthbert' found browsing on the site of Durham Cathedral when they there ended their long series of wanderings

THE DUN COW, DURHAM, *circa* A.D. 1300.[1]

with his shrine in 995. A cow was certainly carved on the north-west corner turret of the Nine Altars or eastern chapel of the cathedral about the year 1300, and gave rise after the Dissolution to the saying—

> The dun cow's milk
> Makes the prebend's wife go in silk.

In the end black once more asserted its predominance. Writing early in the seventeenth century, Gervase Markham says: 'Of our English cattle those bred in Yorkshire, Derbyshire, Lancashire, and Staffordshire are

[1] Hutchinson's *History of Durham*, 1785-1794, ii. p. 226. In Grimm's drawings, vol. ii. MS. Add. Brit. Mus. 15,538, fo. 135, the cow is shown as a mere crumbled fragment. Carter's sketch is still more vague.

generally all black of colour, and they whose blackness is purest, and their hairs like velvet, white with black tips, they are of stately shape, big, round, and well huckled together in every member, short-jointed and most comely to the eye, so that they are esteemed excellent in the market.'[1] So prevalent, indeed, was the black colour in the North of England and the South of Scotland, that bulls, cows, and oxen were given the generic name of 'black cattle.' Originally the Scottish thieves appear to have called the 'black cattle' they were driving off their 'black-mail' or 'black-rent'; the term being afterwards applied to the money paid them for foregoing these exactions in kine.

The two other principal breeds of English cattle in the seventeenth century were those of Lincolnshire and Somerset. The Lincolns were 'for the most part pied, with more white than the other colours (black or red), their horns little and crooked, of bodies exceeding tall, long and large, lean and thin-thighed, strong hoved, not apt to sorbate, and indeed fitter to labour and draught.' The oxen in Somerset and Gloucestershire were 'generally of a blood-red colour, in all shapes like unto those in Lincolnshire, and fitter for their uses.' It was considered bad to mix the red and black races together 'for their shapes and colours were so contrary that their issues were very uncomely'; the most that was permitted was to cross the different sub-species of each, the very most to cross the pied Lincolns with the red Somersets.[2]

In spite of this, by the middle of the eighteenth century, although the general colour of Yorkshire cattle was still black, a considerable mixture had taken place and there was a large sprinkling of other colours. In the South of England black became the distinctive mark of Welsh cows and Welsh and Scotch runts. Some of the Welsh cows proved as good milkers as many of the larger English sort, giving four quarts of milk a day throughout the whole year. It was thought in that case

---

[1] *Cheap and Good Husbandry*, 14th ed. 1683, pp. 59, 60.
[2] *Ibid.* p. 60.

'good husbandry' to keep them and propagate their breed by an English bull.[1] The 'runts' fed and fattened on turnips almost like sheep, and this enabled the Norfolk and Suffolk farmers to improve their poor sandy lands. A small polled Scotch breed of a dun colour was also established in the eastern counties, and was found 'to live on a short bite, milk better than some others of the same bigness, and fat with lesser meat than the horned sort.'[2]

The red English cow was kept by the farmers of the Vale of Aylesbury and the Chiltern hills for dairy purposes, because being of a large size it gave the greatest quantity of milk 'in the time of high grass.' 'But of all the cows in England,' wrote that most intelligent husbandman and delightful gossip, William Ellis of Little Gaddesden, in 1744, 'I think none comes up to the Holderness breed, for their wide bags, short horns, and large bodies, which render them (whether black or red) the most profitable beasts for the dairyman, grazier, and butcher. Some of them have yielded two or three gallons at a meal.'[3]

A new breed, consisting chiefly of white cows, had now been imported from abroad. According to Ellis it was not kept in great numbers, being reckoned very tender, and requiring a longer bite of grass than many had to give. 'These great Dutch or Flanders cows,' he continues (for from these the white breed generally came first), are seldom kept but by gentlemen who can best afford them meat enough, and content themselves with the sight of a fine cow. When this breed is mixed with some of our best English, I have known them to give a great deal of milk. I once had a heifer that I bred from a Dutch white cow, that I sold for four guineas at Dunstable Fair, in her second year, which proved the best milch cow in the country. Another I sold to a gentleman farmer, who, after having kept this large white beast

---

[1] *The Modern Husbandman*, by William Ellis, 1744, vol. ii. June, p. 143.
[2] *Ibid.* p. 147. This would seem to be the origin of the polled Suffolks.
[3] *Ibid.* p. 146. Holderness is the extreme south-east corner of Yorkshire, between Hull and Beverley and the sea.

several years, at last lost her by means of her drinking wash out of a hog tub.'[1]

Notwithstanding that the then unrepealed statute 18 Car. II. had, in 1666, prohibited the 'importation of all great cattle' as a 'common nuisance,' and that no official records have been found of cattle landing from Holland,[2] the fact that they did so land hardly admits of doubt. The introduction of the terrible murrain or 'hyanstriking' into England about 1744 was attributed to traders bringing over to Essex some calves which they had purchased in the Low Countries.[3] In 1756 we read: 'The fine Dutch breed have long legs, short horns, and a full body. They are to be had in Kent and Sussex, and some other places where they are still carefully kept without mixture in colour; they will yield two gallons at a milking, but in order to this they require great attendance, and the best of food.' The Dutch and Alderney cows of that day were classed together as being very like one another in shape, and especially in the shortness of their horns, which distinguished them from 'the English breeds.'[4]

Still, there is little room to doubt that a native race of short-horned cattle had continued to exist on the banks of the Tees, as well as in Holderness. According to a tradition handed down from father to son, the Smithsons had been in possession of an excellent breed of shorthorns at Stanwick, ever since they obtained that estate from the Cathericks in the middle of the seventeenth century.[5]

---

[1] *The Modern Husbandman*, p. 147.
[2] *History of the Short-horn Cattle*, by Lewis F. Allen, Buffalo, 1872, p. 23.
[3] *The Compleat Cow Doctor*, by Joshua Rowlin of Hollins, Lamplough, Cumberland; Glasgow, 1794, p. 249.
[4] *A Compleat Body of Husbandry*, London, 1756, p. 215.
[5] Considering the great interest that the Smithsons took in agriculture, especially cattle, the only reason for disputing the fact that there was an old breed of shorthorns at Stanwick, as well as at Studley and Newby, appears to be that the first Duchess cow came from Stanwick. Even the Rev. W. Holt Beever (*Leading Short-horn Tribes*, p. 62) thought it necessary to write: 'As regards the legend that the Duchess breed had been kept intact for two centuries in Stanwick park, it is assuredly incredible, considering the intrusive ways of the mosstroopers alone.' 'Two centuries ago,' from 1846 take us back to 1646, about the time the Smithsons acquired Stanwick, and legends of

Some thirty or forty years later Sir Edward Blackett of Newby did much to improve a valuable herd of shorthorns which he is said to have inherited from his father. He acted in concert with the Aislabies of Studley Royal; and the entrance hall of the splendid mansion he built from the designs of Sir Christopher Wren, at a cost of £32,000, was hung round with portraits of his celebrated shorthorns.[1] Their descendants remained at Newby till the sale of the property in about 1750. The Aislabie herd was kept up for a considerable time longer.

The great improvement which was effected with these early shorthorns is universally attributed, both by the purist and the alloy factions, to the use of bulls from Holland. The former, however, suggest that the Dutch bulls in question were the descendants of some shorthorn cows said to have been sent as a present by James II. to William of Orange at the time of the Stadtholder's marriage with his daughter Mary.[2] In that case the improvement the Dutch bulls effected would be the result of a change of climatic influences like that accomplished by the shorthorn re-importations from America.

---

mosstroopers in Yorkshire after that date would furnish an entirely new and unsuspected chapter of Border history. While ridiculing *a priori* the traditions of the Northumberland family, the Rev. W. Holt Beever did not hesitate to accept as authentic the story of the wild cattle of Chillingham. According to him, while it was impossible to keep a herd of shorthorns in Yorkshire under the Commonwealth and the later Stuarts, on account of the mosstroopers, there was nothing improbable in the maintenance of the Chillingham herd close to the very Border during five centuries of Scottish rapine. Professor M'Kenny Hughes, in his valuable paper *On the More Important Breeds of Cattle which have been recognised in the British Isles in Successive Periods*, printed in *Archaelogia*, lv. pp. 125-158, has disposed of the Rev. J. Storer's deduction of the Chillingham cattle from the *Bos urus*, but his own view of their descent from Italian cattle brought over by the Romans must be considered as still *sub judice*.

[1] Bell, p. 24.

[2] James II. was of course only Duke of York in 1677; the old jingle also has—

> What will rhyme with porringer?
> What will rhyme with porringer?
> The King, he had a daughter fair,
> And gave the Prince of Orange her.

Edward IV. in 1480 had allowed his sister, Mary of Burgundy, to export 1000 cattle and 2000 rams from England to Flanders, Holland, and Zealand.—*Rymer's Foedera*, ed. 1727, xii. p. 137.

It is not amiss to remember that some of the so-called 'Dutch bulls' may have come from the district of Holland in Lincolnshire, and not over sea from the Netherlands. On the other hand, the alternative 'Holstein or Dutch breed' points clearly to an origin along the continental shores of the North Sea.[1] It should also be recollected that, as regards the improvement of English

HOLSTEIN OR DUTCH BULL, 1792.[2]

horses, the Earl of Huntingdon, on the conclusion of his embassy to the States-General, did, as a matter of fact, bring back with him a team of black coach-horses, most of which were used as stallions by his tenantry; and it was to the same source that Bakewell resorted for mares upon which to found his breed of Leicestershire cart-horses.[3]

If the 'Dutch' bulls did indeed come from the Netherlands, they must not be confounded with the modern

---

[1] *History of Quadrupeds*, by T. Bewick, Newcastle, 1792, p. 26. It is worth noting that, during the interval when the importation of foreign cattle was permitted in 1801-14, the cattle from Rotterdam were black and white or dun and white, and those from Hamburg 'a good red and white fleck.'—*Sockburn Short-horns*, p. 46. Bates recorded that the owners of Rudchester, two miles west of Heddon-on-the-Wall, 'went over to Holland and bought cattle, very good.'—Holt Beever, *Leading Tribes*, p. 22.

[2] Bewick's *History of Quadrupeds*, 1792, p. 26.

[3] *Live Stock*, by George Culley, p. 32.

Dutch race of black and white cattle. This, as has been well remarked, bears no resemblance to the cattle in the pictures of Paul Potter, Rubens, Cuyp, and Teniers, and everything points to a fresh race having been introduced after the murrain of which 200,000 cattle died in Holland in 1745.[1] It is the red-coloured cattle of East Friesland that resemble most nearly in outline the 'Holstein or Dutch breed,' as drawn by Bewick in 1792.[2]

HOLSTEIN OR DUTCH COW, 1792.[3]

The most successful early improver of shorthorns in the county of Durham was Michael Dobinson of Witton Castle, who went to Holland to buy bulls in the early part of his life. He and his neighbours continued to be noted, even in George Culley's youth (1740-50), for having the best breed of shorthorn cattle, and they sold their bulls and heifers for very great prices.[4] In or about the year 1730, as has been said, John Bates of Aydon (grandfather of Thomas Bates of Kirklevington), being at the October Great Fair in Yarm, saw a cow of this breed, belonging to Dobinson's brother, who lived at the Isle, near Sedgefield. He called at Witton Castle on his

---

[1] *La Race Short-horn dite Durham*, par M. le Marquis de Chauvelin, pp. 5, 6; Housman, *Live Stock Journal*, No. 869. p. 531.
[2] Cf. sketch of an East Frisian cow in Dr. O. Rohde's *Rindviehzucht*, Berlin, 1875, with woodcut of 'Holstein or Dutch cow' above.
[3] Bewick's *History of Quadrupeds*, 1792, p. 26.
[4] *Live Stock*, by George Culley, p. 42.

way home, and bought six cows and a white bull. From this stock he continued to breed till his death in 1777, when the cattle passed into the hands of his second son, William. Bates well remembered how extraordinarily good they were at Clarewood down to the year 1800,[1] and they continued in high repute at Chollerton till 1820.[2]

Other importations of bulls from Holland were made by Sir William St. Quintin of Scampston, on the Yorkshire Wolds, near Malton, whose experiments in crossing shorthorns with Alderneys are recorded in the *New Farmer's Calendar* of 1802.[3] Both he and Sir James Pennyman, another baronet who took a great interest in breeding, are said to have derived their shorthorns from the Aislabie herd at Studley. Sir James had estates and residences at Beverley Park and at Ormesby in Cleveland. To induce his agents to pay attention to the stock, he frequently made small wagers as to whose oxen would make the greatest weights and the best prices. He was greatly respected by all of them, and they spared no exertions in his interests. The farm accounts at Ormesby, commencing before 1745, regularly recorded the sales of the Pennyman shorthorns, with their weights and proof in tallow, for they were often sold by weight. As the soil there is a strong clay no turnips were grown, and the cattle were kept in winter on only hay and straw. Notwithstanding this, the five-year-old steers generally averaged about 1960 lbs. There were several entries of payments of half-guineas for cows sent to Studley, showing that the original connection was frequently renewed.[4] The steward at Ormesby repeatedly assured Major Rudd that Sir James told him his breed was a cross between the old shorthorns and the Alderneys.[5]

Mr. Milbank of Barningham, between Richmond and Barnard Castle, appears to have obtained some of his

---

[1] Bates MS.   [2] *Newcastle Courant*, 8th April 1820.
[3] *An Account of some of the Stock of Short Horned Cattle of Charles and Robert Colling*, by Bartholomew Rudd, 1821, p. 7.
[4] Bates MS.
[5] *Origin and Pedigrees of the Sockburn Short-horns*, 1822, p. 55.

stock from the Blacketts of Newby.[1] Among his purchases from Studley was the celebrated red and white Studley bull (626), calved in 1737, and possessing wonderful girth and depth of fore-quarter, very short legs, a neat frame, and light offal. His steward's sons long remembered the fine animals his oxen were as they went to be slaughtered, with ribbons flying from their horns. Mr. Sharter, one of the steward's family, entering on Mr. Milbank's farm at Chilton, took with him the Studley bull (626), and some cows from Barningham. A cow he bred by the Studley bull, weighed at twelve years old, and after having produced several calves, upwards of 110 stones. The weight of a five-year-old ox, bred and fed by Mr. Milbank, and killed at Barnard Castle in April 1789, was 177 stones 1½ lbs. (the fore-quarters 150 stones 4½ lbs., tallow 16 stones, hide 10 stones 11 lbs.). This was considered to mark an improvement, since the weight was about the same as that of each of Sir Henry Grey's seven-year-old oxen, got by a bull of Sir James Pennyman's, and fed fat at Howick in 1787.

Bates had no doubt of the excellence of the ancient shorthorn blood, but although they got to great weights, a less number of cattle were, as far as he could learn, kept per acre in those days, and a greater quantity of food consumed.[2]

The Studley bull (626) was the ancestor of many of the most celebrated shorthorns. Pedigrees in those early days were kept exclusively in the male line. True to the Horatian doctrine—

>'Tis of the brave and good alone
>   That good and brave men are the seed;
>The virtues, which their *sires* have shown,
>   Are found in steer and steed.[3]

the bull was everything:—

[1] Bates MS.   [2] Bell, p. 54.
[3] Fortes creantur fortibus et bonis;
est in iuvencis, est in equis patrum
virtus.

—Q. Horatii Flacci *Carminum* lib. IIII. iiii; *Works of Horace*, by Sir Theodore Martin, i. p. 166.

MALE DESCENT OF DUCHESS AND RED ROSE

Studley Bull (626)
Red and white, owned by Mr. Sharter of Chilton, calved
1737, "possessed of wonderful girth and depth
of fore-quarters, very short frame,
neat frame, and light offal."[1]
|
Mr. Lakeland's Bull
"Great size, good back, sold into Northumberland."[2]
|
Mr. William Barker's Bull (51)
|
Mr. James Brown's Old Red Bull (97)
Bred by Mr. John Thompson of Girlington Hall,
"good fore-quarters and handle; huggins and rump not good;
excellent getter"[3]—"one of the most celebrated
bulls of his day."[4]

| Mr. Appleby's Duchess, | Mr. Watson's Red Rose,[5] |
| sold to Mr. Charles Colling, 1784. | sold to Mr. Robert Colling. |

The important part played by the female in cattle-breeding came, however, as no new discovery. The Romans were well aware of it. Virgil had sung—

> The generous youth who studious of the prize,
> The race of running coursers multiplies,
> Or to the plough the sturdy bullock breeds,
> May know that from the *dam* the worth of each proceeds.[6]

The earliest regular shorthorn pedigree in the female line is that of the Princess family. It begins with the celebrated cow bought by Thomas Hall of Haughton-le-Skerne, in 1760, to which Mr. Wastell of Great Burdon gave the unromantic name of Tripes:—

---

[1] *Improved Short-horns*, by Rev. H. Berry, Liverpool 1824, p. 16.
[2] MS. *Coates's Herd Book*; Bell, p. 367.
[3] *Ibid.* Bell, p. 366.　　[4] Bell, p. 27.　　[5] *Ibid.*

[6] Seu quis Olympiacae miratus praemia palmae
　　pascit equos, seu quis fortes ad aratra iuvencos,
　　corpora praecipue matrum legat.

—Pub. Vergilii Maronis *Georgicorum* lib. III. 49; translated in Dryden's Works, ed. Scott, xiv. p. 76.

THE PRINCESS (LATE BRIGHT EYES) FAMILY[1]

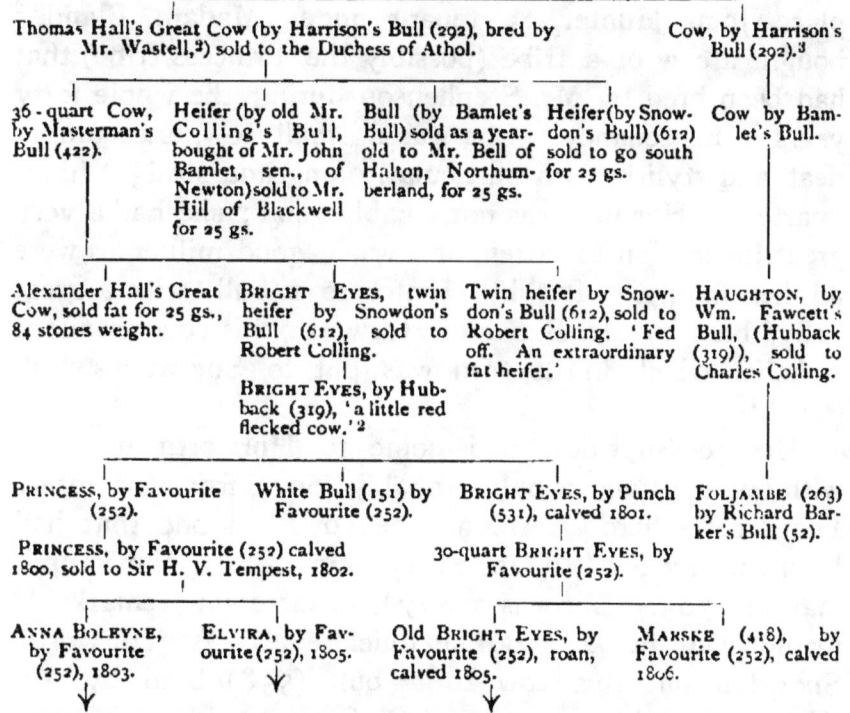

TRIPES, bought in 1760 by Thomas Hall from Charles Pickering of Foxton, near Sedgefield (grandfather of Christopher Mason of Chilton). How bred not known. 'She had a wildness and unruliness about her.'

In some books, but on what evidence does not appear, Tripes is said to have been herself the daughter of Studley Bull (626), and her dam to have been bred by Mr. Stephenson of Ketton in 1739.[4] Mr. Stephenson had

---

[1] See Appendix. Note A.

[2] The error of the Herd Book in identifying this bull with 'Waistell's or Mr. William Robson's Bull' (669) (apparently the same as Robson William's Bull (558) by Masterman's Bull (422)) has made the early pedigree of the Princess tribe appear an instance of close in-breeding among shorthorns before the days of the Collings.—*La Race Shorthorn dite Durham* par M. le Marquis de Chauvelin, Paris, Masson, 1892, p. 4; Housman, *Live Stock Journal*, No. 828, p. 145. Coates also carelessly transposed in the Princess pedigree the order of Wastell's Bull and Masterman's Bull, which he gave correctly in the Bright Eyes.

[3] See Appendix. Note B.

[4] Bates wrote (16th February 1832): 'The original cow (of the Bright Eyes family) came from Chilton and might be by the Studley bull; the first of the tribe came from Holland.'—Holt Beever, *Leading Tribes*, p. 22.

removed from near Ormesby, in Cleveland, in 1731, when the Pennymans had already purchased stock from the Aislabies and were also using Studley bulls. His son went to practice as a surgeon at Hawick, in Scotland, in 1769.[1] At the sale which consequently took place John Hunter, 'a tenant under Madam Bland,'[2] bought a cow of a tribe (possibly the Princess tribe) that had been bred by Mr. Stephenson during the whole forty years of his tenancy. She was a small cow, exceedingly neat and stylish, with particularly long and straight hind-quarters. Her hair was remarkably good; she had a very great inclination to fatten, and was a good milker, as were all her family.[3] Put by Hunter to a bull 'with a great belly,' bred by Mr. Banks of Hurworth, this cow produced a heifer which in its turn was put to Snowdon's Bull (612).[4]

George Snowdon had come to Hurworth in 1774 with six cows and a bull, probably (601), from Sir James Pennyman's herd.[5] Among the cows was one that had been sent by Sir William St. Quintin to Sir John Pennyman in 1765. She was a very handsome cow, remarkable for her wide hooks and fine quick eyes. At Hurworth, Snowdon put this cow to a bull (558) bred by Mr. Wastell from his 'great cow' Barforth (by James Masterman's Bull (422), a son of Studley Bull (626)); the issue was the bull-calf, Snowdon's Bull (612), 'a neat good beast,' for which the father of Mr. Ostler of Audleby, near Grimsby, offered 50 guineas. He certainly had a 'fine fore-end and good crops, but drooped in his hind-quarters and was low-sided.'[6] For some reason or other Mr. Hutchinson of Grassy Nook did not wish to believe that Snowdon's Bull (612) was descended from Sir James Pennyman's stock. 'Had it been a race-horse or a game-cock,' he wrote, 'I should have been more inclined to have given credence to honest George's evidence.'

---

[1] Bell, p. 36.
[2] *Origin and Pedigrees of Sockburn Short-horns*, by J. Hutchinson, p. 56.
[3] Bell, p. 36.   [4] *Ibid.* p. 58.
[5] *Ibid.* p. 26.   [6] *Ibid.* p. 59.

Hutchinson's incredulous cross-examination of Snowdon, about three months before the latter's death, completely failed, however, to shake his testimony.[1]

In 1777, from the union of Hunter's Stephenson heifer and Snowdon's Bull (612), was born that most famous of all shorthorns, the bull (319) which eventually received the name of Hubback:—

HUBBACK

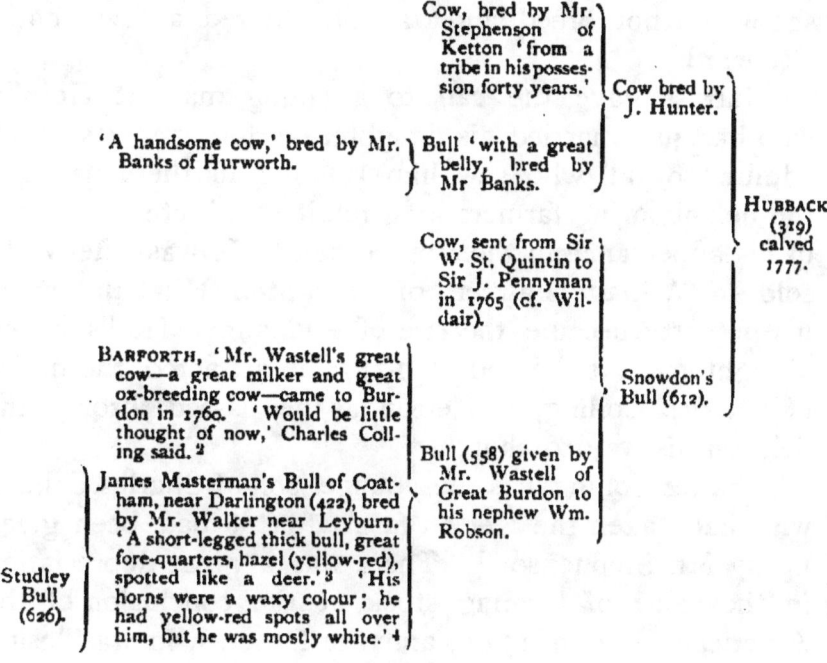

The dam proved to be a good milker, and was on that account retained by Hunter when he gave up farming and sold off the rest of his stock. Those who remembered

---

[1] *Origin and Pedigrees of Sockburn Short-horns*, pp. 48, 49.

[2] MS. note by T. Bates on the Ketton Catalogue. Barforth came, no doubt, from the herd of Mr. Croft of Barforth. He and Mr. Milbank of Barningham were considered to have the purest breed of shorthorns on the Tees in about 1740. The colours of their cattle were red and white; and white with a little red about the neck, or roan.—Bailey's *Survey of Durham*, p. 227.

[3] MS. note on the Ketton Catalogue, 'from information given by Charles Colling.'

[4] Bell, p. 60.

her going in the lanes at Hurworth (for Hunter had no land), described her as a small, remarkably good handler, her carcase near the ground, and all her points very fine. Hunter took the cow with the calf at her foot to Darlington market and sold them to a quaker, who the same day resold them to Mr. Basnett a timber-merchant. On his way home, Basnett called at the door of a blacksmith named Natrass, at Harrowgate (near Darlington), and sold him the calf for a guinea; the cow he kept, but growing quickly fat, on the good land near Darlington, she would not breed, and was slaughtered a few months afterwards.

Natrass gave the calf to a young man at Hornby who had just married his daughter. Here he was much admired by all who saw him running in the lanes, and the neighbouring farmers long retained minute particulars of his appearance in their memories. He was afterwards sold to William Fawcett of Haughton Hill, and while his property became the sire of Alexander Hall's heifer, Haughton. At the end of 1783 he attracted the notice of Charles Colling, of Ketton, as he passed through the field on his way to church.

Charles Colling was the younger son of Charles Colling, who had taken the Ketton farm after it had been given up by Mr. Stephenson.[1] There was a great depreciation in the value of farming stock at the conclusion of the American War in 1782, and the father, who was losing considerably, gave up the farm to his son Charles, retiring himself to Skerningham. The elder son, Robert, had been apprenticed to a grocer, but being in delicate health he returned home and joined his brother Charles in partnership until, in the spring of 1783, he took the Barmpton farm, having previously resided at Hurworth. It was old Mr. Colling who laid the foundation of the Ketton herd by the purchase at Yarm Fair of a cow called Cherry.

Neither of the brothers had received anything more

---

[1] Bates noted on the Ketton Catalogue that J. Walker, a former tenant, left Ketton in 1757.

than an ordinary education, and their portraits do not give the idea of men possessed of a high order of intellect. Acting possibly on the advice of George Culley, Charles Colling had paid a prolonged visit to Mr. Bakewell at Dishley in 1783. Bakewell was a regular practical physiologist. He cut up and dissected the carcases of his sheep and cattle, examined their flesh, bones, and sinews, put them in pickle and afterwards hung them in his laboratory. His great principle was in-and-in-breeding, quite contrary to the ideas then prevalent, according to which the coupling of animals nearly related was sinful in the extreme and sure to produce debilitated offspring. To a Scottish farmer who visited Dishley in 1771,[1] Bakewell's idea that cattle and sheep, as well as horses, were to be distinguished and valued by their blood seemed quite a new doctrine. Pasture, Bakewell admitted, had a great effect on the size and value of cattle, but blood and 'goodness of kind' should, he held, be the chief aim of a breeder. Profit, he owned, was his immediate object, as he thought it should be that of every farmer. 'If,' said he, 'with the same quantity of grass, I lay on my longhorns a finer flesh, more valued by the butchers and more relished by their customers, I have more profit in rearing them.' 'Try different kinds of cattle on your farm,' was his advice, 'and propagate those which thrive and fatten best; dispose of the rest, and by degrees you will improve your stock as far as your pasture can produce it.'

He brought up no calves by hand, but kept common cows for nursing them. The calves were always allowed to suck six months, and this, he said, would have drained his best cows too much, and have been apt sometimes to hinder them taking the bull. Another reason for the use of nurses, he appears to have wished it to be believed, was that his best cows yielded more milk than was necessary for their calves, though George Culley did not hesitate to declare that the cows at Dishley were invariably bad milkers.

[1] *Annals of Agriculture*, xxviii. p. 588.

Bakewell's bulls began to do business when fourteen months old. They were kept together by themselves in strongly fenced enclosures. A deep narrow ditch with the earth thrown out into a mound sufficed to deter them from attempting the hedge. Frequent handling by boys and servants from the time they were calves had trained them to perfect docility. It was amusing to see two little fellows, one of them not yet five years old, mount on the backs of the bulls, and with the gentle touch of a willow wand turn and direct them as they pleased. Bakewell considered that the legs could not be too short, and naturally wished to see the flesh on back and sides rather than on the legs.

The Dishley herd consisted of 150 'black cattle,'[1] all longhorns of the Lancashire breed with the exception of a few shorthorns of different ages. These—his 'patches'[2] as he called them—were kept by Bakewell on the same pasture in order to show the superiority of his own breed. This was unquestioned by his visitors, though his shorthorns were very good of their kind and no worse than the ordinary cattle of the district.

He sent few animals of his breed to the ordinary markets. They were in such reputation, being visibly excellent, that he received commissions at extraordinary prices from all parts of the kingdom and from Ireland. He had been at great expense in carrying out his improvements. Many farmers in the neighbourhood treated him as a whimsical man and gave out that he was bankrupt. In fact before he had succeeded in improving his stock, his finances were very low. For one season's use of a bull it was his rule to take half the value.

All his cattle were housed from November to March, so as to effectually 'winter-hain' his fields. This he considered to be of great importance, as preventing the ground being poached and producing rich grass in the spring. What he called 'housing' was not in close byres, but in 'shades' round a straw yard, which held forty

---

[1] *i.e.* neat cattle, see p. 26, above. [2] *i.e.* 'party-coloured.'

cattle. Every stall was about six feet long and held two animals tied to wooden pillars on each side; the place where the straw was laid before them was laid or lined with brick. He fed his cattle through the whole winter on nothing but straw; only the young cattle sometimes got a few turnips.

Owing to his 'housing' all his cattle in winter, he had no occasion for 'shades' in his fields. He was not so much in favour of shelter as many farmers were. He thought it seduced cattle to delicacy and idleness, and that for the sake of ease they would neglect their food. At any rate, in his opinion, it was a great error to make 'shades' or shelters near the hedges, as this caused cattle to hurt the fences and withdraws them from the best pasture. He therefore recommended the use of simple 'shades,' in the form of an L or two sides of a square, that could be easily shifted, in the middle of the fields. He placed rubbing posts in every enclosure to save the fences and afford a natural exercise to the cattle.

Bakewell obtained a permanent success with his Leicester sheep, but the fame of his long-horn cattle rapidly declined after his death. This had been foreseen by Arthur Young, who considered his exertions 'beyond all comparison more successful in sheep than they had been in cattle.' Young rightly complained that 'in the article of quantity of food consumed by a given weight of beef no experiments were offered to his consideration, which was not the case with the sheep, which had been carried through some very interesting trials of this sort much to their advantage.'[1] George Culley had convinced himself that shorthorns offered a better field for improvement than did longhorns.[2] Colling, as has been already stated, said that the finest herd of shorthorns he ever saw was that the Culleys took with them into Northumberland from Denton in 1767.[3]

The Christmas following his visit to Dishley, Charles

---

[1] *Annals of Agriculture*, xvi. pp. 602, 603. [2] *Live Stock*, p. 82.
[3] Bell, p. 40.

Colling dined with his brother Robert. Mr. Robert Waistell of Elly Hill was also there. After dinner Robert Colling asked Charles if he knew of a bull that would do to serve his cows and Waistell's until a large bull calf they were rearing was fit for use, ideas of the merit of cattle being then entirely dependent on their size. Charles told him of Fawcett's little bull (319) which he had seen in the field going to church. He thought it might be bought for very little and sold again without loss when they were done with it. His brother and Waistell then asked him to buy the bull for them, and he did so for eight guineas.

Charles Colling married soon afterwards Mary Colpitts, in June 1784, and when Christmas came round again he and his wife dined with Robert. Waistell was also of the party. Dinner being over, Robert said that the large calf being now in service he and Waistell no longer required the little bull (319) Charles had bought for them, and asked him if he could find a customer for him. Charles asked what price they wanted. Both said they would be glad to take prime cost, whereupon he said they might send him to Ketton. Some time passed and no bull appeared. Charles Colling began to think his brother and Waistell had found out the little bull's value and intended keeping him. One day, however, on returning home from an absence of some duration, he found the bull had arrived. He took Mrs. Colling out to see him, and did not hesitate to pronounce him 'better than any bull he had ever seen.'

Soon after Charles Colling had received the bull, Waistell sent a cow to be served by him. Colling told Waistell's man that he might leave the cow, and go home and tell his master that he might have her bulled if he paid five guineas, as he would take nothing less for service by that bull. The man returned to say that Waistell would not pay five guineas for having a cow served by a bull for which he had only received four guineas as his half share.

On 14th June 1784, Charles Colling bought, for £13,

in Darlington market a cow from Mr. Appleby of Stanwick.[1] Sir Hugh Smithson, as a young man, before he was High Sheriff of Yorkshire and married the heiress of the Percy family, had kept up the celebrity of the Stanwick shorthorns by paying the greatest attention to their breeding.[2] He used regularly to weigh his cattle, and the food they ate, so as to ascertain the improvement made for the food consumed, which was the first authentic account of this being done. Sir Hugh had been created Duke of Northumberland in 1765,[3] and in consequence of this Charles Colling gave the name of Duchess to his Stanwick cow. She was a massive short-legged animal of a beautiful yellow-red flecked colour; her breast was near the ground and her back wide; she was, too, a great grower. Charles Colling considered her handling very superior, and no one was a better judge. He even went so far as to say that he considered her the best cow he ever had or ever saw; and confessed that he could never breed as good a one from her even by his best bulls, which improved all his other cattle. This is good evidence of the prepotency of the Stanwick herd, even in females, and that a clearly distinctive character still attaches to the Duchesses, in spite of their numerous viscissitudes, enforces the fact that they must have had their root in very careful breeding. The most extraordinary stories that diseased imagination or jealous spite could invent were circulated, many years afterwards, with respect to

[1] Letter of Charles Colling to Thomas Bates (Ketton, 30th December 1810). Mr. Thomas Appleby was tenant of lands at Stanwick in 1756, paying an annual rent of £190 : 10s.; in 1785 his rent was £304 a year. It is very probable that he occupied much of the land now known as 'Park House Farm.'—(Information supplied by the kindness of the Duke of Northumberland, K.G., November 1895.)

[2] Bates MS. fo. 28. The reference to the long-forgotten fact of Sir Hugh Smithson having been High Sheriff of Yorkshire in 1738, is just one of those minor details the accuracy of which materially corroborates the general truth of the narrative in which they occur.

[3] Persons envious of the good fortunes of this handsome and courteous nobleman referred to him as a 'Yorkshire grazier.' Hearing of this the duke replied by ordering his piper to strike up the tune of the 'Yorkshire Grazier' alternately with the more romantic 'Chevy Chase,' and the custom was continued at the Alnwick rent-dinners into the present century.

the cattle kept in Stanwick park[1] by people who, notwithstanding the careful researches they professed to have made, never even heard the name of Mr. Appleby. According to these the park was a regular bear-garden of every variety of cattle in existence (with the single exception, of course, of shorthorns), the bulls and cows all running wild, and allowed to commingle and calve at random.

The great antiquity of another shorthorn family, that from which the cows Marcella and Mantalini appear to have descended, is shown by an old pedigree with a pencil-note appended in Bates's handwriting:—

Little Hollon was got by Mr. Chapman's Bull of Dinsdale which was got by a Bull of Mr. Colling's of Barmpton. Her Dam by Mr. Thomas Weatherill's Bull of Oxenfield, which was a son of the noted Dalton Bull, and her Grandam by Mr. Hill's White Bull of Blackwell, which was out of a cow sister to the noted Blackwell ox, which weighed when killed 151 stone and 10 pounds the four-quarters,[2] *and the family was in the possession of Mr. Hollon's[3] family for 135 years.*

Mr. Christopher Hill's father was the first who gave as much as £10 for a five-year-old draught ox in the Darlington country. His judgment was considered so good that about 1760 his son's herd at Blackwell was by far the most eminent in the north. Three two-year-old heifers were sold out of it to a Mr. Pelham in Lincolnshire for £50 each, a price till then unknown.[4]

A little later—before 1777—Mr. Hutchinson removed from Smeaton, where he had been best known as a sheep-breeder, to the rich peninsula of Sockburn on the Durham side of the Tees. He possessed himself of 'a large yellow cow with some white, most remarkable for her mellow handling,' but of unknown pedigree. She received the name of Old Sockburn. In answer to his brother's

---

[1] The assertion that the Duchess came 'out of Stanwick Park' seems to be an elaboration of the story—*Saddle and Sirloin*, p. 146. It is probable that the park wall was not completed until after 1797.

[2] Killed at Darlington, 17th December 1779.

[3] Of Stresham, near Blackwell. [4] *Sockburn Short-horns*, p. 47.

inquiry, Mr. Thomas Hutchinson, who went to reside at Sockburn in 1783 wrote, in 1821: 'She might indeed have been descended (for anything I know to the contrary) from the old woman's propitious dun cow found at Durham (some time back now), which directed the monks attending the remains of St. Cuthbert to that seat of ease and magnificence. They were nearly of the same colour, I

THE NEW DUN COW, DURHAM, A.D. 1784.[1]

believe, and the old woman's was a short-horned cow (as you will find her to be if you take the trouble to examine her effigy upon the west corner tower of the east transept of Durham Cathedral), but whether she was of Dutch extraction or not I really cannot say.'[2] The statuary in question (really on the turret attached to the northwest corner of the eastern chapel known as the Nine Altars)

---

[1] *Grimm's Drawings*, vol. ii., MS. Add. Brit. Mus. 15,538. 'The new dun cow' has since crumbled like its predecessor. The drawing of it, given in the American *Herd Book*, vol. viii., and Allen's *History of the Short-horns*, p. 21, is quite inaccurate.

[2] *Ibid.* pp. 11, 12. It is said in *Saddle and Sirloin*, p. 144, that Old Sockburn was 'good enough to be modelled for the cathedral vane.' This is one of the many slips in that taking volume.

was a 'restoration' made in about 1778 of the Gothic original; it represents 'a good pointed cow of the short-horned breed attended by two portly dames in the costume of the reign of George III.'[1] The exigencies of sculpture caused the legs to be unnaturally coarse; the horns were made this time of lead, lest she should ever again be reduced to the condition of a polled beast. Old Sockburn's grandson, by Christopher Hill's Blackwell bull, won the Durham premium of 1787, beating a bull of Robert Colling's.

About the same time that he acquired the Stanwick Duchess, Charles Colling also bought a cow named Daisy, at Brafferton, a village about a mile from Ketton. She was 'an animal very neat in shape and very inclinable to make fat.' Colling had no doubt that she had a cross of Masterman's bull (422, 670).[2] Her tribe was second only to the Duchesses for milking properties. From Alexander Hall, Colling obtained the cow called Haughton, of the Princess tribe.

In March 1785 Hubback (319) gained the premium for the best bull at Durham. The following year Gabriel Thornton, who had lived with Mr. Maynard of Eryholme since 1774, entered Charles Colling's service as bailiff. Thornton mentioned the fine cattle at Eryholme to his new master, and on the 30th of September Colling and his wife rode over to see Mr. Maynard's herd. A seven-year-old cow, by Ralph Alcock's bull (19), rivetted their attention. After some negotiations, Maynard, who knew the family to which this cow belonged were bad milkers, agreed to let them have her for twenty-eight guineas.[3] They were to take her second calf by Dalton Duke (188) the next spring at ten guineas. Accordingly this heifer-calf, which received the name of Young Strawberry, came to Ketton on 13th February 1787. The dam had previously been known as Favourite, but she was called Lady Maynard by Colling. At the time of her purchase she was again in calf to Dalton Duke (188), and gave birth to

---

[1] Raine's *St. Cuthbert*, p. 55.   [2] Bates MS.   [3] *Ibid.*

the Crooked-tail Bull at Ketton in 1787. The Collings bid in vain fifty guineas for a heifer from Old Sockburn,[1] so their Eryholme purchases ought to be regarded as cheap.

Lady Maynard came of a tribe which had long been established at Eryholme.[2] The first of them Mr. Maynard remembered was a 'gray-coloured cow' which his father had when he was a schoolboy before 1750. The terrible murrain, 'vulgarly called Hyanstriking,' was then devastating Yorkshire, and there was a strict cordon drawn and a watch kept to prevent any cattle passing the Tees northwards; nor was any Yorkshire butter admitted to Darlington market. The cream from Eryholme was therefore taken across the river in a boat and churned on the Durham side.[3] The grey cow was the dam of 'a black cow with a white belly and white legs to the knee,' and the grandam of a red cow named Necklace. These three were all of them great milkers, indeed they had always to be milked before calving.

Necklace, 'far from being the best of cows at that day,' was the only one sent by Mr. Maynard to Jolly's Bull (337), 'a dark red with black brindled intermixed,' who left a powerful impress on all his descendants and especially on Lady Maynard. He was bred by Mr. Wastell of Great Burdon.

In 1770 Mr. Jolly of Worsall, then a youth of seventeen, often saw Mr. Wastell at Darlington market. Having always heard of the superiority of his cattle, he took the opportunity one day, as Wastell was getting his horse, to ask him if he would kindly show him them. 'I am going home, young man,' was Wastell's ready reply, 'and if you will fetch your horse and accompany me, I shall be glad to do so.' On reaching Great Burdon,

[1] *Sockburn Short-horns*, p. 1.
[2] See Appendix (C). 'The pretensions of Mr. Maynard's cow and heifer to be called pure short-horns' were called in question by Hutchinson who declared that 'Lady Maynard' had horns 'nearly a yard and a half long.'—*Sockburn Short-horns*, p. 49. The old Teeswaters, *circ.* 1744, had rather long horns 'turned gaily upwards.'—Berry, p. 14.
[3] Bell, p. 93.

THE FAVOURITE (OR LADY MAYNARD) FAMILY

'A gray-coloured cow' at Eryholme, *circa* 1745.

'A black cow with white belly and white legs to the knee.'

Necklace. 'A red colour.' 'Mr. Maynard = Jolly's Bull (337). 'A dark red with black only sent one cow to Jolly's Bull and she was brindled intermixed.' far from the best of cows at that day.'—Bates MS. fo. 19.

Strawberry = Jacob Smith's Bull at Givendale (608). 'A yellow-red with a white back and white legs to the knee.'

= Ralph Alcock's Bull (19), bred by Michael Jackson of Hutton Bonville, 'remarkable for his handling and lively looks; all his stock were like himself.'

Favourite Cow, called afterwards Lady Maynard, red roan.

```
(Dalton Duke) (Dalton Duke) (Dalton Duke) (Hubback) (Foljambe) (Lame Bull)  (Bolingbroke
    188           188           183          319       263      357         or Favourite)
 Miss Lax.     Young Straw-  Crooked-tail  white heifer, Phoenix.  Lady May-  Mason's White
                 berry.         Bull.        died.                nard's Bull.  Bull, 1796.
                                   
                (Foljambe)                  (Ben, 70)  (Grandson of  (Favourite)
                   263                       Venus.    Bolingbroke)     252
                Bolingbroke                               280         Young
                 (86), 1788.                           Lady, 1796.    Phoenix.
```

(Manfield, 404, (Son of Hubback) (Favourite) (Mason's (Son of Fav- (Favourite) (Favourite) (Comet, 155) (Favourite) (Favourite) C.C.
 or Barning-   calved, 1788.   252   White Bull) ourite, 253)   252   252   Petrarch, 488,   252   252   died.
  ham, 56).   Cow, 'a great feeder Lily, 1797. b.c. 1799. Cupid, 177, Windsor, 698, Mary, 1804. Alfred, 23, *Alley*   Comet, 1804, North Star,
                of large frame and   *Mason*   1799.   1803.   1808.   1809.   *blood.*   155.   1805.
                good handling.'   *blood.*

Chapman's   (A son of Maynard's
Grandson of = Maynard's Old = Hubback Old Yellow Cow)
 Punch?   Yellow Cow   calved 1788. Mr. Maynard's last
       Bull = Cow   Favourite cow, calved
              Starling.   1798.

Wastell took Jolly straight to the cow-house. There Jolly beheld a young bull-calf the like of which he had never seen before. He asked Wastell if he would sell it. 'Yes,' he said, 'if you will give thirty guineas for him.' Jolly did not hesitate: 'Then I'll give you thirty guineas,' and Wastell clenched the matter with 'Young man, he is yours, and you may fetch him whenever you like.' Jolly's father was exceedingly angry when he heard his son had bought a bull-calf at so high a price, and declared he would never see the money again. Being intimate, however, with Wastell, he gave his son the thirty guineas the next morning, and told him to bring the calf home.

Jolly's Bull (337) was used for some years on the farm at Worsall. One day, as young Jolly was down in the 'holms' along the bank of the Tees, a gentleman rode up and asked if he still had the bull he had bought from Mr. Wastell as a calf. Jolly said he had, and in answer to a further inquiry, added that he had no objection to sell him. They returned to the farmhouse together; Jolly held the horse while the stranger went and examined the bull. On his return he asked the price. 'Fifty guineas,' said Jolly. Hereupon the other pulled out his purse and told fifty guineas into the young man's hand, disclosing himself at the same time as Jobson of Turveylaws, a farm near Wooler, in Northumberland. Strange as it may appear, though he lived to be ninety, Mr. Jolly never bought another good bull.[1]

Possessed now of Duchess, Cherry, Daisy, and Lady Maynard, the four best short-horned cows in existence, Charles Colling proceeded to put them to Hubback (319).[2] Four beautiful heifers were the result, but the white one from Lady Maynard, calved in 1788, which Colling especially admired as exceeding her dam, met with an accident and died as a two-year-old.[3]

By an extraordinary error of judgment—his first and most fatal—Charles Colling, discouraged by public opinion

---

[1] Bell, pp. 17, 18.   [2] *Ibid.* p. 52.   [3] *Ibid.* p. 60.

running rather on size than quality, sold Hubback (319) then ten years old, on 26th October 1787, to 'Mr. Hubback near Newbiggin in Northumberland' for thirty guineas.[1] Colling had long offered him for sale, to Mr. Thomas Booth of Killerby among others, before any one would give that price.[2] Yellow-red and white in colour, with clean waxy horns, mild bright eyes, and a very pleasing countenance, Hubback (319) was one of the most remarkably quick feeders ever known. He retained his soft and downy coat long into the summer; his handling was superior to that of any bull in his day.[3] George Coates admitted that his quarters were long and straight, and his breast, flank and twist of great size, but he considered that his pens were rather down and flat, not formed to the hooks from the rump, and that, soft as it was, his coat was not very long.[4] Some of these criticisms scarcely tally with other notes of his: 'HUBBACK, head good; horns small and fine; neck fine; breast well formed and firm to the touch; shoulders rather upright; girth good; loins, belly, and sides fair; rump and hips extraordinary; flank and twist wonderful.'[5] 'Mr. Hubback' never had a good bull before or after that which immortalised his name. The bull was used generally in the district during the remaining three or four years of his life. His issue, even that from inferior cows, was much admired by breeders—prepotency all the more remarkable when it is remembered he was not in-bred in the least himself. All agreed that they never had such stock as he produced. None, however, paid attention to it, or reared bulls by him, and very soon no trace of his blood was to be found in the district.

It was the opinion of all good judges of shorthorns in Bates's early days that had it not been for the bull

---

[1] Bates MS. fo. 36, 58. Probably Mr. William Huggup of Spital House, near North Seaton, whose surname was pronounced locally 'Hubback.' Bates saw the vigorous old bull and calves got by him in 1790, the year before his death; see Allen's *History of the Short-horns*, p. 40. Mr. George Hubback of Cowpen Bewley, near Hartlepool, was elected a member of the Durham Agricultural Society in 1800.—*Sockburn Short-horns*, p. 29. There seems no tradition of Hubback (319) left in Northumberland.

[2] Bell, p. 34.   [3] *Ibid.* p. 36.   [4] *Ibid.* p. 63.   [5] *Ibid.* 367.

Hubback (319), and his descendants, the old valuable breed of shorthorns would have been entirely lost, and that where Hubback's blood was wanting the stock had no real merit. Bates was himself of the opinion that no stock ought to have been entered in any shorthorn herd-book that had not Hubback blood in its veins.[1]

So far from approving as yet of Bakewell's system of in-and-in-breeding, Charles Colling seems purposely to have sent the neat fine heifer, Haughton, by Hubback, which he had bought of Alexander Hall, to Richard Barker's Bull (52), in order to avoid a second cross by Hubback. A more complete contrast than this bull to Hubback and his offspring, it is difficult to imagine. He was a large, well-shaped, but coarse and wiry-haired beast, which could not be made fat, and his flesh not of a good quality. He was of a dark red flecked colour, with large head, dark horns, and a black nose.[2] His dam was a big coarse cow; Hill's Red Bull (310), his sire, apparently the first-prize bull at Durham in 1784,[3] was sold to go to Ireland. The produce of Haughton and Richard Barker's Bull was Foljambe (263).

FOLJAMBE

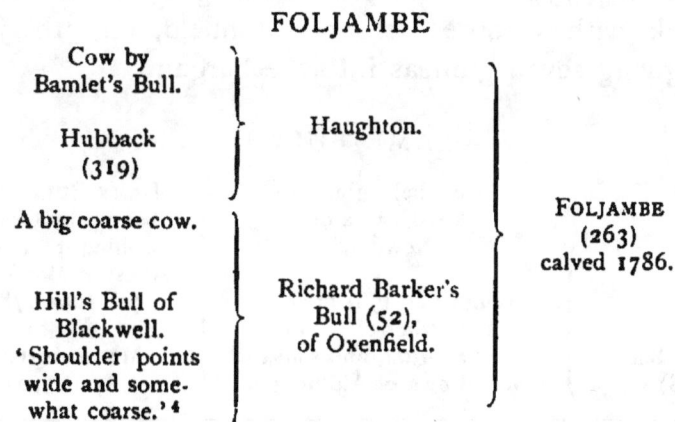

Foljambe was white with a few red spots. According to George Coates, he was a thick useful beast with good handle and dark face.[5] Charles Colling told Bates that

[1] Bell, p. 37. [2] *Ibid.* pp. 54-9. [3] *Sockburn Short-horns*, p. 22.
[4] Bell, p. 56. [5] MS. *Coates's Herd Book*; Bell, p. 366.

he was not of good quality and could not be made fat;[1] Coates, on the contrary, declared he heard Colling say that 'this beast did him most good.'[2] Bates was more in Colling's confidence than any one else, while Coates was a buyer.[3] The 'most good' cannot have been much, since Colling sold Foljambe to Coates for Mr. Foljambe for fifty guineas, on 14th December 1787, when he was about twenty months old.[4] He had only served a very few cows at Ketton. One of these was the beautiful Lady Maynard and another her coarser daughter, afterwards known as Miss Lax. As if to show the uncertainty of breeding, Mrs. Colling's cow, Phœnix, Foljambe's daughter from the former, took his character, which was the very reverse of her dam's; while Lord Bolingbroke, his son from the latter (86), blood red with a little white, calved 12th November 1788, was 'the best bull Coates ever saw,' resembling Hubback in being 'cleanly and neat,' though 'inclined to be podgy or big-bellied and rather low-sided.'[5]

Serious as was Charles Colling's negative error in selling Hubback, it was nothing compared with the positive harm done to his herd by the introduction of Lame Bull (357), which he exchanged for a son of Hubback with George Best of Manfield, on 7th June 1788, giving seven guineas into the bargain.

LAME BULL

Mr. Maynard's Dalton Bull (188).	Cow belonging to Mr. Charge of Newton.[6]	LAME BULL (357), bought by Charles Colling from Geo. Best of Manfield, 7th June 1788, sold to Mr. Robertson, April 1791.
	White Bull (98) bred by Mr. Barker of Kipling near Scorton, and sold to Mr. Brown of Aldbrough.	

---

[1] Bell, p. 55.   [2] MS. *Coates's Herd Book*; Bell, p. 367.
[3] Mr. L. F. Allen suggests that this eulogy on Foljambe was the result of Colling's chagrin at having parted so prematurely with Hubback.—*History of Shorthorns*, p. 45 n.
[4] Bell, pp. 55, 63, 367.   [5] *Ibid.* p. 55.
[6] 'Charge's White Bull (save tips of ears) was by Mr. Aislabie's Bull of Studley.'—MS. note by Bates on Ketton Catalogue.

Charles Colling put Lame Bull (357) to a cow he had bought from Mr. Fairfield of Coatham, and so bred Johanna.[1] Lame Bull was also, Colling believed, the sire of Colonel's dam, sold to Colonel Simson in Fifeshire. Johanna's family, though of a neat form, were a hard-fleshed tribe of shorthorns, with bad wiry hair quite the reverse of the Hubback blood.[2] After this Charles Colling sent his cows to his brother Robert's bulls, Broken Horn (95), Punch (531), and Ben (70). These three bulls did their herds a great deal of harm.[3] Their true pedigree appears to be[4]:—

BROKEN HORN, PUNCH, AND BEN

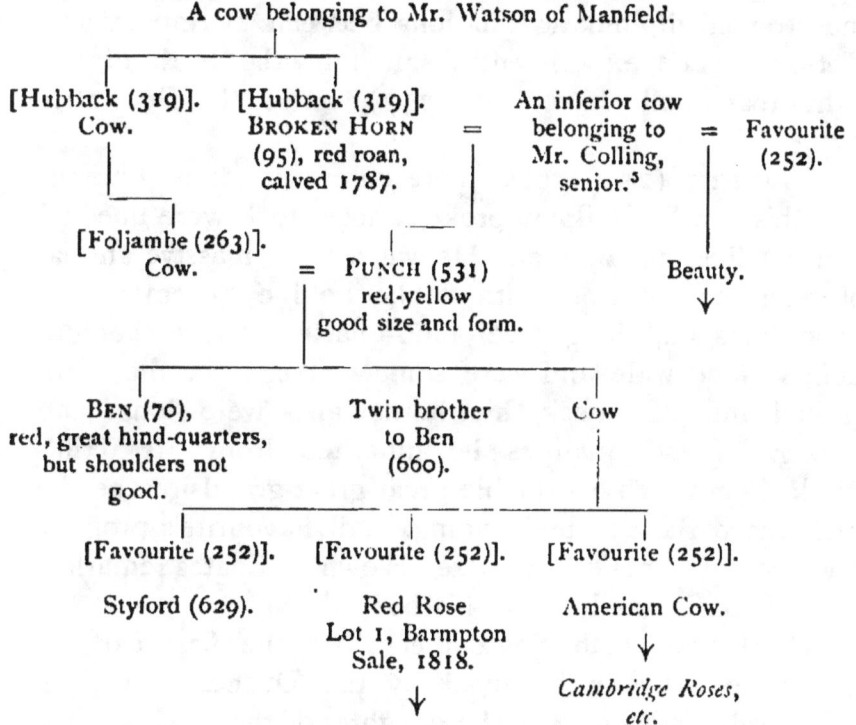

Although Lord Bolingbroke (86) was such a good bull, Charles Colling used him very little, and Robert Colling not at all, as there was a 'shyness' between the brothers at the time. Indeed, this 'shyness' became so

---

[1] Bell, p. 61.　　　　　　　　[2] *Ibid.* p. 82.
[3] *Ibid.* p. 43.　　[4] *Ibid.* p. 211.　　[5] *Ibid.* p. 38.

developed that when Mrs. Colling sent her cow Phœnix a second time to Ben (70), after having dropped Venus to him, Robert Colling told her servant, 'I wonder your mistress should send a cow to my bull.' The man gathered that she was not to be bulled by Ben (70), and took her back to Ketton. On her arrival, in order that she might have a calf of some sort, she was put to Lord Bolingbroke (86), although he was both her nephew and her half-brother. It was thus not till about ten years after Charles Colling had paid his celebrated visit to Bakewell in 1783, that he put in practice a system of in-and-in-breeding among his shorthorns, and then his doing so, instead of being intentional, was the accidental consequence of the strained diplomatic relations between Barmpton and Ketton.[1] The experiment resulted in the birth of the light-roan bull Favourite (252), on 15th December 1793.[2]

Favourite (252) took more after his dam Phœnix than his sire Lord Bolingbroke, whose stock were unequal and not like one another. He was a large massive animal of good constitution, with a fine bold eye, remarkably good loins and long level hind-quarters. His shoulder points stood wide and were somewhat coarse; they protruded into the neck, his horns also were long and strong. These qualities he inherited from Mr. Hill's stock, from which his double great-great-grandsire, the sire of Richard Barker's bull, sprang. All Favourite's progeny had shoulders more or less like his own.[3] Coates remarked that his back was low and his body down.[4]

On Friday 11th September 1795 the first calf by Favourite (252) was calved by the Duchess cow; on Wednesday 5th October the daughter of the Stick-a-Bitch cow gave birth to his second calf, which grew to be the famous Durham Ox.[5]

Notwithstanding Favourite (252) was so good a bull, his dam, Phœnix, was never again put to his sire, Lord

---

[1] Bell, p. 43.   [2] Ibid. p. 63.
[3] Ibid. pp. 55, 56.   [4] Ibid. p. 366.   [5] Ibid. p. 62.

Bolingbroke (86). She was not inclined to breed, and after many bulls had been tried without success, Charles Colling had recourse in 1795 to the yearling Grandson of Bolingbroke (280), three-fourths a shorthorn and one-fourth a Galloway. A red and white heifer-calf called Lady was the result.

The introduction of this Galloway strain into the Ketton herd was perfectly accidental. Colonel O'Callaghan of Heighington, one of Charles Colling's neighbours, had purchased two red Galloway heifers from George Coates, and agreed with Colling in 1791 to have them served by Lord Bolingbroke (86), on the condition that if there was a bull-calf Colling was to have it. O'Callaghan's Son of Bolingbroke (469) accordingly came to Ketton, and as the old cow Johanna had not bred for two years, Charles Colling tried the experiment of putting her to him as a yearling. The produce in 1794 was Grandson of Bolingbroke (280). Bates saw this young bull exhibited at Durham, 31st March 1797, when he took only the second prize; Charles Colling never showed anything again.[1] Bates objected more to the bad shorthorn blood in Grandson of Bolingbroke than to the Galloway alloy.[2]

Lady, the offspring of Phœnix and Grandson of Bolingbroke (280), was a fat-rumped cow, sleek from giving very little milk and being well kept, but lacking hair and handling. Indeed all the stock descended from her wanted good hair and proper handling, but being bad milkers, they showed fat on the tail-head, which is called 'the fool's piece.' Bates never knew any of them that were good milkers and lived to see the very best shorthorns ruined by the use of bulls of this strain. Whenever two crosses on both the sire's side and the dam's were of Lady's blood, he never knew even a decent bull bred, however good-looking were the cows.[3]

Charles Colling sold Lord Bolingbroke (86), when eight

---

[1] *Sockburn Short-horns*, p. 28. Hutchinson was wrong in supposing the bull shown at Durham to have been Favourite (252).
[2] Bell, p. 43.   [3] *Ibid.* p. 43.

years old, to William Jobling of Newton Hall, Northumberland, on 9th January 1797, for seventy guineas.[1]

IMPROVED HOLSTEIN OR DUTCH BULL, 1800.[2]

Vigorous to the last, the old bull was killed at Newcastle on 12th of May 1800, being sold at one shilling per

IMPROVED HOLSTEIN OR DUTCH COW, 1800.[3]

pound.[4] The stock he left with their white faces and red bodies bore a great resemblance to Herefords.[5]

[1] Bell, p. 61.　　[2] Bewick's *History of Quadrupeds*, 1800.　　[3] *Ibid.*
[4] MS. note by Bates in Berry's *Improved Short-horns*, p. 17.
[5] Bell, p. 118.

The complete transformation that the 'Dutch or Holstein breed' underwent in Northumberland in the last years of the eighteenth century is attested by the woodcuts of the Improved Breed which appear in the edition of Bewick's *History of Quadrupeds* printed in 1800. The celebrated artist did not hesitate to ascribe this marvellous improvement to the practice of in-breeding, as in the case of Bakewell's longhorns. Indeed the passage relating to in-breeding as applied to the longhorns in the edition of 1792 is transferred bodily to the account of the shorthorns in that of 1800. It concludes:—

It has long been an established maxim that to improve the breed it is necessary to cross it with others of an alien stock; under an opinion that continuing to breed from the same line weakens the stock. This idea, however rooted it may have been in the minds of former practitioners, is now entirely set aside by the modern practice of breeding, not from the same line only but from the same family. The sire and the daughter, the son and the mother, the brother and sister, are now permitted to improve their own kind. This practice is well known under the term of breeding *in-and-in;* and in this way, the improvement of the several breeds has advanced rapidly to a height unknown before in any age or nation.

CHAPTER III

HALTON CASTLE

1800–1809

HALTON CASTLE is little over a mile to the north-east of Aydon Castle, but the situation is very different—quite open to the south, with a splendid panorama of the Tyne valley from an elevation of 500 feet above the sea-level. Luxuriant pastures on the limestone rise at the back to the site of the old Roman fortress of HUNNUM on the line of the Wall. The fourteenth-century tower, with the arms of Halton carved in a stone panel on the east wall, and round bartizans at the corners of the battlements, was raised, in popular parlance like many other Northumbrian fortalices, to the dignity of a castle by the addition of a comfortable Jacobean dwelling-house. In front is a delightful old-fashioned garden, well sheltered with walls; and close to the entrance the ancient chapel which probably marks the scene of the murder of the saintly King Alfwold in the eighth century.

Thomas Bates took up his residence at Halton at May Day 1800. With a view to stocking the extensive farm he purchased his first shorthorn cows from Charles Colling, giving him for one of them the first hundred guineas the Collings ever sold a cow for.[1]

It was this price, then deemed incredulous, that brought the Colling stock into great repute. Every newspaper was trumpet-

[1] Bell, p. 100.

ing the marvellous wonder, and every gentleman who wished to begin improving his herd went to the Collings for a bull, till it came to be considered a great favour to get one from them. Previously Mason and the Joblings had been looked upon as the great shorthorn men, but this was immediately altered. Everyone who talked of Improved Shorthorns now set his affections on those of the Collings. Mason still had a party that stood up very stiffly for his breed, but the 'dons,' who were able and willing to give high prices, flocked to Ketton and Barmpton. If they could be served there, well; if not, they had to fall back on Mason or Jobling.[1]

The heifer in question seems to have been a daughter of Venus by Ben, and own sister to Mary the roan two-year-old which Colling sold to General Simson of Fifeshire for three hundred guineas in 1806. From the aversion Bates afterwards expressed for Ben's blood, we may be certain that his heifer did no good. Bates also hired Daisy Bull (186) from Charles Colling till the October, when he went into service with Mr. Thomas Jobling at Styford on Robert Colling's Styford (629) returning to Barmpton.

After viewing the celebrated Ketton ox, now four years old, all breeders who recollected the best shorthorns before Hubback's day, and were well acquainted with his descendants, admitted to Bates that the most perfect animals would be those bred from Hubback cows to a Favourite bull; and that all other crosses coming between Hubback and Favourite had done the greatest harm. This remark took deep root in Bates's mind; he always acted upon it in his breeding, and to it, he considered, his success was entirely due.

On 19th October Bates purchased some grazing steers by Favourite (252) from Robert Colling, but although the Barmpton herd was then at its very best, he found them inferior to the twin steers by the same bull which he had

---

[1] Bell, p. 183. Mr. William Charlton, the venerable writer, got confused between the first hundred-guinea heifer Bates bought from the Collings and his first Duchess bought at the same price. A similar confusion led to the *canard* that his first Duchess never had a calf.

bought from Mr. Arrowsmith of Ferryhill the previous autumn. Mr. Arrowsmith was accustomed to feed his animals fat at two years old; in 1801 he sold two steers and two heifers at that age for £25 each, beef being then 10s. a stone.

Finding that there were very few really good shorthorns fit to breed from, and knowing that the best West Highland cattle were small consumers and good grazers, Bates bought the best of these he could obtain to put to the bulls he hired from the Collings. By his directions Mr. Moorhouse, a great importer and a good judge, who lived near Skipton, selected twenty heifers in 1800 and another twenty in 1801 from extensive purchases made from M'Dougall of Lorne and other gentlemen in the West Highlands who had paid great attention to their native breed. Ten out of the forty were nearly equal to the best shorthorns in quality and gave very rich milk. By the cross with the Colling bulls their issue became superior to all except the very best shorthorns; the milk equalled that of the shorthorns in quantity, while retaining the quality of the West Highland.

This cross-bred stock, moreover, proved almost more hardy than the pure West Highlanders. There was on the Wark Eals estate an open pasture of 200 acres which had formed part of Broadpool Common. This lay very high (750-850 feet), near the water-parting between the Solway and the Tyne. Except in the actual mountains there are few more exposed places in Scotland, and the winters in Northumberland were particularly long and severe. The pasture in question was let in 1794 at one shilling an acre, and after being partially surface-drained was valued at only thrice that amount, when sold during the highest war-prices in 1806. Here Bates kept twenty-four of these cross-bred cattle, winter and summer, for the six years 1802 to 1807 inclusive. They were tied up at night in winter in the byres, and went out in the day-time to top the heather and eat the coarse herbage. They had a double coat of hair, one long and the other a close fur or down thicker than that of the Highland breed. The calves were sent up there late in the autumn, were wintered and summered there for two years, and then brought down to the tillage at Halton as two-and-a-half-year-olds. There they received straw and a few turnips the first winter, and were then grazed on inferior land during the summer and on the fogs in the autumn. The last winter, when they were three-and-a-half years old, they got turnips and were sent to market in April, bringing from £26 to £27 a head in 1805 and 1806, when beef was about 7d. a pound for the best West Highlanders and 6d. or less for ordinary shorthorns.[1]

Bates kept all his hay at Halton for the milk-cows in April and May after calving. No other class of stock ever got hay from him in Northumberland, unless it were a little coarse hay from bog-land.[2]

Glimpses of the happy family life passed between Aydon and Halton Castles continue to be afforded by the indefatigable correspondence of Miss Joyous Moore, whose military enthusiasm was aroused by Colonel Browne of Mellington, a distant cousin, then at Newcastle in command of his regiment, riding out to call at Aydon Castle, in full regimentals:

AYDON CASTLE, 29*th August* 1800.

My eldest nephew, John Bates, is one of the cavalry to guard ten miles round Hexham. He serves without pay. When he has his helmet and scarlet jacket he has quite the look of a soldier. He uses the broadsword in the parlour sometimes, but I have never seen him on horseback; he is reckoned to sit a horse very well. Thomas is neither so tall or lusty; we got him excused being a soldier. On Thursday the 14th inst. the colours were presented to the cavalry and the infantry by Colonel Beaumont's lady. My nephew Thomas drove me and his mother to Hexham in the chaise to see the ceremony. We were very near and heard all she said. It was a very fine speech and she spoke it very well. This you may well suppose when I tell you that the officer of the cavalry, who was to receive the colours, attended so closely to it as to forget what he had to answer. He said, 'Sir, Madam,' and was then so confounded that he could utter

---

[1] Bell, pp. 88, 89. [2] *Ibid.* p. 88.

no more. He attempted it again, but not to the full purpose. He received the flag and bowed. He is a very sober man and should have taken a glass to give him courage. The officer belonging to the infantry spoke his answer with great spirit. I saw Colonel Beaumont's party leave the town in a 'sociable.' I never saw one before; they are rightly named, and are very clever. This one took nine persons in the middle, and three in one end.

AYDON CASTLE, 20*th March* 1801.

Many persons come into this country and buy fat cattle and sheep. They take them to Liverpool, Manchester, and other places to sell. My two nephews sold some about a fortnight ago; one received £440, the other £560. Thomas Bates breeds fine sheep and lets out tups for the season.[1]

AYDON CASTLE, 10*th November* 1803.

I hope when the French hear how well prepared the English are to protect their country, that it will deter them from making an attempt upon it. When the emigrant priests were in England, an old priest who lived about four miles from here used to come to teach my nephew, John Bates, French. Bonaparte, he said, was nobody; others there were who contrived and then Bonaparte acted what they would have him to do.

There is every preparation making in this country for its defence. On Tuesday last an army agent bought six hundred and forty bushels of oats from my nephew Thomas Bates. They are all to be sent in this week and the next to Hexham, about four miles from the place where he lives. He has a threshing machine that goes by water; his father has one here that is worked by horses which turn the wheel. At quickest motion they can thresh sixty Winchester bushels in an hour. On a dry day they take in a stack, and on a wet or frosty day, when the men and horses would not otherwise be employed, they thresh under cover.

AYDON CASTLE, 22*nd February* 1804.

The first week of this month threw Newcastle into a great panic, an express from the north having arrived with the news that the enemy had either landed or intended to land. Before

---

[1] In April 1803, Bates, being at Morpeth market, dined with Admiral Collingwood. He produced a good ring which had been found on removing some stones near the Roman encampment at Halton, and as the Admiral expected daily to be called to town, requested him to take it to Sir Edward Blackett there.—W. Clark Russell, *Collingwood*, p. 110.

this alarm, two thousand of the army of reserve out of Yorkshire marched from Newcastle to Hexham and Corbridge on their way to Carlisle by different lines to meet at those places till they were ordered further.

With this letter referring to the mistaken firing of the beacons along the coast, celebrated in the pages of Sir Walter Scott's *Antiquary*,[1] Aunt Joyous's correspondence closes. She preserved her lively disposition to the very last, and was accustomed to walk over to Halton Castle every fine day to see her favourite nephew. In March, when Thomas Bates spent several days at Aydon Castle helping his father, she had as great a flow of spirits as ever. She was highly pleased with beating George Bates and her sister at cribbage in the evenings, and repeated many of the verses and maxims with which her memory was stored without any variation. One Saturday she complained of finding the staircase too steep. As she would never hear of a doctor coming to see her, Thomas pretended to be unwell himself and asked one to call. The doctor did not think her case serious. She dozed on an easy-chair in an upper room for the greater part of two or three days, while Thomas assiduously tended her with warm handkerchiefs. Early on Wednesday morning the 28th of March 1804, she passed tranquilly away in her eightieth year.[2] *Ultima suorum*, the last of a name so long held in honour on the Welsh Marches, Joyous Moore, in accordance with a wish she had expressed to her nephew, was laid in the Bates family vault in Corbridge church.

It was principally with a view to crossing his West Highland heifers that Bates had bought Daisy Bull (186) from Thomas Jobling of Styford in August 1802 for thirty guineas, and had hired Styford (629), Red Rose's own brother, from Robert Colling at Christmas 1803. These were the two bulls he had led out to show to

---

[1] Waverley Novels, *The Antiquary*, ed. 1831, p. 338, note to chap. xxiv.
[2] Letter of Thomas Bates to William Holmes, Halton Castle, 2nd April 1804.

Alderman Blackett of Newcastle, a younger brother of Sir Edward Blackett of Matfen, who visited him at Halton Castle in 1804. The predecessors of both were by Hubback (319), and they both showed his character in their looks, hair, and handling in a very striking manner. After examining both bulls very attentively the alderman said they reminded him greatly of his father's cattle at Newby in his early years, and pointed out the distinguishing characteristics.[1] Bates put Styford (629) to forty cows of his own and near twenty of his father's, but there was not a calf of his get that did well, while a severe loss was sustained in 1804-1805 by many of the cows casting their calves. Bates, indeed, had been most unfortunate with his cattle, losing no less than a hundred head in four years.[2]

The very substantial legacy that Bates received under his Aunt Joyous's will now aided in accomplishing what had been his heart's desire since the Darlington Great Market of 1799. For 100 guineas he purchased from Charles Colling the following Christmas Duchess-by-Daisy-Bull (186), calved, Thursday, 12th March 1800.[3] He bought her daughter at the same time, apparently for sixty guineas, but Mrs. Colling importuned her husband to request the return of the heifer, and Bates had reluctantly to consent.

Milk occupied an important place in Bates's mind at the time. A few days before he had gone over from Ketton to Barmpton to dine with Robert Colling. On his giving him an account of the celebrated 48-quart cow at Stamfordham,[4] his host rejoined: 'When I came to Barmp-

---

[1] Bell, p. 24. Bates's pocket-book contains the memoranda :—' March 10, 1804, measured White Bull at Halton, six years old, girth 7 ft. 10 ins., length 6 ft. 2 ins., weight by measure 90 stones (of 14 lbs.); Red Bull, six years old, 7 ft. 9 ins. by 6 ft., weight 84 stones; two-year-old Kyloe, 6 ft. 7 ins. by 5 ft. 7 ins., weight 60 stones. Old Red Bull came yesterday and two-year-old Kyloe went to the Eals to-day. November 15, 1804, Red Bull went away, girth 7 ft. 14 ins., length 6 ft.' It would seem that the White Bull was Daisy Bull, the Red Bull Styford, and the two-year-old kyloe Chieftain (135).

[2] Bell p. 206. [3] *Ibid.* p. 62. [4] See above p. 8.

ton, after the death of Mr. Harrison, I had no thought of becoming a breeder of shorthorns, and only kept dairy cows. My sister, who then lived with me, said that the milkmaid had repeatedly told her that one of the cows gave an extraordinary quantity of milk. She happened to mention this again one Sunday evening before the cows were milked, and I said " Let us go and see the cow milked and we can then measure the milk and see what she really gives." The servants knew nothing of our intention '—here Colling turned to his old servant who stood behind his chair with, 'I believe, Joshua, the milkmaids do not get up early on a Sunday morning, therefore twelve hours could not have transpired between the morning and evening milking.' 'After seeing the cow milked,' he continued, 'we had the milk measured and found that there were twenty-six quarts and a half, ale measure, that evening alone.'[1]

It happened that Styford (629), who had been returned by Bates and made a seg, was then at Ketton[2] together with the Grey Bull (also belonging to Robert Colling) which Mr. Hustler and his tenant at Acklam had hired the year before. No man ever had better fingers than Charles Colling, but as to the handling of these two animals Bates for once ventured to differ from him. After a long discussion, he begged Colling to re-examine them and say whether he bided by his first opinion. Colling did so, and then openly acknowledged his mistake, saying:

'The Grey Bull has precisely the same handling that Hubback had, and better than that of any other animal except the cow you bought of me yesterday evening. Her handling I consider the best, and all her family have had the same.'

Bates then asked how the two bulls in question were bred. Robert Colling replied that they were the offspring of two own sisters.

'What bull was each by?' continued Bates.

[1] Bates MS. fo. 32.   [2] *Ibid.*

'Styford was by Favourite (252); the Grey Bull by the White Bull (151).'

'What bull got the White Bull?'

'Favourite.'

'And the White Bull's dam?'

'Hubback.'

'Have you any cow now bred as the White Bull?'

'None now in my possession.'

'Have you any other bulls that are bred as the White Bull, direct from Hubback to Favourite?'

'None.'

Bates then told Robert Colling that he had bought the only cow Charles then had that went direct from Hubback to Favourite, and that he would give him his own price if he would let him send this (Duchess) cow to his White (Princess) Bull.[1] This Robert for some reason refused point-blank. It was in vain that Bates offered a hundred guineas for the service, assuring the Colling brothers that the result would be superior to anything they had ever seen. Subsequently, he repeated the same offer but it was again rejected.[2]

As some mitigation Charles Colling promised Bates that Duchess-by-Daisy Bull should be served by her own son Duke (224), a roan two-year-old then going to Mr. Gibson's at Stagshaw Close House within a mile or two of Halton. Mr. Gibson also was a party to this arrangement. In the meantime a Leicester ewe on which Bates set great store, having paid ten guineas for her service by Mr. Donkin's tup, was chased into the mill-pond at Halton by Gibson's pointers, which had been allowed to range by themselves. The drowning of the ewe was aggravated by its turning out that she was big with ewe-lambs whereas she had only bred tups before. Gibson declined to make

---

[1] A note in Bates's pocket-book at the time of this visit to Ketton refers to the hire of Leicester rams and of a bull for 1806:—'26th December 1804. Took a three-shear tup by Points (let two years to Mr. Barker) for thirty guineas and another at forty guineas, and to have the first offer every year afterwards if I wish it. To have the bull Mr. Seaton has had for 1806 at forty guineas.' This was probably the White Bull.

[2] Bell, p. 33.

any compensation. In an access of Welsh passion, Bates seized a horse-whip and rushed to chastise him. John Moore Bates hastened to Stagshaw and with great difficulty dragged his brother back to Halton. For this Thomas never forgave him. From the first the comparisons between the brothers had not been altogether in Thomas's favour. Popular tradition says that both unfortunately fell in love with the same young lady, and that the good looks and brave demeanour of the elder carried the day.

In the end the matter of the ewe was submitted to arbitration, but Bates only obtained butcher's price for her without any reference to the breeding value. Gibson now refused to allow him to have the use of Duke (224), as he had promised, though he was used as a hack bull by every one else in the neighbourhood. 'Contrary to his natural disposition,' Charles Colling declined to interfere. At the same time he entreated Bates to let him keep Duchess-by-Daisy-Bull, then in calf to Favourite (252), as well as her daughter. After the treatment he met with respecting Duke (224), Bates not unnaturally refused. 'Ask what sum you like,' insisted Colling, 'and I will give it you.' 'No sum would induce me to part with her.' 'I acknowledge you are right; you have in her the best cow in England.' Bates told Colling he had long known that as well as he.[1]

At a very large dinner party at Wynyard in 1807, Bates reminded Charles Colling of his breach of faith with respect to the use of Duke (224). Colling rose from his seat and acknowledged the impropriety of his conduct. He deserved, he said, the reproof he received so publicly and hoped Bates would forgive him. Bates immediately went and shook hands with him. Robert Colling, a man of a high sense of honour, then beckoned to Bates and told him that he could never have thought such a thing of his brother. This was the only instance Bates had ever to find fault with Charles Colling, and they continued ever afterwards on the best of terms till Colling's death in 1836.[2]

[1] Bell, p. 39.   [2] Bates MS. fo. 39.

During the summer of 1805 Bates visited Ayrshire, where he was struck with the resemblance the cattle bore to the old inferior shorthorns on Tyneside.[1] He continued his tour to the remotest parts of the Highlands, and found the spirit of agricultural improvement equally alive there as in more favoured countries. The Highland farmers, however, were greatly at a loss which species of cattle to select. They were then breeding from the coarsest and largest animals. These they found most acceptable to the dealers, who took them to Yorkshire, in consequence of the absurd emulation among the graziers there who should sell his cattle at the greatest price per head, without attending to the cost incurred in making the animal so high-priced. Thus it happened that the Highlanders were for some years losing by neglect the valuable qualities their stock possessed. Bates was astonished that this should be the case with a people so well-informed as the Highlanders, and who showed such quickness of perception on other subjects.[2]

In 1805, Duchess-by-Daisy-Bull produced a bull-calf, to which Bates gave the name of Ketton (709). In high dudgeon at being unable to have her served by Duke (224), he put her to Chieftain (135), the son of Daisy Bull (186), from a West Highland cow, and in 1806 she gave birth to Laird. Still in a dilemma, Bates now had recourse, against his better judgment, to Mr. Mason's St. John (572). This bull, born in 1804, was the eighth calf of Fortune, and was by Favourite (252). Fortune was of doubtful origin. She is entered in the Herd Book as bred in 1793, by Charles Colling, sire Bolingbroke (86); her dam may have been by Foljambe (263), as there stated, but in that case her grandam was certainly not by Hubback (319), nor was her great-grandam bred by Mr. Maynard, for Colling never bought any females from Maynard except Lady Maynard and her calf Strawberry.[3] Fortune was kept by Mason with Lily, the daughter of Miss Lax and the grand-daughter of

---

[1] Bell, p. 76.   [2] *Ibid.* pp. 131, 132.   [3] *Ibid.* pp. 95, 96.

Lady Maynard. The constitutions of both were impaired by high feeding or other causes.[1]

'The Agricultural Society of Tindale Ward' was founded in the autumn of 1804. Bates was the only person not resident in the parish of Ovingham who attended the preliminary meeting; and it was agreed that all meetings of the Society were to be held in the 'town of Ovingham.' The movement had probably been suggested by the Duke of Northumberland who became the patron. For some time all Bates's suggestions were adopted and acted on. The principal promoter of the Society came over to Halton Castle one day in 1806. Bates gave him a hospitable welcome and showed him his farm, explaining that he then kept on a less quantity of grass land more than double the stock that the farm had ever carried before. Then enlarging on the national importance of improving live stock from the greater returns good animals made for the food consumed, he continued:

'All other farmers have only to follow out what I have proved and they will reap the same advantage.'

'They must first possess your stock or they cannot do it,' retorted his visitor.

'True, but I am ready to furnish them with my stock for the purpose.'

'Consider what a fortune you will make,' said the other in a rage. 'I have several friends to whom I wish to give the advantage of disposing of their stock at high prices, and when your opinions are proved to be correct, as I have no doubt they will be, you will receive the benefits I wish them to have. I shall, however, take care to oppose your views so that it may not be known how superior your stock is, and I possess the influence that can accomplish this.'[2]

On the 7th of June 1807, Duchess-by-Daisy-Bull dropped a heifer calf to St. John (572). Bates named this Baroness. Though she pastured that summer with nineteen other cows and was kept in the same way in every

---

[1] Bell, p. 97.   [2] Bates MS. fo. 9.

respect, getting no hand-food whatsoever, Duchess-by-Daisy-Bull gave fourteen quarts of milk twice a day. Each quart of milk, when set up and churned separately, yielded 1½ oz. of butter or 43 oz. a day. The butter was made up for the Newcastle market in half-pounds of 10½ oz., which were sold at one shilling each. The skim milk was bought by the labourers at a penny a quart, and allowing two shillings for the subtraction of the cream this made 14s. 4d. a week. Altogether, therefore, the cow brought in more than two guineas a week.[1]

A sale by auction of ten shorthorn cows and heifers, thirty-five West Highland ones and fifteen of the mixed breed, held at Halton, 23rd October 1807, did not at all answer Bates's expectations, owing to the novelty in the North of this mode of doing business, and the sinister influence already alluded to. The catalogue gives little information as to the shorthorns then at Halton. Of the West Highlanders, some were from the island of Islay; and others, bred by the Duke of Argyle, by M'Gibbon, by Laird M'Dougall of Kerera, and by M'Neil of Oransay. All the cows had been hand-milked except five West Highland ones which had suckled their calves.

On the other hand, Bates had the satisfaction of seeing his views on the national importance of improving the live stock of the kingdom adopted not only on Tyneside, but also by the Bath Society:—

ROBERT HARRISON *to* THOMAS BATES

BATH, *3rd December* 1807.

The meeting of the Bath Society concluded last night. We (I have enrolled my name as a member) assembled on Monday morning in full force and proceeded to business, which consisted of discussions on agricultural topics. This was arranged in the most orderly manner; every member in giving his opinion must rise and address himself to the chairman (Mr. Hobhouse, M.P. for Wiltshire). The consequence is that no person stands up who has not some tolerable observation to make; less nonsense

---

[1] Bell, p. 21.

is heard here than in any society I know. I communicated to this Society the resolutions of the Tyneside meeting in October, urging in the best manner I was able its warmest co-operation and support. My observations were received with the most marked attention, and the Secretary was immediately ordered to write to Lawes of Prudhoe.

The show of live stock was poor indeed; three cows, three heifers, one bull, and two oxen, all of the Devon breed—one of Mr. Gray's cows was the best—were the whole of the cattle. One solitary pig, three pens of Southdown ewes, one pen of Anglo-merinos, and three fat wethers, two, three, and four steer made the sum total of the exhibition. I was appointed one of the judges, and had much difficulty to make up my mind to give the best ox the premium, for he fell far short of his two-thirds.

The meeting was well attended; all the peers, baronets, and M.P.s of the district were here, and a most respectable body of gentlemen agriculturists. Sitwell, Curwen's son, and a Colonel Cunningham (a gentleman jobber that you met at Workington) were the only Northern lights. We spent an evening most agreeably, and wanted only a few Tynesiders to have made the thing complete.

On receiving this letter from Bath, Bates, to expedite matters, drew up himself a memorial on the subject:—

AN ADDRESS TO THE BOARD OF AGRICULTURE AND THE AGRICULTURAL SOCIETIES OF THE KINGDOM ON THE IMPORTANCE OF AN INSTITUTION FOR ASCERTAINING THE MERITS OF DIFFERENT BREEDS OF LIVE STOCK.

The Past is the most valuable source of instruction for the Future. The lessons it teaches are often dearly purchased; but nations, as well as individuals, have borne ample testimony to the salutary effects that result from an attention to them. Let us consider the Agricultural History of our own country. Until very lately manufactures and commerce were looked upon as almost the sole sources of national wealth. They were, accordingly, almost the only objects of legislative attention and favour. Under these discouraging circumstances, Agriculture, except in a very few illustrious instances, was practised only by the illiterate and unenlightened. With small capitals, and with little inclination for prosecuting such inquiries as are harbingers of improvements, these plodded on without deviation from the path of their forefathers. Uncherished by the protecting hand of power, the intelligence and capital of the nation being in a great measure directed to other objects, Agriculture languished, or at least made no considerable advance.

From an apathy so fatal to its best and most substantial interests, the country was roused by years of scarcity. It was at length discovered that some little attention was necessary to be paid to the interests of Agriculture. Foremost in the list of those who directed the efforts of the legislature to this desirable purpose, stood Sir John Sinclair, whose patriotic zeal is entitled to the highest praise. The establishment of the Board of Agriculture; the consequent survey of every county in the kingdom; the enthusiasm for the profession imparted to all ranks from the nobleman to the mechanic; the formation of provincial societies: all these are the effects of the national genius being directed towards Agriculture, and they have already produced returns of incalculable benefit.

It is a fact worth whole volumes of abstract reasoning, that some farms in Northumberland pay now a greater sum in direct taxes on the land than the amount of their rentals during the American War; and yet, from the improved cultivation of the soil, and the still greater superiority of the live stock, the present occupiers realise much greater returns for their skill and capital. Does not this account in a great measure for the depression the country felt at a period when taxes were trivial in comparison to the present? Had Agriculture not been improved since then, would it have been possible for us to maintain the arduous contest in which we are now engaged?

Much has been done in some districts, but a wide field for improvement still lies before us. If ever there was a moment in the History of Britain, which called upon her sons for every kind of exertion, it is the present. Cut off from foreign supplies in the time of need, ought we not to cultivate with increased diligence our own resources, and improve and extend them to the utmost? Our extensive colonial possessions will always ensure us a supply of articles of luxury, but ought we not to inquire with more anxiety and earnestness how far our country is able to furnish from its own soil, that absolute necessary of life, bread for its inhabitants? Now, Great Britain has, in the best of years, supplied barely a sufficiency for its population, and when, from unfavourable seasons, its produce has been diminished, we have been forced to depend on foreign sources of supply. It becomes a question of most serious urgency, what means can be devised to obviate the dangers of such a situation.

To this great end, I conceive that nothing can contribute

more than the skilful selection and improvement of the breeds of live stock. There is no doubt that the various breeds differ very materially, both in the length of time, and in the quantity of food which are requisite to fit them for slaughter. Hence, by choosing the proper species of stock, a part of those extensive tracts of land which are at present set apart as pasturage, may be taken into cultivation, and by adopting alternate rotations of corn and green crops, we may raise a greater quantity of grain, while still maintaining as great a proportion of live stock.

But, in order that the best varieties of each species of stock may be universally adopted, it is necessary that accurate experiments should be made to ascertain the comparative merit of each variety, with a fair statement of expenditure and return. Were all farmers made sensible of the real difference which there is in stock, their own interest would prompt them to adopt the best; and what will so soon convince them of this difference as accurate experiments, on the impartiality of which they can rely? And where is impartiality to be looked for, if it cannot be found in a National Institution, where there can be no interest in any deception?

If but one pound more in ten could be obtained of butcher's meat (and more than double this increase might be expected), how much would it contribute to the advantage of the farming interest? And, if live stock were universally improved in this proportion, how great an addition would be made to the prosperity of the Nation?

The high importance of the subject is evident from the observations of those more immediately interested in attending to it. All who have made the trial acknowledge that they find a great and material difference in the variety of breeds, not only in the quickness of thriving, but also in the consumption of food. It is only of late, however, that any minute attention has been paid to the selection of breeding stock. Indeed, in any records now extant, we find no person who was eminent in these investigations until the time of Mr. Bakewell. Having shown how much might be done by such trials, he excited breeders, throughout many parts of the kingdom, to make similar observations, either upon the stock of the district where they resided, or upon such varieties as judgment or fancy led them to bring from other districts. To forward this great object, amongst others, agricultural societies have now been established in almost every district.

But what information can be gathered from all the observations hitherto recorded, to regulate the selection of any one variety as more suitable than another, for any particular purpose?

Excepting the race-horse, whose speed and bottom are pretty accurately known, we have made but few precise inquiries into the merits of animals, many of them perhaps more deserving of our attention. We have discovered in general that there is a material difference in stock, but what, or how great that difference is, we have not yet been informed with precision or certainty.

This important information, I fear, has not been much advanced by the premiums offered by the agricultural societies in consequence of the respective merits of the different breeds of stock not being sufficiently ascertained. It is true that persons supposed to possess the greatest knowledge are generally appointed judges in awarding the premiums; but what one set of judges think right one year, another often think wrong the next, or, as is generally the case, the judges determine that stock to be the best to which they have been most accustomed, or which they possess themselves. The long-horned or the short-horned are favoured by those who breed the one or the other species, and so in other cases. No discovery is made of the different merits in the stock exhibited, but only of the prejudices, or at best of the sentiments and opinions of the judges. In fact many individuals are discouraged from introducing breeds different from those which are prevalent in their districts, on account of the difficulties, hazard, and prejudices, which they have to surmount.

A striking instance of prejudice occurred when Messrs. Culley and Thompson first introduced the Leicestershire, or Dishley breed of sheep, into the northern part of Northumberland. The dealers from Yorkshire, who bought the cast stock of that district in the autumn, refused to purchase this breed. Messrs. Culley and Thompson were obliged to feed them on turnips in their own neighbourhood, until the Yorkshire dealers were convinced that the sheep which they had rejected were quicker thrivers and more profitable. So great has been the change since the prejudice was removed, that scarcely a vestige now remains of the original stock.

Some societies determine the comparative merit of animals by weighing them alive, and when dead, taking the smallness of the offal as the criterion by which the best may be distinguished. But no accurate judgment can be drawn from this, since poor keep when young greatly tends to the increase of offal (tallow excepted). Other modes, it is true, are adopted by more enlightened judges, such as fineness of bone, true symmetry of form, and quality of flesh, ascertained by the animal being peculiarly mellow and free to the touch, which indicates a strong disposition to fatten. In many districts, however, these circumstances are not attended to; and as the looks and handling of animals depend greatly upon their keep, and their degree of fatness, nothing can be determined with precision, unless the stock has been for some time kept on the same food.

Even under the best regulations, the premiums also of agricultural societies often go into the hands of a few individuals. This discourages those who see no prospect of succeeding; and in a few years the societies are either much neglected or wholly discontinued. Besides, the expense and trouble to which individuals are put in the keep of their stock, in order to gain the premiums, exceeds sometimes the value to be obtained.

One great good, however, has resulted from these societies, and that is the creating and diffusing of a spirit for improvement and free inquiry, which appears, at present, almost universally to prevail, even in the remotest parts of the Highlands of Scotland. It behoves well-wishers to the cause to keep up that spirit, and institute experiments, under every variety of climate, soil, and situation, in order to determine the merits of each species of live stock that may be thought worthy of trial, by ascertaining the quantity of food consumed, as well as the improvement and return produced.

The great importance of a general Institution for carrying on these experiments, is what I am anxious to submit to the consideration of the Board of Agriculture, and of all the Agricultural Societies. It is evident that by this means the intrinsic worth of every particular breed, and its comparative value with other breeds suitable for the same purposes, would be accurately known. The best breed for any particular purpose would at last be discovered—a discovery which would be universally beneficial. Even the plain common farmer would know the real merits of the stock in his possession, and adopt what was best. The individuals possessed of the best breeds would receive in the prices of what they disposed of, or let out, the sure and true reward of their merits, together with the high gratification of having rendered an important service to their country. The purchasers or hirers, would be sensible that they paid no more than an adequate price. The whole farming interest would be well rewarded for their exertions; and, as they could then afford to pay a greater rent, the landed proprietors would be particularly interested in the result. At the same time, the public would be furnished with an additional quantity of food, which would support an increasing population, or, at least, be sufficient for our present numbers, without dependence upon foreign aid. If, as an eminent writer has observed, 'he who causes two blades of grass to grow where only one grew before, is deserving better of his country than all the race of politicians put together'; surely they are not less deserving who raise a greater proportion of animal food, from a given quantity of vegetable produce than has been hitherto done.

One very great evil would be prevented by such an Institution. Breeders whose stock is in general estimation, and who are in the habit of letting out males by the season, or who breed with an intention of disposing of their females, keep them very high while young, and use all methods of forcing them forward in condition. Many who wish to have their stock in repute keep all their breeding stock high, a practice which evidently tends to weaken the constitution of the animals themselves, and, of course, to debilitate their offspring. From this cause, more perhaps than any other, we find breeders complaining that they are disappointed in the produce of their stock. Indeed, many animals that are called high-bred, produce very puny, weak, and tender stock, and are not prolific. Now, if the merit of every breed were ascertained there would be no necessity for such a practice. An enormous expense would be saved to the breeders, and the public would derive a benefit, since the keep given to those forced animals would be applied to feed off such as are intended for slaughter, and thus produce a greater quantity of animal food.

The more the subject is investigated, the greater becomes our conviction of the extent of improvement of which it is susceptible. But improvement in this department is not so quickly accomplished as in cultivation, where a more beneficial mode of practice than what we knew before is easily seen and readily adopted. A new or improved variety of grain, or seed, when once selected, is soon diffused, and multiplies much faster, as well as much more than do animals. This consideration, however, should excite us to lose no time in setting about our plans for their improvement.

It is true that some agriculturists have been trying experiments on these subjects for their own information; and the public are highly indebted to a few distinguished patriots who have communicated their experiments for general benefit. Among the foremost of these I cannot help mentioning Mr. Curwen, whose liberal communications of the experiments which he carries on at Workington are highly honourable to his name, and must render the society over which he presides the resort of distinguished agriculturists from all parts of the United Kingdom. The

Duke of Bedford, too, pursues with much zeal and intelligence the various plans of agricultural improvement carried to so great an extent by his lamented brother. Lord Somerville's exertions are also well known to the public. But till these experiments are carried on with different species of stock, and upon a much more extensive plan than a few individuals, however eminent, are competent to accomplish, they cannot be productive of accurate results. If a public Institution of the nature now recommended were once established, there is every reason to believe that it would create a new and important era in agriculture.

I can see no objections to this plan, except such as may be made by individuals, who are in possession of breeds, from which they derive, at present, great emoluments, and who might be apprehensive of their diminution, if the comparative merits of different breeds were accurately ascertained. These persons may be cool at the first in adopting such a plan, and in allowing their stock to be brought to the test. Yet, I would gladly hope that they possess public spirit sufficient to run the risk of sacrificing their private advantages for the general good. In case of the success of their stock, it would establish their reputation without any future dread of disparagement, and consequently increase the demand. Moreover, as the refusal of those who possess the best varieties of any breed, would render that breed liable to disrepute, from the experiments being tried upon the inferior varieties, they would readily see the risk to which they are exposed, by not acceding to this plan. Those breeds, too, which for any length of time have been attended to, have evidently an advantage over those which are at present neglected, or have but lately been brought into notice. If the persons possessing the former hesitate in such a case, little doubt will remain to what cause their conduct is to be attributed.

The longer this desirable Institution is delayed, the greater will be the number of prejudiced individuals; for the possessors of a breed which happens to be in repute, by putting other persons in possession of a few of their stock, will link them to their cause, and render them instrumental in preventing an accurate investigation into the merits of the breed.

To draw up any plan in detail appears premature and unnecessary, until it be known what encouragement will be afforded by the Board of Agriculture and the Agricultural Societies. The extent to which the Institution can be rendered efficient, depends upon the support which it meets with, and the funds which are raised for carrying it into execution. It is sincerely to be wished, should the Honourable Board approve, that Parliament would see the expediency of granting the funds which may be necessary for an undertaking so highly conducive to the public welfare. Were one great National Institution established, the Agricultural Societies in the different districts would the more readily lend the assistance requisite to try such experiments as the peculiarity of their situations may afford. THOMAS BATES.

HALTON CASTLE, NORTHUMBERLAND,
    19*th December* 1807.

For anything that has really been effected since, this Address might be dated 19th December 1895. The questions of the intrinsic value of stock and of the best modes of promoting sensible breeding, have made next to no progress during the interval. It is impossible for any candid person to read the Address without being struck at the depth and breadth of the views of a man whom his ignorant enemies charged with attempting to ram dogmatic fancies down their throats. With Bates everything was founded on the results of careful experiments in the first instance, and his anxiety that others should participate in the benefits of his conclusions was mistaken for mere rant by those whose prejudices rendered them incapable of scientific inquiry and deaf to logical proof.

A remarkable six-year-old brindled ox of the shorthorn and West Highland cross, was exhibited by him at Newcastle in February 1808 for the benefit of the Infirmary. It measured :

> Height at shoulder, 4 ft. 10 ins.
> Length from nose to setting on of tail, 9 ft.
> Girth at ribs, 9 ft. $2\frac{1}{2}$ ins.
> Girth of fore leg, $7\frac{1}{2}$ ins.
> Breadth between fore legs, 1 ft. 3 ins.
> Breadth of back at ribs, 2 ft. $9\frac{1}{2}$ ins.

The fore-quarters weighed 51 stones (of 14 lbs.) 13 lbs.; hind-quarters, 45 stones 2 lbs.; tallow, 14 stones 10 lbs.; and the hide 6 stones 12 lbs. It was sold to Mr. William Robson, the principal butcher in Newcastle, for £75 : 7 : 5.

In Bates's opinion the block-test was insufficient to

pronounce on the merits of an animal; confirmation by the palate was indispensable:

GEORGE HARBOTTLE *to* THOMAS BATES

LONDON, *2nd March* 1808.

Having had the honour of a card to Lord Somerville's Annual Dinner, which was held yesterday, at the Freemason's Tavern, I was agreeably surprised to hear from his Lordship, that a piece of your fat ox had been on the table. Presuming you would be gratified on hearing the reception it met with, I am induced to address a few lines to you.

After disposing of the prizes, Lord Somerville begged leave to inform the company that a piece of beef of the Crossed Breed (describing it) had that day been on the table. It was, he said, obligingly sent by a Mr. Bates of Northumberland, whom he had not the pleasure of knowing, yet, he trusted the gentlemen present would join him in drinking Mr. Bates's health. This was received with unusual approbation, and without flattering, I assure you I was not the most silent with my knuckles on the table. The beef was not on the table where I sat, consequently I am unable to tell you how it ate.

The party to dinner amounted to about 400, Lord Somerville in the chair, and at his right hand was the Duke of Clarence. Amongst others I noticed the Duke of Bedford, Sir John Sinclair, Sir Thomas Sebright, and Sir Thomas Carr. These gentlefolks were at a cross table at the head; the rest of the company were disposed in four tables down the room. The show was inferior to what I have seen.

George Harbottle had left Anick Grange near Hexham for the fine farm of Remenham on the Thames; the servants he took with him came north again 'on account of the want of great coal fires.'[1] He afterwards went to Russia as agent to Prince Demidov. Lord Somerville's dinners developed into the Smithfield Club.

Bates's promising young bull, Laird (by the cross-bred Chieftain (135) out of Duchess-by-Daisy-Bull), measured when just two years old, 28th March 1808, 7 ft. in girth, and 5 ft. 6 ins. in length, equal to a weight of 62 stones, a gain of 30 stones in twelve months. In May the cele-

[1] Bates's Letter Book, fo. 31.

brated Comet, then two or three months older, was two inches less in girth.

The importance of having portraits of cattle accurately drawn to scale so that the dimensions of each part of an animal might be at once ascertained, was fully realised by Bates. Robert Pollard received an order to execute in

BRINDLED OX, HALTON, 1808.

this manner, for £50, a hundred coloured prints, 23¼ ins. by 17¾ ins., of the brindled ox shown at Newcastle from a painting by Wilson.[1]

The print itself is very badly executed, but the particulars will give breeders an idea of the animal. The ox was bred from a West Highland kyloe heifer not near a third of his weight. I fed her after breeding him, and when fully fat she weighed only 28 stones (of 14 lbs.) the four quarters. Till three years and a half old, he was kept upon Broadpool Common in the parish of Simonburn in the summer months, and was wintered upon straw

[1] 'T. Wilson, portrait and animal painter, No. 15, west side of Dean Street, Newcastle-on-Tyne.'

at Halton. The Common consisted of land which I sold to Mr. Ridley two years ago at 75s. an acre for the freehold, and it is not worth more than 3s. an acre to rent by any tenant at the present day, a great proportion of the herbage being heather. He had my feeding pasture from September to November when he was put to turnips. The April following, I sold his partners of the same age and breed for £26 : 10s. each at Skipton market. This ox was under 61 stones, the weight of the others at that time, and I kept him at the desire of a Newcastle butcher to kill at their races in June. On turning him out to grass, however, he made so great an improvement that I kept him on till he was six years old. During that period he had no other indulgence but what I gave my other feeding stock—grass in summer, and in winter, turnips and straw in a loose fold.[1]

CHARLES GORDON GRAY *to* THOMAS BATES

STRATTON HOUSE, CHILCOMPTON,
14th November 1808.

I only received the print of your fat ox a few days since and beg to return my sincere thanks for the honour. . . . The account of the ox proves the excellence of your judgment in particularising two animals, each good of their kind, and in blending them so that their best properties are reproduced. I think, however, that the characteristics of the ox have not had justice done them by the painter—the fore-quarters are not sufficiently extended, the rear fore-leg is much too high to the shoulder point, the off hind-leg is badly expressed, and the point of the rump and tail is unnatural. The head, countenance, and neck are good and show well the Highland extraction, but the form generally is not well executed. I regret much not having seen the ox himself.

I read with attention and pleasure your pamphlet on the methods of testing the merits of the different breeds of live stock. It accords much with my own ideas, and proves most forcibly that were the hints you give, and the strong facts you produce, generally acted on, it would be of the utmost benefit to the community. Unfortunately, we have as yet too much prejudice in agricultural pursuits for men to adopt them. I shall present the pamphlet to-day at our general meeting in Bath, to the President, Mr. Hobhouse, Dr. Perry, and the Society.

Though asked to stay at Melmerby Hall on his way,

[1] Letter Book, fo. 48.

Bates gave up going to Mr. Curwen's sheep show on Belle Isle in July 1808, as he could get no companion in consequence of its being the Newcastle race week. An invitation to meet a party of ladies at Warden confirmed him in this abstention. 'No Cumberland scenery,' he replied with patriotic gallantry, 'not even the lake of Ulleswater, can please me so much as the beauties of the Tyne. When this is adorned with sprightly faces and charmed with melodious sounds, who can help exclaiming, Can aught on earth surpass it?'[1]

By way of making out the week, he went to the Newcastle races for the first time in his life, though he had been born and bred within sixteen miles of them. On the Grand Stand he unexpectedly met Misses Jane and Sophia Atkinson from Temple Sowerby in Westmoreland. In the evening he was again much struck with Miss Sophia's beauty as she entered the playhouse. The races were well attended. Cardinal York, the property of Mr. George Hutton and the winner of the cup on the Thursday, carried off also a sweepstake on the Monday and the hundred-pound plate on the Tuesday, a thing unknown before. On his return to Halton, Bates went immediately to Park End. As it was a rainy evening, he set to work to dance in hard earnest, and so nearly cured the bad cold he had caught on the road from the races. Much to his chagrin, the ladies from Park End could not come over to Stagshawbank Fair owing to their being unable to procure proper conveyances, all the chaises being bespoken. Notwithstanding their absence, he entertained a bevy of eight-and-twenty to tea, including Mrs. Fenwick of Bywell, an heiress of about twenty-five with more than a hundred thousand pounds, and a widow lady in full bloom with as much or more at her disposal. The next day he was engaged to go to a house-warming by two ladies, whom the gentlemen of the county wished to encourage. After such honours, he feared to become a very vain fellow, and had not his heart and soul been fully occupied

[1] Letter Book.

in admiration of the beauties of North Tyne, he would certainly have made a bold attempt at a wife.[1]

Stagshawbank Fair was very thinly stocked with cattle fit to graze that year; most of them were very poor and lean. Sheep were a little lower than the previous year; the horses made a bad show, and though trade ruled dull the prices were fully equal to the value. The 'filly' show, on the contrary, was pretty good, the young 'sprouts' very promising, though in the tent where Bates and his friends dined the country girls were a little annoyed by their superiors looking on, and he only saw one very warm squeeze given. He was recalled from Gilsland, where 'a happy harmony pervaded the company,' by the serious illness of his mother.[2]

Diana Bates, or 'Madam Diana' as the workpeople called her, had become thoroughly naturalised in Northumberland, and quite participated in her husband's love of agriculture. She had no patience with the Northumbrian idea that to make good hay it is absolutely necessary that it should be soaked with rain. On a fine day she would go herself with a rake into the meadow, and with ' Now, hinnies, we'll tuck it up to-day, and knock it out to-morrow, and stack it the next day,'[3] urge the labourers to greater haste. Always an eager collector of specifics recommended by 'Aunt Blayney, Cousin Rogers of the Home,' or Mr. Oakley of Oakley, she would send directions 'to make the blue bottle for sheeps' feet' and other pastoral remedies to Halton with a note—

My dear Son Thomas—I have sent you some good recipes which may be of use to you when I am gone.—Your loving mother,
                                                                DIANA BATES.

A dangerous operation, to which she submitted most courageously at her already advanced age, preserved her life for another fourteen years. She died in her ninety-

---

[1] Letter Book. [2] *Ibid.*
[3] Memo. of conversation with Wm. Hunter of Boggle House, 27th June 1882.

third year, in 1822, in consequence of falling off a chair in taking down her Bible from the book-case.

As the result of a faction in the Tyneside Agricultural Society attempting to prevent his receiving the credit due to his proposals for experiments in feeding cattle, Bates did not go up to London at the end of April 1808 as he had intended. The opposition which the measures passed at the general meeting of the Society in October encountered from those in authority had only, however, the effect of increasing his exertions. He transmitted the Address to the Board of Agriculture, which he had drawn up, to Sir John Sinclair, and also wrote to him by the same post. Most wonderfully strange did he think it when neither Sir John nor the Secretary of the Board acknowledged the receipt of either the Address or the letter. Indeed, Sir John afterwards sent him his own address to the Board at the conclusion of the session without the slightest allusion to having received that from Halton. Bates could only conclude that his letter had been lost in the post. Subsequently he was glad to find that Curwen of Workington had prevailed on the Board to admit the propriety of the measure. The experiments that would be carried out in consequence of the premiums offered would, he felt sure, convince the Board and the public of the importance of the National Institution he had suggested.

He sent Mr. Curwen a cheese from his dairy at Halton for the Workington meeting in October, trusting that it would convince the longhorned breeders that the mixed breed—shorthorn on Highlander—'which he had dared to recommend as paying the most for what they consumed, possessed the property of pleasing the taste, not only in beef but in dairy produce.' Two other cheeses were sent by Bates at the same time to the Mayor of Newcastle and to his old friend Lord Collingwood with the two notes:—

Mr. Thomas Bates presents his compliments to Mr. Mayor, and begs his acceptance of a cheese from his own dairy. The beef Mr. Mayor has from Mr. Robson is from the same breed of cattle, which Mr. Bates submits to the judgment of the public whether it exceeds other breeds in flavour as much as he is convinced it exceeds them in producing a greater quantity of animal food from a given quantity of vegetable.

Mr. Thomas Bates sends his most respectful compliments to Lord Collingwood, and begs his acceptance of a cheese made from his own dairy.
To the Lord Collingwood,
Commander of His Britannic Majesty's Fleet off Cadiz.

HALTON CASTLE, 22nd August 1808.

For his own part, Bates continued 'to keep up his amusements rarely,' being invited out almost every day that he did not entertain large parties at Halton. Sometimes as many as a dozen 'beautiful damsels' gathered round his bachelor tea-table.

During this time Bates was conducting an experiment at Halton to determine the respective increase made by soiling and grazing cattle. It lasted from 14th May and ended on 1st October, a period of 140 days; his soiled kyloe left a profit of $10\frac{1}{2}$d. a day, his grazed kyloe $7\frac{3}{4}$d. He requested Mr. George Atkinson of Staingills, near Temple Sowerby, to let Mr. Curwen have his red kyloe cow for the experiments in feeding which were to be conducted at Workington. He also tried to procure two two-year-old shorthorn heifers for the same purpose from gentlemen in his own neighbourhood, but without success.

THOMAS BATES *to* JOHN CHRISTIAN CURWEN

HALTON CASTLE.

It is as I have ever said impossible to induce the leading breeders to send stock for such an experiment unless they be induced either by interest or by hazarding the risk of losing the estimation of their stock by keeping it back. The reception I have met with plainly convinces me that no endeavours, however praiseworthy, will be supported on their own merits when there is so much private interest to struggle against, and when many thousands are yearly made on the supposititious idea of particular stock being better than other without the risk of competition.

I was in hopes that Mr. Mason would have procured you two heifers long before this. He told me at midsummer that you had asked him. With so extensive an acquaintance among the leading breeders of short-horned cattle over the whole of the county of Durham and all Northumberland to Tweedside, I thought he would have found two which would have been a credit to the breed, particularly as many of those which are called high-bred are apt not to be breeders. In that case, as they never come a-bulling, they lose no time in feeding, and grow quicker than open heifers. Mr. Robert Colling has at present a heifer of this description which surpasses any animal ever yet seen for fat. This I thought would make Mr. Mason choose those of the same kind. If you have not heard of any from him, I have here a heifer of Mr. Mason's purest and best stock, belonging to Mr. Temple of Hilton Castle, near Sunderland, which is at my bull (of the mixed Highland breed), but has never yet been a-bulling. I shall go to Sunderland to-morrow to see if Mr. Temple will part with her for your purpose. I would much rather you obtained these shorthorns through any other hand, as however fairly the experiments are conducted, or however proper the stock for them, as I prefer another breed, every advantage would be taken of saying unhandsome things.

I have put the Address I sent the Board of Agriculture in the press (as I have never heard from the Board), hoping thereby to rouse the landed interest for a measure so highly conducive to their benefit.

THOMAS BATES *to* JOHN SPEDDING [1]

HALTON CASTLE, 15*th September* 1808.

I am very glad your stock pleases. As No. 58 is not in calf,[2] I could wish you would send her to Mr. Curwen's to be fed at his experiments, viz. to have her food weighed and the improvement she makes properly ascertained. She is the only tolerably good one of the mixed breed you got. If No. 27 has not taken the bull, as she is from the same kyloe as No. 58, she would make a fair comparison—the pure Highlander against the mixed breed. Although you say it is contrary to the Board of Agriculture's wish to encourage mixing of breeds, yet if Mr. Curwen will

---

[1] Of Mirehouse, near Keswick.

[2] No. 58 at the Halton sale, October 1807, was by Styford (629). No. 27 was 'bred from Mr. M'Neil's stock of Oransay.' Mr. Spedding bought a dozen cows and heifers there, at very low prices.

allow No. 58 to be treated as the other stock, I should have no doubt of the result, provided she is in health.

THOMAS BATES *to* GEORGE HARBOTTLE

HALTON CASTLE, 30*th September* 1808.

I am sorry to say our Agricultural meeting at Ovingham makes slow progess in anything that can be beneficial. They are just following the old track which all other provincial societies have got into. We are quite lost for want of a head; the Duke of Northumberland is the patron.

Notwithstanding the last year has been rather unfavourable for the sale of stock, yet farms are taken higher than ever in this country. The Greenwich Hospital have let a number lately even higher than what Dilston was let for in 1805.

Mr. Curwen gives me a very flattering account of his crops. He is at the head of the Cumberland Agricultural Society, and is a real ornament to it. Perhaps no man ever took more pains to give encouragement to the tenantry of a county. I was at Workington last year and was highly pleased with the reception he gave us.[1]

I am still pursuing the cross breed of cattle. I become more attached to it every day. I am convinced that there is no other breed I have ever tried which pays so much for what it consumes. I keep double the number of cattle on Halton that was kept by my predecessors, and have as many sheep and a third more in tillage than they had. I get as much per head as they did for their shorthorns, the extra price per stone making up the deficiency in the weight.

After the Workington meeting in the autumn of 1808, Bates made a tour of the Lakes. He was much pleased with the scenery, 'the tints of the woods appearing to great advantage in October.' On his return to Halton, he wrote:

I began on the 1st of November to cut twenty acres of turnip-tops for feeding stock and kept my whole stock upon them and

---

[1] Workington Hall was then the Holkham or Woburn of the North. The visitors 'had two days' farming at the Schooze, and dined in a large wooden booth. Mr. Curwen was M.P. for Cumberland, and the gathering had rather a political tinge about it. Every one rode through the fields and saw the ploughs at work, and scanned the turnip drills, and then came back to finish the business portion of the day among the cattle, in the yards, or at the sale of shorthorn heifers.'—*Saddle and Sirloin*, p. 88.

the rusted wheat straw for a month. This will make me strong for fodder this winter, having housed as many turnip roots during last month (November) as will serve all my stock for three months to come—a most excellent practice on such strong soils. The tops are worth from 2s. to 3s. per acre, and to have the turnips carted off and laid up clean and dry, and the land ploughed before winter are no small advantages.

In December Mr. Mason came to spend a few days with Bates. They were dining at Mr. Donkin's when a servant brought the news of an outbreak of fire at Halton. They all instantly repaired thither. A spark from a candle had caught some straw while the maids were milking. The whole premises were in danger, but providentially the wind abated when the fire was at its height, and the active exertions given by the neighbourhood soon got it under. Two stables and two byres adjoining the castle were entirely destroyed, but the stock was got out without loss. The property was fully insured.

At Christmas Bates found that 'his bachelor habitation was less comfortable each succeeding winter, and wished that some good-tempered fair one would come and sit herself down and take possession of his fireside.'

Thomas Bates *to* Matthew Atkinson[1]

HALTON CASTLE, *3rd January* 1809.

With regard to procuring West Highland cattle to breed from, experience is the great requisite. Even when you know how to choose the females, it takes no little skill afterwards to associate them with proper males. If either side is not properly selected, the produce is sure to disappoint the expectations of the breeder. It was not without considerable expense that I obtained those from which I have been breeding. Had I not had the benefit of Mr. Moorhouse's[2] friendship in allowing me to select whatever I pleased out of his large droves going south (and he bought at that time all the best lots), I could not have obtained such as I got. Even with that favour, every calf I have bred from a real kyloe has laid me in at least £5 when calved; for many of those I selected as quick grazers were bad milkers, and when they

---

[1] Of Temple Sowerby.   [2] See above p. 60.

proved so on trial I fed them and their produce, keeping none as breeders which were not good dairy cows. It was only by this means that I obtained a lot of breeding cows to my mind; this I can now keep up by those I rear. I keep 40 cows and all their followers till three years old.

THOMAS BATES *to* REV. J. PATTENSON[1]

HALTON CASTLE.

I keep my young stock very poor. This enables me to keep double the quantity this farm ever carried before. I find six months good keep will make them ready at any time for the butcher. I sold 29 steers on Tuesday for £20 a head; I could not have obtained more than £4 a head twelve months ago. The stock I breed will feed at any age; I have had a-year-old which weighed 40 stones and a two-year-old which weighed 70 stones.

I am well aware of the prejudice against stock of a mixed breed, because they seldom have been known to succeed after the first cross. But experience has convinced me of the propriety of it when done with judgment. If anything is wrong on either side the produce will inherit it; but when male and female are both in perfection their produce is superior to either. I have now animals with various of blood, some with only one-sixteenth of the original Highland breed, and find no falling off from the perfection of the first cross. Almost everything depends upon the male of the short-horned breed; I have only seen improved calves from four bulls.[2]

THOMAS BATES *to* JOHN SPEDDING

HALTON CASTLE, 25*th July* 1809.

My bull Ketton has grown so heavy this summer that two-year-old heifers cannot bear his weight. If your cows are done with the bull you have, I should be obliged by your sending him off with a careful driver, directing him to come by Carlisle, Brampton, and Chollerford.

Soon after you were here I let Laird for this season to go near to Sockburn into the very heart of the short-horned country which will give the mixed breed a trial.

I sold the whole lot of three-year-old steers you saw at £20

---

[1] Of Melmerby, near Penrith.   [2] Letter Book, fo. 48.

each the beginning of June. They made uncommon progress on going to grass. This is one of the greatest advances on a whole year's breed of stock that I ever remember. Mr. Mason offered me £4 a head for them the year before. As lean stock then sold, they were not worth so much. Their summer's keep was very indifferent, and they got only a few turnips with straw (no hay) during the winter.

Thomas Bates *to* John Christian Curwen.

HALTON CASTLE, 5*th August* 1809.

My stock is all doing well. You remember the small steers you saw when last here and which I hoped to get, worth as much as the runts Mr. Donkin bought at £12 : 15s. a head. I have fully accomplished this, as they were sold two months ago at £20 a head. They were computed to weigh 46 stones at 8s. 9d. per stone or 7d. a lb., sinking offal; they gained in the last twelve months more than 30 stones according to the computation of several good judges. You will recollect Mr. Mason saying that he offered me £4 a head for them the spring before last and that they were not worth so much in the market. An increase to five times the original price in a twelvemonth with only indifferent keep is no common thing, and places no small value on stock capable of doing it. If Mr. Mason sends you an account of the experiments he has been making on his short-horned cattle which we saw, for insertion in the Agricultural Report of Cumberland for this year, I wish you could view the ground on which I have kept this mixed breed of stock from my first commencement to the present time, and take down a report from those who have seen how they have been kept together with the improvement they have made, as also the butcher's report of the beef with the prices obtained each year. I have also the dairy account of last year, kept exactly for a twelvemonth. It is these joint considerations which show the intrinsic value of any breed of stock.

Thomas Bates *to* Baron Hepburn

HALTON CASTLE, 10*th August* 1809.

I have received your kind invitation and shall do myself the pleasure of paying you a visit. As the weather seems now more settled and I have not seen the Lothians, I shall come on horseback, sending my portmanteau by the coach.

Thomas Bates *to* John Christian Curwen

Halton Castle.

I have spent a fortnight under the hospitable roof of Baron Hepburn at Smeaton. I never passed a more pleasant time in my life. The country is the best cultivated and most productive I ever saw. The soil and climate are excellent for grain-growing, and the great quantity of lime they use forces such crops as no land produces elsewhere which has been so long in tillage. Peas and beans seem peculiarly adapted to this country. The Baron's farm is a perfect garden. His Swedish turnips are the best I have seen this season, and I never saw a better crop at this time in any year. His white turnips are equally good and his corn very promising. You would be highly delighted with the country. In other districts you may see a farm here and there as well cultivated, but nowhere the general husbandry of a whole district so well attended to. The farmers are well-informed, shrewd men, and very liberal in communicating their practice. To live stock they have paid no attention as their soil does not require to be laid to grass in order to bear white crops. But at the Baron's the greatest connoisseur would relish the well-fed Scotch beef and mutton on the table.[1]

Thomas Bates *to* John Christian Curwen

Halton Castle, *4th October* 1809.

I have spoken to the person who got the last short-horned cows for you to be looking out for eight more that are younger and promising for milk, by the time you come here. I hope you will make it convenient to spend a few days with me before the Ovingham meeting on the 26th. Most of my neighbours go from it to Newcastle Fair. I should wish to have some conversation with you before the meeting to see if some plan could not be arranged for trying the merits of each breed of live stock upon their proprietors' own farms, and having the results accurately ascertained and reported to the public. I see no difficulty in it if all are willing. What part do you think the Board of Agriculture would take in such a plan? As far as an individual ought to go, I am ready to offer my assistance and support.

[1] Letter Book, fo. 57.

Thomas Bates *to* Baron Hepburn

HALTON CASTLE, *6th October* 1809.

My servant takes two couples of ducks to Newcastle in the morning and will forward them by the carrier to Mrs. Hepburn. I have got comfortable accommodation made for them and hope they will arrive safe and well.

A numerous party from this neighbourhood and the county of Durham attended the Workington meeting this year. All were highly gratified with the reception they met with from Mr. Curwen. Nothing was omitted that could render it agreeable and instructive. During the day the examination of the farm and his numerous experiments occupied the attention of every one that had any relish for Agriculture. Those that prefer the sports of the field were also accommodated, it being a Hunt as well as an Agricultural Meeting. The brilliant display of beauty and fashion exhibited at the balls and routs in the evenings could not fail to charm every admirer of the fair ones.

My neighbour, Sir John Swinburne, is appointed one of the stewards next year, when a still more numerous party of us will attend. He is president of the Tyneside Society this year.

Our party were delayed a day on the road on account of the floods which had taken away many bridges and rendered the roads impassable. We were at Carlisle the evening of the 18th. The next morning we went up on the Castle to view the rivers near the town and a more awfully grand sight I never saw. The whole surface of the waters was covered with grain. The lower part of the town was all under water and all the surrounding country. The Caldew was higher than ever was remembered.

Thomas Bates *to* John Spedding

HALTON CASTLE, *October* 1809.

I have determined to leave home on Thursday the 2nd of November to pass the winter in Edinburgh. I should be glad to have the pleasure of seeing you and Mr. Brown of Tallentire here before then.

With regard to a bull, as Laird is kept for the entire season by the person who took him last year, I shall have none to spare except the one you had last season. Should the gentleman you mention wish to have him, I will take twenty guineas for the use of him till the 1st of October next, provided they do not bull any other person's cows for less than 10s. 6d. a cow.

During his tour in Ayrshire in 1805, Bates had noticed the practice of forming compost heaps with twelve-inch layers of peat between six-inch layers of fresh manure, and turning the heap over a few weeks before applying it to the soil. On his return to Halton he adopted the same system and derived great benefit from the first trials. He tells us, however, that this benefit did not invariably accrue:—

In a distant part of the farm there was a very deep peat moss. I prepared the compost in the usual way in winter, but on turning it in the spring, I saw no trace of the manure. The whole compost applied to the turnip crop proved a complete failure. The barley sown after the turnips was not half the crop I had previously had on the same field, nor were the clover and the seeds that followed any better. I then applied fifteen chaldrons of lime per acre, and ploughed it in for oats. The crop was a very great one and the field continued afterwards very productive. Not having then studied chemistry, I could not account for the deterioration and the subsequent improvement. It was this that induced me to go to Edinburgh to study chemistry.

It turned out that the peat used in my first trials was free from oxide of iron, while the presence of it in large quantities destroyed the manure in the last case and rendered it useless. The application of a large dose of lime removed the bad effects of the oxide of iron and converted it into a beneficial manure.[1]

[1] *Farmer's Magazine*, 1850, i. p. 4.

CHAPTER IV

EDINBURGH AND HALTON

1810–1812

AT the age of thirty-five, then, Thomas Bates, a Northumberland tenant-farmer, set out in November 1809 to enter his name as a student at the University of Edinburgh. His note-books contain careful epitomes of the lectures he attended. The course was not confined entirely to practical agriculture, but comprised mental and moral science. Dr. Ritchie on Logic, Professor Brown on the Mind, and Professor Stewart on Moral Philosophy, alternated with Dr. Hope on Chemistry, Professor Playfair on Natural Philosophy, Mr. Jameson on Natural History, and Dr. Coventry on Agriculture. Judging from the syllabus Bates made of his lectures, Dr. Coventry must have been a most able exponent of the principles of agriculture in all its branches, and many of his aphorisms on the breeding and management of cattle must have sunk deep into the mind of the young practical farmer in the trite form in which he preserved them :—

Sheep fatten better on new grass, cattle on old pasture. In the same breeds the coarser animals consume the more food. Rich food given at an early age gives cattle an inclination to quick and early maturity. Pushing them early forward brings them sooner to maturity but does not get a greater size. Rich feed induces a delicacy of constitution, and in early life gives a tendency to old age. . . . Pampering an animal renders it delicate. . . To increase size, let an animal not be put on too rich land. . . . Degenerate progeny is owing to want of food. Lean stock turned into rich pasture make no improvement and acquire bad habits. Put lean stock on coarse pastures. The former habits of animals must be attended to. Highland stots will not rest indoors in summer owing to the heat, yet short-horned cattle do very well in the house. Twenty-four acres subdivided into three will feed twenty-seven cattle if shifted every fourteen days. . . . Cattle vary in size and quality of flesh; accurate knowledge is required to enable us to choose those best adapted to any one particular purpose. By striving after one good quality in a breed you may loose many valuable ones. Form, quality of flesh, hardiness of constitution, and prolificness, all deserve equal attention. (Reports to the Board, Mr. Lawrence on Live Stock, and models by Garrow). Length of legs aids cattle in travelling and on rough pastures and for hilly ground. . . . A Leicestershire sheep would not do on a Highland hill, more owing to want of food than to cold. . . . A breed that attains to earliest maturity degenerates the soonest. . . . In choosing a stock the best formed should be preferred. Mr. Donaldson found that the best formed animals always bore travelling best. Colour may indicate qualities in a breed. Dark coloured cattle are supposed to be the hardiest. As animals get larger they turn lighter coloured. Of animals of the same size, the lighter coloured appear the larger. . . . The bull should be chosen to improve the deficient points of the cows; a bull with a defect is not objectionable if he answers in this particular. Calves should not be born too late for rearing on grass. Mashes and succulent food should be given to cows for three weeks before they calve, and till the grass comes, or else the milk will fall off. Ninety calves have been lost from twenty-three cows in five years through their casting calf. For abortion the twenty-first week and the second calf are most dangerous. Long milking of pregnant cows was found to cause abortion in Cambridgeshire. . . . Some cows keep their milk from one calf, if two are put on them; fear prevents them letting their milk down. Twelve quarts of milk a day is fully sufficient. . . . Weakly calves have eggs and boiled fish given them. . . . Turnips and hay will support calves in a few weeks; a 6-lb. turnip bruised will serve a calf daily during the spring. They should have good grass in summer and 10 lbs. to 12 lbs. of turnips a day the next winter. . . . Ferns and leaves may be used for litter, and are better than none, as was Mr. Bakewell's plan. . . . Water should be given oftener to wintering stock than is usually the case, and on cold days it would be better to give it in a trough in the house. ... The gains of dairying are steadier than those of the cattle market. Milk cows are never equal to the demand. ... Giving hay to coarse young cattle in winter will not pay. ... Pennant and others suppose the cattle at Cadzow and Chillingham to be remains of the old wild breed. Dr. Walker and others think they were introduced from abroad as curiosities. Long-horned cattle are said to have come from Ireland; the Craven breed, improved in the Midlands, were originally great milkers, but changed in this respect when better fed. The short-horned or Dutch and Lincolnshire breeds are supposed to have come from Holland, but the characters are different. The Suffolk duns, said to have come from Poland, are without horns, and give much milk of a poor quality. ... The colour and disposition of cattle change by domestication. The wild cattle vary in size, shape, hair, and direction of horn more from pasture than climate. ... The Highland and Welsh breeds are the same, and at no distant period occupied the whole island. Within one hundred and twenty years this was the only breed in Yorkshire, according to Marshall. In well cultivated districts the face and legs become white, and the cattle turn lighter-coloured by better treatment. ... The Kyloes are the hardiest breed; they received their name from the district of Kyle. They came originally from Kintail. They are of useful form, good meat, and mature early. Bakewell and Marshall thought them the best breed in the island. The breed has been much injured by attention to bone. If improved like shorthorns, they would not do on poor keep. ... Rich food gives early maturity and delicate habit; it is only profitable when you want to spread a breed. The fixing of breeds is much more difficult than is imagined. No stock should be put in a situation below its keep. Some breeds vary as milkers,—perhaps a variety of Kyloes will leave more milk for a given quantity of food than any other breed. ... The Earl of Marchmont's factor had four cows and a bull of the Teeswater breed, and by crossing enlarged the size of the old Ayrshire breed. The Teeswater cows coupled with the Ayrshire bulls made the greatest improvement. An Alderney cross was also introduced.

His friendship with Baron Hepburn of Smeaton introduced Bates to a wide circle of Scottish acquaintances. He was soon on intimate terms with his lecturers, Dr. Coventry and Mr. Brown, and with Lord Woodslie, Mr. Graves, General Dixon, and Mr. Rennie of Phantassie.

It was impossible that a man who threw himself with such energy into advanced agriculture could resist the temptation of acquiring a suitable freehold of his own for his experiments and his improvements. Rennie wrote to tell him of an estate of 1400 acres of fen land in Lincolnshire to be sold the following March.

Thomas Bates *to* George Rennie

EDINBURGH, *February* 1810.

I should have gone to see the estate in Lincolnshire if more time had been allowed. I suppose other parcels will be put up afterwards, and I shall wait till the summer to see the land there when vegetation is at its strongest. I have been here two months attending the University, and am so much pleased with the studies I am engaged on, that I have determined not to leave before the classes break up, which will be about the beginning of May. I will do myself the pleasure of calling at Phantassie on my homeward journey.

My fat sheep at Halton were sold last week at $9\frac{1}{4}$d. per lb., sinking offal. The Newcastle corn markets are much depressed in consequence of the immense arrivals of grain. However, the price of fat stock leaves farmers such profits that they can afford to sell their corn lower; we never have both very low at the same time. I wish you could send me to Halton four bushels of seed beans of the sort you most approve of by themselves, and also a mixture of beans and peas, such as I saw great crops of at Phantassie last summer.

A letter from the vicar of his parish posted Bates up in Tyneside gossip, including the first introduction of steel pens:

Rev. George Wilson *to* Thomas Bates

CORBRIDGE, 21*st March* 1810.

From your various pursuits in Edinburgh, you cannot have much leisure time upon your hands, yet I am willing to believe that you will find a few minutes to spare for the perusal of what comes from an acquaintance in this part of the country.

Though you have been removed this winter, perhaps a degree and a half to the northward of this latitude, I do not apprehend you have felt a colder climate, than what you might experience upon Halton Downs. Sunday, 25th February, was one of the severest days I ever travelled in to the chapel, and even scared my friend Dobinson, of Great Whittington, from attendance. My congregation was consequently chiefly composed of your own township.

Last Sunday was almost as cold though a westerly wind, but the men of Whittington ventured to face it as well as a lady to be churched. Upon my return home, I called at Aydon Castle, and was happy to find both the master and mistress of the mansion in good health. They were preparing for dinner upon a roast turkey, to partake of which I received a most cordial invitation; and had not my afternoon duty stood in the way, I would have 'devil'd' for them both the rump and the gizzard, for the keenness of the air had set me very sharp. Your father said he thought he had caught a little cold, but as to your mother she looks like a four-year-old, and I'll be bound to say, notwithstanding her years, there is not a fairer or finer skinned woman in all Edinburgh.

Mr. Bell[1] informed me everything was going on to his satisfaction. A rascally fox had been paying you a visit more than once, but had committed no depredations. It is supposed that he had even the audacity to take up his abode in your garden, for he had been paddling upon the top of the hen-house, and had leapt over the wicket that leads toward the chapel. He had also attempted a burglary upon a house where you had some lambs, but found it too well secured; in short, he has hitherto been baffled in all his efforts, but it is necessary to be upon guard against him and shut up the troops early in the evening.

Soon after you left the country, Mrs. Giles[2] sent us down three turkeys, a present I understand to Juliana, which she will be happy, I am sure, to thank you for when she sees you. Unfortunately they all proved hens; I took the liberty, therefore, for my daughters have been absent all the winter, of placing one of them upon the spit, and have since sent the other two to see their old gentleman acquaintance again at Halton. After remaining there nearly a week, they returned on Saturday, and I suppose everything is now as it should be.

Yesterday I called upon Miss Donkin at Sandhoe, where I saw Mr. Donkin who gave a good account of his wife. Miss Donkin has received a number of the patent pens, any of which, if you choose, you must mention soon, as I understand there are many applicants for them. I have taken three of them, each different as to its mountings. To one handle you may have as

---

[1] The steward at Halton. [2] The housekeeper at Halton Castle.

many nibs to fasten on as you please, the brass nibs at 2s., the silver at 3s. You may choose a hard nib, a soft nib, or a middling one. I am now writing with one.

To your late friend, Willy Jobling, I have not heard that any successor as bailiff to Greenwich Hospital has been appointed, but I understand that Wailes's brother is the most likely to succeed him. His farm is to be let, and Willy Hunter and young Michael Brown, who married Miss Hall, are talked of as tenants.

This letter was soon followed by another from Jobling's brother:

ROBERT JOBLING *to* THOMAS BATES

NEWCASTLE, *9th April* 1810.

Presuming a catalogue of the stock of Horned Cattle belonging to my late brother and your particular friend may be acceptable to you, I now hand you one. I shall be glad, if you are in Northumberland at the time of the sale, to see you at Newton.

The sale took place at Newton Hall, on the 24th and 25th of April. The thirty-eight cows and heifers made £987, an average of £25 : 19 : 5, the nine bulls, £310 : 5s., an average of £34 : 9 : 5. Bates does not appear to have been present. His only purchase was the yellow and white lot 10 of the heifers for £14 : 10s. In the printed catalogue this heifer is said to have been the daughter of lot 12 of the cows (Bella, by Traveller, dam by Bolingbroke). The Winifred family are supposed to spring from Bella, who was sold to Mr. Winship of Newcastle for £26 : 5s.; while the white two-year-old heifer (afterwards called Snowdrop), sire Duke, dam Darling, lot 21, sold to Mr. Ryle of Biddick for £21, was ancestress of the Bensons.

Spring passed and summer came; Bates still lingered at Edinburgh in his lodgings at Mrs. Ramsay's, 5 College Street.

His return to Northumberland was anxiously awaited by the talented, but high-strung, Walter Blackett Trevelyan of Netherwitton. Mr. Trevelyan had tried the Devon breed on his estate.[1] His opinion that the farmers of Scotland were too apt to overlook and underrate their own breeds of cattle was quoted as authoritative by Sir John Sinclair.[2]

WALTER BLACKETT TREVELYAN *to* THOMAS BATES

NETHERWITTON, 17*th June* 1810.

Appearances of my neglect towards you are certainly much against me. I wrote a full answer to yours of April, but had not, upon re-perusing it, the resolution to send it. Being fitter for after dinner than before, I will preserve it for your reading here. If you will do me that particular favour, on your way home, this house being on the shortest and best road from Edinburgh, give me a previous line lest I might be out of the way.

This day week I dined at Matfen—Ladies Collingwood and Blackett being present—and met your brother John. He is not an agriculturist nor a horticulturist, although he speaks of having a garden of twenty acres, for the Newcastle markets, on some ground on the north of the Tyne.

The drought has been severe and continuous, so I have been looking to my dung pits, that the manure, for the drill turnips, may be as moist as can adhere to the forks.

Last week I went from Newcastle up Tyneside to Haydon Bridge. I do not believe that in any part of England, this county excepted, there is so great an extent of horticultural rather than agricultural tillage to be seen. We certainly take the lead nearly out of eye-reach. An experimental farm, once my hobby, would place us amongst the celestials. You and I can pull together—and if sadly jeered at by the over wise, we shall not be altogether derided.

When Bates finally did tear himself away from his academic studies, he unfortunately missed seeing Rennie at Phantassie:

GEORGE RENNIE *to* THOMAS BATES

PHANTASSIE, 11*th August* 1810.

I am sorry I had not the pleasure of seeing you on your way south. I shall give you my experiences of what I saw in the

---

[1] *General View of the Agriculture of Northumberland*, by Bailey and Culley, p. 122.
[2] *Account of the System of Husbandry in Scotland*, 1812. For a notice of Mr. Trevelyan see *Gentleman's Magazine*, April 1818.

North. My first expedition was to Kinross Fair, where there was a great show of very fine cattle which sold from 10s. to 11s. per stone. I also attended Forfar Fair, where there was likewise a full market of very fine cattle, which if anything sold higher. The principal buyers at Kinross were the Williamsons from Aberdeenshire and the Armstrongs from the South. I leave you to judge what beef must sell at in the spring, so as to pay the feeders.

Bates was very pleased with Lincolnshire, he told Rennie, but did not find an estate there to his fancy:

I was highly gratified in seeing the richest and most fertile soil covered with the most productive crops of grain I ever beheld, particularly oats, then nearly all cut and the quality of the grain superior to anything I have seen since 1803. I found a large variety of shorthorns like those which formerly prevailed along the coast of Northumberland, especially about Boston where there is a quantity of grass-land deep and rich, not unlike that around Haggerston near Berwick.

Shortly after his return to Halton, he complied with Mr. Curwen's request to give him an account of his method of storing turnips:

THOMAS BATES *to* JOHN CHRISTIAN CURWEN

HALTON CASTLE, 14*th September* 1810.

My best land is in grass and the great numbers of cattle and sheep, which I annually breed, rear, and fatten, require in proportion to my farm, more green food to maintain them properly in winter than I can possibly raise. I am obliged therefore to grow as many turnips as I can provide manure for. By keeping my cattle in fold-yards in winter on turnips, and on clover in summer, I can sow nearly all my fallow with turnips. The tillage portion of my farm has a wet spongy sub-soil, and it is impossible to cart off the turnips in winter without doing material injury to the succeeding crops. This induced me to try carting them off before winter. I first tried to keep them in the old barns which I had no use for after the erection of my threshing-machine, but I found that they began to heat when piled together, and could not be kept beyond three weeks or a month.

After trying various methods, I find it best to put them into long stacks on the surface in a dry and airy situation. An old stack-yard adjoining my byres and fold-yards is well adapted for the purpose. The turnips are laid down in a long ridge, seven to eight feet broad at the base and piled up in the middle to about the height of six feet, like the roof of a house. The whole is thatched with straw and fastened with ropes. The cover does not require to be thicker than on corn or haystacks. Sometimes I build the sides straight up to the height of about four or five feet, placing hurdles round the outside. If holes are made through the heap, the turnips are still better preserved. A considerable exposure and free circulation of air are the grand requisites to preserve them from the extremes of vegetation or putrefaction.

In an early season, and if my turnips have been sown soon, I begin storing them about the middle of October, but generally it is the beginning of November. If the weather continue dry and favourable, I keep stocking the bulbs as fast as my stock consume the tops. They are taken up on a dry day in order that as little earth as possible may adhere to the roots, and they are immediately made secure in the store heap so as not to be injured by frost. At this time of the year turnip-tops are more nutritious than the best after-grass. In a favourable season I have supported one hundred and sixty head of cattle on them for four weeks with only rusted wheat-straw, and they made much improvement. The greatest advantage, however, of this practice is that it enables me to sow land unfit for barley with wheat after the turnips, and in some seasons the wheat thus sown is a better crop than where the land has been bare fallow. Turnips thus stored keep well to the middle of March, when I always have Swedish turnips which carry me through to the grass. They are, however, not liked by stock, nor do they improve them so much as those led fresh from the field. Had my soil and situation been more favourable for growing turnips, I certainly should not stack so great a proportion, nor so early as I do; for in some seasons their further growth continues till Christmas, and their juices are at the same time matured. But as my situation is cold and exposed, and frost and snow occur frequently, by storing my turnips I am rendered independent of the weather, and have always at hand a constant supply of sound and nutritious food. Previously to trying this method, I incurred a greater expense in raising three weeks' consumption of turnips in a hard frost than my whole crop now costs storing.

If properly stored turnips do not decrease much in value till February. In some seasons it is even later before they become much shrivelled. Though they certainly grow lighter by being kept till the middle of March, yet I think on an average of years there is a greater loss by frost, when they are left in the field till that season. There is the further advantage that the loss by storing can be foreseen and provided for, whereas the destruction of a crop by frost places a farmer who has a large stock to support in a very embarrassing situation.

In some seasons I have stacked from five to six hundred tons of turnips, and having been properly 'topped and tailed' when taken up, they kept perfectly well till all consumed. Much depends on the expertness of those employed in this operation. The cut should be made directly at the junction of the top with the bulb, and the same caution is necessary in striking off the fibres. If either cut is made in turnip itself rottenness ensues, and the juice oozing out renders the root unpalatable to stock and occasions a griping and looseness very prejudicial to their thriving. If too much top is left vegetation is resumed on the first heat. It requires a careful person constantly to overlook the operation. Should any turnips be improperly cut or bruised, they should be separated and used first, as they would spoil any they were mixed with.

I never stack my Swedish turnips in the autumn as I find they increase in succulence when left growing till the middle of April. When all my spring seed is in, except where they grow, I cut, cart, and stack them in the same manner as the white turnips in the autumn. I have had them keep very well to the end of May and the beginning of June.

It took in Bates's opinion a capital of five times the amount of the rent to farm profitably. At Halton he employed a capital of £7500, one half of which he had sunk in permanent improvements, of which he had only the benefit during the remainder of his twenty-one years' lease. More and more impressed with the importance of possessing a freehold, Bates had purchased the Hall Garth estate near Durham, but the title proved defective in respect of tithes. He now bought a moiety of the manor of Kirklevington near Yarm in Cleveland from Mr. Waldy. The price was £30,000, of which Bates paid down £20,000 in hard cash. About a thousand acres in extent on the new red sandstone, one half of the estate was excellent land with good old grass fields, and the other half poor cold clay. There was no mansion, and Bates eventually resided in the long low farm-house of the Town Farm. The farm-buildings were on the south side of the house, so that the principal rooms looked out on the lane to the north. The manor of 'Lenton,' as Kirklevington is still called by the natives, was the king's land at the time of the composition of Domesday Book. Granted to the Bruces at the beginning of the thirteenth century, it was given by Adam de Brus, Lord of Skelton, in marriage with his daughter Isabel to Henry Percy on the condition that the bridegroom and his heirs were 'to repair to Skelton Castle every Christmas Day, and lead the lady of the castle from her chamber to the chapel to mass, and thence to her chamber again; and after dining with her, to depart.' After the revolt of the first Earl of Northumberland, the manor was confiscated by Henry IV. and granted to Roger Thornton, mayor of Newcastle, in compensation for the losses he had incurred in the king's service. Subsequently it came into the hands of the king's son, John, Duke of Bedford, and his widow, Jacqueline of Luxemburg. In consequence of some flaw in an entail, it was not returned to the Percies with the bulk of their estates, and they continued vainly to petition for its restoration down to the reign of James I. The name of Bates had no foreign sound in Cleveland; William Bates of Easby had contributed £50 towards the national defence against the Spanish Armada.[1]

To the ordinary observer the district round Kirklevington offered anything but a tempting prospect at the time of its purchase:

Nearly the whole of the extensive flat Vale of Cleveland was a cold tenacious clay, resting chiefly on the blue lias. This impervious stratum spread over the entire district a degree of exhaustion, wetness, and sterility, that gave it a bleak and barren

---

[1] *History of Cleveland*, by Rev. John Graves, Appendix No. viii. The connection of the Cleveland branch with Northumberland appears in the suit of Leonard Bates of Welbury in *Calendar of State Papers, Domestic, Addenda*, 1566-1579.

aspect, especially the lower portions which were full of stagnant water.

Once abundant in grass and famous for its cheese and horses, the vale had seen its pastures converted piecemeal into tillage. This was ploughed as long as it would grow a corn crop—little or no manure being brought back—until it was lost in wet and adhesiveness. Little stock was kept; less and less manure was made; and every third year came a bare fallow. The great mass of farms were totally undrained, and when once wet seasons became prevalent, starvation spread over nearly every parish.[1]

In Bates's own words:

At the time of my entry in 1811, almost the whole of the tillage portion of the Kirklevington estate was under the three-course system—1. Bare fallow; 2. Wheat; 3. Oats—still so prevalent in the district and so deteriorating. No farm-manure was laid on any of the crops. A ton or cartload of lime then cost twenty-eight shillings, with the expense of leading, and consequently little or none was applied. The tenants were bound by their agreement to lay what manure they made on the grass land. The gentleman from whom I bought the estate told me that he would never depart from this system, and he was advanced in years, and had lived all his life in the district.

I began by applying all the farm-manure to the tillage, and as far as was possible for turnips. Where the land was too strong, I applied it to beans, drilled like turnips, in rows twenty-seven inches apart. On the wheat crop which followed the turnips, I sowed grass seeds to lie one or two years; this refreshed the tillage. Then as this was a slow process, I bought as much manure as the farm used and put it on the bean crops, taking wheat after the beans. After a wet spring followed by a sudden drought the land could not be well wrought for beans, and they failed as well as the subsequent wheat crop. I accordingly changed the system and after turnip fallow and wheat sowed red clover. Contrary to my experience in Northumberland, I found that on our Cleveland strong lands red clover would stand two years. Following after the second year's clover, I had most excellent crops of wheat without any manure. On repeating this, after an interval of nine years, the first year's clover was good, the second year's very inferior.

I therefore finally adopted a twelve-course rotation, and finding that feron grass in wet seasons, and couch, when any

[1] Milburn's *Prize Essay on the Agriculture of the North Riding of Yorkshire*, R.A.S.E.

was in the land, began to increase by taking two crops after two years' sheep pasture, I modified it to this extent :—first year, bare fallow; second year, wheat (on which was sown 5 lbs. of cowgrass, 5 lbs. of white clover, 2 lbs. of hop clover, and 2 lbs. of parsley with ¼ bushel of Italian rye-grass, per acre); third year, sheep pasture; fourth year, sheep pasture; fifth year, fallow (turnips if possible with 3 tons of lime per acre); sixth year, wheat; seventh year, red clover; eighth year, beans; ninth year, turnips (if possible, and again limed) tenth year, wheat; eleventh year, beans; twelfth year, oats. The only bare fallow is once in twelve years, and coming after two white crops, with an intervening bean crop, properly horse and hand-hoed, as well as manured, leaves the land in high order for the commencement of the next twelve years' rotation.[1]

His views on the proper manner of applying lime, and his reasons for settling at Kirklevington are enforced in the following letter:

THOMAS BATES *to* WALTER TREVELYAN

EDINBURGH, *8th December* 1810.

I am much obliged for your interesting dissertation on lime and morasses. Your sentiments are in unison with mine. I have been fully convinced of the utility of lime from actual experience. I have laid nearly twenty fothers an acre over my tillage land at Halton during the last ten years. The improvements on that farm are principally due to draining and liming, matters despised by my predecessors. They used to assert that lime was of no use and for many years applied none, though they had it for the burning. I cannot give you a fairer proof of the alteration in my crops than by stating the fact that last year my corn tithe was (£92 : 10s.) more than thrice the value they generally paid (£28). Great as this disparity is in the value of the grain crop, it is still more manifest in the turnips and clover. To one who knows how much the future improvement of a farm depends on the green crops for raising manure, it is needless to say more.

Lime operates in many different ways. Perhaps the proper time for applying it has not been attended to so much as it ought to be. I apply it to the clover crop, and where I have been able to get it spread immediately after the first cutting of the clover,

---

[1] *Farmer's Magazine*, 2nd series, xxi. pp. 4, 5.

the effects on the after-growth are abundantly evident. A field which had the lime so applied in 1809 was left for pasture last year, and during the dry spring and autumn last year exhibited the finest verdure, carrying double the stock it would otherwise have done, as was fully confirmed by the adjoining new leys of the same year which had not been limed.

But it is in vain for a tenant to expect any countenance from his landlord for his improvements. The only return I have received is being told that the farm is not one shilling better for what I have done to it. To provide against this unpleasant situation I have purchased a place in Yorkshire to retire to, where most probably I shall live in future. This purchase will require my attention when I leave here in May, and prevent me profiting from your kind invitation.

Duchess-by-Daisy-Bull had run more than a year without Chieftain (135) getting her with calf. Donkin of Sandhoe, who had married Bates's aunt Catherine, offered to buy her, and with much reluctance Bates let her go. Petrarch (488) proved more successful than Chieftain, and Bates implored Donkin to let him have his favourite cow back again. A blind bargain was struck: Bates was to take Duchess and pay a premium of twenty-five guineas, and Donkin was to have the option of buying her calves at twenty-five guineas a-piece. She had four more calves: in 1810, Marquis (407) by Petrarch; in 1811, the twins Romulus and Remus by Sir Oliver (605), and in 1812 another bull calf by this last named bull. Donkin insisted on taking them all. The twins turned out worse than useless, he complained, though they cost him so much money.[1] Sir Oliver (605), Bates considered to be one of the worst constitutioned bulls he ever knew, worse even than his sire, St. John (572). All his produce were like him in this respect; few of his calves lived many years; those that did had delicate calves, so that it would have been better if they had been dead-calved themselves.

It is said that the Ketton and Barmpton herds showed evident signs of degeneration after the slaughter of

[1] Letter of William Donkin to Thomas Bates, Sandhoe, 3rd January 1820.

Favourite in 1806, and that this caused Charles Colling much anxiety.¹ However this may be, he resolved to take advantage of the prestige he had attained as a breeder, and of the inflation of prices consequent on the great European wars, by a public sale of his shorthorns at Ketton, in October 1810. At this the bull Comet (155) made 1000 guineas, and the whole forty-seven lots averaged more than £150.

DUCHESS 1st, 1815.²

Donkin had told Bates that he did not intend purchasing any of the females at Ketton. Bates proposed that they should buy between them the two-year-old heifer, Young Duchess, by Comet (155), the only grand-daughter of their Duchess-by-Daisy-Bull; but his aunt's husband declined the offer. Knowing full well that there was a disposition to run him up for anything he really wanted, Bates, who was accompanied by his friend Mr. Smith of Haughton Castle, agreed with the auctioneer that he was to go on taking his bids as long as he held up his umbrella. To Bates's surprise and indignation, Donkin was the only person who bid against him, from 121 guineas to 183 guineas, at which figure Young Duchess became his property. Asked immediately by a gentleman, what he

¹ Bell, p. 46.
² From the original coloured drawing by Mr. James Fawcett of Scaleby Castle, formerly at Kirklevington.

would take for his bargain, he replied, 'Not a thousand pounds.' Nor was this mere bravado: he posted over to Sandhoe and offered Donkin 500 guineas more for the grandam Duchess-by-Daisy-Bull, that is to say, to buy her outright, free from Donkin's claim to her calves, but his uncle-by-marriage would not consent.

Young Duchess, when she reached Halton, was pronounced 'a shabby animal' by the whole neighbourhood. Old George Bates especially ridiculed his son's purchase of her, and said that he had many better cows in his own herd which he would have given him for nothing. The Aydon Castle cows were indeed considered the best in the district, and were very fine large ones, much larger than those at Halton, but they certainly did not show the same high breeding that Young Duchess did. Bates took great care to point out to the young farmers who flocked to Halton to see her, the properties cattle ought to have. He went so far as to blindfold them, before they handled the animals, in order that he might be sure that they understood and appreciated properly the good qualities of each.[1]

The competition for Young Duchess destroyed the cordial relations existing between Halton and Sandhoe. Up to that time Bates had given Donkin his own price for anything he wished to part with, and had let him have at his own price anything he asked for. There happened to be a great dearth of provender that winter, and Donkin had asked Bates to buy some of his cows. Bates offered to take them at thirty guineas each, but eventually it was agreed that he was to have the calves they produced for their winter's keep. During his absence again in Edinburgh that winter (1810-1811), his steward Robert Bell paid every attention in his power to these cows. Before calving they had the same quantity of turnips and straw as the feeding cattle, and nothing but hay and turnips after calving. Donkin, however, went over to Halton and

[1] Bell, pp. 180, 181, from information given by the late Mr. G. H. Ramsay.

found them in a miserable condition without the vestige of anything near them but a little straw. On his remonstrating with Bell, the latter said they had very few turnips to spare for anything, and the appearance of the cows and byre spoke sufficiently as to their treatment. Bell subsequently admitted that the cows had shrunk considerably after calving. In the spring Donkin had a great loss of calves, and Bates offered to let him have those his cows had dropped if he paid whatever he liked for the winter's keep. From what Bell and the two men who waited on the cows told him, Bates made out an account of £7 : 10s. a head, considerably less than the profit his feeding cattle left for the same food; but in sending Bell over to Sandhoe, he told him to take whatever Donkin chose to give. Donkin paid the whole amount, but let the matter rankle in his mind for nine years and then charged Bates with gross imposition.

In passing through Newcastle, Bates happened to catch sight of an advanced copy of the *Agricultural Survey of Durham*, just written by John Bailey who had been the joint author, with George Culley, of the *Survey of Northumberland* in 1797. The Durham report was a perfunctory performance, encumbered with dissertations serving to exhibit Bailey's skill in mechanics and arithmetic. There was a needless reference to Bates's Highland crosses in Northumberland, to the disparagement of their milk and butter properties,[1] and Bates at once took up the cudgels:

---

[1] *General Survey of the Agriculture of the County of Durham* by John Bailey, 1810, p. 237. The composite character given to cattle in Durham at this time appears in the following passages :—' In the autumn of 1805 Mr. George Taylor of St. Helen's, Auckland, went purposely into Devonshire to procure the best of the Devon breed he could find; for this purpose he traversed the whole county and selected ten cows and a bull. . . . Mr. Shafto of Whitworth has also procured five cows and a bull from the north of Devonshire: the bull is of a good straight form and a better handler than most of the breed. Mr. Shafto has also a small kyloe cow, bought of Mr. Bates of Halton, which for real good handling is scarcely to be surpassed. About three years ago Mr. G. Baker of Elemore got some Devon cows from near Bath, and a French bull said to have been an excellent one; but he is now breeding from the improved shorthorns. Mr. J. Hopper of Witton

Thomas Bates *to* Joseph Walton[1]

HALTON CASTLE, *22nd October* 1810.

If Mr. Walter Johnson brought a correct message from you, I observe an error with regard to what you have done in ascertaining the quantity of butter from a quart of milk. As I am sure Mr. Bailey will be obliged to both you and me for putting him right, may I request you to try an experiment similar to mine by putting up this evening a quart of new milk taken from the whole quantity milked from each of the six cows you tried before. At the same time please measure how many quarts each of the cows gives. The cream from each of the quarts put up I should like you to see churned yourself in separate bottles on Wednesday morning, having stood three meals, and I will thank you to put down the time when each of the six cows calved and when they were last bulled, together with their pedigrees. Then if you will put the whole of these memoranda in your pocket-book and come and dine with me on Thursday to stop over the Ovingham meeting the next day, I shall feel particularly obliged by your company. Mr. Curwen comes into this neighbourhood on Wednesday, and will probably dine with me on Saturday. If so, I shall expect you will take up your quarters with me as long as you can stay in Tyneside. I have a thousand things to inform you of that I have learnt at Edinburgh.

Thomas Bates *to* John Bailey

HALTON CASTLE, *22nd October* 1810.

You say Mr. Robert Colling tried crossing with the best Kyloe cows he could procure.[2] This is incorrect. He never had a good Kyloe cow in his life as he has repeatedly told me, and yet he might have procured good ones. He has also repeatedly told me in private (and would perhaps not wish it said again without his leave, and I mention it relying on your discretion), that he believed the best breed of cattle might be produced by crossing

Castle procured twelve cows and two bulls of the Herefordshire breed, but finding them bad milkers he sold them to Mr. William Salvin of Croxdale.'—*Ibid.* pp. 238-241.

[1] Of Stanhope, co. Durham.

[2] The whole passage ran :—' Mr. Robert Colling has frequently crossed with the improved short horned bulls, and the best kyloe cows he could procure : the produce made very fat, and much earlier than the pure kyloe ; but he has now given it up, finding that the pure improved short horns are more profitable.'—Bailey's *Survey of Durham*, p. 238.

if good Kyloes could be obtained suitable for crossing with short-horns. Mrs. Charles Colling has a two-year-old heifer in calf out of a West Highland Kyloe I gave her which Mr. Charles Colling refused a hundred guineas for on the evening of his sheep sale. I saw her for the first time the day following, and in my opinion she was worth more than any animal sold at his sale except the heifer I bought.

I have written to Mr. Walton and other eminent short-horned breeders asking them to make experiments similar to my own. I shall have one of the oldest breeders of short-horns with me to-morrow. He means to test himself my dairy-cows' produce in milk and butter, including the short-horns, and the Kyloes, and the mixed breed between them. His experiments, as well as my own for six years previous to your being at Halton at my sale, and for the three subsequent years, are at your service for publication.

I could also give you a much longer account, and a very correct one, of the short-horned breed than what you have given in the Report, if I can obtain leave of the gentlemen who have given me it. An enlarged pedigree of the stock sold at Ketton would be a most valuable acquisition to the report of the county of Durham. If you will promise to insert it in the appendix, I will immediately set about drawing one out. I would devote all the time I have unengaged to accomplish this, and bring certificates signed by those who give me the information for your perusal previous to its being published. The attention required from you would be slight, and I myself would pay the additional expense incurred in the publication, if the Board of Agriculture made any objection.

I have found in the experiments I have made a very great difference between cows of the variety of the same breed with respect to the quality of their milk. Some Kyloes yield only $\frac{3}{4}$ of an ounce of butter from a quart of new milk, others $2\frac{3}{4}$ ozs. for one quart. Some short-horns give only $\frac{1}{4}$ oz., others $2\frac{3}{4}$ ozs.

I have an instance this year; it is a heifer (Baroness) in her first calf, out of Duchess (grandam of Young Duchess which I bought at Mr. Colling's sale) and got by Mr. Mason's St. John, which yielded upon trial, 20th September last (although she calved 25th November, and I have always found the milk yield most butter when the heifers calve first, and it diminishes as they become older calved), $2\frac{3}{4}$ ozs. of butter for one quart of new milk. She gave but three pints of milk, 20th September last; for some time after calving, while on white turnips with hay and straw, not getting what turnips she could have ate, she gave nine pints of milk per day.

I never did try all my cows in winter, while on hay and straw with a few turnips, but my housekeeper says that they generally give one-third more milk when they are put to grass in May than when kept in the house, but that they produce no more butter.

My head servant, Robert Bell, had a short-horned heifer of the common breed of this neighbourhood which calved in February last. He tried her milk in the beginning of April while on white globe turnips with hay and straw, and each quart, he says, yielded 1¾ ozs. of butter. In May, after turning her to grass, his wife tried her again, and had then ¾ oz. of butter; in June she was joined to my dairy cows and yielded likewise ¾ oz.

The result of the average of trials of heifers of the first calf and the same when become cows having had calves, in point of yielding the greater quantity of butter for a quart of milk is in favour of the heifers.

I should not have written so much upon this subject, had I not invariably found that the different individuals of each correct breed inherit the milking properties of their progenitors as much as their propensity to fatten and other qualities.

Thomas Bates *to* John Bailey

Edinburgh, 19*th November* 1810.

I saw Mr. Robert Colling lately. He desired me to request you would add a short note in your appendix to the Durham Report in reference to the opinion he gave you on crosses between the improved short-horned bulls and Kyloe cows, to the effect 'That he has seen much better Kyloe cows since he tried his crosses than he ever had himself; that he could not tell where to procure the best Kyloe cows when he made his trials of crossing, and that he by no means considers his trials sufficient to decide the superiority of the pure short-horns as being more profitable than the crossed breed, but only as being more profitable than the crosses which he made.'

Mr. Walton of Stanhope has informed me by letter that his experiments (on the quality of butter from a quart of milk) were made while his cows were kept on good hay and had only calved a short time. Therefore no comparison can be drawn between these experiments and those I made. As your appendix will be short, the following will satisfy me, and I beg you to charge me with the expense of inserting it:—

'The trials I mentioned as made by Mr. Thomas Bates with Kyloe milk were the average result of his whole dairy during the summer months, when being on grass cows yield considerably less butter from a quart of milk than when fed on hay, in some instances even less than half, consequently no comparison can be drawn between his and Mr. Walton's experiments and their respective breeds. Mr. Bates also says that cows after their first calf generally yield milk producing more butter than when they are older; and that he has found it not incompatible to unite in the same breed of cattle a great propensity to produce both milk and butter, and a great inclination for growth and fattening at an early age, and that all these qualities are inherited by their progeny.'

The grandam of Young Duchess (No. 38 in the catalogue of Mr. C. Colling's sale) was purchased of Mr. C. Colling, December 1804. She gave not only a great quantity of milk, but milk of a superior quality for yielding butter to any short-horn cow I ever knew. For several weeks after calving her last calf, she made 14 lbs. of butter, and even six months after calving she made 7 lbs. of butter per week, and generally milked to within a short time of her calving. A heifer of this Duchess breed, being the first calf got by Old Favourite, weighed when little more than three years old within 6 lbs. of 100 stone, 14 lbs. to the stone, and was allowed to be a greater curiosity than the Ketton ox of the same age when shown with him at Darlington in the spring of 1799. I was informed of the fact of the weight of this Duchess heifer by Mr. and Mrs. Charles Colling since the sale, and Mr. Robert Thompson of Chillingham Barns can inform you of his opinion of her as he examined her with me when shown at Darlington.

The pedigree of Young Duchess as I received it from Mr. and Mrs. Colling is thus: By Comet, dam by Favourite; grandam by Daisy (a son of Favourite); great-grandam by Favourite; great-great-grandam by Hubback; great-great-great-grandam by Mr. Brown's famous old bull of Aldbrough. And what adds to the value of this pedigree is that the cow by Mr. Brown's old bull was as good as any of the tribe since, without her of course being improved by those bulls that have so much benefited the other tribes of short-horns. Mrs. Colling assured me that this tribe has always been the best milking tribe. This Duchess tribe is the only instance now remaining of the produce of Hubback being put to Favourite without some other bull intervening; which circumstance, added to their being a great milk and butter tribe, gives them a pre-eminence over any other tribe of short-horns. But this Duchess breed falls short of the return made in milk and butter by the mixed breed in my possession, produced by crossing Kyloes with short-horn bulls, when the food consumed is taken into consideration.

It is well-ascertained facts that can alone determine the merits of different breeds as well as of the different tribes of the same breed.

The heifer I mentioned in a former letter by Mr. Mason's St. John gives me reason to think that tribe produce much butter though but little milk, and by judicious breeding, keeping that object in view, a tribe giving more milk without diminishing the quality might be produced, which would be a great acquisition in the present advanced state of our knowledge on those subjects. The more the subject is discussed, the greater will its importance appear to every reflecting mind.

With that utter want of logic which is often calculated to drive a correspondent to despair, Bailey, after having gone out of his way in the first instance to give a disparaging account of the butter-making qualities of the Halton kyloes though they were in Northumberland, and after having received Bates's offer to draw up full pedigrees of all Colling's stock, and to bear the expense of printing them, now chose to understand that he was asked to give the pedigree of Young Duchess only. Treating Bates as if it were he who was dense, he wrote:

In the body of the work I have given no pedigrees. If I added anything on this subject, it ought to be a *full pedigree of Charles Colling's stock*, and not of a *single cow*, and that belonging to a person *residing out of the district* of which I have to give an account of the agriculture.[1]

THOMAS BATES *to* JOHN BAILEY

EDINBURGH, 3*rd January* 1811.

I found your letter on my return here. I saw Mr. Robert Colling while in England. He informed me that you had sent him the paragraph I had drawn up which he said was quite correct. I am therefore surprised why it is not inserted but instead thereof a further assertion, viz.: that 'he has never seen any Kyloe cows so good as the best short horns.'[2] From what I know of Robert

[1] Letter of John Bailey, Chillingham, 20th December 1810.
[2] 'Mr. Robert Colling informs me that since he made his trials in crossing (related p. 238) he has seen better kyloe cows than those he experi-

Colling I think if he had considered the matter, having had so few opportunities of seeing any good Kyloe cows, he would not have made the assertion, and would even consider himself obliged to you for leaving it out of your Report. If you wish to give a faithful report of the practices in the county of Durham, you ought to give Mr. Charles Colling's account of his crosses, which are the reverse of what his brother states. Mr. Charles Colling has repeatedly assured me, even within this week, that the produce of a Kyloe cow Mr. Colling had from me was equal to any short-horn he ever knew in quickness of grazing, and that the Kyloe cow herself got fat on land which would not make a short-horn fat.

I am quite astonished how you could draw up the paragraph you have sent from what I stated. You only need refer to our correspondence to be convinced of the impropriety of it. I never said that the short-horned cows I experimented with were of Mr. Charles Colling's breed nor of the Duchess tribe, neither was one of the Duchess tribe included in the experiments.

You state that you are confined to relate only what you found practised in the county of Durham, and yet you did not scruple to draw up an unfair statement of my stock in another county without my knowledge. Surely, therefore, you will allow me to give a correct statement which I shall confine to the same number of words as you have used:—'Mr. Bates informs me that his experiments were made on the average of his short horns and the average of his Kyloe cows while the cows were on grass.'

I was induced to take the trouble I have done in order to avoid any future altercations. I am only sorry I have failed.

Bates had, it seems, urged Charles Colling to give the pedigrees of the cattle sold instead of merely a bare list with the dam, sire, and price, on a broadsheet, as was the custom after a sale at the time. The idea of a Herd Book was, however, so foreign to the mind of breeders in those days, that Colling completely failed to grasp it, and chose, like Bailey, to imagine that Bates wished for the preposterous absurdity that the pedigree of Young Duchess should be the only one given:

_{mented with; but he has never seen any kyloe cows so good (in his opinion) as the best short horns: and if the doctrine holds good that "*like produces like*," he is inclined to think that no improvement is likely to be made in the improved short horns by crossing them with kyloes.'—Bailey's *Survey of Durham* (Supplement), p. 410.}

Charles Colling *to* Thomas Bates

Ketton, 17th October 1810.

I am sorry to acquaint you that it is not in my power to comply with your request in particularly giving Duchess's pedigree in the catalogue of the prices of my sale. I have had a number of applications of the kind from the different purchasers, which I was obliged to refuse, and the insertion of yours singly would have given great offence to those gentlemen that have made a similar request.

After this unsatisfactory letter, Bates made a point of going over to Ketton himself at Christmas. Mrs. Colling asked him whether it was true that he had said, after buying Young Duchess, that he would not take a thousand pounds for his bargain. He replied that it was.

'How far would you have gone for her?'

'I did not tell any one that.'

'Had I but known how far you would have gone, you should have paid the uttermost farthing.'

'I knew that very well, and took the precaution of not bidding openly myself, so that it was not known till after she was knocked down that I was the purchaser.'

The results of his conversations at Ketton, especially with Mrs. Colling, Bates noted down on the sale catalogue, and these notes proved afterwards one of the main foundations of the Shorthorn Herd Book. It still required a great deal of personal pressure before Charles Colling would give him the pedigree of Young Duchess under his own hand:

Charles Colling *to* Thomas Bates

Ketton, 30th December 1810.

According to your request I send you the Pedigree of Young Duchess which you purchased at my sale. She was got by Comet, her dam by Favourite, her grandam, which you bought of me was got by Daisy, her great-grandam was got by Favourite, her great-great-grandam by Hubback, her great-great-great-grandam was bred by Mr. Appleby of Stanwick, and got by Mr. Brown's famous old Bull of Aldbrough.

Daisy Bull was got by Favourite, his dam by Punch, and grandam by Hubback.

The three-year-old heifer shown with the Durham ox at Darlington, weighed within a few pounds of one hundred stones at that age, and was the first calf old Favourite got, and out of a heifer got by Foljambe, and her dam was the Duchess cow that was got by Hubback, the grandam of the cow sold to Mr. Bates.

After the failure to obtain proper recognition of the experiments he had been so long conducting on the relation between the quantity and quality of milk and butter, and after the impossibility of getting Colling and Bailey to comprehend what he meant by 'an enlarged pedigree' of the Ketton stock, Bates had to submit to a third, and for him a yet greater disappointment:

JOHN CHRISTIAN CURWEN *to* THOMAS BATES

WORKINGTON HALL, 5*th January* 1811.

What with indisposition, and what with the effects of vexatious and untoward circumstances, I have been very ill and uncomfortable for the last three months. This will account for my suffering your letter to remain so long unacknowledged.

I hope you continue pleased with your purchase of Kirklevington. I think what you purpose can scarcely fail of answering. Land will, I consider, advance yet greatly in value.

I have been compelled to take a hand in helping other people to breed short-horns. I have got a bull from Mr. Mason—for the country, not myself; my situation precludes it, a calf at three months old costs me five guineas. If I can encourage other persons, I shall be satisfied.

I had 95 gallons, 3 quarts (wine measure) churned yesterday. It gave 30 lbs. (of 16 ozs.) or 17 ozs. on three gallons. This, considering the severity of the weather, is not doing ill.

My turnips keep well, thanks to you. Several farmers have imitated the example set them. The being able to have them unfrozen is a great matter for dairy cows.

I hope you will very shortly receive the Report. It will require great allowances. My patience has been exhausted by the printer; I wish yours may not be by the author. I do not know how you will like what I have said respecting stock. I have spoken my real sentiments, pretending to no knowledge on the subject. Mr. Maynard jun. has surprised me. I hope the result may justify him.¹

I see Sir John Sinclair is supporting the Regent, though he was made Right Honourable by Mr. Perceval. Gratitude in politics would be a monster, and, like honesty, somewhat out of place.

I fear I must go to town soon. Mrs. Curwen is much obliged

WHITE OX, HALTON, 1810.²

for your present of a book. She is very much pleased with it, and I know of no better judge.

Meanwhile Bates had fed another remarkable ox at Halton, the portrait of which (by Wilson) shows a considerable improvement on the brindled ox of 1808.

Thomas Bates *to* John Christian Curwen

EDINBURGH, *8th January* 1811.

I have just returned from England, where I have been for a fortnight. I found everything going on well, except the loss of

¹ The result did justify Mr. Maynard: Mr. Whitaker, as will be seen, was astonished at the excellence of Laird's stock.
² From the original oil-painting formerly at Kirklevington.

six calves, which have died of the quarter-ill. One of these, from the yellow Kyloe, was my greatest favourite.[1]

My white ox turned out excellent beef, but you are no longer an unprejudiced observer of the improvements in live stock. I was surprised in the autumn to find you so great an advocate for the short-horns, although your judgment was formed on no other proof except the *ipse dixit* of men who take every pains to undervalue the real merits of other stock and offer no proof of the value of their own except their own assertions. Never was more art used than in puffing off Mr. Mason's stock in the Durham Report.

Had you accepted my offer of Mr. Colling's cow and two of my mixed breed for your experiments, and given a faithful statement of the food consumed and return made, you would have rendered an important service to the country. It would not have cost an hour's labour in the day for a man to have done it, and I am convinced you would not have been any longer an advocate for the short-horns.

I have at length got Duchess's pedigree under Charles Colling's hand. This tribe of short-horns is the only one directly bred from Hubback to Favourite without some other cross intervening, and the red bull (Laird) Mr. Maynard has taken is an improvement on this tribe, being in greater perfection than Comet, though two years younger. But prejudice must have its day.

JOHN CHRISTIAN CURWEN *to* THOMAS BATES

WORKINGTON HALL, 12*th January* 1811.

I rejoice to hear from all quarters so favourable an account of your purchase.[2] I hope it will do more than realise all your expectations.

I could almost scold you for supposing me so prejudiced that I should not be prepared to make a fair report. Of your cross I have no experience, I must therefore speak with diffidence. I have, to the best of my belief, stated the question fairly as to the objects you have in view.

That the short-horned are an admirable breed of cattle cannot be questioned; that they are at the utmost point of perfection, no prudent man would contend. As milch cows I do not believe they have their equal. With respect to butter, that is a

---

[1] No place suffered more from quarter-ill than did Halton. It was not until 1815 that the practice of setoning was introduced as a preventive; it soon became very general.—Letter of William Charlton, 26th October 1872.

[2] Of the estate of Kirklevington.

question which it will require more experiments to decide than have yet been made. I can promise 20 ozs. of butter from 12 wine quarts on particular food; 17 ozs. is the least. I did not try the experiment at the height of the clover. Oil cake makes a great difference. I am disposed to think that the short-horned eat more than the long-horned. I have not a milch cow which is not fit for the butcher. Here is an advantage I never saw in any other cattle—to milk, work, and lay on fat at the same time. I hope when you have seen the Report, you will be of opinion I have dealt fairly, though I have no hesitation in declaring that I am strongly of opinion your cross will never reach the short-horned.

Though you are not in the county of Durham, your experiment in cross-breeding may be of such public utility that it ought to have every fair play.

My contention is not with the Kyloe, but the long-horned, the worst breed of cattle in the island. Here, then, the matter must rest till you have seen the Report, when I shall be obliged for your candid sentiments.

John Christian Curwen *to* Thomas Bates

Workington Hall, 11*th February* 1811.

The Report is at length finished. I shall be very anxious to know whether I have stated your objects fairly.

As a cow-keeper, I am very partial to the short-horns. When there is plenty of good food and no exposure, I do not think for that object they can be surpassed. Situation makes a great difference.

The country appears much pleased with Mr. Mason's bull. I am satisfied we shall agree in the opinion that the long-horned cattle are the worst breed extant. Many of my farmers are getting rid of them. I have sent Mr. Gibson's man an order for sixteen heifers for my tenants.

I beg you not to spare the Report but to tell me your real sentiments. I shall be happy to correct anything that is erroneous.

Curwen's 'experimental cattle' consisted of a couple of shorthorns, Herefords, Glamorgans, Galloways, and long-horns, and a solitary Sussex. The greatest profit, £8 : 10 : 1, was on one of the shorthorns, which increased in weight from 90 stones to 115 stones. The next best balance, £6 : 16 : 5, was left by a Hereford which began at

61½ stones and gained 28½ stones. In the first case, the food, in which 6½ stones of oilcake was the only artificial stimulant, cost £7 : 17 : 7 ; in the latter, £7 : 19 : 11. Each animal was purchased at 4s. and sold at 6s. per stone. The Board of Agriculture awarded Curwen a premium of £50 for these experiments.

Thomas Bates *to* John Christian Curwen

Halton Castle, *June* 1811.

I found your favour on my return from taking possession of Kirklevington. I am sorry to find you suspect me of want of charity, as our mutual possession of that virtue was the only ground I had for expecting your forgiveness of the freedom with which I have always expressed myself when our opinions differed. The sole reason why I did not comply earlier with your request and give you my opinion of the last year's report of the Workington meeting was that I found that the prejudices you had contracted last summer were not to be removed by recalling to you those first principles which, in common with me, you had adopted as the true criterions of the merit of live stock. You may be offended at my use of the word 'prejudice,' a thing so foreign to your general character, but I know no other word that applies to opinions formed on no foundations except the assertions of others.

Had you viewed Mr. Maynard's conduct in taking a bull of my breed (Laird) with an unprejudiced mind, your conclusions would have been very different from those exhibited in the Report. According to the declaration of a champion of shorthorns, Mr. Maynard had for forty years sold his cattle to the butcher for the most money of any man in England, while at the same time he attended to his dairy. The cows he sold to Mr. Colling and Mr. Mason were (as they themselves stated to me) parted with solely from being bad milkers. On seeing his cows milked, I expressed to him the great difference there was between them and my mixed breed, on which he agreed to come and see my dairy milked. He had the milk put under lock and took charge of the key himself till it was churned. It was through his consequent conviction of the great superiority of my mixed breed both in milk and butter that he hired my bull, which he had previously examined at different times, and which he was convinced was equal in every respect to Comet though Comet was two years older.

Had you tried my stock, and those in the county of Durham

by the same criterion as Mr. Maynard, your Report would not have contained opinions founded on the assertions of interested individuals, but on facts by which the merits of stock can only be decided. Had your mind been unprejudiced, you would have given Mr. Maynard the credit he deserved as a man of sound judgment, capable of altering his opinions when he saw convincing proof of superiority.

Should you wish to return to the first principles we held in common, and to judge by facts not opinions, I will with pleasure give you a meeting at the tup-shows in the county of Durham, and if the breeders there will allow, we will visit their dairies and you shall have the same opportunities here, though my dairy cows this year are quite out of condition from an infectious disorder last winter which has injured their milking exceedingly.

I trust my freedom will convince you that my sole aim is truth, and that I would rather offend you by contradicting your hastily formed opinions than suffer you to be influenced by prejudice.

Having now expressed the points on which I differ, allow me to return my sincere thanks and those of numerous friends to whom I have lent the Report, for the pleasure we have received in the perusal of it on account of the liberal spirit with which it is in general drawn up.

JOHN CHRISTIAN CURWEN *to* THOMAS BATES

WORKINGTON HALL, 13*th November* 1812.

Whatever may be my sentiments on the comparative merits of the two breeds of cattle, it has no influence with our Agricultural Society. Mr. Spedding received two premiums. Had you sent your stock here, they would have had full justice done to them.

I should feel miserable could I suppose I could in the most remote degree have injured you. The meeting here was much pleased with Mr. Spedding's stock. I wish you had been here, and if you had shown I was wrong in anything advanced by me, I should have been happy to have corrected it. You must allow me to entertain that opinion which my judgment dictates, but facts must be correct, and it has ever been my anxious desire to have them so.

An injury does not authorise a declaration of war by one state against another. It is the refusal to make reparation. I have just referred to what I said last year. I do not see anything you have to complain of. If you can point it out, your letter shall appear in the Report. The only thing I request at your hands is to acquit me of any intention of meaning to injure you or to serve those who would not thank me for so doing, so little store would they set upon my opinion.

The interest Curwen and Bates both took in the spread of religious knowledge fortunately ended their estrangement :

JOHN CHRISTIAN CURWEN *to* THOMAS BATES

WORKINGTON HALL, 12*th January* 1813.

Though we may have differences of opinion as to the Kyloe or the short-horned, we can have none on the subject of enlightening the minds of our fellow-creatures. I highly applaud your zeal.

My opinion is that had we met, we should have amicably settled any apparent differences. Should you be tempted to visit Spedding, I should be truly happy to see you. To hear from you will give me great pleasure for I cannot forget your attention and kindness to me.

In his correspondence with Curwen, Bates had insisted on the superiority of his cross-breed over the general average of shorthorns; the value of the shorthorn blood in the cross became more apparent when it was contrasted with the pure West Highlander.

THOMAS BATES *to* DUNCAN CAMPBELL[1]

HALTON CASTLE, 13*th July* 1811.

As I told you was likely to happen, we had a great and sudden advance in price of fat and lean stock in the beginning of June. This continued nearly a month, and now we have as great and sudden a fall again. I had all my fat cattle in hand when the advance took place, and immediately sold everything to the butcher that was ready. I got from 9s. to 10s. a stone for my fat cattle. The Duke of Argyle's three-year-old stots which I bought at Newcastle Fair the 29th of October last at £15 : 4s. a head, I sold fat at £19 each, and the steers of my own breed, now three years old, for £24 : 10s. The latter were not so good as the Duke's stots last November by more than £3 a head. It is the opinion of several gentlemen who saw them together at

[1] Of Ardnave, Islay.

that time and again when sold, that my own had gained above three times the weight of the Duke's in the time of feeding.

I entered on my new purchase in Yorkshire at May Day. The other part of the same Lordship has been on sale since, and from what they ask my bargain is worth £7000. There has been no depreciation either in the value of estates or in the annual value of land in this neighbourhood or in Yorkshire, but in North Northumberland there is a considerable falling off and considerable arrears of rent unpaid.

I am keeping a bull calf off the grandam of the white ox, the portrait of which you saw in Edinburgh. He is all red except a small white mark under his belly and a few white hairs on his tail-head. The West Highland cow he is off has bred me ten calves and they have all been of the same colour, except one which was grizzled. The bull calf is very promising, and is, I think, the best calf she has ever had.

THOMAS BATES *to* DUNCAN CAMPBELL

HALTON CASTLE, *3rd September* 1811.

I have procured a customer for half the twenty Highland cows you mentioned to me in Edinburgh that you would have to dispose of this autumn, and will take the other ten myself.

I will send a driver to meet them at Dumbarton, about a week before the Falkirk market in October, so that they may reach this before the droves from Falkirk overtake them on the road. Your drivers must be particularly careful in boating, and drive the cows short journeys every day.

I shall have great pleasure in showing you here the improvements made from the stock I got from you in June 1800.

THOMAS BATES *to* DUNCAN CAMPBELL

HALTON CASTLE, *9th November* 1811.

I have deferred writing until I could give you a correct account of Brough Hill Fair. It is the market which principally regulates the sale of Highland stock in the autumn. It was a very large market this year, upwards of 10,000 cattle, chiefly Highland stock, for the most part in good condition. Upon the whole it is considered to have been a good Fair, though many lots remained unsold. Of those which were bought at the late markets in Scotland, few can have done more than paid their expenses. Highland heifers were sold for six to nine guineas, and stots from ten to sixteen guineas a head. Buyers were numerous, but the fact that the Irish cattle offered were so much larger, induced many to try them.

I am glad to find that I shall have the pleasure of seeing you here this autumn. I have got all my stock housed except two of my fat cattle, a cow and a steer. These I should like you to see before they are killed. Comparing them with the stock in a lean state would give you an idea of the improvement the latter are capable of making.

Thomas Bates *to* Duncan Campbell

Halton Castle, December 1811.

The weather this week has set in so very severe that I should not advise you to come into this country before the spring, but before you leave Edinburgh, I should have great pleasure in showing you and General Simson my stock and farm.

I can meet your wish to have a bull of my stock at Whitsuntide or sooner. One of my last year's calves out of a Highland cow is (as I believe I mentioned before) very promising and likely to make one of the best bulls I have yet bred; it is quite adapted for the Highlands in colour and every other respect. Although we generally use bulls at a year old in this neighbourhood, it is in my opinion too early. I could let you have the use of an older bull which I have used myself and think highly of, until the other is fit for use. He has been poorly kept and is not too heavy for the smallest Highland cow. The bulls could come together, or the calf be sent when two years old. Your cast cows got by my bulls should be worth fully five guineas a head more than real Highland cows to any grazier.

The experiments I have made require only to be generally known, for stock thus bred, to be even more run after than Bakewell's Leicestershire sheep were when first introduced. It is said that a heifer of the short-horned breed was sold in this country last summer for £1150. I am ready to bet the seller that sum that I have better heifers descended originally from Highlanders. Should you commence breeding from my stock I shall always be ready to give you my ideas how to proceed. I wish you could have seen a cow I now have, eight years old, out of a Highlander. She gave milk till this year, and though she had twice a sore udder in the summer, is at present quite a phenomenon for fat. She is to be killed with a steer at Christmas, and I will send you particulars of their weight.

I am sorry I cannot get to Edinburgh this winter, having so

much to do, but beg you will give my best respects to General Simson and all my other friends.

THOMAS BATES *to* DUNCAN CAMPBELL

HALTON CASTLE, 11*th January* 1812.

The cow and steer were killed for November market this week. Two of the kyloe stots, bred in the Island of Islay, were killed also, and all four carcases exposed for sale to-day on the shambles.

One of the Islay stots, six years old last spring, was without exception the best Highland stot I ever saw. He came from the Highlands when four years old, in the spring of 1809, and was bought out of a drove, by a gentleman a few miles from here, for the purpose of feeding to as great a perfection as possible. He was put for a short time on a rough pasture and then passed two summers on the best grazing land in Northumberland. Each winter he has been fed on hay and turnips, and no expense spared.

The other stot, five years old last spring, was bought in a lot of forty, October gone a year. He was fed all last winter on turnips, grazed on the best of land all summer, and has had hay and turnips since.

The forty Islay stots in question cost the same price as those of the Duke of Argyle's I bought, and formed together two of the best lots of stots I ever saw grazed here. Several good judges gave it as their opinion that both lots were worth above £3 a head more than the two-year-olds of my own breed in October 1810, yet both in June, and now, these Highland stots fall far short of mine in weight.

The six-year-old Islay stot weighed 71 stone 1 lb. the four quarters (14 lbs. to the stone); that five-years-old-off weighed 48 stones.

My steer, three-years-old-off weighed 64 stones. He had only one year's feeding; in June 1810 he was looked upon as the average of that year's breed and valued at £7, being then both lean and small, having been very poorly kept while young.

What, however, particularly struck the attention of all the breeders and graziers present, and was also admitted by all the butchers, was the very apparent superiority of the beef of my breed over that of the real Highlanders, which till now has always been looked upon as the best beef in the kingdom. The grains of my beef were much finer and better marbled; the fat not only finer but of better colour. This superiority was still more marked in the coarse joints than in the fine ones. So great was the difference that the beef of my steer was not only sold for 2d. per lb. more than that of either of the Islay stots, but was nearly all sold before any of the others was taken.

I wish you had seen my steer and the cow alive. The latter would have astonished you. She was about twice the weight her mother (a real Highlander) would have fed to, and yet the bone of her leg was smaller than her mother's. I should have sent you her sirloin, but it was so much fatter than any beef I ever saw in Edinburgh that I feared it would have been thought too fat for eating. It is only a year last autumn since this cow was milked, and when in milk she was the leanest cow I had.

In order that you may judge for yourself that three-year old steers of my breed are better flavoured and more delicate eating than any pure Highlander, I take the liberty of desiring my butcher to send you a few ribs of the fore-chine of my steer by the waggon, and beg your acceptance of it. I hope you will desire the friend you know of most partial to the Highlanders to procure the fore-chine of a kyloe stot to be tried with it. The beef will, no doubt, be much injured by the carriage, yet notwithstanding the advantage that beef killed in Edinburgh must have, I court the trial in order that you and your Highland friends may judge for yourselves.

Never did any exhibition make so many converts as that of to-day. I have opened a mine of wealth for Highland gentlemen, if prejudice does not prevent their deriving the benefit. No unprejudiced grazier in this country would now feed a pure Highlander if he could procure grazing stock of my breed.

Pray inform me how many cows you will cast for sale this autumn. I will provide for taking them all myself, and will send a bull down in order to have their calves all of my own breed.

Four years had now passed since the publication of the Address to the Board of Agriculture. During these, Bates had not attended the meetings of the Tyneside Society. Hoping that at length 'reason would have some sway,' especially after the evident fiasco at Workington, he now brought the subject of feeding trials again before the Society. He did not wish to put it to any expense, and merely begged that two of the body might be deputed to see the experiments which he had commenced carried definitely through. This was accordingly conceded:

THOMAS BATES *to* DUNCAN CAMPBELL

HALTON CASTLE, *6th March* 1812.

Several unforeseen circumstances have prevented my intended visit to Edinburgh this spring. I trust, however, that you will be able to favour me with your company here, and judge for yourself of the system of breeding I am pursuing.

I put three steers, of my mixed breed, two years old last spring, which had gone on my sheep pasture all summer, and had no fog in the autumn, into a fold on the 9th of December, to feed on white globe turnips and straw. They were computed by good judges to weigh 22 stones each on the 5th of June; on the 9th December they were estimated at 34 stones each, and according to our markets at the time were worth £12 to kill. They are now calculated to weigh 49 stones each. Two gentlemen are at present weighing the turnips they consume, in order to compare accurately their increase in weight with their consumption of food. The result will be of great importance both to myself and the public. The procedure is most satisfactory to me, as I neither knew nor had ever seen the gentlemen until they came to my house, and they were sent by gentlemen who are themselves breeders of short-horned cattle, and have no object but to ascertain facts from which the public can draw their own conclusions.

Should General Simson be in Edinburgh give my best respects to him, and I should be glad to see him with you, if he can make it convenient.

Mr. Mason of Chilton conducted an experiment about this same time, to ascertain the weight of beef gained by the food, principally turnips. His three steers, under three years old, gained twenty stones each in twenty weeks, and averaged, when fat, seventy stones each. Bates's three, kept in a similar manner, consumed only two-thirds of the food Mason's did, and gained a third more weight in the same time. They were by Ketton 1st (709) out of a half-bred Kyloe, and so were really three-quarters shorthorns.[1]

This proved that the Halton steers left more than double the money the best shorthorns in England had ever done on any occasion whatsoever.

[1] Berry's *Improved Short-horns*, pp. 34, 35, and MS. note by Thomas Bates.

William Nelson *to* Thomas Bates

LILBURN, 20*th May* 1812.

I wish it was in my power to see your cattle before they go to the butcher. I hope to have the whole particulars as to their age, weight, breeding, the quantity of each kind of food consumed, and also the increase of their dimensions. I should not have expected them to eat so few common turnips. In few experiments I have ever made have the cattle eaten less than one-quarter of their supposed dead weight, indeed most have eaten one-third of it; but I always find that the quantity of food consumed diminishes as they advance in fatness.

The rye grass and roota baga only reached Lilburn a few days ago. Let me assure you of the great value I place upon both. One field of eighteen acres sown with the produce I got from Halton some years ago is the admiration of all the country.

The experiments in cattle-feeding were still in progress at Halton when a letter had arrived from the Professor of Mineralogy in the University of Edinburgh, which was destined to lay the foundations of one of the few unclouded friendships that Bates ever possessed:

Professor Jameson *to* Thomas Bates

EDINBURGH, *March* 1812.

I find it will not be in my power to get the length of Halton Castle this summer.

I enclose a letter from my particular friend young Mr. Mackenzie of Applecross, one of our great Highland proprietors who is very desirous of making himself acquainted with the husbandry of your part of England.

I have known Mr. Mackenzie for several years, and have found him a man of the strictest honour, of most amiable manners, and deeply versed in all the sciences in which you feel an interest.

Should your domestic arrangements admit of an additional inmate, I am convinced the only inconvenience you will experience will be the regret on parting with a gentleman of Mr. Mackenzie's talents and worth.

I hope you have not renounced your mineralogical pursuits. They will always prove an agreeable relaxation, and to you may afford means of improvement. Mr. Mackenzie is well skilled in mineralogy. I expect that together you will be able to give me a correct idea of the structure of the North of England.

THOMAS MACKENZIE *to* THOMAS BATES

EDINBURGH, 31*st March* 1812.

My view in wishing to pass some time this summer in Northumberland was solely the acquisition of so much practical knowledge in husbandry (which I have sought in books in vain) as might enable me to carry on judiciously the improvement of a large Highland property (at present nearly in a state of nature), which my father leaves entirely to my management.

From having seen you occasionally at Dr. Coventry's class, I had the highest character of your talents from my friends Mr. Jameson and Mr. Erskine. I immediately thought of you as a person not only eminently qualified to advance my agricultural knowledge, but from whose general information I should derive both pleasure and instruction. Professor Jameson obligingly offered to communicate my views to you, and I consider his having introduced me to your acquaintance as far from the least of the many favours I have received from him.

I think I shall be able to arrange my affairs here, so as to be at liberty about the middle of next month. I shall then do myself the pleasure of visiting you; and should it not suit you to gratify my original wish of residing some months under your roof, I shall at least have the advantage of your advice as to the best means of attaining my object.

Of Mackenzie's sojourn at Halton, Bates subsequently wrote :

Mr. Mackenzie was the only agricultural pupil I ever had, and I can truly say I never had more pleasure in society than I had during his residence with me. Every day was fully occupied, from leaving our beds till we returned to them again in the evening. We made several excursions to view farms and stock in Durham and Yorkshire as well as in Northumberland. There was not a vacant moment for several months, till he thought he had gained all the information he needed, and that he would go and put it into practice on his own estate of 130,000 acres in Ross-shire.[1]

In July 1812 I had a large party at Halton Castle for the

[1] Bates MS. fo. 10.

meeting of the Tyneside Society at Ovingham. Mr. Mason of Chilton, and Mr. Thompson of Stamford in Bamburghshire, were present. Short-horns became the topic of conversation. I appealed to Mr. Thompson, who remembered well both Lady Maynard and her daughter Phœnix, if ever he knew a descendant of Lady Maynard that lived to maturity equal to herself. He immediately answered, 'None were equal to herself.' I asked Mr. Mason if he could say anything to the contrary. He remained silent as did the rest of the party. Here then, in addition to my own view, was the uncontradicted opinion of Robert Thompson, who was much older than myself and whose judgment of short-horns and Leicester sheep was never called in question.[1]

The Ovingham meeting was one of the largest ever held by the Society. It was attended by shorthorn breeders from all the principal districts of Northumberland, Durham, Yorkshire, and Cumberland, while Scotland was ably represented by Mr. Mackenzie. The facts which the Halton feeding experiments had established were communicated to this influential assembly, and there was no getting over them.

It was on this or a similar occasion that Mason of Chilton called to breakfast at Halton. Barbara Giles, the housekeeper, had just put the week's butter in readiness for the Newcastle market on the Saturday, and Bates told him that however ready he was for breakfast, he should have none until he had counted the butter. There were three hundred half-pounds to go to market, besides what was used in the house and sold at home. There were then thirty cows which had calved, and the butter sold for above one shilling the half-pound. This left more than ten shillings for each cow in butter alone, besides the value of the old milk otherwise sold, while all the calves were reared by the pail, and none allowed to suck. Had all the milk been creamed and made into butter, there would have been twice the number of pats. Mason, thrown off his guard at this display of dairy produce, confessed to Bates: 'You can go on breeding shorthorns,

---

[1] Bell, p. 55.

because they pay you in milk, butter, and beef, but we cannot do so unless we sell them at high prices to breeders.'

Mason, as Bates plainly told him, was keeping at the time three sets of cows: one to breed calves and then get dry (which was no hard matter) in order to attract notice by their high condition; a second as wet-nurses to rear the calves; and a third to supply his family with milk and butter. 'This,' Bates added many years afterwards, 'is a system that would ruin any man, if he had the land rent free and no outgoings to pay; yet many continue to pursue this reckless course in order to gain premiums, attract public attention, and gratify their vanity at the cost of their pockets.'

Bates's relations with Mason were not improved when in August 1812, Baroness, the daughter of Duchess-by-Daisy-Bull by the Chilton bull St. John (572), dropped down in the byre after milking, and never rose again. John Charlton of Heddon-on-the-Wall, the cow-doctor, said, on seeing her, 'This is what I have met with before in cattle having St. John's blood; I am certain that it is hereditary in it.' Bates had never before heard of hereditary disease in cattle, and laughed at the idea.[1] Dear-bought experience taught him that Charlton was right. Many who had St. John's blood lost whole fortunes by it—Arrowsmith, as he told Bates in 1837, no less than £10,000; while Raine acknowledged to him in the presence of Mr. Black's brother that he had nearly given over breeding as hopeless.

At the hour of parting, which, as Professor Jameson had predicted, occasioned feelings of deep regret on both sides, young Mackenzie of Applecross cordially thanked Bates for all he had learnt during his sojourn at Halton, but asked leave to tender a little advice on his part. 'You have,' he said, 'been giving your time and attention to receiving whoever comes to your house, and the return they make is to laugh behind your back at the disinterested kindness you invariably show. They are the most undeserving

[1] Bell, p. 91.

persons I ever saw or ever met with. Instead of benefiting from your kindness and hospitality, they have all, except one gentleman, said everything they could to injure you, and would stop at nothing to accomplish it. I cannot leave Halton without telling you this. I shall always be glad to receive you and Mr. ——, who has behaved as a gentleman, but there is no one else among the many hundreds I have seen at your house with whom I would ever have any further communication. I hold them in contempt and abhorrence.' Bates told Mackenzie that he had long been aware of the facts he mentioned, but that he was not going to alter his habits on account of the ingratitude of his visitors; nor did his relations to his neighbours undergo any change while he remained in Northumberland.[1]

[1] Bates MS. fo. 10.

CHAPTER V

HALTON: AGRICULTURAL DISTRESS

1813–1820

BATES had exhibited cattle at the Tyneside Agricultural Society's meetings, held sometimes thrice a year, from its first formation in 1804. Though he never exhibited his best animals, he had been successful at every show. At the last show of 1812, he sent better samples than ever before, but received no prize. One of the judges told him afterwards that those who influenced the proceedings had decided, before going out to examine the stock, that he was never to have another premium from the Society, however excellent the stock he exhibited. This was confirmed in 1819 at Berwick-on-Tweed by a shorthorn breeder who had removed to the northern part of Northumberland from Tyneside. An early visitor of Bakewell and of high standing in the agricultural world, this breeder was dining at the same table with Bates at the ordinary at the 'Hen-and-Chickens.' Before a very large company, he asked Bates, after dinner, how his stock was going on. Bates replied that he had not exhibited since 1812 in consequence of the resolution passed by the Tyneside Society. 'You ought not to reflect on me seven years afterwards,' continued the other; 'I avow myself to have been the proposer of that resolution. We were perfectly right in doing so; we had none of your blood, and after giving you premiums for so many years, it was time to put a stop to it, and in our own interests to help the sale of our stock.'[1] This jealousy was largely the result of the evident superiority Bates's herd developed through the use of Ketton, and of his refusal to allow any cows except his own to be put to him; besides which he reared few bull calves except for his own use.[2] After the treatment he received in 1812, Bates never exhibited again till the York meeting of 1838. He continued his subscription to the Tyneside Society until the evil course it had taken led to its dissolution in 1821.

For these petty annoyances a cheery letter from Smeaton must have proved a most excellent cordial:

Baron Hepburn *to* Thomas Bates

SMEATON, *7th July* 1813.

I rejoice to hear such flattering accounts of your young stock. It always affords me real pleasure to see genius and industry rewarded. I always told you that the cross of the Scotch breed would answer your expectation; and you may recollect I advised you to try it on yourself, and now that you have proved it so successfully in one species of stock, I trust you will listen to my advice and carry it one step further. Do come down to us when you can spare as much time and cast your eye over our fair ones, and to the lass you like best, sing our Scotch song, which begins thus—

> Tell me, lassie, will you marry me now,
> For I canna come every day to woo.

I have sent your compliments to my neighbours, Rennie and Brown, but I delayed writing until I could tell you of my farm, and I have the pleasure to say that in every particular it would do your heart good to see how promising it really is. My autumn-sown wheats are well planted, and have been now about a week in bloom, so I think they will cut by the second week of next month. My oats are strong and showing the ear. My peas and beans completely cover the ground—a vigorous growth and full of bloom, and setting remarkably well. My roota baga is far advanced, and my turnips generally in the rough leaf, and both fully planted; my hay harvest somewhat advanced. Plenty

---

[1] Bates MS.   [2] Bell, p. 181.

of clover plants but late and soft; I expect a great second crop, and I have plenty of old straw to mix up with it. My practice is to carry the straw to the field and to mix it with the clover when cut. This is a little more expensive, as I must carry the straw out and home again; but the new and the old incorporate so intimately that my horses are fonder of it than of pure hay. Twenty head of cattle paid me of nett return £25 a head for their keep, and my sheep netted £1 : 19s. a head. Since 1798 I have not had so good a crop of all corn as 1812 has yielded, and the prices last year nearly double. What I say of my own, for men generally are fond of speaking of themselves, I may truly say of all my neighbours'. So we are contented, quiet, and happy, and I thank God we have no rascally drunken discontented weavers amongst us, 'our pleasures to alloy.' Our staples are wheat and bairns. I wish you would begin to raise recruits in the matrimonial way, and give the king some good healthful subjects. Try the cross and fill up the waste of war. Your butter last year was super-excellent, and my wife is your humble suitor this year for a single kit if you can spare it. I congratulate and rejoice with you on the glorious news from the Peninsula.

The prosperity of East Lothian was marvellous:

WILLIAM NELSON *to* THOMAS BATES

LILBURN, *26th July* 1814.

I fully intended being with my friends in Tyneside ere this, but having gone into Scotland for a few days, ran out both my time and money. I went through East Lothian in the end of May, and saw one or two farms in that delightful district. Mr. Rennie's at Phantassie is in the highest state of cultivation, and is worthy of being possessed by a good farmer, who appears proud of the trouble of showing his management. Their bean husbandry I was much pleased with. Indeed such land with so many local advantages of manure, climate, soil, and markets, is rarely to be attained at any price. I saw also a Mr. Cuthbertson's farm at Linton Mains, nearer the sea by Haddington, which was well managed. One of the finest rides I ever had was up the Esk river.

I want no more convincing proofs of the value of the mixed breed of cattle than what I have seen at Halton, and your statement is another proof. Here we cannot drive the people to buy my lean half-bred cows at their real value, and I have not yet got a few acres of grazing to finish my stock with.

I have kept for you a catalogue of Mr. Culley's sale, filled up with the prices, which I procured from himself. The stock did not sell high. The Wark ewes there and at Chillingham Barns were really useful, and met the worst sale of the year.

If you would marry and had a daughter I would wait for her. Why do you sport so long with time? A single man knows nothing of the real enjoyments of life.

Somewhat late in the day, the Honourable Board of Agriculture repented them of their contemptuous disregard of Bates during six long years:

COMET (155), 18th JUNE 1815.[1]

SIR JOHN SINCLAIR *to* THOMAS BATES

21st August 1813.

It is mentioned in the Durham report that you would favour the Board of Agriculture with the result of some experiments for ascertaining the proportion between milk and butter. It would be obliging in you, therefore, to send that information to me as soon as may be convenient, under the cover of the Earl of Caithness, Edinburgh. I should be glad also, that you would make comparative trials of churning the whole milk, both sweet and sour, against churning the cream alone; also on the different value of the butter-milk of each, for feeding pigs. Any information from you will always be acceptable to the Board.

On the memorable 18th of June 1815, while Napoleon

[1] From the original coloured drawing, formerly at Kirklevington.

and Wellington were grappling in the throes of war at Waterloo, Bates took a last look at Comet (155) whom he had seen soon after his birth. He was accompanied by his friend Mr. James Fawcett of Scaleby Castle, near Carlisle, who took a faithful likeness of the famous old bull. Comet seemed in health at the time but weighed only 80 stone. A week later he was dead. Although he was the sire of Young Duchess (afterwards called Duchess 1st), the heifer bought at Charles Colling's sale in 1810, Bates always

KETTON (709), 1814.[1]

considered Comet the worst blood in the Duchess pedigree.

That 'Emperor of all bulls,' Ketton (709), was now 120 stone and too heavy for bulling. Tommy Thompson, the cowman, said he never got a middling calf all the seven years he was in service. More than sixty years afterwards Mr. William Charlton, who had lived near Bates, and ultimately settled at Sutton in Essex wrote:

I think I can see the grand old animal standing in the bull park with his fine head and placid countenance, his beautifully arched neck, his deep and roomy chest, his short and widespread legs, his handsome shoulders and full crops, his long, straight, and level back, his heavy flank and deep ribs, his well-

[1] From the original coloured drawing, formerly at Kirklevington.

formed beautiful quarters and heavy thighs, and his tail so nicely set as to give symmetry to the whole frame. How oft on my youthful mind was impressed the idea that I should never see his like again. His image was so imprinted on my memory that whenever I began to examine a prize bull, Ketton came full in view, and then many defects were soon prominent. Still, although Mr. Bates used Ketton for so many years, a Duchess heifer or bullock could easily be picked out of his herd. There was something in their very countenance, and in their prominent gait, and, above all, in their superior touch like none else. In that last quality they had no equals.

The Duchesses now received some curious visitors from the East:

ROWLAND FAWCETT *to* THOMAS BATES

SCALEBY CASTLE, *8th April* 1815.

My brother mentions the arrival, in one of their ships, of a cow and bull calf of the Abyssinian breed (much the same as the Indian). He has shipped them for Newcastle. He begs you to accept the cow with his kind respects, the calf he has given to me. They were to leave the port of London on the 5th inst., and may probably be at Nichol and Ludlow's wharf at Newcastle by this time. As you are much nearer that port than I am, I must beg you to make inquiries so that they may be forwarded to Halton with as little delay as possible. Please inform me when they do arrive that I may send a man to bring the bull home. I hope they are good stock as they will afford us an opportunity of making a trial of the cross.

Bates does not appear to have risked the experiment:

ROWLAND FAWCETT *to* THOMAS BATES

SCALEBY CASTLE, *17th April* 1816.

James desires me to return his best thanks to you for all your care and trouble about the Abyssinian cow. I do not think the calf will make a bull, it had therefore best be cut. The year old seems to get on very well, but they are not in any respect at all equal to the Guzerat cattle.[1]

From 1793 to 1813 there had been twenty extraordinary years, good for farming beyond all precedent.

---

[1] Another letter dated 'Scaleby, 4th August 1820,' says 'Our Indian bull begins to thrive well.'

From the autumn of 1794 to 1814 the value of live stock had nearly trebled; the price of mutton had advanced from 3½d. to 10½d. per lb., that of beef from 3s. 6d. to 10s. 6d. per stone. Rents and labour had also trebled during the same period. The tendency now was for all these things to fall back to double their value in 1793, but the interest on mortgages and the amount of taxation remained the same as during the prosperous period.

The end of the French war brought an unprecedented amount of distress on the agricultural community. Bates did all he could to obtain relief by remissions of taxation. He conducted a voluminous correspondence on the subject, and urged his views on most of the leading members of Parliament:

THOMAS BATES *to* The Right Hon. NICHOLAS VANSITTART[1]

HALTON CASTLE, *30th January* 1815.

At the present price of grain I am confident the crops in the counties of Durham and Northumberland will not pay the expenses incurred in their cultivation, without one shilling to pay rent, or any profit for the farmer's skill and capital. The want of money among farmers is extremely great in consequence, and a considerable part of the agricultural labourers have been out of employment for some months. Many farmers, I am convinced, have had their capital reduced one-fourth this last year, and so much is the agricultural spirit of the country depressed, that if redress be not given before March, a very large portion of the tillage lands will be laid to grass.

The duty on horses kept for husbandry operates as a great check on the cultivation of inferior soils. On a considerable proportion of the tillage lands of this district, it amounts to fully two shillings per quarter on the wheat the farmer has to sell; in many cases it comes much higher.

The last duty on malt has had a bad effect in causing an increased use of spirituous liquors. I hope a considerable addi-

---

[1] Chancellor of the Exchequer from 1812 till the beginning of 1823, when he was created Lord Bexley. His nephew, Henry Vansittart of Kirkleatham, was an early breeder of shorthorns, purchasing Alexina, Trifle, and Peg Woffington at the Wynyard sale of 1813, but Pedigree is the only animal entered as bred by him in the *Herd Book* (ii. p. 467), before vol. ix.

tional duty will be laid upon these and on foreign wines, and other articles not necessaries of life.

Wages in this part of the country, where the labouring class depend upon their own industry and not on parish relief, or an increase of their families, have doubled within the last twenty years, when the average price of wheat was 44s. per quarter, while for the fourteen years up to last May it was 88s. per quarter. The agricultural improvements that have been introduced into general practice within the last twenty years in this part of the country have made the land more productive. As the tenants reaped the benefit of this in the first instance, and the proprietors afterwards in the advance of rent, so the labouring class should now have the full advantage of it, and this has been gradually taking place with the advance of wages. The labourers, as well as every other class of society, now live much better in this country than they did formerly.

I conclude this hastily written letter in urging you to press the bill at all hazards.

SIR JAMES GRAHAM *to* THOMAS BATES

March 1816.

I have not been able to obtain a sight of your letter to Mr. Vansittart, but from your letters to me I believe you must have repeated to him nearly what I did in a letter I wrote to him previous to his stating that he should continue the Property Tax at one-half the present rate. I stated that if he would confine the tax to 5 per cent for two years only, and on rent actually paid into the hands of landlords, and give up the tax upon occupiers entirely, and also the tax on agricultural horses, I believed it might pass, although the extent of the landed distress was beyond all belief. But if he proposed to continue the smaller portion of the tax on occupiers, and on agricultural horses, I assured him it would be universally considered as so unjust, unequal, and oppressive, that he would lose the tax on property and income, and I have repeatedly pressed this upon the Ministers, but they would persist, and have been defeated, as I foretold them six weeks ago. I perfectly agree with you that the prices of all sorts of produce will in a year increase very much, and perhaps be in a very few years as much too high as they are now too low, and also that trade and commerce and agriculture must go hand in hand. One cannot be depreciated without affecting the other, and if a Corn Bill had passed in 1813, preventing importation when wheat was at 72s. a quarter, some part of the evil we now feel would have been averted; but the low prices have not been produced by importation, as only a fortnight's consumption was imported in the year 1815. The present Corn Bill, with the reduction of taxation, will do everything for the agricultural interest, I should hope, and the less Parliament interferes the better.

THOMAS BATES *to* SIR JAMES GRAHAM

HALTON CASTLE, 18*th March* 1816.

I trust both you and Mr. Vansittart now see that all the distress which the country is labouring under, has arisen almost entirely from the Corn Bill of 1814 not having been passed. I do not mean the unjustifiable proposal to raise wheat to 105s. per quarter, but that to keep it stationary at about 72s. per quarter, so far as corn laws can have that tendency. I presume you now see the effect of the Corn Bill passed last session. I was so fully convinced that it had the tendency to keep down the price of grain that I offered my crop of wheat at 72s. per quarter, though had the Bill not passed I would not have taken 88s. per quarter, which was the average of the fifteen years preceding. Will you ask Mr. Vansittart for the letter I wrote to him at the commencement of this session, before I heard of any of his measures? If advice directly contrary to a man's own interest has any weight in the House of Commons, that letter should have. I hoped Mr. Vansittart would have paid more attention to my suggestions as to the remission of the Property Tax on tenants and of the tax on horses used solely in husbandry.

I told Mr. Vansittart in 1814 that if the bill as then amended were not carried, no after-measures could rescue the country from the distress into which it would be thrown. My most urgent request now is that Parliament should not in any way interfere with the corn trade, but leave grain to find its own level. This it will soon do. The worst is past, and the pendulum is already reverberating. One extreme begets the contrary, and before six months are past, wheat will, I am sure, be 72s. per quarter if Parliament does not interfere. The House of Commons can do nothing in the measure without doing harm both to the grower and the consumer. The farmers and all tradesmen connected with agriculture have suffered more than was ever the case before in the same space of time: the agricultural depression and the fall in the value of stock makes a farmer's capital not worth half the sum it was two years ago, but no redress can now be given by Parliament except under the head of taxation. The injury is done; time alone with increased skill and industry can provide the remedy.

I trust you will attend the House of Commons and oppose every measure that is suggested to raise the price of grain. All half measures do harm. The agricultural and commercial interests of the nation are so interwoven, that every attempt to serve the one at the expense of the other will recoil upon that which is unnecessarily relieved. Except the remission of the Property Tax and the Horse Tax, grain needs no further support than the bill passed last session. To raise the price by legislative interference at a time when manufacturers, dependent on agriculture, are out of employment, would be to increase the misery already produced by not passing the Corn Bill a year sooner. Natural causes will produce an advance of prices all in good time.

Sir James Graham *to* Thomas Bates

April 1816.

Your ideas respecting the present state of the agricultural interests of the kingdom and the relief to be given, are so sensible and coincide so entirely with my own opinions upon the subject, that I have sent them to Mr. Vansittart, and urged his adoption of them. You will have the pleasure of seeing that some of them have been adopted. My opinion is that everything has now been done for the relief of the agriculturist, except the repeal of the tax on the agricultural horses, and on horses used in single-horse carts carrying coal and lime for hire, which is the most unequal and hard ever imposed, and I hope will be repealed, as a friend of mine, Mr. Burrell, has given a notice to that effect the week after the House meets. Mr. Vansittart has proposed a reduction of the duties on horses on small farms, but that is not sufficient in my mind. The other relief to the agriculturist is an extension of time for steeping barley grown in the Northern climates, as the time now allowed is not nearly sufficient for coarse barley, or such as has very thick husks, and which is the case with the Northern barley: if twenty hours more were allotted, it would be a great encouragement to the malting Northern barley, and of course to the grower. I have pressed this upon Mr. Vansittart, but he says that it cannot be allowed without great injury to the Revenue, as frauds would be practised, and he will not allow that your plan would prevent these frauds; however, I shall again and again urge him on this subject, and also to agree to give up the tax on horses as I have stated, and then I am sure all reasonable agriculturists would be satisfied.

I find agricultural produce rising in all parts of the kingdom. I perfectly agree with you, and have for many months stated my opinion to every one, that no doubt all produce will be as high before April 1817, as any fair and honest agriculturist could wish it to be, and the less interference by Parliament the more beneficial for the agriculturist.

Ever grateful for the benefits he himself had received from his residence at Halton Castle, Mackenzie of Applecross wished to have his bailiff instructed there:

THOMAS MACKENZIE *to* THOMAS BATES

APPLECROSS HOUSE, 25*th April* 1817.

I return my sincerest thanks for your kindness in allowing me to send the young man, who is to be my bailiff, to Halton.

I shall briefly mention my object in sending him to you. For the first year he will have little to do with me, as the out-going tenants have the way-going crops, but afterwards he will have the whole charge of the culture and improvements of the farm under my direction. The buying or selling of stock I generally contrive to manage myself. Were I determined to keep only Highland cattle I could myself instruct him in the detail of their management, but as I have a great inclination, when my farm is in order, to try the mixed breed, I wish him to be exactly acquainted with the minuteness of their management, of which, you know, I am ignorant, and to gain some knowledge of stock. I am, above all, desirous that he should see and imitate the regularity of your system, not only in work, but in the habits of your workmen, and form himself on Robert Bell's model as nearly as he can. You know me so well as to be aware how little I could tolerate a servant who had any high notions. I think he can in no way be better guarded against them, than by seeing your system under Robert Bell. As he has not been accustomed to have the command of workpeople, he might run the risk of mistaking presumptuous domineering for proper authority. Perhaps you will also permit him to see how Robert keeps his books.

Mackenzie was invited to contest the representation of Ross-shire. Bates urged him not to canvass a single voter nor to be at one shilling of expense himself. 'No member of Parliament ought,' he wrote, 'to be put to any personal expense in his own candidature. His duty is an arduous one, indeed, if performed properly no employment is so arduous. The post is an honourable one when obtained on these terms, but not otherwise.' Mackenzie followed Bates's advice, and was returned by his county after an expensive contest for his opponent, and his position was never afterwards challenged. He never opposed the Government but once, and then they were defeated by a large majority, the only time indeed that they were so.[1]

BARON HEPBURN *to* THOMAS BATES

SMEATON, 18*th September* 1818.

This conveys my kind wishes which indeed always attend you, and I have the pleasure to inform you that my barn-yard is shut with an abundant crop. I have nearly as much bulk, with 20 per cent more corn and 20 per cent of superior quality. I have this day sown nearly half of my stiff land fallow with wheat, in a dry bed and very cloddy.

My cows have not buttered well this season. Some dropped too early, some too late, so I hope you have received my commission for butter. It will relieve Lady Hepburn's anxiety, if you will be so obliging as to let us know about it.

I hear they are sending from Berwick potatos and even turnips to Covent Garden market. I think wintering cattle will be low at the next great Falkirk Tryst, where I always lay in stock. When your leisure will allow, I would like to shake hands with you here.

Mr. James Fawcett of Scaleby Castle left interesting memoranda of the long visits he paid to Halton:

Enclosed within the precincts of the large old-fashioned screen, in the Great Room of the Jacobean house appended to the Border Tower, with a blazing fire in front and copious libations of tea, with old Barbara's cream and currant cake, we cosily read Davy, Marshall, and Culley, and discussed short-horns and everybody and everything connected with them till the small hours of the morning. The days were spent in looking over the farming operations, in visiting the paddock, with the old ash-trees and shed, where the elephantine Ketton the First and his successors stalked about, and the rich pastures to the north

---

[1] Letter to *Farmer's Journal*, Kirklevington, 22nd March 1844.

where the renowned herd roamed in luxurious plenty, and in riding out to see the stock and farms in the neighbourhood.

On one occasion we visited Mr. Donkin of Sandhoe, who had purchased from Mr. Bates, a few years before, the famous Daisy cow (then called Duchess) grandam of Duchess 1st. She was then in her twentieth year and not having bred for some time, was about to be sent to the butcher. I was so much struck with her fine clear eye, extraordinary breast and shoulders, and

DUCHESS-BY-DAISY-BULL, 1815.[1]

general appearance of an 'old has been,' that I made a drawing of her.[2]

On another occasion our ride was extended to the Tees valley, and the principal herds of the famed district were visited; among them that of Robert Colling, at Barmpton, which was shortly after disposed of. We spent a pleasant day with that gentleman, and there was much interesting discussion about his cattle. He pointed out to us a favourite five-year-old cow, which had, the week before, produced three living calves at a birth, having been previously barren. This he attributed to the effect of ploughing on the farm the foregoing spring. We also

[1] From the original coloured drawing by Mr. James Fawcett, formerly at Kirklevington.

[2] Mr. Fawcett, writing from memory, seems to have been mistaken as to the exact age of Duchess-by-Daisy-Bull at this time; according to Bates the drawing, with the spire of St. John Lee Church in the distance, was made in her 16th year, *i.e.* in 1815. It must have been her grand-daughter Duchess 1st that Mr. Thomas Bell remembered in 1820 'going in the Peafield at Halton the summer before she was sent to the butcher.'—Letter to Thomas Bates of Heddon, 15th March 1870. Mr. Bell referred enthusiastically to Halton Castle as 'a place that I like above all others in the world.'

saw Charles Colling, then living at Croft. Mr. Bates's criticism on the herds was by no means sparing where he could see faults.

The character of the Duchesses at this time was that of good and handsome wide-spread cows, with broad backs, projecting loins and ribs, short legs, and prominent bosoms. The head was generally inclined rather to be short and wide than long and narrow, with full clear eyes and muzzle, the ears rather long and hairy, the horns of considerable length, but of free waxy quality. They were good milkers, and had for the most part a robust healthy appearance. The colour was almost uniformly red, with, in many of them, a tendency to white about the flank. They had also generally what Mr. Bates called the Duchess spot of white above the nostril. A strange anomaly occurred in the case of Duchess 6th. I recollect her being calved.[1] She was very handsome, and of the most orthodox colour, but with a round spot of several inches on the flank, of the deepest black. Whether this indicated a harking back to some ancestral Highland alloy, or a freak of the cow's imagination, is a curious question.[2]

Mr. Bates had at this time several other good tribes besides the Duchesses, and a number of beautiful cows of more or less affinity to Argyleshire heifers, which he had procured from Scotland at considerable cost and trouble. The produce of these bred to the short-horn bulls, Ketton and Ketton 2nd, seemed to answer admirably. After a few crosses they had all the appearance of first-class shorthorns.[3]

Being an intimate family friend, Mr. Bates was often at Scaleby, and induced my father to try a cross of the new breed on his long-horns. For this purpose he insisted on furnishing us with a Ketton bull, which answered so well that no more long-horns were got, and by the purchase of a few well-bred shorthorns, many of them recommended by Mr. Bates and some from his own herd, a complete transformation was effected. Among these were some descendants of the cow Princess, whose daughter Elvira was brought into Cumberland and produced some good animals.[4]

[1] This was in the autumn of 1819.

[2] Mr. Thomas Bell knew himself of a case in which a pure shorthorn cow, after rubbing against some black cart-grease and making a black mark on her skin, produced a calf with nearly the same black mark on the same place.—Bell, p. 180, *n*.

[3] Bell, p. 179.

[4] Mr. John Grey too, wrote, 'That well-known breeder, Mr. Thomas Bates, when in Northumberland, brought this cross to great perfection.'—*Farmer's Magazine*, p. 94, August 1841.

In order to enforce the importance of good female descent in the case of cattle, and to found regular families, as well as to avoid the necessity of everlastingly searching out new names, Bates had begun numbering the produce of his Duchess 1st and her heifers in the order of their birth, a system which recalls that adopted in male descents by the sovereign houses of the two charming miniature principalities of Reuss in central Germany. Every prince in both branches is christened Henry, but is distinguished by a number in succession to the last born, which number he still retains if called to the throne. Less exacting, however, than shorthorn-breeders, the princes of Reuss relieve their lieges from further feats of arithmetic by commencing a new notation on the verge of five-score; a Henry XCIX is followed by a Henry I.

At Robert Colling's great sale at Barmpton, 29th September 1818, Bates met Lord Althorp who purchased there:—

	Guineas.
Lot 4. DIANA by Favourite, dam Wildair by Favourite, grandam by Ben (70) bulled by Lancaster (360)	78
Lot 20. ROSETTE, 4 years old, by Wellington, dam Red Rose by Favourite, dam by Ben (70)	300
Lot 23. NONPAREIL, 5 years old, by Wellington, dam Juno by Favourite, grandam Wildair by Favourite, great-grandam by Ben (70)	370
Lot 61. REGENT (544), 3 years old, by Wellington, dam Rosebud by Windsor, grandam Red Rose by Favourite (252), great-grandam by Ben	145

The twelve-year-old bull Marske (418) was the only animal there that Bates himself would have cared to have purchased. He was sold to Mr. Maynard for 50 guineas.

Lord Althorp appears to have paid a visit to Halton shortly afterwards. Bates impressed on him the great value of the Duchess blood.

In returning from Mr. Witham's sale at Lartington in the autumn of 1819, Bates saw Robert Colling for the last time. Their conversation turned chiefly on the pedigree of the Princess family.[1]

Mr. Hustler of Acklam died that same year. He had bought the cow Old Daisy from Charles Colling previous to 1807. He matched Daisy when on fog in the autumn against one of the cows of his tenant, Mrs. Appleton. On the milk of one meal being measured, Mrs. Appleton's cow was found to have given 15½ quarts, and Daisy 16 quarts. Mrs. Appleton therefore lost, and by the terms of the wager had to pay Mr. Hustler a new hat; if she had won he was to have found her a new gown.[2] Mr. Hustler had bred from Daisy the cow Fairy, 'one of the best cows by Duke (224),' after the Ketton sale of 1810. He had also brought back from America Robert Colling's Red Rose heifer by Favourite (252), known as the American Cow, and bred from her Acklam Red Rose by Yarborough (705), for which he repeatedly refused 400 guineas. At the sale which now took place at Acklam, young Mr. Waldy, who was an agricultural pupil of Mr. Jolly's, bought this cow for twenty-six guineas and the auctioneer's fee. Notwithstanding this excellent bargain, his father waxed very wroth, and insisted on his selling her again in the autumn. Bates bought her as he also did Fairy. It would seem that he paid sixty guineas for the pair. Soon afterwards, still in 1819, Bates met Charles Colling in Darlington. On his mentioning his purchase, Colling told him, 'My brother Robert and I breakfasted with Mr. Hustler in going to the Marton exhibition just before his death, and we consulted together after seeing his cows. Neither of us ever bred so good a cow as the Acklam Red Rose. She has exactly Hubback's handling.'

At Lord Althorp's request, Bates bought, what he considered were two good cows, Spot and Sparkles, both probably by the Duchess bull Duke (226), from Mr. Compton of New Learmouth, for £50. In acknowledging

---

[1] Bell, p. 97.
[2] Thomas Bates to Mr. Priestley, Kirklevington, 14th December 1844. *The Pencraig Herd of Shorthorn Cattle*, Monmouth, 1877, p. 6; Bell, p. 22.

this amount Compton added, 'I am sorry you have such a bad opinion of Carham (114). I hope his stock will prove better than you expect.'[1] Both Spot and Sparkles were sent, probably against Bates's advice, to a very promising bull called Young Star (619), purchased by John Grey of Milfield for 120 guineas—the top price—at the Pitcorthie sale, 9th October 1818, after General Simson's death. Young Star by North Star (458) was out of Mary, the heifer bought from Charles Colling for 300 guineas. Mr. James Fawcett of Scaleby Castle was repeatedly at Pitcorthie with Mr. Bruce, General Simson's nearest neighbour, who once offered 600 guineas for this bull.[2] 'Some of the Fifeshire farmers pleasantly assured John Grey when he bought him, 'Ay, man! what a price for nowt! but he's a bonny beast an he had been black.' 'If he had been black,' said Grey, to their speechless amazement, 'I'd not have carried him home.'[3]

Considering how largely Ben (70) figured in the pedigrees of his costly purchases at Barmpton the year before, Lord Althorp was naturally much concerned when Bates energetically warned him against that bull's blood:

If the Ben (70) blood is as bad as you seem to think, it is very difficult to avoid it, for, if I remember rightly, it comes into Windsor's (698)[4] blood, and therefore into Star's. Pray what is the disease, and what cattle of this blood are subject to it? I know that Mason's St. John (572) sort have weak constitutions, but I do not know of any particular disease. Though I am very eager to see Spot and Sparkles, yet I shall not be over at Wiseton till the end of September. I think it will be best for me to accept your kind offer to keep them for me until the cool weather. I think you have made an excellent bargain for me, for the prices are very little more than butcher's prices.

It was not necessary to wait long for a fatal example

---

[1] Letter of Mr. Compton, New Learmouth, 8th July 1819; Thomas Bates's MS. Correspondence, IV. fo. 11.

[2] MS. note by Bates on the Pitcorthie Catalogue.

[3] *Saddle and Sirloin*, p. 123.

[4] Windsor (698) (by Favourite (252), dam Venus by Ben 70) was own brother to Star's (619) dam Mary.

of the disease in the very animal that Lord Althorp had singled out. Bates's vaticination proved only too correct:

JOHN GREY *to* THOMAS BATES

MILFIELD HILL, 1*st September* 1819.

I intended to have written to you in a post or two to acquaint you of the time when Sparkles was served, and to arrange about her going home, when I had the pleasure of receiving yours from Yorkshire informing me of the chance of Lord Althorp doing me the honour to call on his return from the Highlands. Should his lordship find time to do so, I shall have much pleasure in showing him anything that this country can afford that he may deem worthy of his attention. I think I might venture to say that I could show him on my farm 100 acres of as good turnips as he would find in any man's possession from the Tweed to the Thames. But it is a subject of no ordinary regret with me to state that I cannot show him Star, as that valuable animal died last week. I found on my return from your neighbourhood that he had been very ill during my absence, and that in spite of every attempt to relieve him, he continued growing worse from day to day and entirely forsook his food, so that he was supported for some time by gruels. He never had any difficulty of breathing, but seemed to suffer much pain internally. When he died at last quite exhausted, we found the seat of his complaint to be in the liver, which was covered in parts with small hard knots of a whitish colour, and also was fastened to his side. His throat, heart, and lungs were quite clear of any appearance of ulcers or disorder of any kind. The complaint cannot have been of very recent origin. It is probable that the extreme heat of the weather, by which he seemed to be excessively distressed, brought it sooner to a crisis. I wish he had lived one year longer that some more varieties might have been produced from so pure a stock,—but all wishes and regrets on that head are now equally unavailing.[1]

Sparkles was served by Star on the 25th June, and again on the 16th July; since which time she has never come in season, from which I presume there is no danger of her. I was going to propose to you, when you wish them to go home, to allow your man, who will no doubt be a careful driver, to bring along with him three cows that I have at Mr. Donkin's, for doing which,

---

[1] A clearer example of the fatal tendency of breeders to disregard the all-important point of hereditary constitution cannot be imagined.

I would of course pay his days' wages and expenses; or if you prefer it, I could send a man from here with Spot and Sparkles who would bring mine home. If you determine on the former plan, you can send them off at your own convenience; but if the latter, be pleased to let me know.

Our turnip crops were beginning to feel the severe drought, but have now abundance of rain, and are generally the best I ever remember to have seen. Everything is much refreshed by the change of weather; although in this stage of the harvest, many would have willingly dispensed with rain for a week or two. I send for Spot to-day.

John Grey *to* Thomas Bates

Milfield Hill, 5th October 1819.

You may have heard from Halton that Spot and Sparkles have been there for a fortnight. The cows that I expected from Sandhoe were not ready to come, but having engaged to send Lord Althorp's to Halton, I despatched them notwithstanding.

Bates also bought for Lord Althorp a cow belonging to Mr. Hutchinson, who was in partnership with Mr. Place as a banker at Stockton. Apparently this cow was Alpha by Alfred, the dam of the Masham heifer belonging to Mr. Thomas Place of Spennithorne, and which, at four years old weighed, the four quarters, 78 stones, tallow 12 stones:

Lord Althorp *to* Thomas Bates

Wiseton, 30th October 1819.

I should very much like to have Sparkles, and the thirty guineas cow of Mr. Hutchinson's home as soon as I can. I am very full stocked, so that I think it better to decline Mr. Hutchinson's old cow, and Mr. Place's heifer, for the thirty guineas cow will give me a cross of all the blood, at a moderate expense. I am very sorry for Spot's accident, but I hope, by rest, she will soon recover; as to your standing to the loss, we will talk about that if it should occur. I must, however, ask you to let her stay at Halton till after her calving, as moving her before will now be impossible, and perhaps you can have her put to Duke.

Parliament is to meet, as you know, on the 23rd of November, which will prevent my Northern expedition before Christmas. If the recess is long enough I will come afterwards.

After Ketton (709) ceased to procreate, Bates felt severely the contrast between his offspring and that of the best bulls descended from him like Ketton 2nd (710) and Ketton 3rd (349).[1] The origin of Ketton 2nd, got by Ketton and calved in 1811, presents problems of much

KETTON 2nd (710).[2]

perplexity. The statement in the Herd Book is that his dam was by a Grandson of Favourite, and his grandam by James Brown's Red Bull, the sire of the Stanwick Duchess, bought by Charles Colling in 1784. It is said that the Grandson of Favourite in question was born of ' a first-rate milker and feeder, own sister to Old Cherry by Lame Bull (357).'[3] The evident satisfaction, however, which Bates felt on Mrs. Colling telling him in 1810 (before he was personally interested in Yarborough (705) by the purchase of his daughter Acklam Red Rose), that the Cherry Cow, Yarborough's dam, had nothing of Lame Bull in her,[4]

---

[1] Bell, pp. 44, 81.
[2] From the original coloured drawing, formerly at Kirklevington.
[3] Beever's *Leading Shorthorn Tribes*, p. 40.
[4] 'Yarborough's dam, Mrs. Colling remembers well, had nothing of the Lame Bull in her, but was of the Cherry tribe by Favourite, and being out of Old Cherry it could be none other than the Cherry Cow by Hubback.'— MS. note of Thomas Bates on the Ketton Catalogue. According to the *Herd Book*, Cupid (177) was the sire of Yarborough's dam and was by Son of Favourite (253) a grandson of Lame Bull (357).

suggests that it was with her that Ketton 2nd was most nearly connected. One thing is certain and that is that, judging from his portrait, Ketton 2nd was an indifferent bull, and any tribe less prepotent than the Duchess would have been impaired by his cross. Yet Bates specially selected this very portrait as one of the illustrations for his intended History of Shorthorns.

Bates now tried the experiment of using Marske (418), the roan thirteen-year-old bull which Maynard had bought at the Barmpton sale. Marske was by Favourite, and his dam, Old Bright Eyes, gave fifteen quarts of milk twice a day. Bates had been for some time in possession of his sister, and though this roan Bright Eyes cow was now fifteen years old he fruitlessly had her put to her brother. Marske was the sire of Duchesses 7th, 8th, and 9th.

While Duchess 1st and Duchess 3rd were at Kirklevington in order to be near Marske (418) at Harlsey, Mr. Jonas Whitaker of Burley in Wharfdale, a quaker cotton-spinner, came over to see Bates in the autumn of 1819, his 'religious character serving as an introduction.' Although he 'never had one grain of knowledge or judgment of short-horns or any other cattle,' Whitaker had for above ten years been endeavouring to procure the best shorthorns as a good business speculation. On seeing the two Duchesses, he exclaimed, 'I now see where I have erred. I must get this blood or I never can succeed. I paid as high as 140 guineas for Golden Pippin at Mr. Robert Colling's sale, and bid 622 guineas for Lancaster (360), but the auctioneer did not observe my nod, and knocked the bull down to the last bidder at 621 guineas. I am now glad that he did so, for Lancaster would have done as little for me as the other bulls I have tried. I wish you would only procure me some good blood that I may have a chance of success.' Bates replied that Lord Althorp had asked the same favour, and that until his first friend wanted no more cattle, he could not assist any one else. As has been said, he received a letter from Lord Althorp at the end of October to say that he was very fully stocked. Soon after this there came a private offer of a number of cattle bred by Mr. Hustler of Acklam. Among these were Alfrede and Red Daisy (of the Old Daisy tribe which Bates considered the best shorthorn blood next to the Duchesses), and Meteora and Young Priestess. The owner knew nothing of their real value, and, wishing to oblige Bates, named 'a low price indeed.' Bates bought them all and then let Whitaker have them at prime cost. He also helped Whitaker to buy two other cows, Young Honer and Yellow Rose, from Mr. Baxter of Acklam. Thanks to Bates's opinion of Whitaker's 'religious character,' the cow with which, and the descendants of which, Whitaker gained most premiums, only cost him five guineas.

After he had got so many cows and heifers from Bates, Whitaker came to Halton Castle in 1820, and admiring there Robert Colling's old cow Bright Eyes, the sister of Marske (418), begged Bates to go with him to see the descendants of her daughter Barmpton by Favourite, at Mr. Baker's at Elemore. Bates always regarded the Bright Eyes as 'one of the best tribes of short-horns'; of this, his own use of Marske (418) on the Duchesses is sufficient proof. On reaching Elemore, however, he and Whitaker found that all the young Bright Eyes stock were by the roan bull Hermit (305). This they both considered a thoroughly bad animal though there was little in his breeding to account for it. Indeed Mr. Baker was a rigid purist and a sworn enemy of 'alloy.' 'No cow or bull,' he wrote in 1821, 'can ever be called a Short-horn with a cross of Scotch blood. I would just as soon put the Duke of Grafton's Penelope to a Scotch pony, to breed a racer, as a cow to the grandson of Bolingbroke, out of a Galloway cow, to breed a Short-horn.'[1] All that Bates knew against Hermit's ancestry was that his grandsire Surplice (634) 'had very low flat sides, and was an excessively hard handler.' This was all the stranger to him as Surplice's dam was by his own first bull Daisy (186). Still there

[1] Hutchinson's *Sockburn Short-horns*, p. 56.

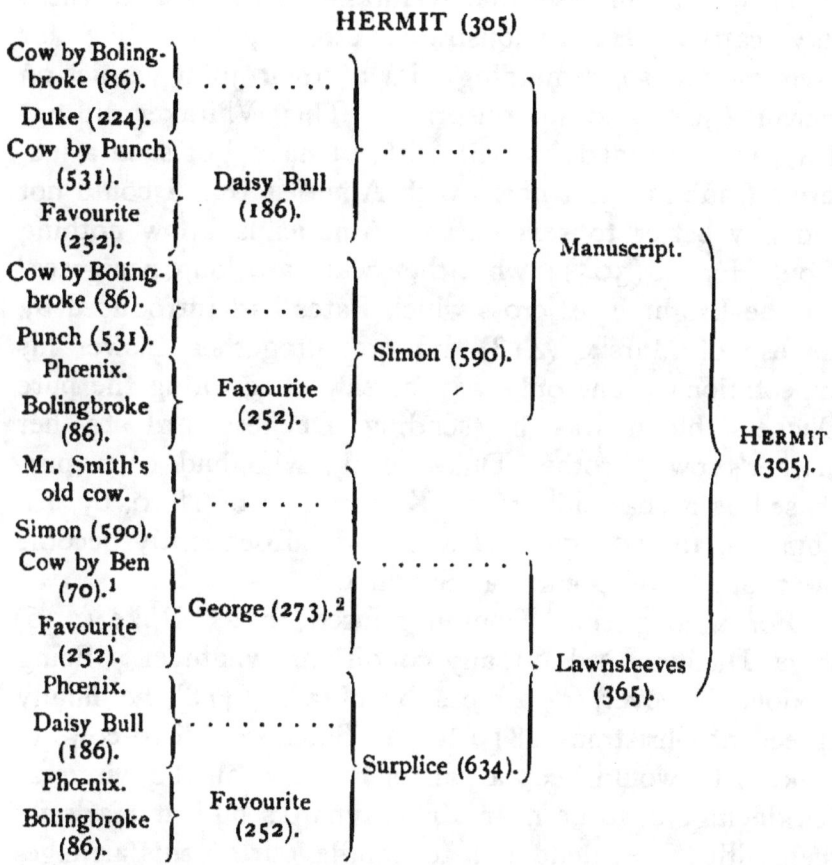

remained the stubborn facts that Hermit (305) was a bad bull, and that his issue at Elemore were 'a ruined stock that had lost all character.' Bates warned Whitaker that he ought to have nothing to do with them. Whitaker expressed his conviction that Bates was right, and promised to follow his advice.[3] Some years afterwards Bates

[1] Bred by Colonel Simson, out of a cow bought by Charles Colling from Mr. John Newby.—*Herd Book*, i. p. 59.

[2] It seems absolutely certain that Lawnsleeves (365), the sire of Buxom, calved in 1813, could not have been the son of Charles Colling's alloy bull George (276), calved in 1810. Besides this, Mr. John Wood stated positively that Lawnsleeves was about eight years old in 1815, see Appendix F. Nor could Lawnsleeves have been by Mr. R. Colling's George (275), dam Lady Grace, said in the Herd Book to have been calved after Barmpton in 1810.

[3] Bates's opinion of the Elemore stock is borne out by the description in the MS. Herd Book of Bonny, by Lawnsleeves, a grand-daughter of Bright Eyes, bred by Mr. Baker in 1815, as having 'bad shoulders and crops.'

found to his surprise that Whitaker had bought these very cattle. He remonstrated strongly with him for having done so, reminding him of their joint verdict on Hermit (305) and his offspring. This Whitaker did not deny, but screened himself with, 'I have got such a lucrative trade in short-horns with America that I could not find any better to send. The Americans know nothing about Hermit (305); what they want are long pedigrees.'

The Bright Eyes cross which Bates had introduced by his use of Marske (418) did not altogether answer his expectations. The only way he saw of restoring the pure Duchess blood was by sending Duchess 3rd to her mother's own brother Duke (226), who had been purchased as a year-old at the Ketton sale of 1810, by Mr. Compton for 105 guineas, and had subsequently become the property of Donkin at Sandhoe.

For a long time Donkin refused to let Duke (226) serve Duchess 3rd on any conditions whatever. Being anxious, however, for a cross by Marske (418), he finally agreed at Christmas 1819 to let Bates send two cows to Duke if he would keep a pair of his cows all the winter at Kirklevington, to be near Mr. Maynard's bull at Harlsey; then with the evident wish to shuffle out of this arrangement he wrote a ridiculous and contemptible letter:

WILLIAM DONKIN *to* JOHN CHARLES MAYNARD

SANDHOE, 28*th December* 1819.

I have agreed to let Mr. Thomas Bates have two cows bulled by a bull of mine the ensuing season, for his having two cows of mine bulled by your bull Marske, for which he says he has made you a compensation; but this proviso was made by me, that I was to be satisfied previously that Marske was living, and in such a state of health as to be likely to serve my cows. I have made this agreement to oblige Mr. Bates, and I shall now be obliged to you for telling me candidly whether you think Marske is in such a state as to get my cows with calf, and if Mr. Bates has made you such compensation, as I will be at no expense myself.

Maynard was a gentleman; he gave Donkin no loophole, and communicated the correspondence to Bates:

JOHN CHARLES MAYNARD *to* WILLIAM DONKIN

HARLSEY HALL, 31*st December* 1819.

In answer to your letter I beg to say that Mr. Bates agreed with me for two cows of yours to come to Marske. This agreement I expect will be fulfilled. Marske, I can truly assert, is in perfect good health, and a sure calf-getter. Your cows shall be well taken care of and free of expense, Mr. Bates having promised to answer the demand.

JOHN CHARLES MAYNARD *to* THOMAS BATES

Herewith you will receive Mr. Donkin's letter and my answer, which I hope will not allow him any plea to be off the agreement with you. I cannot with convenience send any cows to your bull this year, having Marske and another bull. Perhaps by another twelvemonth you may have a favourite at Kirklevington, which will be a consideration to me as the distance is not near so great.

Being unable to plead Marske's death or incapacity as a reason for voiding his agreement with Bates, Donkin now courted a rupture by referring to 'the gross imposition that had been practised on him in the keep of his heifers at Halton in the winter of 1810.' This, contrary to his calculations, brought a remarkably temperate answer from Bates who offered to do any possible thing to oblige Donkin on any terms he chose—a soft answer, which instead of turning away wrath only added fuel to the fire. Donkin wound up an ill-conditioned rigmarole of imaginary grievances with, 'I deny ever having had a favour from you. For this I am truly grateful, as you are one of the last men I would choose to owe an obligation to. Our intercourse ceases. I have other employment than attending to your visionary chimeras. I will read no more of your letters, and if unavoidably anything is to be settled between us, my clerk shall speak to Robert Bell.' With feminine inconsequence, he added the postscript, 'If you are determined upon having your cows sent here, Robert Bell may attend to see that they get the right bull.' Forced by circumstances to put up with the hallucinations of a cantankerous uncle-by-marriage, Bates had Duchess 3rd served by Duke (226) in February 1820.

Had he known at the time that there were still some of the pure Princess blood left, bred direct from Hubback to Favourite, his patience with Donkin would probably have been exhausted long before. This he did not discover to be the case even when a month later he was making inquiries to assist old Mr. Coates in preparing the first Shorthorn Herd Book.

The compilation of a Herd Book is said to have been mooted at the Wynyard Agricultural Meeting in the autumn of 1812, but the death of Sir Henry Vane Tempest which occurred soon afterwards delayed the matter indefinitely. Somehow or other, George Coates was always looked upon as the proper editor. Originally a butcher at Haughton-le-Skerne, he had hired Robert Colling's prize tup in 1784.[1] From Leicester sheep he passed, like so many other breeders, to shorthorn cattle, and gained the premium for the best bull at Darlington in 1791.[2] His jealousy of the subsequent success of Charles Colling is said to have caused a coolness between them.[3] Taking a farm at Great Driffield, he bred and fed there a celebrated cow.[4] He afterwards fell into reduced circumstances, and the compilation of the Herd Book was suggested as a means of affording him a livelihood. The subject was brought up again by Colonel Trotter on the evening of Robert Colling's sale in 1818. A subscription was started, and the list largely signed by the breeders present. Names, however, are not cash; nothing further was done, till, thanks to the zeal of Bates, who had entered enthusiastically into the scheme, a meeting was held at the King's Head at Darlington in February 1820. After this, Bates often accompanied Coates to see Charles Colling at Croft, taking with him

---

[1] *Sockburn Short-horns*, p. 23.   [2] *Ibid.* p. 26.
[3] Letter of 'S' in *Country Gentleman*, 27th July 1871; Allen's *History of the Short-horns*, p. 233.
[4] A coloured print of this cow, dedicated to Richard Langley, Esq., was published by Coates in 1804.

the written statements he had jotted down at the dictation of Charles and his brother, many years before the Herd Book was thought of. Up to this time, very little attention had been paid to the pedigrees of cattle except by Bates himself. Most of the breeders who had any pedigrees of their stock had got them through him. The Collings never volunteered any information on the subject, but they never refused to answer Bates's questions. Unless, however, he had known all their stock and their descendants from close application, he would never have been able to draw the brothers, as he did when once he began to write out the pedigrees in their presence. After the death of Robert Colling, which took place on the 7th March 1820, he vainly hoped that some memoranda of importance might be discovered at Barmpton:

CHARLES COLLING *to* THOMAS BATES

MONKEND, 19*th May* 1820.

Mr. William Robson has looked over my brother's papers and has not been able to find anything like a pedigree of his stock that goes further than the catalogue of his sale which will be the only sure ground for Mr. Coates to go upon.

It was on the 13th March 1820, that Bates in trying to keep Coates straight, obtained from Mr. Maynard the remarkable account of the lineage of the cow Lady Maynard which plays so important a part in shorthorn history.[1] Of scarcely less moment was the statement made to him nine days later by Alexander Hall, giving the account of the origin of the Princess family.[2]

Lord Althorp now sent Rosette to Halton in order that she might be served by Duke (226) at Sandhoe. A comparison with the Acklam Red Rose going in the same pasture fell out very much to Rosette's disadvantage, although they were out of own sisters.[3]

---
[1] See above, pp. 47, 48, and Appendix A.
[2] See above, p. 35, and Appendix B.   [3] Bell, p. 292.

Lord Althorp *to* Thomas Bates

DUNSTABLE, 26*th April* 1820.

I am on my way up to town, and as I must on getting there drive short-horns out of my head, to replace them by politics—certainly a more disagreeable subject, and perhaps in the present state of things a less useful one—I write you a parting letter to say that I suppose Rosette[1] will have arrived when you receive this, and she may be put to the bull, the first time that she is ready. I think I told you that I wished you to bleed her when she arrived.

Sparkles has had a bull calf; I have not seen him, but I hear he promises well. I cannot, however, be very sanguine about him, for I think her coarse shoulders must come out in a bull, but I shall not have him cut till I see him in the end of July. By the bye, I do not know whether you would like to see our Cattle Show at Doncaster, which will take place about that period; if you should, I need not say I shall be most happy to see you at Wiseton, and I will let you know the day in time.

Lord Althorp now proposed to hire His Grace (311), 'a grand animal, the very image of his father Ketton, with that fine fleecy coat that so remarkably distinguished the Duchess tribe.'[2] His Grace was dropped by Duchess 2nd on the 13th of April 1818.[3] He bulled his first cow, at thirteen months old, and measured the next day, 14th May 1819:

Length, 4 ft. 8 ins.
Girth, 5 ft. 8 ins.; at ribs, 6 ft. 6 ins.; at hooks 6 ft. 2 ins.
} Estimated weight, 35 stone 8 lbs.

Measured again on 27th December 1819, his weight was 54 stone, a gain of 8 stone in seven months; indeed, Robson the butcher considered him and a two-year-old bull off Younger Duchess of similar proportions to be each 56 stone at the time. At two years old, April

---

[1] See above, p. 149; *Herd Book*, i. p. 475.
[2] Letter of Mr. William Charlton; Bell, p. 181.
[3] His Grace's *signalement* in the Halton service-book is: 'red with white stripe down face, and red spot in it above far eye, red nose, two white stripes on near shoulder, a white spot behind far shoulder, some white on crop, buttocks, and legs, and a white spot on throat.'

1820, His Grace measured :

> Length, 5 ft. 6 ins.
> Girth, 7 ft. ; at ribs, 7 ft. 11 ins. ;
>   at hooks, 7 ft. 4 ins.
> Hooks, width, 2 ft.
> Length of hind quarter, 1 ft. 10 ins.
> Bone below the knee, 7½ ins.

Estimated weight, 65 stone 8 lbs.

In eleven months, then, between a year and two years, he had gained 30 stone, the same weight that Laird had gained in twelve months. This meant an increase of a stone in every ten and a half days, while the best of the other animals measured took twelve days to make this. Robson the butcher declared His Grace weighed fully 72 stone, which would represent a gain of a stone each week.

LORD ALTHORP *to* THOMAS BATES

ST. JAMES'S PLACE, *1st May* 1820.

You may say what you please, but I will say that I am very much obliged to you for the inclination you show to accede to my proposal about His Grace. I am thus situated: I have a young bull that I expect will be good enough to put to all the common Regent heifers next year, but I hardly expect that he will be good enough to put to a bull breeder, and therefore if Nonpareil should have a heifer calf, I shall not exactly know what to put her to, because my young bull will not be worthy, and Regent is not correct enough in his shape to be put to his daughter, but even in this case she will not be fit to put to the bull before the 1st August. I would therefore propose to take His Grace for a twelvemonth, or to the end of the year 1822, from the 1st August 1821, if Nonpareil has a heifer calf; if she has a bull calf, then I will take him for the year 1822.[1] I do not consider myself bound to allow Champion the use of a bull I hire, as this is a very unusual thing to be allowed, and I should also rather prefer that you did not allow me to take in cows to His Grace. This being the nature of my proposal, I shall be obliged to you to let me know what you will ask for the hire of him for the periods I have mentioned.

[1] Nonpareil, as it turned out, had a bull-calf Sirloin by Regent on 22nd May 1820.

Lord Althorp *to* Thomas Bates

St. James's Place, 15*th May* 1820.

I write you a line to say that I have bought Premium,[1] and I have desired Mr. Grey to send her to Kirklevington. She has only calved about a month, so she will be very fit to travel home with Palm-flower.[2] When she arrives, please write to Hall, at Wiseton, and he shall send for the two. I only gave £31 : 15s. for her, so she is very cheap.

His buying Premium, Spot's mother, shows that Lord Althorp must have been very satisfied with Bates's purchase of the daughter for him, and it might have been better taste if he had employed him again as intermediary. Palm-flower, a sixteen-quart cow, was descended from the celebrated Old Sockburn.[3]

John Hutchinson *to* Thomas Bates

Stockton, 17*th July* 1820.

You will see Major Rudd and I am corresponding in the *Farmer's Journal* on shorthorns. I think we shall, before all be done, raise a laugh amongst the breeders. I think I see you smiling at us already. I have heard nothing of Palm-flower how she performed her journey, nor from Lord Althorp.

By His Grace (311), Palm-flower afterwards produced a heifer calf, Eliza, which Lord Althorp told Hutchinson 'promised to be quite capital, being very large, well-formed, and very healthy.'[4]

In response to Lord Althorp's invitation Bates stayed at Wiseton for the Doncaster meeting of 1820. As the party were leaving the dining-room after dessert, Lord Althorp, turning to one of his friends, said of Bates, 'Wonderful man! wonderful man! He might become anything, even Prime Minister, if he would not talk so much.'[5]

While at Wiseton Bates alluded to the fact of Duchess

---

[1] *Herd Book*, i. p. 441.   [2] *Ibid*. p. 429.   [3] See above, p. 44.
[4] *Sockburn Short-horns*, p. 40 n.
[5] *Ex inf.* the late Mr. George Burdon of Heddon House.

3rd being in calf to her uncle Duke (226) as 'the only hope of the shorthorns.' 'I will give you fifty guineas for the chance, calf or no calf,' said Lord Althorp. 'I would not take two thousand guineas for the chance,' was Bates's reply. The future history of the Kirklevington herd proved that Bates was right in the high value he placed on the embryo Earl (646).

At Doncaster, Bates met the Rev. Thos. Harrison the breeder of Firby (1040), a bull that had been much used by Lord Althorp, and did not hesitate to tell him of his error in putting Styford (629) in the pedigree, since Colonel, who followed as the next cross, was slaughtered by Thomas Jobling at Styford, before Styford (629) came there in 1799-1800. A fracas appears to have resulted from this disclosure, but Harrison apologised the next day for his conduct in the presence of the Rev. James Armitage Rhodes and Mr. Jonas Whitaker who happened to have been with Bates at the time.

Blindly imitating the Collings, Lord Althorp crossed in-and-in till he sacrificed constitution. His cattle had thin quarters and no breasts.[1] Many of them were deformed and unhealthy.[2] He was for ever doctoring and dosing them, and thought all other animals should be treated in this same way:

LORD ALTHORP *to* THOMAS BATES

WISETON, 16*th November* 1820.

I arrived here last night and found His Grace arrived safe, but not quite well, as he coughs a little. I have therefore ordered him to be blooded pretty copiously, and to have a dose of physic. If this does not remove the cough, I shall apply the blister which succeeded so admirably with Spot.

I like him very much indeed, and a great deal better than I expected. Indeed, I can only find one point to criticise, and that is that the arm of his foreleg is coarse and too muscular. This, however, is of no great consequence, and I have no doubt from the perfection of his shape and quality in every other respect that he will do me the greatest service. I use the word

---

[1] *Saddle and Sirloin*, p. 171.   [2] Bell, p. 162.

perfection, because I think it most appropriate to the merits of His Grace.

I wish you joy of the birth of Earl Percy.[1] Duke has scarcely more of his own blood than the young Earl has, and therefore if he turns out well he will be very valuable indeed.

I am afraid your plan of a ministry composed of men of all parties will not succeed. The members of the Administration would have no confidence in one another, and the public, who would attribute their union to interested motives, would have no confidence in them either. This has always been the case in all coalitions which have ever taken place, and I think probably always will be. The first thing to give energy to a Government is that its members should have confidence in the honour and integrity of one another; the second is, that they should agree in the main principles on which their political opinions are founded; and without these two qualifications no Government can be good for anything. The Ministry which you propose would probably not possess the first of them, though perhaps it ought, and certainly would not possess the second. For these reasons I cannot agree with you in wishing the experiment to be made.

Lord Althorp *to* Thomas Bates

WISETON, 20*th December* 1820.

I am happy to tell you that His Grace's cough is nearly well. He has never appeared ill at all with it, and has shown such a constitution as I never saw in any beast before. I was able only to relax his body a little by three pints of castor oil; salts, the medicine I usually give, had no effect upon him. Upon consideration, I thought it best not to apply a blister; I was afraid it might irritate him, and I am always very cautious about a bull's temper. I like him better every time I look at him. I think I should say he is the finest bull I ever saw in my life, not excepting Duke himself.

Champion has seen him and admires him very much, but has, as I expected, refused your offer. You will not be surprised at this, when I tell you that Empress is now in calf to the young brown bull with a black face, which he showed at Doncaster. He is determined to have a new sort of his own. I told him that they might probably be a better sort of cattle than the improved short-horns, but that most undoubtedly they would not be improved short-horns, but something else. I have not seen

---

[1] The Earl (646), calved 2nd Nov. 1820, 'red with white stripe down face.'

Champion's young bull, who is matched against Sirloin (604), but from Champion's manner, when he saw Sirloin, I think the match is won already. I hope Earl Percy (646) is well and thriving.

Things were desperately bad, but never, throughout the troubles of a lifetime, did Bates allow his spirits to sink. Hope always buoyed him up:

THOMAS MACKENZIE *to* THOMAS BATES

ST. JAMES'S STREET, 12*th April* 1821.

I am not nearly so sanguine in my hopes of the return of a better time to Agriculture as you are. The loss of capital and the depreciation of the value of stock is such that many must, I fear, sink before any advance takes place.

I went this week to look at the stock shown for the prizes of the Agricultural Association, and was much disappointed. The swine were good, but the cattle (mostly short-horns) very indifferent indeed.

I do not find my residing for so large a portion of the year here much in favour of my farming operations in Scotland. However, I have raised the name of my stock (West Highlanders) so high that I sell my two-year-olds at the price others get for their two-year-olds. For this I consider myself indebted to the information acquired at Halton.

LORD ALTHORP *to* THOMAS BATES

WISETON, 12*th September* 1821.

I have had one calf dropped to His Grace since I saw you. It is a heifer out of Rosabella, who was got by Lancaster out of Rosette. Although it came a month before its time, and although Rosabella is far from being a very strong constitutioned cow, the calf appears uncommonly well, and promises to be very clever. I do not know how his having got a heifer calf out of this cow agrees with your system.'[1] I am afraid you must either give up your system in this case or His Grace's constitution, and I am sure you need not give up the latter.

[1] This is said to have been founded on the idea that if a cow is only served once she will produce a heifer, but if twice a bull.

Lord Althorp *to* Thomas Bates

LEAMINGTON, 16*th September* 1821.

I have no wish to use His Grace to any more cows, as I have so many in calf to him that I have hazarded enough upon one bull, therefore he will be ready to go whenever and wherever you please. The only cow I should put to him by any chance would be the Lancaster heifer, who has had so good a calf by him [1] and this, of course, would depend upon how the calf looks when I get back to Wiseton; and this is of no consequence, as I think Regent likely to get as good an animal out of her.

I cannot quite give way about the Lame Bull. Charles Colling writes me word that he was got by a bull of Brown of Aldborough's, so that on his sire's side at least he was well bred. Then nothing can show better constitution than Peeress, Cecil, and Ketton.[2] Spot is a fine cow, and Premium, her dam, a much finer cow, and these are Lame Bull all over. Premium is got by May Duke (Lame Bull), dam by Cupid (Lame Bull), grandam by Favourite, out of Johanna, who was got by the Lame Bull himself, and I think you told me that nothing could be better than Johanna. Cecil's horns are quite waxy, and so will those of the heifer out of Harrison's cow by him be, and so are Premium's; and, as to the shoulders, in all these three they are the most remarkable points about them; and I should prefer Cecil to any bull I know for a cow who had coarse shoulders.[3]

When I told you that the calf out of Rosette was amiss, I did not mean to describe him as actually diseased.[4] He had a scouring upon him for some time, and as this is the disorder most prevalent among the short-horned breed of cattle, I am determined never to use a bull whom I can ascertain to have been affected in this way; that is to say, as long as I can find one that has never been so. This is my only objection at present to Radical.[5] The public admire him very much.

[1] Rosabella, roan, calved March 1819, by Lancaster, dam Rosette, etc., produced, 28th August 1821, a roan cow-calf Thalia by His Grace.—*Herd Book*, i. 465.

[2] Ketton (346), by Comet, dam Cherry, 'bad in girth,' according to the Dean of Bangor.—Beever's *Leading Tribes*, p. 40.

[3] The MS. Herd Book characterises Cecil (120) as 'sides too flat' (Bell, p. 366); while Coates did not hesitate to pronounce him 'as plain a dog as I ever saw.'—Beever's *Leading Tribes*, p. 40.

[4] See Bell, p. 205.

[5] The roan bull-calf Radical (535), calved 26th February 1821, by Duke (226), dam Rosette, died in the December following (*Herd Book*, i. 475). Bates wrote, 'Lord Althorp blames Duke, when the disorder in the calf was from Rosette's tribe,' see below, p. 179.

Daisy, the daughter of Sally, belongs to Mr. Simpson and not to Mr. Smith. Sally herself came to Smith's share, and I saw a very promising bull calf out of her by Ketton (Lame Bull again), at Dishley in the summer.[1]

It was eventually settled that His Grace should go from Wiseton to Greenholme, where Mr. Whitaker was not satisfied with Ketton 3rd (349):

JONAS WHITAKER *to* THOMAS BATES

BURLEY, *2nd March* 1821.

Mr. Coates says nothing when the book will be ready. I suspect he has not got Mr. Mason's list. Priestess has produced two quey calves,[2] Alfrede, a quey and a bull, both very small and nearly ten days before her time was up; and Penelope, the heifer bought of Mr. Waldy, a quey calf very large; all to Ketton 3rd, their colour like the sire. Cleveland (146) is doing well, but the cows do not hold to him. You had your doubts of his being a calf-getter; may I ask your reasons? Golden Pippin calved a quey to Harold (291), which died a few days after its birth. From the Lancaster heifer I have a bull calf, which is to be superior to most if Mr. Hutchinson's opinion is to be taken. It is certainly a fine large calf with a coarse head. Mr. Briggs was over last week; he talks of keeping Ketton the 3rd another year, or of buying him out. One of the heifers you bought of Baxter is an extraordinary good one;[3] another, which would not breed, I am feeding.

Mr. Curwen called here on his way to London. He expressed great surprise at seeing, as he said, so many good animals, saying, 'I informed you at the Barmpton sale that I considered my stock equal if not superior to any in the kingdom; but I find yours equal in symmetry and far superior in size.' I suppose he wished to pay me a compliment.

JONAS WHITAKER *to* THOMAS BATES

BURLEY, *15th September* 1821.

I thank you most sincerely for your offer of His Grace. Cleveland is certainly better. Several of our cows do not hold to him, yet I do not think the fault is in him, as several do. I

---

[1] Tippoo (649) red and white, by Ketton (346), dam Sally.
[2] Young Priestess, calved 1817, bred by Mr. Hustler, produced in 1820, November 1, Priscilla and Parentia by Ketton 3rd.—*Herd Book*, i. 442.
[3] This was Yellow Rose by Major, dam by R. Colling's Gray Bull (a son of the White Bull).—*Herd Book*, i. 561.

have arranged to have Harold (291), whose stock are now looking well. You will not be offended when I say Ketton the 3rd's are too much like himself—coarse in the offal and bad in their hindquarters. I wish you had never known him, as the transmission of his coarseness to his stock is quite evident. I hope His Grace's are exclusively his own superior shape. Rosebud, out of your Red Rose, has slipped her calf.

Jonas Whitaker *to* Thomas Bates

Burley, *3rd October* 1821.

His Grace has just arrived, much fatigued, although a careful man has been on the road with him since last Thursday morning. He has a cold upon him, but a little rest and good nursing will, I have no doubt, bring him soon round. The wet has been against him. I hope, from the report that accompanies him, that our cows not in calf to Cleveland will hold to him; and hearing a bad report of Harold as a calf-getter, I am inclined to accept your kind offer of retaining him until you want him at Christmas. I may now pronounce Cleveland nearly well. On looking at your letter I see Mr. Briggs is to see Leven[1] before he determines which of the two bulls he will take. Is it pounds or guineas I am to pay for Cleveland?

Meanwhile, on the expiry of his lease of Halton Castle at May Day 1821, Bates had removed his herd to Ridley Hall, an estate of 200 acres, on the South Tyne, between Haydon Bridge and Haltwhistle, which he had been led to purchase in 1818 under peculiar circumstances.

The Rev. Nathaniel John Hollingsworth, who had vacated his fellowship at St. John's College, Oxford, on marrying a daughter of the Margaret Professor of Divinity, was presented by the Bishop of Durham, the Hon. Shute Barrington, to the living of Haltwhistle in 1809. The accuracy of an account of the 'Reformation' given by the bishop in a charge to his clergy was severely impugned. Hollingsworth volunteered to defend him, and published anonymously a small volume bearing the attractive title of *Three More Pebbles fresh from the Brook, or the Romish*

---

[1] Leven (374) red and white, calved in 1819, got by Ketton 3rd, dam Duchess 2nd by Ketton.

Goliath slain with His Own Weapon. He carried the Low Church cant of the day to the extent of advertising in the Newcastle paper for 'a schoolmistress possessing genuine piety.' The Bible Society movement afforded him another welcome opening. Early in 1812 he asked Bates to grant him an interview at an inn in Hexham before a meeting to be held for establishing a Bible Society in Tindale Ward.[1] He soon ingratiated himself into Bates's favour, and stayed for days together at Halton Castle, often expressing his regret that Bates was not his parishioner. From time to time Bates assisted him in raising money and managing estates with which he was connected, without looking for any remuneration. In February 1818, returning from Cumberland in the company of Mr. Nicholas Burnett of Black Hedley, Bates called at Haltwhistle vicarage. Hollingsworth was so importunate for him to remain that he did so, and let Burnett continue his journey alone.[2] The vicar then disclosed the deranged state of his affairs, and the awful effect it had on his mind. In the words of William Robson, his tithe-proctor, he 'was fallen from his clothes'; his sleep was gone, and he could get no rest at night.[3] His wife, he said, had just written to Lady Barrington to see if an exchange could not be effected to prevent his living being sequestered for the benefit of his creditors. Bates said that such public exposure would never do, and asked if there was no other mode by which he could be relieved. Hollingsworth replied that Miss Lowes had consented to sell Ridley Hall in order to assist him, and wished to know Bates's opinion of its value. He had already offered it in vain to the adjoining landowners, and did not wish to advertise it lest it might 'unhinge Miss Lowes.'[4] Bates had valued Ridley Hall three years previously at £12,000, but the value of land had fallen greatly in the interval, and the old pasture by the river

---

[1] *Letter to the Lord Bishop of Durham in consequence of the Letter of the Rev. N. J. Hollingsworth, relative to the Sale and Purchase of the Ridley Hall Estate*, by Thomas Bates, Newcastle, 1830, p. 4.
[2] *Ibid.* p. 34. [3] *Ibid.* p. 67. [4] *Ibid.* p. 38.

side had been ploughed out, and the soil carried away by the floods. £12,000, Hollingsworth said, would so increase Miss Lowes's income as to free him from all care.

As Bates was about to leave, Mrs. Hollingsworth urged him to ride down with her husband to Ridley Hall and return to dinner. On looking at the house Bates suggested that it might be made into an institution for clergymen's daughters if he could obtain the Bishop of Durham's sanction. The next morning the vicar's wife so worked on his feelings by the representation of their distressed condition that he agreed to take the estate himself, provided that he was ensured that there could never be any claims made for any tithes not then paid. Hollingsworth declared that a *modus* paid for hay, hemp, and reek covered everything. Bates reminded him that when the Rev. Dr. Scott made fresh claims upon the estate of Wark Eals in the parish of Simunburn, his father would not contend against a clergyman, however unjust his demands, neither would he himself.[1] Again Hollingsworth gave the strongest assurances that there could never be any fresh claims made by any vicar of Haltwhistle against the Ridley Hall estate, the tithes from which averaged under £8 a year.

Bates hastened to London to see the bishop who declined to sanction the proposed institution; then, though for seven years he had experienced the unpleasantness of two farming concerns, one in Yorkshire, and the other in Northumberland, he yielded to Hollingsworth's solicitations, and signed the agreement to purchase for £12,000, without consulting his own legal advisers.

The next time he went up to Haltwhistle, the curate met him with, 'I do not know what you have done, Mr. Bates, but you have quite recovered the vicar, he was all despondency before you came; he is all life and spirits now.'[2]

[1] *Letter to Bishop of Durham relative to Ridley Hall*, p. 40.
[2] *Ibid.* Appendix, p. 10. For a French reader it is well to explain that in England a *curé* who has only the lesser tithes of a parish is called a vicar, while a *vicaire* is called a curate.

In proceeding to Darlington, Burnett was overtaken by William Armstrong, a cattle dealer in Haltwhistle parish, who told him 'Mr. Bates, whom you left at the vicarage is still there; but if Hollingsworth does not requite him for his kindness in helping him to let his tithes, then say I am no judge of mankind.'[1] Even with the *modus*, Burnett considered that Ridley Hall was not then worth £8000, being in very bad condition, and not of a good tenure.[2]

Although he entertained a very high opinion of his estate at Kirklevington, Bates, when it came to the point, was probably extremely loath to quit Northumberland.

[1] *Letter to Bishop of Durham*, p. 35 n.   [2] *Ibid.* p. 70.

CHAPTER VI

RIDLEY HALL

1821–1830

THERE is no more lovely situation on the South Tyne than that of Ridley Hall. The Allen, issuing from a gorge which with its woods and crags is quite Tyrolese in character, unites with the larger stream in a stretch of excellent haugh-land, and on a little higher ground are the house and gardens, admirably sheltered, with rich pastures leading westwards to the ancient chapel of Beltingham.

On removing to this second paradise from Halton in May 1821, Bates imprudently drove the calf The Earl (646), 'the hope of the shorthorns,' all the way, and turned him out on the fresh grass for the first time after the day's journey. The Earl swelled in consequence, and frequently did the same when on green food, clover, or turnips.

Bates often tendered advice and rendered assistance to his shorthorn friends, especially to young breeders anxious to commence a herd. Prominent among these was the Rev. James Armitage Rhodes, then of Carlton, near Pontefract, whom he had met at Doncaster in 1820.

REV. J. A. RHODES *to* THOMAS BATES

20th December 1821.

I am happy to hear you have recovered your health, and that your farming concerns are going on satisfactorily, although your prognostication as to the rise in the price of wheat is rather tardy in being realised. I have this morning a most kind and affectionate letter from Mr. Coke of Norfolk, in which he invites me and my wife 'to spend several weeks with him at Holkham.' His opinion of the state of agriculture is of a gloomy kind, and he cannot apparently raise his mind upon those buoyant spirits which always support you.

Mr. Whitaker has quite recovered. In order to console me for the loss of the Herefordshire tour, we have been making a little trip into the county of Nottingham, to see Lord Althorp's, Mr. Champion's, and Mr. Simpson's stock. We were much gratified, and Mr. Whitaker has added many new particulars to his history of short-horns. I have recommended him to publish, promising to use my influence with Mr. Hutchinson to write him a poetical preface in the same classical taste as the specimen, which was lately inserted, with other accurate and delightful observations, in the *Farmer's Journal*. Warren, 'hide thy diminished head'! Day and Martin,[1] creep into insignificance and obscurity! For who will hereafter read the praises of Japan blacking when stockbreeders shine, and Furioso, Orlando, and Norman Willy dance amidst all the flowers of poetry through the wilderness of fancy and romance?

I have been particularly unhappy in my first attempts to acquire celebrity as a breeder of short-horns. I bought Marchioness and Cecilia of Mr. Whitaker, at Ferry Bridge. I sent Marchioness to the house of a friend, for fear of infection from the cows which had here miscarried. On the 3rd she presented me with a beautiful dead calf, about half-grown. Cecilia was taken ill without any previous indisposition on Monday noon last, and before midnight she was dead. Her complaint was the quarter-ill in a most virulent degree. Two of my purchases at Doncaster kick in the most persevering manner, and must be parted with forthwith; but the red cow[2] is well, quiet, and likely in all appearance to produce a calf in the proper course of nature. The preceding misfortunes have given me pain, but, as it was impossible to prevent them, it is fruitless to repine. A bountiful Providence has given me many other sources of happiness, and to them I must gratefully turn.

Young Daisy[3] is well, and we think much improved. We

[1] The celebrated manufacturers who kept laureates to puff off their blacking.
[2] Kate, red, calved 1817, by Mr. Pilley's Wellington, dam by Mr. Pilley's Eclipse (*Herd Book*, i. 353). Bought for 37 guineas by Mr. Rhodes, and sold to Mr. J. Hunt for 35 guineas. (G. Coates.)—Bell, p. 203.
[3] Young Daisy (947) roan, calved May 1820; bred by Mr. Whitaker, by Orpheus (473), dam, Red Daisy by Major (398), grandam by Windsor (698), great-grandam Old Daisy, etc. etc.

believe he has been the cause of the miscarriages by a trick he has of butting the cows. We have therefore got him a pair of spectacles to peep through.

Of the three bulls to which Mr. Rhodes alluded as bred by Mr. Hutchinson, the Stockton banker, at Grassy Nook on the Tees, Furioso (270), a twin by Mason's Son of Blaize (75), was sold to Sir Charles Morgan, M.P., in Monmouthshire :

> I pray you to Tredegar go,
> Where *in bovili otioso*
> You will see a Furioso—
> (Got by a bull of Mr. Mason)—
> A fatter never grazed in Basan![1]

Orlando (471), own brother to Furioso (270), was purchased by Mr. Wrightson of Cusworth, near Doncaster, after leaving, so Hutchinson tells us, five hundred of his offspring, principally heifers, in the Yarm district.[2] The yearling Norman Willy (456) he intended to show in London :

> right away for Piccadilly,
> Shall he go, my Norman Willy![3]

Hutchinson certainly succeeded 'in raising a laugh amongst the breeders' by his *Origin and Pedigrees of the Sockburn Shorthorns, with Remarks, an Appendix, and a Supplementary Essay: the whole intended to elucidate many points necessary to be known by south country breeders and amateurs. In Prose and Verse*, 1822.[4] The lines with which he interlarded this ill-balanced production would not always scan or rhyme. It was with some justice that

---

[1] *Sockburn Short-horns*, p. 40. [2] *Ibid.* p. 8. [3] *Ibid.* p. 40.
[4] London: Printed for Evans and Ruffy, 9 Bridge Row, by W. Robinson, Stockton. On the title-page the Latin motto, inculcating due care in calf-rearing :

> Tu modo, quos in spem statues submittere gentis,
> praecipuum jam inde a teneris impende laborem.
> P. Vergilii Maronis *Georgicorum*, lib. iii. 73.

was followed by the English distich,

> To breed the best let all who may aspire,
> Investigate the pedigrees of dam and sire.

he described himself as

> a loreless Wight,
> Whose bow-string twangs his gander plume
> To an advent'rous height.[1]

He had been strongly advised to strike out all 'his foolish verses as tending to make the book ridiculous'; but if he were laughed at it would be in the company of his brother-in-law Wordsworth, and

> For a musical chorus, did Handel or Hooke
> Ever equal the short-horns of green Grassy Nook?[2]

All the same it is impossible not to admire the enthusiasm which led John Hutchinson to buy back the Sockburn (Blanche) blood that had been so long in his family, and, though no practical farmer himself, to raise it again to a high position among English herds. His bull Sir Leoline (603) was awarded by the Board of Agriculture the first premium at Westminster in 1821, albeit a Durham gentleman (apparently Mr. Baker of Elemore) pronounced him to be 'no more a short-horned bull than he himself was.'[3] Like Bates, Hutchinson cautioned Agricultural Societies that they ought to prevent, as far as possible, all partiality or private interest from influencing the decisions of the judges. In his jingling jargon, he recommended to the Honourable Board of Agriculture

> That Herefordshire judges
> Decide the Durhams' prizes,
> *Atque vice versâ*
> So Equity advises,[4]—

a view that was to a great extent shared by Bates.[5]

Mr. Rhodes was not the only person who had bad luck with his cattle; but to crown his misfortunes, when Kate, the red cow that Bates had recommended him to buy at Doncaster, did 'produce a calf in the proper course of nature,' it had a black nose, and then promptly died:

---

[1] *Sockburn Short-horns*, p. 53.
[2] *Ibid.* p. 51.   [3] *Ibid.* p. 41.   [4] *Ibid.* p. 42.
[5] 'The Hereford and Devon breeders know the best shorthorns better than do the breeders of short-horns in general.'—Bates MS.; Bell, p. 51.

THOMAS BATES *to* JONAS WHITAKER

RIDLEY HALL, 19*th February* 1822.

His Grace reached Kirklevington the evening before I left and had borne his journey well considering the effects of the severe cold he had had, and which I fear, from his thin appearance, he is not likely soon to get quit of. Ketton the 3rd also got safe up, and was so for some days afterwards, when he began swelling, and they had to tap him frequently while I was there.

I am really sorry to hear our friend Mr. Rhodes has been so unfortunate with all his cows, and the calf having a black nose, and Lord Althorp saying Pilley's cattle are all so, has surprised me not a little. I did hope our friend had been well laid in as a beginner at Doncaster, but the other cows having all cast calf, and this calf with the black nose being also dead, hurts my feelings very much, particularly as I found Mr. Rhodes so very tractable and willing to follow advice. The other cows could all be got to pay for feeding without any loss, but this cost, I think, about the double of butcher's price when fat, as markets are at present. I do assure you I thought well of her at that time, and had it not been to serve our friend would have bought her for myself, and notwithstanding all you say, and I have no doubt justly, as his lordship's report may well be depended upon, and notwithstanding that his lordship will have no more of Duke's stock, if our friend Mr. Rhodes chooses to part with the dam of the black-nosed calf I will yet give him prime cost, and if he likes to part with her would send for her and put her to The Earl, son of Duke, dam out of the own sister of Duke, and the first bull calf she brings shall be at his service, if he chooses to keep it as a bull *solely* for the use of his neighbours' cows in his own parish and his own, thus promoting the patriotic object he has in view and a truly laudable one it is. Do write and inform him of my offer, and keep up his spirits, and tell him if he is not quite disheartened, and will trust me again to choose for him, I will keep looking out for another which I hope will be more fortunate and never bring a black-nosed calf. To The Earl she will never breed a black-nosed calf, if I have any judgment whatever of cattle. I do really wish to have her to redeem both the cow's character and my own judgment, and I trust our friend will not let his delicacy overcome him if he wishes to part with her. If he does I should wish to have her here immediately before she takes the bull, therefore will thank you to let him know my sentiments on the subject, and give him your advice what you think he ought to do without any reserve whatever.

You were very desirous I should sell him the heifer out of Duchess the 4th, by Ketton 3rd, when at Kirklevington last spring. She (Duchess 6th) is now in calf to The Earl, bulled the 19th December last, and as fat as when you saw her in June. If you thought the offer of her for our friend at fifty guineas, and the dam of his black-nosed calf, which cost about thirty-seven pounds would do I will give him the choice. For this heifer I would not take one hundred guineas; it would be therefore giving above fifty for his cow, notwithstanding the black-nosed calf she has brought. I would be at the expense of sending the heifer to him and bringing his cow here. Advise him as you think best for his interest. I make the offers through you, that you may give him your advice, but desire him also to follow his own after all.

I have now two bulls (The Earl and Duke 2nd) by Duke, out of (Duchess 3rd) the dam of Ketton 3rd, and a heifer by Marske (Duchess 7th) out of the same cow, and bulled by The Earl, and for the three I would not take 3000 guineas, bad as times are for farmers.

Lord Althorp blames Duke, when the disorder in the calf[1] was from Rosette's tribe, a discovery I made from having Red Rose's own brother, Styford, hired in 1804, of Mr. R. Colling, and you may remember a cow, the third generation therefrom, at Kirk-levington, when Mr. Ellis was with you from Redcar, after Major Rudd's sale in 1819. I had not one of that bull's (Styford) get, which did well, and I put forty cows of my own, and near twenty of my father's, and they also were the same. Old Ketton's stock were the up-making of me, and now that I have again got the blood, pure of other mixtures, shall never again part with it for any other tribe of short-horns I have ever seen. You would, I am sure, spend a pleasant time at Wiseton with Lord Althorp. He is a truly humble[2] gentlemanly man.

I did purpose on taking up my pen to have given you my pedigrees for Coates. If I knew when he had all ready prepared, I would give him a meeting, either at your house or elsewhere, and give him every information in my power. Previously to his publishing, it would be right to show the whole to Mr. C. Colling, and if you would come there (to Croft), I will meet you at any time, by having timely notice. *To have the whole read to Mr. Colling, two persons being present, or perhaps our friend Mr. Rhodes would make a third, and hearing Mr. Colling's remarks, would prevent any after disputes about pedigrees. If this is not*

---

[1] Radical (535), by Duke (226), dam Rosette, see above, p. 168.
[2] *Sic* in MS.

done, they will be endless. Hutchinson is sure to attack this book. Notwithstanding all I have said to him, he has published quite the contrary to what I had told him I knew and had seen myself.

I am truly sorry that you have been so unfortunate with your cows; it is very disheartening. I lost in four years 100 head, but the last seventeen years, since I bought the first Duchess, my stock have done well, and though this place had the character of stock being often ill, I have not had anything unwell as yet; no farrier or cow-doctor has ever been here.

On Whitaker's communicating this letter to Mr. Rhodes, the latter, one of the few real gentlemen with whom Bates had any dealings, at once wrote a double reply:

REV. J. A. RHODES *to* THOMAS BATES

HORSFORTH HOUSE, 22*nd February* 1822.

Mr. Whitaker has this day forwarded me your letter respecting Mr. Pilley's red cow, and I beg you to accept my thanks for the offer you have made, so that I should be no loser by following the advice with which you favoured me at Doncaster. But, my dear sir, I should be ashamed of myself if such a consideration could ever find admission into my mind. I have the fullest confidence that the opinion you then gave me was perfectly justified by appearances; and the misfortune which has since happened of a black-nosed calf, does not lead me for a moment to question the accuracy of your judgment. I beg you therefore to believe me very much your debtor for this as well as many other acts of kindness. If the cow upon further trial should still show an alliance to the Black-a-moors, I must try again to find something more likely to answer the purpose which I had originally in view, and any small pecuniary loss can be of no consequence whatever. Had I been admitted to more extended acquaintance with you, I flatter myself that you would not have given yourself a moment's concern on the subject. I have ordered a horse-hoe to be made, of which I beg your acceptance as a small token of my esteem and gratitude. We were much gratified by a little tour into Nottinghamshire, and did not meet with anything like Major Rudd's cautious conduct.

REV. J. A. RHODES *to* JONAS WHITAKER

Mr. Bates appears to me to take up the subject of the cow under an erroneous impression, as if I were disappointed and disgusted with her. The fact is that the circumstance of the black nose to the calf, may in a great measure be anticipated on account of the known defect in its size. The frame of the cow is, as far as my imperfect knowledge can enable me to decide, very superior, and if so, the price I gave was not worthy of consideration. But if I had in truth been 'hung on' with her, as it is vulgarly termed, I trust I should be one of the last men to wish to be relieved from any pecuniary loss by 'taking in' any body else with her. Except the circumstance of being torn behind, she has no defect that I know of, and she is likely to be the mother of excellent milkers, as her bag is large, and she milks very well indeed. She has become so quiet that I can lead her anywhere. These circumstances, naturally, have acquired her a sort of individual favour and attachment with us, and we are therefore quite indisposed to part with her on any idea of imperfection or insufficiency. But, at the same time, I have received so many marks of favour and regard from you and Mr. Bates, that I am willing to accede to his wishes, and to be guided by your judgment. If you think that I have a wrong opinion of his notions, and that he really estimates the cow in the same way that I do, I will yield to your suggestions, and put the matter into your hands, on only this promise that Mr. Bates, and not you, shall make the compensation you think equitable, that no advantage shall be taken of Mr. Bates's constitutional ardour, and that the payment shall be in cows and not in money.

Whitaker, always anxious to keep himself before the public, asked Bates to find him a prize-winning heifer. Bates left no stone unturned to oblige him:

JOHN GREY *to* THOMAS BATES

MILFIELD HILL, 28*th February* 1822.

I cannot as yet give you any information about such a heifer as Mr. Whitaker is in want of. She should be something superior for such an exhibition. Having been confined entirely to the house for the last ten days by a severe sprain in my leg, I have had no opportunity of making any inquiry on the subject. I expect to see Mr. W. Jobson to-day, to whom I will communicate your message, as he may perhaps know of something likely. I believe Hunt of Thornington, Curry of Brandon, and some others, have heifers of an age that would suit to show for the sweepstake at the Tankerville Arms,[1] in the first week in April,

[1] At Wooler.

but they will not probably be visible until that time. I wish you would give us the pleasure of seeing you at that time, and choose for yourself. At any rate, if I hear of anything likely, you may depend on my letting you know. I fear these times will operate against the general improvement of the breeds. Money is scarce with the generality of land occupiers, which will have the effect of throwing a monopoly of the good things into the hands of yourself and such as can afford to go on under any circumstances. I have a bull by Duke[1] that I have used one season, and intended to continue one or two, but am not at all sanguine about breeding, having lost 16 out of 38 of my last year's calves by the quarter-ill during the present winter.

I wish Ministers had taken your advice and given the bonded grain to feed the destitute manufacturers, or to sell to them at a moderate price, instead of enriching a few speculators at the expense of the agriculturists.

THOMAS BATES *to* RAWDON BRIGGS THE YOUNGER

RIDLEY HALL, *12th April* 1822.

I have this morning sent off my year-old bull Enchanter, the one Mr. Whitaker wished to have. He is by His Grace, out of Fairy, the white cow you saw at Halton going with Red Rose. He appears promising, his colour a yellow roan like Comet. Being perfectly gentle he has never had a ring in his nose, but it is advisable to have one put in before he gets too strong, and I have desired Mr. Whitaker to have it put in at his place if his men have ringed bulls before. I have given Mr. Whitaker liberty to send what cows he likes to him. I added that if he sent his year-old bull, out of Alfrede, to serve your cows until Enchanter came, you could send your cows over to Greenholme on Enchanter's arrival there. This you can arrange among yourselves. I have desired Mr. Whitaker to ask Mr. Rhodes to send the cow (Kate) he bought from Mr. Pilley last year at Doncaster, if he desires it.

I am sorry you did not sooner inform me of Cleveland's disability. I feared it when I saw him last summer, and said so to Mr. Whitaker, but his cows being no better to His Grace made me think that the fault was in them. I have fixed for my man to be at Catterick Bridge next Thursday night, and have written to Mr. Whitaker to send Cleveland there.

I have had better health this winter than for seven years

---

[1] Fitz-Duke (259), by Duke, dam by Phenomenon, grandam Red Rose.

before, which I attribute much to this place. You thought it gloomy last October—we certainly had a very unfavourable day to see it in—yet if you and Mrs. Briggs can arrange to take both Cheltenham and the sea-side before August, I think you would be delighted with it in that month. The scenery is then very pleasing, and up the Allen river very romantic.

I have preserved the game at Kirklevington, and hope next season you and young Mr. Greenwood will come and thin it. The north part of the township which belonged to the Messrs. M'Leod has been sold to Lady Amherst. When I am certain of your coming, I will ask leave for you to shoot over it also, giving leave to her Ladyship's friends to shoot over mine in the winter.

Jonas Whitaker *to* Thomas Bates

Burley, 24th April 1822.

My best thanks for your prompt compliance with my request of exchanging bulls for Mr. Briggs. I was much grieved to see Cleveland in so bad condition. He left here in fine order, and returned in appearance only about half the size. Your servant would inform you of his miserable looks. Mr. Briggs's cowkeeper must have been guilty of shameful neglect, and I think you had better write to Mr. Briggs to Millbank to desire him to order his servant to take good care of Enchanter. I will order our servant to tell his men that it was your particular wish that he should be well attended to and put into a loose place, where he will be well screened from the weather. It would be most grievous that so fine growing an animal should be thrown back and perhaps injured for ever for want of a little attention and good keep. I am glad to say he has arrived safe with the exception of a little lameness upon the far hind foot. The shoe appears to have hurt him; our man has taken it off, and in a few days I hope he will be quite well and fit to travel forward to Millbank. Mr. Briggs requested that he might be sent on without loss of time as many of his cows were coming on. Alfrede's bull-calf was to have gone for a few weeks, but Mr. Briggs changed his mind, and I am now glad of it as it is doubtful whether he is a calf-getter.

I may well despond; disappointments attendant on breeding have crowded upon me from every quarter. You will be concerned to hear that Golden Pippin, Wildair, Alfrede, Meteora, and Young Honer and Yellow Rose (the two last bought by you of Baxter) are not yet in calf, notwithstanding that we have reduced them so much as to lose the benefit of their milk.

I am very greatly obliged for your continued tokens of friendship, and would gladly have accepted your kind offer of putting cows to Enchanter, could Mr. Briggs have done without him for a little time. I am afraid to send them to Millbank, where there is such poor accommodation; and without grass fields to run in they would entirely lose their milk.

I informed Mr. Rhodes that Enchanter was expected here on Tuesday, and that it was your wish that the Pilley cow should be put to him. I am happy to say that I have not communicated the infection of despondency to Mr. Rhodes. He is in good spirits and very desirous of benefiting his neighbours. With the assistance of The Earl I trust his benevolent intentions may be accomplished; indeed I incline to believe that a bull less ennobled would be of essential service for the purpose, for I confess Enchanter has many perfections and would in my opinion be fully adequate.

I will now attempt to answer your behest to examine closely and report honestly. Enchanter, I consider, good in colour; handle fair; apparently likely to be a great grower; head, good; horn, turned well but too thick and too high at the roots; neck, fair; shoulder, fair but a little too upright; breast, not prominent enough, but likely to make fat; knuckles or shoulder points, very fair; crop, back, sides, and flesh, good; girth, good; huggins, not very good at present; hind-quarter, fair but droops too much; rumps, good; twist, moderate; offal or legs, rather too strong; hocks, turned in too much; feet turned too much out. I think he will get good stock; with Cleveland's, I am much disappointed. They are nearly all bowed in their fore and hind legs, bad in their shoulders, and round in their huggins; indeed very much inferior to Harold's stock.

Having stated my opinion of Enchanter, I should be glad to know in what points or parts the Lame Bull you speak of has transmitted his faults to him. We will put a ring in his nose.

Your pedigrees will no doubt be inserted in the *Herd Book*, but when is doubtful. I have written very urgently to Coates, but I do not expect he will be able to submit any part of his pedigrees for inspection next month.

You say so much of The Earl that I have an itching upon my fingers to touch him. If I was once half-way to Ridley Hall I think I should break through the embargo and, notwithstanding your prohibition, shake you by the hand. I cannot subscribe to the justice of your system of not seeing The Earl till next year. I anticipate the pleasure of seeing you and him both this year and the next. By examining him in the progress of his growth, I shall be better able to judge of his superior merits.

JONAS WHITAKER *to* THOMAS BATES

BURLEY, 24*th June* 1822.

Mr. Briggs has been so kind as to let Enchanter come here for three weeks (on my lending him Alonzo (28) out of Red Daisy by Cleveland, and my taking four of his best cows to put to Enchanter). I am glad to inform you he has much improved in condition. We brought and shall return him in a cart, a conveyance perfectly easy to him.

Meteora, Alfrede, Young Honer and several more cows continue to take bulls regularly, although they have had nearly half a dozen to serve them. We have tried without success bleeding, working, and the starving system.

Mr. and Mrs. Rhodes and myself returned last Friday from an agricultural tour of twelve days through Derbyshire, Warwickshire, Worcestershire, Gloucestershire, Herefordshire, Leicestershire, and Nottinghamshire. We were not gratified with the appearance of good management. We found the Herefords a useful even stock for the grazier and the butcher, but miserably bad milkers. Take the average of them and they are superior to the average of short-horns, but their best are not equal to the Improved Short-horns. The breeders are strongly prejudiced in favour of their own cattle, and claim that the butchers of London will confirm that the meat of a Hereford is finer and much superior to any shorthorn.

Herefordshire is a most delightful county, and occupied principally in breeding cattle, the stocks of which are immensely large. We saw the best. Many of the cattle were extraordinary, but so materially different are their breeders' opinions to ours of the perfections of an animal that I was frequently at a loss to reconcile opinions which to me appeared erroneous. I must, however, confess that in forming their judgments, Hereford breeders appear to be actuated more by the profit of the thing than we are. Elegance and symmetry are, in my opinion, too much esteemed by us, at the expense of usefulness.

Coates has not, to my knowledge, made any further progress with his book. I have said and urged all I can.

Kate (Pilley's cow) is here to be served by Enchanter if she comes again; she is a great favourite.

Bates continued to take the very keenest interest in

all political measures affecting agriculture, and especially in the remedies proposed in Parliament for dealing with the great distress that the fall of prices had brought on tenant-farmers and, to a far greater extent, on the poorer landowners. He was too well read in political economy to believe in the efficacy of many of these nostrums.

THOMAS BATES *to* JOSEPH BUTTERWORTH [1]

KIRKLEVINGTON, 5*th March* 1822.

On passing through Newcastle on my way from my present residence at Ridley Hall, I was surprised to find from a corn-merchant who has found means to influence the decisions of Ministers on every question respecting the Corn Trade, that Lord Londonderry had proposed to the Committee on the Agricultural Questions to allow wheat to be imported at 67s. per quarter on paying a duty of 15s. per quarter. I mentioned this to some friends at Darlington market yesterday, and at their desire I rode down this morning to see Lord Stewart at Wynyard in order, by a personal conference, to try to convince him of his brother's error in the hope that ere too late the plan might be abandoned. On my road I learnt that his Lordship had passed through Stockton yesterday on his way south.

You are, I believe, not on the Agricultural Committee. I know Lord Althorp, but he has taken up a party view, and I cannot communicate with him in any hopes of doing good. I have, on different occasions, written to Sir James Graham, and in 1814 and at other times he handed my letters to the Ministers and they had some attention paid them, but he is not I believe on this Committee, and what I want to find is an active intelligent man, with a mind unprejudiced by party considerations, who would seriously consider the subject. I regard the present distress as the prelude to greater prosperity than this or any other country has ever known if only sound rational policy be followed. I am no party man either in politics or any other thing. I think for myself on all subjects, and Truth is the one object I keep in view. Disregarding all consequences to myself, I follow wherever Truth leads, and when fully convinced that my object is a proper one I leave no stone unturned to promote it.

The welfare of Agriculture should be the prime consideration

---

[1] M.P. for Dover.

of every state that has land capable of cultivation. I told the Committee on the Poor Laws in 1818 that I knew from experience in cultivating land that, without foreign grain, this country might be made to maintain four times its present population. I convinced Mr. Sturges Bourne[1] at the time that my views on the Poor Laws were correct, but he would not act on his conviction. A majority of the House, however, agreed with me afterwards, though Mr. Bourne told me that previously to my seeing him at his own house,[2] the Committee had been unanimously of a contrary opinion. I am convinced that foreign grain would now be sold even at a duty of 40s. per quarter, although the averages are lower, I believe, than for thirty years back. If you will examine Dr. Chalmers on this subject you will be struck with the justness of his views which correspond almost exactly with mine. I shall be glad to enter fully on the subject either with you or any member of the House you may recommend.

On the introduction of a Ministerial bill with a sliding scale of duties, his friend, Mackenzie of Applecross, then M.P. for Ross-shire, wrote him a letter which shows that the representatives of agriculture in 1822 were in precisely similar case to those in 1896, so curiously does history repeat itself:

THOMAS MACKENZIE *to* THOMAS BATES

LONDON, *May* 1822.

Your letter on the Corn Bill is very much in unison with my own ideas. I have unfortunately been prevented, by ill-health, from taking any share in Parliamentary business this session. It is not exactly the influence of the moneyed interest in Parliament, as you seem to suppose, that prevents more efficient steps being taken to protect agriculture, but various causes combined. There is a strong party of real agricultural members in the House, but not half a dozen of them ever agree on the same plan. Those who are mere theorists and talk of free trade are, of course, united and against every one of the plans of the agriculturists. So are various other interests. Then there are a great many who won't think on the subject at all, or who cannot devote time to it, and, as the simplest way of settling the matter, determine to support whatever the Minister proposes. The

[1] The Rt. Hon. William Sturges Bourne, M.P. for Christchurch.
[2] 15 South Audley Street.

Minister himself, when he comes to make his proposition, has to consider, not merely what is the best plan, but what he can carry through the House, so that perhaps that which is ultimately adopted, may not, even by the Minister himself, be deemed the best—and multitudes of agriculturists vote for it, merely because they think it the best they can get. If I am correct in this view, it follows that the great evil is want of union among the agricultural members themselves—one well worth cure. My own opinion is that Lord Londonderry's bill will pass nearly as it is.

Mackenzie's opinion that Lord Londonderry would carry his bill was confirmed by Lord Althorp's views:

Lord Althorp *to* Thomas Bates

THE ALBANY, 1st *June* 1822.

I certainly do not think the bill proposed by Lord Londonderry is the best measure which could have been adopted for the regulation of the Corn Trade, but I think it so great an improvement upon the present law that I cannot do otherwise than support it. I think the best system under which the farmers of this country could live would be with open ports, a duty of 20s. upon the import of wheat, and a bounty of 15s. upon its export, and other grain in proportion, as I am convinced that the vain attempt to keep up the price of corn by a high import price has been the cause of all our present distress. Without the power of exporting corn in years of super-abundance, there must be frequent recurrence of distress, and the only way in which we can have any chance of exporting is by means of a bounty. I am afraid you will not agree with me in opinion, but it would require a longer letter than I ever wrote (except about cattle), to give you my reasons. I think, however, you will be led to doubt whether I may not be right, if you will consider the subject. These are the grounds: A country that grows its own consumption in average years, must grow more than its consumption in good years, and if it cannot export the superfluity the price must fall seriously low, and the increased quantity cannot make up for the deficiency of price.

It does not seem to have occurred to Lord Althorp that it would have been a simpler procedure to strike the balance, and place a 5s. duty on the import of wheat.

The Londonderry bill passed the Commons; Bates only redoubled his opposition to it:

SIR THOMAS BUCKLER LETHBRIDGE[1] *to* THOMAS BATES

You have now nothing more to do than to send up all the petitions you can to the Lords. I will name some of them, Carnarvon, King, Suffield, Lansdowne, and Holland, and all the other noble Lords, who are generally opposed to the Government. I should hope and believe they will not only receive your petition with care, but present it with proper representations. You may rely on my support. I am daily and hourly ever here to serve the cause of the landowners and occupiers who are now so unjustly used.

Bates posted up to town himself; his friend Mackenzie had already left for the Highlands in consequence of failing health:

THOMAS MACKENZIE *to* THOMAS BATES

APPLECROSS, *3rd August* 1822.

I cannot tell you how much I regret your not having arrived in town before I left, as few things would have afforded me such sincere pleasure as to shake hands with you once more. Perhaps, from my acquaintance with Parliamentary business, I could have saved you some trouble. *If all who are interested in agriculture had had a small portion of your zeal and activity, matters would be very different from what they are. I dare say you have seen enough to satisfy you of the apathy of great men, in regard even to what most dearly concerns their own interest.*

The plan which you sketched out for my journey northwards would have been quite delightful, but I fear would not have suited my state of health. I came from London to Leith by sea by my physician's advice, and thence here by very easy stages, where I am living in perfect retirement.

When you reach Ridley Hall I shall expect a few lines from you, as I shall be anxious to hear that you have not suffered from the fatigue which you have undergone. Our crops are beautiful, and promise to be early; no demand for cattle at any price, wool very low. I know not what is to become of tenants and landlords unless times soon mend.

[1] M.P. for Somersetshire.

Disappointed and disgusted with what he saw at Westminster, Bates hastened back to Ridley Hall.

Jonas Whitaker *to* Thomas Bates

BURLEY, 30*th July* 1822.

I was much disappointed I had not the pleasure of seeing you in your flight from London through Leeds, and now the only expectation I still had appears blasted by the sale of your heifers. I must therefore visit you, as I have no hopes of your coming hither.

I do not blame the bulls for the incontinence of our cows, without it was in some manner owing to Cleveland when he became non-effective. Subsequently they have had Sir Henry Ibbetson's bulls as well as ours and Enchanter, yet Golden Pippin, Meteora, Ruby, Alfrede, Dandy, Young Honer, and Yellow Rose, still come regularly every three weeks; I am glad to say Kate appears to hold to Enchanter.

Coates, in your absence from home, called here on his way to Mr. Mason's. I gave him a letter to Mr. Colling, who was very unwell and quite unable to enter into the various pedigrees he wished to submit for his inspection. On his return he called again and found Colling rather better, and had some conversation with him. I believe Colling gave a warranty as to the correctness of Coates's history of his short-horns. Coates goes upon a plan approved by Lord Althorp. The materials of the book are now with Baines, the printer of Leeds, and it is expected to be finished in about two months. All your additions were put in, and if you have any more I shall have much pleasure in taking them to Baines, provided they are here in three weeks. Coates has got pedigrees from Mason, Simpson, Smith and many more who were backward in making out their lists. Champion will not make any return.

Jonas Whitaker *to* Thomas Bates

BURLEY, 14*th August* 1822.

I was at Doncaster, and last Saturday I had the pleasure of seeing Lord Althorp here, but I will tell you nothing in hopes of your anxiety having a little influence in bringing you here.

Mr. Coates has agreed to have the book printed at Otley. He desires me to say that if you have a wish to have the likenesses of any of your stock introduced in it, he will be glad to receive them as soon as possible. I hope you will send the water-coloured drawings of Ketton the 1st and his dam.

No prints will be introduced into the book but those paid for by the owners. A general application will be made by Coates to the principal breeders to send what they please. The impressions are made by a stone operation, and are not expensive. Mr. Hulmandell, Marlborough Street, London, will charge about three guineas for engraving and printing 600. Lord Althorp has ordered that number of Nonpareil; I have of Moss Rose and of Miranda; Coates has of an old Bull called Barningham, and of Patriot; Mason, of Gaudy, and I believe of St. John. Sir Henry Ibbetson will send likenesses of two of his breed, and no doubt many more will be glad to avail themselves of the opportunity of exhibiting their superior animals upon paper. The impressions cannot give the colours of the animals but the book will describe these.

RAWDON BRIGGS THE YOUNGER *to* THOMAS BATES

HALIFAX, 21*st August* 1822.

We dined with Mr. Whitaker on our way from Swarcliffe on Monday, and we saw Mr. and Mrs. Armitage Rhodes at Harrogate. Both Mr. Whitaker and Mr. Rhodes had been at the Doncaster Agricultural Meeting. They gave a very poor account of the whole affair. Several animals were offered for sale, but there was not a single bid at any one, so all were withdrawn. Everything was flat and gloomy. Where the mischief will end, I dread to think! The next session of Parliament may bring the landed interest to their senses; if it does not, the fundholder and ready money man will become the proprietor of most of the landed property.

Enchanter is with Mr. Whitaker. He is a sure calf-getter, as far as my cows are concerned. Lord Althorp has bought Bright Eyes of Mr. Whitaker. I think you bought some of her stock for him in the North, and he likes them so well that he wishes to have the old cow herself. His Lordship has been shooting on the Duke of Devonshire's moors at Bolton. He spent a day with Mr. Whitaker on his way there. We are thinking of trying another sale this autumn at Ferrybridge. Sir Henry Ibbetson wishes to join us.

Mr. Jolly, who, it will be remembered,[1] bought the celebrated bull called after him, from Mr. Wastell, retired

---

[1] See above, p. 47.

from farming in 1822. Although he had the wisdom to use this bull (337) for many years, he never bought another good one. At the sale which now took place at Worsall near Kirklevington, Bates never saw a more ordinary lot of cows, not two alike. Jolly had made money by farming, but he might have made ten times as much if he had kept up his herd.[1]

After Coates had collected all the pedigrees that were to be inserted in the Herd Book, Whitaker wrote them out in a neat clear hand, inserting at the same time in red ink Coates's remarks on the animals he remembered having seen. These are his laconic criticisms of the first nine Duchesses :

Duchess 1st. Fair.
Duchess 2nd } Sisters { Crops and hind-quarters fair.
Duchess 3rd }   { Fair.
Duchess 4th } Sisters. Very good.
Duchess 5th }
Duchess 6th } Good.
Duchess 7th }
Duchess 8th } Sisters. Both good.
Duchess 9th }

This is considerable corroboration from a not over-friendly source of Bates's assertions that his original Duchess-by-Daisy-Bull was superior to her grand-daughter Duchess 1st by Comet, and that by the use of his three Ketton bulls, he much improved the latter's descendants.

Before going to press the manuscript of the Herd Book was altered in form and shape, and the parts in red ink, which included some observations furnished by Bates, were struck out.

JONAS WHITAKER *to* THOMAS BATES

BURLEY, 16*th February* 1823.

Coates's book being now finished, except a few prints which are expected from London this week, I beg to ask you if you will have yours forwarded. There are several subscribers in your neighbour-

[1] Bates MS. fo. 19.

hood, and if Mr. Coates could send them all together it would be less expensive to you. If one person could receive and remit the money I should be glad as I have advanced the poor old man a little money on account, and I should feel much obliged if you could procure him a few more subscribers. The printer having omitted Ketton 1st and 2nd, you will find them in the appendix. This cannot make any difference, as the index refers to them. Besides the prints expected this week from Mr. Arbuthnot, there will be several more (one of which will be Comet) which will be forwarded hereafter. I should be happy to hear the book has your approbation and sanction, and hope the old man may have something to support him in his old age. *The book is merely a copy of the pedigrees furnished by the subscribers, so that, although errors may appear, Coates cannot be blamed.* I have attentively perused the book and think favourably of it.

Bates had striven in vain for the principle that the Herd Book should include the names of no animals that were not likely to be of future importance. Equally futile was his insistence on the absolute necessity that all pedigrees intended for insertion should be passed by a committee chosen for the purpose, before the book was published. If the Herd Book complacently registered everything that anybody chose to send to the editor, it must cease, he rightly contended, to afford the slightest guarantee for purity of blood. His only reward for the great trouble and expense he incurred in launching the work was to have his two principal bulls relegated to the limbo of an appendix.

Charity and Efficiency are two birds not readily killed with one stone; Coates, 'poor old man,' allowed himself to be influenced by parties who got him to print whatever they pleased, in order to puff off their stock,[1] and Whitaker was preoccupied with getting back the money he had advanced. Coates even altered some of the pedigrees which Bates had procured for him with considerable difficulty from Charles Colling.[2] As a test case, Bates directed his attention to the pedigrees of Brokenhorn (95) and Punch (531). Colling had explicitly told Coates

[1] Bell, p. 96.   [2] *Ibid.* p. 97.

in Bates's presence that neither he nor his brother had ever put a daughter to her sire until they used Favourite (252); yet in the Herd Book Brokenhorn was stated to be by Hubback, and his dam by Hubback, and Punch by Brokenhorn, dam by Brokenhorn. Coates admitted that these double crosses by Hubback and Brokenhorn were inaccurate, but said, 'I had some good friends who had the Punch and Brokenhorn blood, and they wished those bulls to have good pedigrees to them.' In addition to this, instead of Punch's grandam being bred by George Best of Manfield, she was, as a matter of fact, out of an inferior cow belonging to Colling's father.[1]

Jonas Whitaker *to* Thomas Bates

BURLEY, 24*th May* 1823.

Mr. Briggs tells me Enchanter will not be fit to go to Doncaster. Although he has had as much cake as he could eat, he does not improve. I am disappointed, because I had great expectations from him, that he would have completely established your stock and of course part of mine. It cannot now be helped. My cows have been very lucky in calving except Meteora, whose calf died. When you have an hour to spare, I wish you would continue the produce of your stock as given in Coates's book, and favour me with it, and I will enter it in my copy. Kate, the cow you bought for Mr. Rhodes at Doncaster, has produced a bull calf.

Bates's prognostic that if Kate were put to one of his good bulls she would not breed a black-nosed calf was fully justified:

Rev. J. A. Rhodes *to* Thomas Bates

HORSFORTH HOUSE, 29*th July* 1823.

The red cow which I bought of Pilley of Lincolnshire has brought at last a bull calf by Enchanter. It is rather lighter colour than she is, and the nose correct.

On the 25th of January 1823, Acklam Red Rose

---

[1] Bell, p. 38. See above, p. 53.

(or Red Rose 1st), then twelve years old, produced a bull calf to The Earl (646) at Ridley Hall. This calf grew into a light-red bull with a lemon muzzle, as perfect in all his points as could be desired, and evenly and smoothly covered with flesh of the best possible quality. Bates gave it the name of Second Hubback (1423).[1] The Collings had said that Red Rose, the dam, had exactly Hubback's handling, and that of the calf revived the impression the first Hubback had made on the public mind. Several good judges of shorthorns were then alive who had seen both bulls, and agreed with Bates in considering the second Hubback better than the first. The comparison of Second Hubback with Ketton does not seem to have been quite so decisive; at any rate Mr. Charlton wrote:

The last time I was at Ridley Hall I was highly gratified at seeing Second Hubback, a bull well worthy of that distinguished name, and which fully came up to my real idea of what a bull ought to be. I compared him in imagination with Ketton, and if prejudice had not been deeply seated, I might have said he was as good as Ketton. Mr. Bates smiled when I told him he was surely next to Ketton, even he would not say he was better than Ketton.[2]

Probably Bates smiled at the notion of Second Hubback being inferior to Ketton. There is little doubt that he himself considered Second Hubback the best bull he ever had, and destined to be quite a regenerator of shorthorns.[3] The stock got by him were all alike in shape, colour, hair, and handling, as well as in countenance, a point which never deceives a good judge of cattle. All the cows by him were good milkers.[4]

In 1823 Bates again went up to London, this time to oppose what he regarded as one of the oppressive

---

[1] The name might have been objected to in itself; but to range this bull in vol. ii. of the *Herd Book* under 'Second' instead of under 'Hubback,' was a slip of the kind that placed the Kettons in the appendix of vol. i.
[2] Bell, p. 182.
[3] Note by Mr. James Fawcett in *Saddle and Sirloin*, ed. 1895, p. 151.
[4] Bates MS.

measures of the Stockton magistrates. He had to stay several months in town. A hundred and sixty members, headed by a Fitzwilliam and a Lambton, came down to the Committee of the House of Commons to oppose him. He could not get a single vote in favour of what he deemed proper in the Lower House, but the bill he objected to was thrown out by the Lords, with only one vote, that of Earl Fitzwilliam, in its favour.

PHILIP SKIPWORTH *to* THOMAS BATES

AYLESBY, *7th November* 1823.

Having had the pleasure of meeting with you at Mr. Colling's sale, I take the liberty of writing to request the favour of some information relative to short-horned cattle. I remember your purchasing Young Duchess at Ketton, and my present object is to know what bulls you have disengaged for letting or selling, that are bred entirely pure to the Colling stock (either Ketton or Barmpton), or whether there are any bulls in the circle of your acquaintance which are purely descended from either place that you would recommend. If so, please to particularise them as to size, colour, and general appearance, and to give me an idea of the price of a good one.

I have been pretty lucky with the cows and heifers I bought at Barmpton. They are all alive and have bred regularly, though I lost two or three of their produce at first. Lily[1] continues to breed, and the stock from her are very promising. I do not know her pedigree beyond North Star, Favourite, Favourite and Favourite. You, I understand, have a further knowledge as to her blood. Will you be good enough to indulge me in the matter? I am particularly desirous of learning it from an authentic quarter.

Say, if you please, what blood is now the most fashionable in your neighbourhood. Have you seen Mr. Mason's herd lately? Are the short-horns making general and progressive improvement in the North? I hope the spirit of breeding does not relax further than the depression of the times may be supposed to affect it.[2]

[1] Lily, lot 8, Barmpton sale, 66 gs.; *Herd Book*, i. 379. The pedigree is elongated in vol. ii. p. 408, but on what authority does not appear.
[2] Correspondence of Thomas Bates, IV., fo. 85.

Philip Skipworth *to* Thomas Bates

Aylesby, 22nd November 1823.

I am very much obliged by your information, relative to the breeding of short-horned cattle. I am going to spend a week or two in viewing the stock of different breeders. It will give me great pleasure to pay my respects to you the first opportunity, but the distance is too great during the winter.

With respect to Enchanter and Young Marske, I think highly of their blood, and know of no objection to either except a little in the price, but I cannot make up my mind without seeing them. I must indulge myself with a ride into the country in the course of this or next month.

The dry and dreary character of the Herd Book became evident before even it was completed; a companion volume was urgently needed to bring out the salient facts of shorthorn history. Bates who, though possessing so many qualifications desirable in the editor, had been passed over, and was then called on to do most of the work for Coates's benefit, was now expected to write a History of Shorthorns with the Rev. Henry Berry's name on the title-page.

Jonas Whitaker *to* Thomas Bates

Burley, 15th April 1823.

A friend of mine and a breeder of short-horns is very desirous of introducing Coates's book into his neighbourhood in Herefordshire. He wishes to make it more interesting by accompanying the book with a short history of the short-horns, and has applied to me to furnish him with materials. This I shall do as far as my knowledge extends. No gentleman with whom I have the pleasure of being acquainted can render such important information on this subject as yourself. If you would have the goodness to communicate to him any particulars you are in possession of—great weights, etc. etc. etc.—I should feel particularly obliged, and if there are any books illustrative of these, please to buy them for me.

Jonas Whitaker *to* Thomas Bates

BURLEY, *22nd November* 1823.

My friend, Mr. Berry, would be greatly obliged by all the information you can give respecting pedigrees of short-horns. No one is more capable, and I shall be much pleased if you will do it the first convenient opportunity.

Mr. Briggs is expecting to hear from you as to the return of Enchanter. Coates advised Mr. Berry to write to the Rev. Thomas Harrison for information. The answer he has received is that Mr. Mason can give the best information, and that he has the best stock. Mr. Berry will not apply to him if he can avoid it.

Rev. Henry Berry *to* Thomas Bates

ACTON RECTORY, *13th December* 1823.

It will probably be thought by many persons, and by Mr. Mason in particular, highly presumptuous in me, with my very confined knowledge of short-horns, to attempt the little work I propose. Many causes have conspired to remove much of the diffidence I entertained. These were chiefly—the promise of material assistance, the apparent disinclination of any one else to come forward, the recent publication of Mr. Coates's book, and above all, my conviction of the superiority of good short-horns over every other breed of cattle. Add to these the fact that in this great breeding county of Worcester and in several adjoining counties such complete ignorance prevails as to short-horns, that they are generally designated by the term *northerns*, and may happen to be black cattle or longhorns for anything the wise people here know. Occasionally they have received ocular demonstrations of the quality and character of a good short-horn, but ignorance and prejudice have led to the conclusion that these were chance good animals incapable of producing their like.

Under these circumstances I have wanted no spur to enlighten my neighbours, and I propose by a comparative estimate of short-horns and Herefords to compel more diligent inquiry. This can do harm to no party, and may do much good. I shall not attempt to deny that I feel the degree of enthusiasm necessary in all similar undertakings for my favourite stock, but I shall not shut my eyes to their imperfections if I have judgment to discover them.

I cannot conceal the fact that this proceeding will not be quite popular, although the Hereford breeders are pleased to express themselves obliged for the impetus given to their exertions by the warfare I have waged with them. I shall certainly neither please the Hereford breeders nor those gentlemen in the north who have designs to forward to which I will not lend myself, so I think it best for me to take the whole affair on myself and not to implicate any other persons. The consequence of this course will be a very deficient historical account, but I do not regard this of great moment. It is not breeders of short-horns I wish to inform, but I am desirous of making my work the means of directing public attention to the respective merits of the two breeds. I shall feel greatly obliged by any information you may choose to afford, or any suggestions that you may make.

Mr. Mason or his abuse I equally disregard. I think he will hardly openly attempt to annoy me. I anticipate that Mr. Hutchinson also will view my proceedings with little complacency. He has done infinite mischief to short-horns by his absurd puffs, and is now at work again. I hope a gentle hint I have given will render him cautious how he proceeds in a general attack on Mr. Colling's short-horns.

Rawdon Briggs the younger *to* Thomas Bates

Savile Green, Halifax, 23rd December 1823.

You will doubtless have heard of the safe arrival of Enchanter at Kirklevington. My man delivered him safe and well to yours at Ripley. That I thought was the best place for the men to meet at, as there was little chance of their missing each other. We had Enchanter shod before he started, otherwise it would have fared hard with him as there was rather a severe frost on the ground.

Jonas Whitaker *to* Thomas Bates

Burley, 28th January 1824.

Mr. Skipworth of Lincolnshire called here on his return from the North. He says he called at Kirklevington and saw Young Marske (419) and Enchanter. The first he liked except his hind-quarters; the latter was too poor to be shown. I sold him a bull calf to be delivered in summer; until he is fit for service, I believe he intends taking a bull of Mr. Booth.

If you see old Garbutt, I wish you would get a full pedigree of the heifer we bought of him at Yarm.[1] He sent me one but too meagre. She is just springing to calving.

---

[1] Yarm, by Symmetry, d. by Meteor, etc.—*Herd Book*, i. p. 560.

I forwarded your letter to Mr. Berry. He is surrounded by a nest of Hereford hornets, and wishes much that short-horn breeders would join in assisting him to face them in a successful manner. It would amuse you to see the *Farmer's Journals* from the last paper in December up to last Monday's inclusive.

Jonas Whitaker *to* Thomas Bates

BURLEY, *24th February* 1824.

I regret exceedingly we did not visit you last autumn, and am sorry to hear of the serious loss you have sustained in the death of The Earl.[1] I think you said you had a brother or two of his.

I pressed Mr. Berry to visit you, and offered to convey him free of expense, but his engagements would not allow of his leaving home. I have heard nothing particular of his History lately.

Jonas Whitaker *to* Thomas Bates

BURLEY, *25th March* 1824.

At Jervaux Abbey, Mr. Claridge asked Mr. Rhodes and myself to look at his stock. This led him to speak of an offer you had made for White Rose out of Fairy. I soon found the cow by her horns, the only characteristic remaining of a good descent. Her shoulders and girth were very bad, and she was apparently labouring under a severe surfeit, her back being much covered with scabs. I said that I expected to find you at my house on my return, and that if he had any offer to make I should be glad to be the bearer of it, although I thought that now you would not repeat your offer. The man who showed her said he would engage his master would take £40 or 40 guineas for her. This I said I would communicate to you. Although I can see nothing tempting in her or in a daughter of her the man showed us, yet I have that reliance on your judgment that I am disposed to treat for her. I must have a few lines from you to say that you decline her, or else Mr. Claridge will refuse to negotiate with me. Please say that owing to her age you decline her, but have recommended me to purchase her by way of experiment.

---

[1] Always liable to swell since his fatal journey from Halton, The Earl (646) had been found dead at the stake by the man who had given him clover while he went to church.—Bell, p. 47.

JONAS WHITAKER *to* THOMAS BATES

BURLEY, 10*th April* 1824.

I shall decline White Rose. Mr. Hall bought Lady[1] for Lord Althorp at Mr. Stapylton's sale[2] last Tuesday for £80. Mr. Hutchinson bought Phoenix.[3] These two cows were bought by Mr. Stapylton of Major Rudd for £400.

JONAS WHITAKER *to* THOMAS BATES

BURLEY, 6*th November* 1824.

Considering the inclemency of the weather, Mr. Rhodes and I have determined to put off our journey to Ridley Hall until next year, not, I assure you, in contemplation of Mr. Donkin's sale but entirely to enjoy the happiness of your company when the beauties of Ridley Hall can be seen.

What was your bull Laird got by? I saw his descendants at young Mr. Maynard's at Marton-on-the-Moor, and they were most extraordinary animals.[4]

Can you inform me the names of the bulls used at Acklam, beginning with the first and downwards in succession. I see the Grey Bull is said in one place to be by a son of Favourite and in another by a son of the White Bull. If I ask the breeders they refer me to Mr. Bates.

Mr. Rawdon Briggs is about to abandon short-horns, his stock is markedly bad. What do you say of Mr. Berry's production? I hope you saw Mr. Charles Colling. I think it must be pleasing to him as he is the chief of the piece.[5] Mr. William Wetherell says the heifers I bought of him were by Colonel Chaloner's Major (401) by Minor.[6]

---

[1] *Herd Book*, i. p. 359.   [2] At Myton, near Boroughbridge.
[3] *Herd Book*, i. p. 435.
[4] That Whitaker used the epithet 'extraordinary' in a good sense, see his remarks on Hereford cattle, p. 185 above.
[5] If Colling was the hero of the piece, Whitaker himself was certainly the understudy, see Allen's *History of the Shorthorns*, p. 62.
[6] Major (401) by Minor was sold as a year-old at Mr. Robert Chaloner's sale at Waterfall, near Guisbrough, 10th April 1817. He was out of lot 6, 'Matchless, 4 years old, out of Jenny (bred by Mr. Charge), by Mr. Chaloner's Old Red Bull.'—*Catalogue of Waterfall Sale*. Coates never even took the trouble to make use of the genealogical information ready to hand in sale catalogues.

Jonas Whitaker *to* Thomas Bates

Burley, *9th February* 1825.

I am very glad to hear The Earl has bequeathed his points and likeness to so many calves, and in the course of the summer I hope to judge of them personally and not by proxy.

I believe you assign the true cause of the quarter-ill,—variableness of climate and not high condition. Poverty, I believe, contributes much to produce the effect.

Jonas Whitaker *to* Thomas Bates

Burley, *5th June* 1825.

Lord Althorp has fixed a sale at Wiseton on the 6th of August, of twenty head; this is preparatory to Mr. Donkin's. I am sorry to inform you that Sir Henry Ibbetson is dangerously ill in London. I fear he will never return to Denton.[1] Your bull (Young Marske) was looking well on Saturday. I look with much pleasure to the time for visiting your beautiful residence.

Jonas Whitaker *to* Thomas Bates

29th December 1825.

I was very concerned to hear of our friend Hutchinson's failure. I am told it will be a bad business, as Hutchinson and Places's notes were in general circulation, and the deposits in their bank very considerable from the farmers round Stockton. I shall rejoice to hear you are not likely to suffer anything. These are awful times.

At Mr. William Charlton's small sale at Bearl, 10th May 1825, Bates had bought two year-old heifers, Philippa (lot 16)[2] and Rosalind (lot 18).[3]

Lord Althorp had a sale of shorthorns in 1825, and

[1] Sir Henry Carr Ibbetson, 3rd baronet of Denton Park, Otley, died that very day.

[2] Philippa by Exmouth (1021), dam Dandy by Wellington (683), grandam Lavinia by Duke (228), great-grandam by Yarborough (705), great-great-grandam by Bolingbroke (86). A portrait of her daughter Philippa 2nd by Second Hubback (1423), taken at seven years old in 1836, while on grass six months after calving, is given on p. 219 below.

[3] Rosalind by Exmouth (1021), dam by Yarborough (705), grandam by Duke (224), great-grandam by Traveller (655).—*Herd Book*, iii. 606.

obtained very high prices. He was always more of a cow-jobber than a true breeder at heart:

LORD ALTHORP *to* THOMAS BATES

WISETON, *9th August* 1825.

I intend to be at Mr. Donkin's. I shall be much obliged to you to let me know where I can see the cows and heifers, descended from Bolingbroke, which you mention. I do not know that I have any peculiar admiration for Bolingbroke's blood, but I do not dislike it as you do. I cannot think that the sire of Favourite could have been other than a first-rate bull. I rather want some in-calf cows or heifers, just now to fill up the vacancies occasioned by my sale, and to enable me to have another next year. I can only give moderate prices with this view, and therefore I should be glad to hear where I am likely to find any, in case I should fail in supplying myself at Sandhoe. I am very glad to hear so prosperous an account of your stock, but I beg you not to reserve any bulls for me. I like two that I have of my own breeding so well that I do not think I would hire Hubback, or even Favourite, if either of them were offered me. The two I mean are Ivanhoe (1131) by Cecil, out of Bright Eyes, of 1817, who is not only a good animal himself, but has proved himself a bull-getter, and Dandy (951) by Sirloin out of Cecilia, who is untried, but I think him calculated both in flesh and shape to get things suited to Smithfield Market.

Donkin's sale took place at Sandhoe on the 29th of August. Lord Althorp attended it and bought several animals. From lot 27, the two-year-old Rosalind, by Hector (1104), then in calf to Scipio (1421), the Rosalind tribe are descended,[1] otherwise Donkin's stock has left little mark.

Without saying anything to Bates, Lord Althorp offered Robert Bell the management of the Wiseton herd at a salary much beyond what he was receiving at Ridley Hall. Bell nevertheless elected to remain where he was, and on his disclosing the matter, Bates was, naturally, not a little indignant with Lord Althorp.[2] Their relations became decidedly cool.

[1] Beever's *Leading Tribes*, p. 203.   [2] Bell, p. 162.

In 1826 Bates parted with the whole of the mixed breed to which he had once been so devoted. Their constitutions had been completely debilitated by his unfortunate use of St. John's blood. From that time, he bred nothing but pure shorthorns.[1]

During the summer, Hollingsworth, the vicar of Haltwhistle, who was in much improved circumstances, asked Bates to sell him Ridley Hall. As he had no motive in buying it except to serve the vicar, Bates at once said that he would sell it him for what it had cost in cash and improvements. £18,000, he found, was the amount; and he said he would take no less.[2] Hollingsworth had at this time commenced proceedings to set aside the modus in lieu of tithes in Haltwhistle parish which had been paid by the Earl of Carlisle. On the Sunday evening before the Commission sat he solemnly assured Bates, as they were standing at the side of the embankment, near the bridge at Ridley Hall, that the issue of the trial would not affect Bates's property in the least.[3] No sooner, however, was a decision given in his favour than he wrote a letter that might serve as a typical example of the conditional mood:

REV. N. J. HOLLINGSWORTH *to* THOMAS BATES

HALTWHISTLE, 26*th September* 1826.

It is my sincere wish and intention, while we continue, as I trust that we ever shall do, on the same friendly terms that we have hitherto done for so many years, that there shall not any alteration take place in the annual sum which you have hitherto paid me for your tithes; though, of course, in justice to my successors, I cannot receive from you, or any one, the modusses for hemp and hay, after their having been decreed in the Court of Exchequer to be invalid.[4]

A fortnight later Hollingsworth called on Bates and said: 'When I told you no new claims for tithes could

---

[1] Bell, p. 80.
[2] *Letter to the Lord Bishop of Durham in consequence of the Letter of the Rev. N. J. Hollingsworth relative to the Sale and Purchase of Ridley Hall*, by Thomas Bates, Newcastle-upon-Tyne, 1830, p. 46.
[3] *Ibid.* pp. 61, 62.   [4] *Ibid.* p. 48.

be made, on your buying Ridley Hall, I told you what I considered to be the truth, what I believed to be the truth, and I had no reason to think otherwise.'[1] Half an hour after the vicar had left, the Newcastle paper arrived, and Bates could hardly believe his eyes when he read Hollingsworth's public admission that so far back as 1809 he had obtained 'the legal opinions of Sir Thomas Plummer and Mr. Hart of the Chancery bar, as being the first tithe-lawyers of the day,' that the modus from Ridley Hall was invalid, and that he was entitled to the tithes of all hay and agistment as well as of turnips and potatos.[2] Hereupon Bates, with good reason, asked him to do what no honest man would have refused, and guarantee him the estate as he had assured him it was, since without such assurance he would never have bought it.[3] Hollingsworth declined to do this, but offered to buy the estate at a valuation to be settled by two arbitrators as if it were only subject to the tithes he had first alleged. In case the arbitrators differed, it was agreed to leave the price to Mr. Rowland Fawcett as umpire. Hollingsworth then said, 'Why trouble the arbitrators at all?' Bates agreed that if Fawcett was satisfied the estate was worth £18,000 there was no need to do so.[4]

Hollingsworth now proceeded to abuse Bates so violently in Fawcett's presence for not having credited the farm, in his statement of account, with the produce consumed in his house,[5] that, his Welsh blood boiling over at this attack on his honour, he told the vicar he might take Ridley Hall at any price he liked if it were only to get out of his parish. Persuaded by Hollingsworth's renewed

---

[1] *Letter to the Lord Bishop of Durham*, p. 48.
[2] *Newcastle Courant*, 9th October 1826.
[3] *Letter to the Bishop of Durham*, p. 47. [4] *Ibid.* p. 48.
[5] This averaged £27:6:6 for three years; and as two maid-servants employed chiefly in the dairy and on the farm, and a man employed entirely on the farm, had their board, washing, and lodging in the house, Bates considered his not crediting the farm with this did not in the least invalidate his statement.—*A Letter to three of the Arbitrators in consequence of the Letter of December 24, 1828, relative to the Sale and Purchase of Ridley Hall*, by the Rev. N. J. Hollingsworth, London, 1829, p. 30.

asseverations that he had honestly believed the modus to be valid at the time of the sale, Fawcett decided that Bates was to be taken at his word and that £16,300 should be accepted for the estate. The balance of £1700 represented, in the opinion of Mr. Burnett and others, the deterioration in value through the new claim of tithes.[1] Once in possession, although land was falling considerably, Hollingsworth asked £19,500 for the estate from Mr. John Davidson to whom he soon afterwards sold it.[2] He was perfectly aware that Mr. Davidson was ready to give a large price when he made a show of compensating Bates by purchasing it himself.[3]

In order to staunch the blood-feud to which these transactions gave rise, the vicars of Newcastle and Bolam, with Mr. Graham of Edmond Castle and Mr. Chapman, a banker in Newcastle, held a court of arbitration in the good old Northumbrian fashion.[4] After examining the two parties,—Bates's speech lasted from mid-day till five o'clock the next morning,[5]—and collecting a great mass of other evidence, the arbitrators found that Bates 'made the purchase of Ridley Hall on liberal terms, under circumstances creditable to himself and with the kindest intentions towards Hollingsworth; that the conduct of Hollingsworth in leaving Bates open to further claims for tithes was at variance with the understanding under which the estate was purchased, and that the circumstance of Hollingsworth having legal opinions of high authority in his possession upon which he had already acted to obtain

---

[1] *A Letter to three of the Arbitrators*, etc. p. 70. [2] *Ibid.* p. 77.
[3] Memorandum of conversation with the late Mr. John Clayton. Hollingsworth had the effrontery to admit this himself in his letter of 1st January 1827.—*Letter to Bishop of Durham*, p. 34.
[4] Cf. 'The Arbitrament between the Storeys and Hebburns' in 1588, printed in *Border Holds*, i. 303.
[5] Memorandum of conversation with the late Dr. Collingwood Bruce, 1st December 1888. Cf. 'I occupied the attention of the arbitrators under examination for about thirty full hours, eighteen on one day and night.'—*Letter to Bishop of Durham*, p. 59. Mr. Matthew Bell during his candidature at the great Northumberland election of 1826, was detained at Ridley Hall by a three-hours speech from Bates. It was impossible to interrupt it, so Bell opened the sideboard unnoticed by Bates, and sought patience in a bottle of madeira.—*Ex inf.* the late Mr. George Burdon.

the tithes of turnips and potatos without informing Bates, called for serious animadversion.' They further declared that the question of the value of Ridley Hall ought not to have been determined by Mr. Fawcett, but should have been referred to the arbitrators originally named for that purpose. Finally they cleared Bates of all blame but a want of Christian meekness, and two of them at least subsequently retracted even this rider.[1] The evidence was all taken down by Mr. Collingwood Bruce—so celebrated afterwards as a Roman antiquary,—who was one of the few proficients in shorthand in the North. The future Dr. Bruce was then engaged in forming a Presbyterian congregation at Haltwhistle. He often walked down to Ridley Hall and dined with Bates, whom he always found pleasant and affable. On one occasion the Rev. Anthony Hedley, who was building his châlet of Chesterholme near Little Chesters, spent two days at Ridley Hall with Mr. John Fenwick and other antiquaries during their explorations of the Roman Wall, and enjoyed some excellent 'Comet port.' Bates's power of concentrating his thoughts on any subject was so intense, that, as he told Bruce, if a cannon were let off close to his ear when he was 'in a brown study,' he would not hear it. Bates gave at this time £1000 to the Bible Society. He bought up all the bones in the country-side and ground them for his fields, before any one else had thought of using them as manure.[2]

Bates had made it a condition of his accepting the arbitration that each party should be furnished with a copy of the evidence. In direct contravention of this stipulation the arbitrators refused to allow the evidence to be disclosed except at the request of both parties. Hollingsworth refused his consent, although he subse-

---

[1] *Letter to Bishop of Durham*, p. 17.
[2] Memorandum of conversation with Dr. Bruce. Cf. 'May 12, 1827—To Michael Elliot, for putting up new Bone-mill, in part of work done £7 : 2 : 0. June 25—To George Chapman for Bone-mill, Collars, and Fitters for Ridley Hall Bone-mill £21 : 7 : 2. To carriage of above from Whitby to Newcastle £1 : 3 : 6.'—*Letter on the Sale and Purchase of Ridley Hall*, by Rev. N. J. Hollingsworth, pp. 46, 47.

quently printed and published a letter to the arbitrators. One of them, the vicar of Bolam, did not hesitate to say that 'the spite and haughtiness of spirit evinced in this composition contrasted with the prayer at the beginning and the prayer at the end were calculated to impress any considerate person with the hypocrisy of the man who could set his hand to them both.'[1]

An equally unfavourable sketch of Hollingsworth's character, as gathered from that piece among his semi-anonymous writings on which he prided himself the most, had been drawn by the master-hand of Christopher North.[2] It so happened that in the early autumn of 1820 the son of the patriarchal Duke of Roxburghe, a boy of four, bearing the courtesy title of Marquis of Bowmont, was sent for change of air to Gilsland. He had for his companion a little girl of the same age, named Hopkins, whose father was in charge of the party. Hollingsworth was staying at the Orchard House at the same time, and was soon on the best of terms with Mr. Hopkins. They rode out together in every direction, visiting the remains of the Roman Wall and the towers and castles in the neighbourhood. Finally, Hollingsworth took the two children down to Haltwhistle to continue in the vicarage garden the diggings that had formed their chief amusement at Gilsland. In due time, as a requital for his kindness, came an invitation to stay with the Duke and Duchess of Roxburghe at Floors Castle. Mr. Hopkins, it seems, did everything he could to make Hollingsworth's visit pleasant, and took him for a number of excursions. In order to confer immortal lustre on these events, Hollingsworth wrote *Fleurs: a Poem in Four Books*, and dedicated it to the Duke 'in the hope that, at some future time, with the blessing of Divine Providence, it might be interesting and useful to his son the Marquis of Bowmont, especially in confirming his attachment to his paternal domains.'

[1] *Letter to the Bishop of Durham*, p. 14.
[2] Professor Wilson, who was afterwards a great friend of Bates's nephew, John Moore Bates, M.D., while at Edinburgh University.

The poem itself was in the most formal blank verse, but the introductions to the second and fourth books, inscribed respectively to Mr. and Miss Hopkins, imitated the rhythm of Sternhold and Hopkins's metrical version of the Psalms. The following stanzas refer to the diggings of the little Lord Bowmont and his playmate :—

> But, whatsoe'er her *Bowy* does,
>   His *Hoppy* straight must follow,
> Howbeit like Phaeton to drive
>   The Chariot of Apollo.
>
> Witness the delves at Orchard-House
>   Near far-famed Gilsland Spa :—
> I thought upon the Labourers,
>   When late their trace I saw.
>
> And, this, alas! is all the trace,—
>   And that well-nigh displaced,—
> Save what remains at ——,[1]
>   And ne'er can be effaced.[2]

Though printed at Newcastle, the poem was to be sold by Mr. Blackwood at Edinburgh. It must, therefore, have been a cruel surprise to Hollingsworth on taking up Blackwood's *Edinburgh Magazine* for June 1821 to read—

Nothing delights our good-natured readers more than a devil'd poet, or a peppered political œconomist; and verily, we are too skilful restaurateurs, not to understand how to cater to their taste. The truth is, that criticism, *selon les anciennes règles*, is neither a pleasing profession nor a thriving one. To separate the faults and merits of a book, and to administer to each a well proportioned dose of praise and censure, is of all tasks the most dull. 'To praise where we may, be candid where we can,' is a recipe from which an amusing article was never concocted, and from which one never will be concocted to the end of time. It is perfect balm to our souls, therefore, when, in the ordinary discharge of our duties, we chance to meet with a work so superlatively worthless and absurd, as to enable us to set all discrimination at defiance, and conscientiously to inflict the severest punish-

---

[1] It is easy to supply 'Haltwhistle.' For an acknowledgment of the authorship of *Fleurs*, etc. etc., see Hodgson's *History of Northumberland*, II. iii. p. 126.   [2] *Fleurs*, p. 111.

ment admissible by the laws of our profession. Such a work we have fortunately now before us, in the shape of a goodly quarto, and under the title of 'Fleurs: a Poem in Four Books.'

It must not be supposed that the author of Fleurs is a bard *sui generis*, or a *rara avis* of some unknown species, delighting the world for the first time with the brilliancy of his plumage, and the music of his song. He is but one of a very numerous and well-fledged class of authors, whose works but seldom issue from the press, and whose ambition is in general amply gratified by the praise and the pudding conferred by a more limited circulation. The chief constellations in this poetical firmament, consist of led captains and clerical hangers-on, whose pleasure and whose business it is to celebrate in tuneful verse the virtues of some angelic patron, who keeps a good table, and has interest with the archbishop, or the India House. Verily they have their reward. The anticipated living falls vacant in due time, the son gets a pair of colours, or is sent out as a cadet, or the happy author succeeds in dining five times a week on hock and venison, at the small expense of acting as toad-eater to the whole family, from my lord to the butler inclusive. It is owing to the modesty —certainly not to the numerical deficiency of this class of writers —that they have hitherto obtained no specific distinction among the authors of the present day. We think it incumbent on us to remedy this defect, and, in the baptismal font of this our Magazine, we declare, that in the poetical nomenclature, they shall in future be known by the style and title of THE LEG OF MUTTON SCHOOL.

To be received as the head of this distinguished body, we think the claims of the Bard of Fleurs stand pre-eminently high. He is marked by a more than usual portion of the qualities characteristic of THE LEG OF MUTTON SCHOOL; by all their vulgar ignorance, by more than all their clumsy servility, their fawning adulation of wealth and title, their hankering after the flesh-pots, and by all the symptoms of an utter incapacity 'to stand straight in the presence of a great man.'

The Bard of Fleurs is one of those obliging persons whose pen is at the service of any man in his neighbourhood with a pipe in his cellar, and a joint at his fire; and he makes it his peculiar care that those who possess every other luxury of life shall not want for poetry. There is a delightful singularity about him. In his imagination, nature possesses nothing of sublime or beautiful, equal to a well decorated spit. The god of his inspiration hangs suspended from a hook in the larder; and were he to invoke a muse, he would inevitably hitch in something about a hind-quarter, or a long cork. To do him justice, however, he is not ungrateful. A good dinner appears to him a benefit he can never sufficiently repay; and his imagination absolutely gloats over the memory of the sumptuous repasts of which he has partaken at Fleurs Castle, with so much satisfaction to himself, and delight to his hospitable entertainers. As he writes, the ghosts of digested haunches, in all their pristine obesity, arise in his prolific fancy; *barons* now no more, come forth at his bidding, from their unconsecrated graves, and smoke again upon the board. He is haunted by spectres of murdered turtles, and apparitions of pheasants, John Dorys, and ducks and green pease. His bowels tremble as he writes; his gastric juice is in a state of fermentation; his liver ceases to be torpid; his palatal glands redouble their secretions; and the imagination of the poet is triumphant over the whole man.[1]

Coarse and disgusting, indeed, must this 'unbecoming review' of his poem have appeared to the saint who had declared in the preface that 'to have written an hundred unpoetical lines would be regarded by him as of far less importance, than to have written a single line injurious to the interests of sound morality and religion,'[2] and who had deplored in his appendix 'the negative morality' of the Waverley Novels.[3]

Bates laboured in vain for two years to obtain the publication of the evidence given in the Ridley Hall arbitration, which would necessarily have convicted Hollingsworth of perjury. Meanwhile, 'the Bard of Fleurs' was promoted by Van Mildert, the new bishop of Durham, to the rectory of Boldon, worth more than £1000 a year, while the bishop appointed his own nephew to Haltwhistle. Indignant at this issue to his attempted reformation of clerical abuse, Bates addressed a letter to the bishop in such unmeasured terms as purposely to invite a prosecution for libel, in the course of which the evidence given before the arbitrators must needs have been disclosed.[4] His effusion only elicited the curt reply:

[1] Blackwood's *Edinburgh Magazine*, vol. ix. April-August 1821, pp. 345-350. [2] *Fleurs*, p. vi.
[3] *Ibid.* p. 186. Instead of patiently enduring his literary martyrdom, Hollingsworth rushed again into print with an anonymous pamphlet, *The Scorpion Critic Unmasked.* [4] *Ibid.* p. i.

AUCKLAND CASTLE, 11*th September* 1830.

The sale of the Ridley Hall estate is a matter in which the Bishop of Durham conceives he can have no concern; and under the circumstances, he must decline being the medium of communication between Mr. Bates and Mr. Hollingsworth.[1]

Bates sent his letter to the press:

Your Lordship must see that I can neither properly vindicate my character nor show sufficient cause for the removal of an unworthy rector, without the production of the evidence the arbitrators hold in their custody. Do not, my lord, suffer yourself to be infatuated like Charles X. of France. We live in an eventful period, we know not what a day may bring forth. The Church of Christ cannot suffer, but the Church of England may and will.[2] The last forty years of my life have been spent in parishes, where if the Devil himself had the nomination and the superintendence of the rector and vicars, he could not have had them more to his own mind as bringing into contempt religion itself by the characters and lives of the ministers. Your predecessor's[3] acknowledgment that he considered himself under greater obligations to me for my faithful reproofs, than to all the world besides—his always seeing me when I called, and his saying I should never call without his seeing me, however much he might be engaged: these embolden me to act faithfully to Your Lordship also. I have done my duty in addressing Your Lordship; I have relieved my conscience of a heavy weight, in having so long kept the truth from your knowledge. I hope Your Lordship will do yours.[4]

The scandals brought on the Church of Durham by the race for wealth among the leading clergy could not be concealed by the indolent reticence of a mitred Gallio. Parliament interfered: Van Mildert was the last prince-bishop of St. Cuthbert's patrimony, and his successors were deprived of much of his ecclesiastical patronage, Haltwhistle among the rest.

From an agricultural standpoint, the only interest in the weary dispute is the fact which was universally admitted that by six years' unintermittent labour[5] Bates

---

[1] *Fleurs*, p. iv.   [2] *Ibid.* p. 84.
[3] The Hon. Shute Barrington, 1791-1826.
[4] *Letter to the Bishop of Durham*, p. 89.   [5] *Ibid.* p. 73.

had doubled the annual produce of the Ridley Hall estate.[1] Before he finally quitted it, all intending tenants said they had never seen 'land so clean and in so great condition.' The 120 acres that were in tillage now produced no less than five times what they did when he first bought it.[2] Notwithstanding the rapid depreciation of landed property the letting value of the whole estate was increased from £267 to £650 a year.[3]

On 31st August and 1st September 1829 Christopher Mason sold off his herd at Chilton. Lord Althorp bought sixteen cows and heifers and one bull, and Captain Barclay of Ury the celebrated Lady Sarah by Satellite for 150 guineas, the top price of the sale, and another heifer for 73 guineas. Bates's elder brother, John Moore Bates, purchased the year-old bull Frederick by Satellite, dam Newby the Younger, for 44 guineas.

Mason had affected to hold the Wynyard herd in no estimation, but eventually he sent a butcher to buy an old bull of that blood, which 'went back direct to Favourite and Hubback, missing the dreaded Punch,' from Mr. Wood for £20. 'When Mr. Wood was at Chilton three years after, and only caught a glimpse of the bull's head, he exclaimed, "Why, there's my old Prince; he was bought to kill!" And sure enough it was Prince, but canonised in life as St. Alban's (1412).'[4] The superiority of the Hubback blood communicated by St. Alban's to cattle which had lost their character, soon showed itself unmistakably in their hair, handling, and looks.[5] Indeed in this case every one admits that pure blood did convey with it nobility of type.[6] The value of this Princess cross was manifest not only in the appearance of the cattle but in the prices they brought. Whitaker gave 140 guineas for one four-year-old cow by St. Alban's, and Lord Althorp

---

[1] This was the opinion of Mr. John Lowes of Allen's Green.—*Letter to the Bishop of Durham*, p. 68. Mr. Burnett of Black Hedley added that he had never seen a place so improved in the time.—*Ibid.* p. 69.

[2] This was admitted by Hollingsworth's land-agent, William Robson of Colmhill.—*Ibid.* p. 66.

[3] *Ibid.* pp. 37, 66; Hollingsworth's *Letter to the Arbitrators*, p. 37.

[4] *Saddle and Sirloin*, p. 152.   [5] Bell, p. 98.   [6] *Ibid.* p. 278.

145 guineas for another. The animals so descended lived and bred in a way very different from those before its introduction.

After the first day's sale, Mason 'drank the health of his party, and wished them luck of their bargains, but added that he must admit to them, he had lost £20,000 by breeding shorthorns.' This frank admission bore out strikingly what Bates had told him many years before, that keeping one lot of cows to breed, another to suckle the calves, and a third to give milk for the house was what few could afford to do, be their circumstances what they might. Bates had good reason to write:

Who that knows the loss that had been sustained in Mason's herd previous to the St. Alban's cross, would run the risk of using such stock again, unless it were persons chiefly owning animals of the blood, and anxious, for this reason, to bolster it up, however unworthy they had found it from dear-bought experience?[1]

Mason himself subsequently began breeding from kyloes, and bred from nothing else to the time of his death.

During his residence at Ridley Hall, Bates took no steps to bring his herd before the public. He rarely let any bulls, and kept no bull-calves except those he thought he might require for himself, or might induce his friends to take for their own herds. He reared and fed his steers, and also his cows and heifers that did not prove in calf within a certain period. Many a noble female was thus sacrificed. He never had calves born during the three summer months. He very seldom sent any fat cattle to the market. The principal butchers in Newcastle and Shields came to buy his stock at home. They were a most respectable and well-educated class. In their wholesale transactions with the shipping trade, they turned over very large sums of money weekly, and they occupied a high position in their respective towns. Mr. Radcliffe, mentioned several times already, was a frequent visitor and purchaser, as also was Mr. Walker of South Shields, who carried on a very large and lucrative business, and

[1] Bell, p. 98.

whose grandson was the first M.P. for that borough. Buyers often inspected the whole herd and gave their opinions on its merits and defects. The breeder-feeder was made fully aware of the qualities of his cattle by those who tested them on the block and in the scales. For well-bred animals properly fed, a higher price per stone was always obtained, than for coarse ill-shapen ones. The breeder and the buyer, brought into close contact, profited by their intercourse in a manner rendered impossible when cattle are sold by auction or by agents without the owner being present.[1]

Considering that there was no hope of the Church of England being better conducted, Bates determined to sell all his property and settle in America near the Episcopal College of Ohio. The Archbishop of York introduced him to the chaplain of the Bishop of Ohio who was then on a begging mission in England. The chaplain promised to come to Ridley Hall, and Bates expected to induce many other settlers to accompany him. The time was most propitious; there was never so much land to let at once in Northumberland, but the tenants were anxious to emigrate to a country not liable to such violent fluctuations of price as had been experienced in England. The industry and intelligence of the people of Northumberland made them especially fitted for successful colonisation, and Bates was well known from having been born and brought up among them. All to whom he mentioned the subject expressed great anxiety to accompany him if he received sufficient confirmation of the chaplain's reports.[2] It says much for Bates's insight into human nature that he was among the first to see through the double dealings of this emissary which culminated in a great scandal.

The unsatisfactory state of the poor-law, and the conduct of the Yarm bench of magistrates had also much to do with Bates's idea of emigrating. Some townships in Northumberland had never had a pauper or a poor rate

---

[1] Bell, pp. 218, 219.
[2] Letter of Thomas Bates to Bishop Chase, 21st March 1831.

for sixty years together. At Halton he reduced the poor rate in ten years to one-tenth. He found Kirklevington 'a poor destitute place eaten up by paupers.' He gave employment and encouraged industry and frugality. Had all the township been his own, or could he have separated his property to support its own poor, there would not have been a pauper left on it. No one who had been in his employment ever, to his knowledge, sought parochial relief. The only pauper he ever consented to send to the workhouse was an old woman whose own daughter would not take her in when offered four shillings a week, and she did not live a twelvemonth afterwards. He wished a law passed to encourage the labouring classes to save, by establishing a fund for their support in sickness and old age. Into this he proposed that they should pay sixpence a week from the age of twenty-one to sixty, while for the first ten years every employer was to pay threepence for every poor man's sixpence.[1]

In all his long experience in Northumberland, Bates never knew the magistrates do wrong; at Yarm he never knew the magistrates do right if they could possibly help it. At Stokesley, on the other side of Kirklevington, Mr. Mauleverer and the other magistrates were resident country gentlemen, who acted for the good of Yorkshire, and were beloved and respected for so doing. All the Yarm magistrates except one were resident in the county of Durham, and the exception had no estate in Yorkshire. They were entirely in the hands of their clerk. A poor orphan boy was tried for some hours at the Northallerton Sessions, his only crime being that the servant girl at a public-house had put his glass of liquor on a sixpence lying on the table. He had been kept in prison several months, and though he was instantly acquitted by the jury, the magistrates' clerk was allowed full costs out of the county rate. The opening prospects of Lord Grey's administration led Bates to hope that the country as well

[1] Letter of Thomas Bates to the Poor Law Commissioners, 7th February 1837.

as the towns might have a voice in choosing the members of the bench.   In this he was disappointed.

The idea of emigration was reluctantly abandoned. Among others who had resolved to throw in their lot with Bates in the New World, was a farmer named Pickering at Ellfoot, near Haydon Bridge. Bates rode over from Ridley Hall and said, 'Well, Pickering, I am not going to America but to Kirklevington, and I must just see what I can do for you.' He procured him the appointment of under-steward with the Trevelyans.[1]

[1] Memorandum of conversation with Mr. Pickering of the Lees near Haydon Bridge, 1st May 1896.

CHAPTER VII

KIRKLEVINGTON

1830–1837

ON May Day 1830, Bates took fifty cows and heifers by Second Hubback, all as like as beans, from Ridley Hall to Kirklevington.[1] The herd being driven through the country, left a great impression wherever it passed. Two old breeders, Mr. Angus[2] and Mr. Lambert, who cannot be accused of great partiality for unalloyed Durhams, repeatedly declared that it was the finest sight they ever witnessed. Among the drove was the young Philippa 2nd, calved in 1829. Her portrait, though taken at seven years old, is of value as showing what a cow by Second Hubback was like before the cross of Belvedere.

Only a small portion of the Kirklevington estate was then in hand, and Bates had to adapt the size of his herd to that of his holding. During the first two years, he lost a great many calves from 'hoodle,' brought on by their drinking stagnant water in hot weather. All the passages in their lungs and up the windpipes to their gullets were filled with small thread-like worms, which caused them to cough incessantly.[3]

In consequence of this, partly intentional, and partly accidental reduction in the number of his cattle, Bates, when he took the Hill Farm into his own hands in 1831, required about fifty more head of stock. He therefore

---

[1] Bell, p. 49.   [2] *Ex inf.* the Lord Bishop of Hexham and Newcastle.
[3] Bell, p. 217. Cf. Clator's *Cattle Doctor*.

began attending sales by auction much more than he had been in the habit of doing, and became a very extensive purchaser of any well-bred animals that went cheap.

On the 19th of April 1831 he was present, in the company of Mr. Thomas Bell, at Mr. Brown's sale at Nunstainton, on the banks of the Skerne, about ten miles north of Darlington and close to Chilton. John Brown and

PHILIPPA 2nd, 1836.

his father appear to have occupied this farm for about fifty years. It was some 500 acres in extent, much of it old grass land. He himself was a man of equal social standing with Mr. Crofton, Mr. Maynard, and other breeders, and was on very intimate terms with his near neighbour Mr. Mason. A little under forty years old at the time of his sale,[1] he had bred shorthorns for seventeen years, and was 'the only farmer who had access to Mr. Mason's bulls.'[2] After the Chilton sale in 1829, he introduced himself to three Irish purchasers, Messrs. Holmes, Adamson, and Fox, and told them he had a very good well-bred herd that had been always crossed with Mason's best bulls. They accordingly went to his farm and bought a few heifers at a low figure. They specially rejected a white two-year-old heifer by Matchem, which afterwards became the mother of all the Oxfords.[3]

[1] Letter of Thomas Bell to Thomas Bates of Heddon, 14th November 1872.
[2] Letter of Thomas Bates, quoted by Rev. John Storer; Bell's *Weekly Messenger*, 11th November 1872.
[3] Letter of Rev. John Storer; Bell's *Messenger*, 11th November 1872.

Mr. Holmes's judgment of men was little better than his judgment of shorthorns: he described Brown as 'a poor little farmer,' and expressed considerable doubt whether his statements could be relied upon.[1] So far from this being true, Brown referred to himself as one of the rare instances of a farmer turning agent of his own free choice, instead of through force of circumstances. He became agent to Lord Hastings and another nobleman, and during his residence at Seaton Delaval was generally respected as a man of high character.[2] He occupied the west wing of the Hall there, that palatial masterpiece of Vanbrugh's art, which after having been the scene of the wild Delaval revels, had suffered so severely from fire. With his umbrella under his arm, he proved a very real personality to Mr. William Howitt. As he did not show the *littérateur* all the attention he expected, Howitt revenged himself by recording a hedgeside acquaintance's description of him as 'just a farmer body fra' the sooth.'[3] He was, as Northumberland farmers remembered, 'a first-rate judge of cattle, always poking about among shorthorns at sales and shows.' He often spoke with pride of the fact that Bates had appreciated and purchased the stock which he had himself neglected to place sufficiently before the public.[4] At Seaton Delaval he again kept a number of shorthorns. These were duly entered in the *Herd Book*, but they appear to have left little mark.[5]

Considering the needless mystery in which the origin of the Oxford tribe has been shrouded—some people even going so far as to declare that Mr. Brown, from whom Bates said he bought the Matchem cow, was a fraudulent myth —it is satisfactory to be able to produce (with Bates's annotations in italics, his purchases initialed T. B.) the

[1] Letter of Rev. John Storer.
[2] Letter of Thomas Bell to Rev. J. Storer, Brockton, 8th March 1873.
[3] Howitt's *Visits to Remarkable Places*, 2nd series, pp. 375-378.
[4] Letter of Thomas Bell to Rev. J. Storer.
[5] They were principally of the same tribe as Mr. Emmerson's Flora—see *Leading Tribes*, p. 87. The first entry is in 1845, *Herd Book*, vii. p. 324, the last in 1851, *Ibid*. x. p. 259.

CATALOGUE OF THE IMPROVED SHORT-HORNED CATTLE,

Belonging to Mr. John Brown of Nunstainton, in the County of Durham,

WHICH WILL BE SOLD BY AUCTION, ON TUESDAY, THE 19*th* of *April* 1831.

MR. J. YOUNG, Auctioneer; W. J. FEWSTER, Printer, Durham.

COWS.

No. 1. Eight-years-old, by a Bull of Mr. Dobson's; dam by a Bull of J. D. Neasham's, Esq.; bull'd by Monarch,[1] 1st December. 16 *Quarts per Meal of Milk.* £9 : 15s.

No. 2. Three-years-old, by Matchem; dam No. 1; bull'd by ditto, 5th December. *White, short-up-horns, T. B.*, £8 : 15s.

No. 3. Three-years-old, by Matchem; dam No. 9. *Mr. Marley,* £16 : 10s.

No. 4. Twelve-years-old, by Wynyard; bull'd by Waterloo, 27th September; *bred by Mr. Parot of Claxton, near Greatham; Spragon,* £15 : 10s.

No. 5. Five-years-old, by Richard; dam by a Bull of C. Mason's, Esq.; bull'd by Monarch, 20th July. *Frank,* £13.

No. 6. Five-years-old, by Fitz Marsk; dam bred by Mr. Hutchinson; bull'd by Monarch, 11th Oct. *Sharp,* £11 : 15s.

No. 7. Six-years-old, by Meteor; dam by a Bull of C. Mason's, Esq.; grandam, by Lawnsleeves; bull'd by Monarch, 15th January. *Spragon,* £16 : 15s.

No. 8. Four-years-old, by Matchem; dam by Richard; bull'd by Monarch, 30th August. *Red roan, some white on shoulders, flanks and udder, T. B.*, £15.

No. 9. Ten-years-old, by a Bull of C. Mason's, Esq.; bull'd by Monarch, 15th January. *Mr. Willis,* £10.

No. 10. Four-years-old, bull'd by Monarch, 13th August. *Mr. Luke Seamor's breed, flecked, up-horns, and bell on brow, T. B.*, £11 : 10s.

No. 11. Four-years-old, by Matchem; dam by a Bull of C. Mason's, Esq.; bull'd by Monarch, 10th February. *Red flecked, white forehead with two red spots, T. B.*, £9 : 5s.

---

[1] *I.e.* Young Monarch (2327), a two-year-old bull of the 'Oxford' tribe, by Monarch (2324), sold as No. 53 for £80, the one redeeming price of the Nunstainton sale. The pedigree of Young Monarch (2327), as given in the *Herd Book,* is an example of hopeless confusion.

No. 12. Five-years-old, by Barmpton; dam, No. 9; bull'd by Monarch, 15th September. *Flecked roan, small white place in forehead, T. B., £15. Barmpton bred by Mr. C. Colling, out of Mr. Crofton's cow, he bought of Mr. C. Colling, 1822, at 60 guineas.*

No. 13. Twelve-years-old, by Wynyard; dam, the dam of No. 7; bull'd by Monarch, 12th March. *Mr. Ryal, £9 : 15s.*

No. 14. Four-years-old, by Matchem; dam No. 4; bull'd by Monarch, 15th December. *White long up-horns, T.B., £11.*

No. 15. By a son of Mr. Robert Colling's George[1]; dam, No. 13; bull'd by Monarch, 11th March. *Red, white forehead with some red spots, T. B., £15 : 5s.*

HEIFERS.

No. 16. One-year-old, by Matchem; dam, No. 15. *Mr. Teasdale, £16.*

No. 17. Bull'd by Monarch, 14th December, *out of No. 10, by Waterloo Bull kept at Wolviston, calved 21st September, T. B., £8 : 10s. Red, with small white spot on forehead, white tail-end, and white on flanks, sides, buttocks, and croop.*

No. 18. Bull'd by Monarch, 11th January. *Baker, £10 : 10s.*

No. 19. By Childers; dam by Richard. *Sharp, £11 : 10s.*

No. 20. By Matchem; dam by Matchem. *T. B., £9 : 15s. Red-flecked, two white spots on forehead.*

No. 21. By Childers; dam by Matchem. *Mr. Ryal, £8 : 15s.*

No. 22. By Matchem; dam No. 10. *Mr. Sharp, Skepton, £9.*

No. 23. By Matchem; dam by Richard. *Mr. Sheriton, £7 : 10s.*

No. 24. Bought in. *Mr. Sharp, £7 : 10s.*

No. 25. By young Rob Roy. *Mr. Sharp, £8 : 5s.*

No. 26. By Matchem; dam by Richard. *T. B., £6 : 5s. Red, with small bell on brow.*

No. 27. By Childers; dam by Richard. *T. B., £9 : 10s. Roan, with white face.*

No. 28. By Waverley; dam No. 13. *T. B., £7 : 15s. White.*

No. 29. By Merrington; dam No. 4. *Mr. Ives, £14.*

No. 30. By Monarch; dam No. 8. *Mr. Spragon, £10 : 10s.*

No. 31. By Merrington; dam No. 7. *Mr. Ryal, £10 : 10s.*

No. 32. By Merrington; dam No. 9. *Mr. Spragon, £10 : 5s.*

No. 33. By Matchem; dam No. 3. *T. B., £10 : 5s. White, with red ears and eyes, brown muzzle and red on low part of legs.*

[1] *I.e.* Robert Colling's George (1066).

No. 34. By Childers; dam by Richard. *Mr. Spragon, £7.*
No. 35. By Matchem; dam No. 6. *Mr. Arrowsmith, £5 : 5s.*
No. 36. By Monarch; dam by Matchem. *T. B., £5 : 10s. White, partly roan, with grizzled eyes and neck, and red ears.*
No. 37. By Waverley; dam by Richard. *Mr. Jones, £4 : 16s.*
No. 38 and 39. Two stot calves. *Mr. Sheriton, £8 : 2 : 6d.*
No. 40 and another stot calf. *John Thompson, £6 : 10s.*
No. 41. By Monarch; dam No. 1. *Mr. Wales, £3 : 13s.*
No. 42. By Monarch; dam by Richard. *Mr. Wales, £3 : 13s.*
No. 43. By Monarch; dam No. 14. *Mr Thomas Crofton, Halliwell.*
No. 44. By Monarch; dam by Richard. *T. B., £4 : 5s. Forehead and breast red with white.*
No. 45. By Monarch; dam No. 11. *T. B., £4 : 5s. White with light red ears, eyes, muzzle, and low parts of legs.*
No. 46. By Monarch; dam No. 10. *Stot calf (see No. 40).*
No. 47. By Monarch; dam No. 12. *T. B., £4.*
No. 48. Bought in. *George Robson, £4 : 15s.*
No. 49. Bought in. *Mr. Frank, £4 : 10s.*
No. 50. Bought in. *Mr. Spragon, £3 : 15s.*
No. 51. By Monarch; dam No. 13. *Stephenson, £6.*
No. 52. By Matchem; dam No. 1. *Sharp, Skepton, £7 : 15s.*

BULLS.

No. 53. Monarch, two-years-old, by C. Mason, Esq.'s Monarch; dam No. 4. *Nicholson, Darlington, £80.*
No. 54. Eighteen-months-old, by Matchem; dam by Young Ackmet. *Mr. Ryal, £15.*
No. 55. Three-months-old, by Monarch; dam No. 15. *Stephenson, £6.*

Bates's seventeen cows and heifers purchased at Nunstainton cost him only £157 : 12 : 6, an average of £9 : 5s. The sale was destined, nevertheless, to exercise a great influence on the fortunes of the Kirklevington Herd, since from No. 14, the white four-year-old heifer by Matchem, with long up-turned horns, called by Bates the Matchem Cow, there came the whole Oxford tribe. No. 27 is the Childers Cow of the *Herd Book*,[1] No. 33

---

[1] *Herd Book*, iii. p. 312. She bred a c.c. to Gambier (2046), but both dam and calf disappear.

the Brown Cow,[1] and No. 45 the Young Monarch Cow.[2]

No. 14, the white Matchem Cow, summarily rejected by the three Irishmen in 1829, was, we see, the daughter of No. 4, a twelve-year-old cow bred by 'Mr. Parot of Claxton, near Greatham.' Bates himself openly stated that this cow had been bought by Mr. Brown at a public sale, and that her breeder had always used bulls of the Wynyard blood,[3] a circumstance that exactly tallies with his statement that Mr. Porritt used Sir Henry Vane Tempest's bulls from the commencement of the Wynyard herd in 1802 till his death.[4] It was absurd to contend that 'Bates did not search after the pedigree of the Matchem Cow because he knew well,' from Mr. Brown using Chilton bulls for seventeen years, 'that it would lead him up to Mason.'[5] Bates naturally allowed that the Matchem Cow had 'no inconsiderable dash of Mason blood,' seeing that Matchem, her sire, was bred by Mason.[6]

Bates must have had some good reason for keeping Mr. Porritt's name out of the Matchem Cow's pedigree, as entered in the *Herd Book*.[7] What that reason was it is not difficult to surmise. In an unguarded moment, he admitted the direct female descent of the Matchem Cow from old Princess: 'Though I came into possession of so

---

[1] *Herd Book*, iii. p. 295. Her heifer Brown 2nd, by Duke of Cleveland (1937), calved 1834, was sold to Messrs. Bolden, Port Phillip, Australia. *Ibid.* v. p. 117.

[2] *Ibid.* iii. p. 517. Her four bull calves by Gambier and Duke of Cleveland were all steered.

[3] Letter of Thomas Bates to *New Farmer's Journal*, August 1842.

[4] Bell, p. 69. Mr. Porritt died, it seems, in consequence of breaking his leg—Memorandum of conversation with Mr. T. Stephenson. 'Mr. John Porritt, Claxton, near Stockton,' makes a solitary appearance in the *Herd Book*, vol. iv., as the breeder of 'Briton (3221), late the property of Mr. Henry Porritt, Togston, Northumberland, got by Mr. Parrington's son of the old Kirkleatham Bull (a son of Coates's White Bull of Driffield), dam bought of Mr. Parrington, near Stockton.' Mr. Henry Porritt, the breeder of Togston (5487), calved 1823, died at Warkworth at the age of ninety. He was a great man for horses, and is said to have lost his wife's two fortunes in farming.—*Ex. inf.* Mr. J. Crawford Hodgson.

[5] Letter of Dunelmensis, *Mark Lane Express*, 19th April 1858.

[6] *Ibid.* Bell, p. 289.   [7] *Herd Book*, iii. p. 494.

many females of the Wynyard blood, descended from old Princess, I had only one (the Matchem cow, bought at Mr. Brown's sale in 1831) whose stock turned out remarkably well.'[1] That this does not refer to mere crosses by Wynyard bulls is clearly proved by the fact that the Waterloos participated in this way in the Wynyard blood more than did the Oxfords, and surely Bates did not intend to deny the success of his Waterloos? Soon after the Nunstainton sale, he discovered, in making his inquiries into the lineage of Belvedere (1706), that Mr. Porritt of Claxton near Wynyard, retained both Young Wynyard (2859), the admitted sire of No. 4 in Mr. Brown's catalogue, and Anna-by-Lawnsleeves, the great-grand-daughter of old Princess, from the time of the Countess of Antrim's sale in October 1818, to his death in March 1820.[2] Anna-by-Lawnsleeves appears to have been the only Princess heifer in Mr. Porritt's possession during this period, which is exactly that in which No. 4, the Matchem Cow's dam, was calved. The evidence seems almost conclusive. After his denunciations of Mason's blood, however well-founded they may have been, it must have been a bitter pill for Bates to have to put Matchem (2281) in the pedigree of his Oxford Premium Cow, but to go behind the printed evidence of the Nunstainton catalogue, in order to further sully it with Lawnsleeves (365), a bull who, though not badly bred himself, was the son of Surplice (635), denounced by Bates as an 'excessively hard handler,'[3] and the sire of Hermit (305),

---

[1] Bell, p. 71. Cf. the singularly worded passage, which has been collated with the original, 'I have also seen cattle bred from a cow, a daughter of old Princess, by Mr. Baker's Lawnsleeves, and I bought descendants of the said daughter some years afterwards, and three heifer calves descended from her, but I have not now (1846) any cows descended from this cow, and her three daughters.'—*Ibid.* p. 69. Lots 15 and 28, bought by Bates at Nunstainton may also have been of the Princess tribe, descended from Lawnsleeves; the denial of possessing 'any cows descended from this cow and her three daughters,' is both too obscure and too specific to convey any meaning.

[2] *Ibid.* p. 70.

[3] See above, p. 156. For Mr. Wood's description of Lawnsleeves, see Appendix F.

anathematised by him as the ideal of a bad bull, would have been superhuman.

That there was something special about the Matchem Cow's breeding is proved by the fact that out of Bates's seventeen purchases at Nunstainton she was the only one that did well,[1] and founded a family—a sufficient substantiation surely of the distrust with which he regarded Mason's blood.

The talented panegyrist of the Killerby and Warlaby herds was forced to admit that 'the Matchem Cow element had the support of a brilliant success. If an illustrious line of ancestry could not be paraded, an illustrious posterity might be claimed.'[2] His strong partisan ties, however, led him to jump to the conclusion that 'it was morally certain that the Matchem Cow was the result of promiscuous breeding, and that all manner of intermingled varieties of aboriginal races must have entered into her composition.'[3] There is no commoner fallacy than the assumption that, because a shorthorn's ancestors are unknown, they must necessarily not have been pure-bred. The absorption of pure-bred cattle into the ordinary farming stock of the country has been enormous, and the generality of farmers are ill-qualified to regularly post their Herd Book entries. For anything that is known to the contrary, the market cows that formed the foundation of the Killerby and Warlaby herds may have been lineally descended from Stanwick Duchesses. One crucial difference between the Warlaby and Kirklevington schools of thought is that if an excellent shorthorn of unknown lineage makes its appearance, the former attributes its merits to the probability that some of its ancestors were not shorthorns, the latter to the probability that most of its ancestors were good shorthorns. Which school has the greater faith in the virtues of shorthorn blood?

The Grove Farm at Kirklevington was occupied in 1831

---

[1] Letter of Thomas Bates to *New Farmer's Journal*, August 1842.
[2] Letter of Mr. William Carr in Bell's *Weekly Messenger*, 1873.
[3] *Ibid.*

by Mr. Atkinson Greenwell, as tenant of Lord Falkland. Greenwell kept a large dairy, and having been interested in cattle all his life, he was constantly inspecting Bates's herd. This was nearly all red at the time, and in order to restore a greater variety of colour, Greenwell mentioned that his cousin, Mr. Stephenson of the White House, Wolviston, a farm situated on thin soil, four miles due north of Stockton-on-Tees, had a very fine yellow roan bull which Bates ought to buy. Bates, however, would not listen to him.

That same year Whitaker took Second Hubback for use in his herd at Burley. The bull left Kirklevington on 21st February in company with Rosanne,[1] a cow by North Star, which Whitaker had bought as a four-year-old at Mr. Wetherell's sale at Holme House near Darlington, 13th October 1828, for 41 guineas. Second Hubback's dimensions at the time, as an eight-year-old, were:—

Girth at crops	8 feet
„ ribs	9 „ 3 inches
„ hooks over thick of flank	8 „ 4 „
Breadth at hooks	2 „ 6 „ plumb
Length from breast plumb to tail	6 „
„ of rumps	2 „
„ from breast to crops	2 „
„ from crops to hooks	2 „
Girth of fore-leg below the knee	8 inches
„ horn at root next the head	8 „
Estimated weight, 92 stones 1 lb.	

Bates describes Rosanne as 'a long spaced cow; short hind-quarter; broad hooks; white brittle horns, not waxy; fine bone; yellow roan; light-lyred; mellow; well-haired; fine eyes; not unlike R. Colling's Punch cow, North Star's dam.' Her dimensions were:—

Girth at crops	7 feet 1 inch
Length	5 „ 1 „
Breadth of hooks	2 „ 4 inches
Length of hind-quarter	1 foot 6 „
Estimated weight, 61 stones.	

[1] *Herd Book*, ii. p. 511.

All cows got by Second Hubback (1423) were without exception capital milkers, even those from Whitaker's cows of the Western Comet blood which gave little milk.[1]

Whitaker sold his three-year-old bull Bertram (1716) of the Old Daisy tribe to Colonel Powel of Pennsylvania in 1831. Bates had the highest opinion of this bull:

THOMAS BATES *to* JONAS WHITAKER

26th April 1831.

Had Red Daisy been in calf to Second Hubback (1423), no other blood would have been so good. To what did you put her last? I hope to Bertram her son. At least I would have done so, as she could not with her lameness bear Second Hubback. If she gets strong again do try her, after Bertram has gone away, to Second Hubback, the produce will be invaluable.[2]

It is said that Bates met Colonel Powel in Whitaker's yard at Greenholme, and on their examining Second Hubback, asked 'poor old' Coates, who happened to be present, if the bull was not quite equal to the First Hubback; but that the author of the *Herd Book*, though willing to oblige his friends, was not the man to endorse this high estimate of a bull bred by his most bitter critic.[3] The whole story is very improbable, as Bates was doing everything he could for Whitaker, and admired Bertram (1716) extremely:

THOMAS BATES *to* COLONEL POWEL

KIRKLEVINGTON, *1st June* 1831.

I think the bull Bertram which you have bought of Mr. Whitaker of Greenholme is the best bull I know of at present to lay the foundation of a good stock of shorthorns in any country. He is descended from one of the best milking and quickest grazing tribes, and one which yielded meat of the best quality, and, as I found by experiments, left the most for the food consumed. I used the Daisy Bull, brother of the great-grandam of Bertram above thirty years ago; the first Duchess cow I bought

---

[1] Bell, p. 49.   [2] Beever's *Leading Tribes*, p. 58.
[3] *Saddle and Sirloin*, ed. 1895, p. 152 n.

of Mr. Charles Colling of Ketton, bred in the year 1800, was by this bull, and going in a lot of twenty cows, on grass alone, she yielded 14 quarts of milk per meal (ale measure) and made 14 lbs. of butter per week of 21 ozs. This is 294 ozs. of butter per week, so each quart of milk (ale measure) yielded 1½ oz. of butter.

The Daisy Cow, sister to Daisy Bull, was sold by Mr. Charles Colling to Mr. Hustler of Acklam, whose tenant, a man of the greatest respectability, saw her milked and told me she gave 16 quarts of milk per meal; Mr. Robert Colling's Bright Eyes cow gave 15 quarts per meal, and these tribes of improved shorthorns were all quick grazers.

I consider Bertram a much superior bull to Comet, which bull I saw sold for 1000 guineas by public sale, and afterwards £1500 was offered for him.

After his arrival in America, Bertram (1716) is described as 'red, with a little white, a compact, massive form, short in the leg, of fine touch, good hair, and altogether an imposing animal.'[1]

Disappointed in a year-old bull out of Duchess 22nd, which he had intended to use, only weighing 22 stones, Bates purchased Bertram's son Gambier (2046), of the Western Comet (or Gentle Kitty) tribe, from Whitaker for 100 guineas. Although Gambier was by Bertram, and his dam Matilda was the highest priced cow (52 guineas) at Mr. Charge's sale at Newton near Richmond, 14th and 15th October 1828, Bates was by no means satisfied with him as a stock bull.

Greenwell now again mooted the subject of Stephenson's roan bull, adding that his grandsire was bred at Wynyard. At the word 'Wynyard,' a complete change came over Bates's countenance. He often regretted the supposed extinction of the Wynyard herd, and would have given anything to obtain the pure Princess blood of which it mainly consisted. He at once told Greenwell that he would go and see the bull if he would accompany him. Accordingly they drove over to Wolviston the next day. Tradition has it that Belvedere looked out of the bull-

[1] Allen's *History of the Short-horn Cattle*, p. 174.

house as they were passing, and that the sight of his head sufficed to confirm Bates's determination to purchase him.[1] On Stephenson telling him how Belvedere was bred, he immediately agreed to pay all the price asked, fifty pounds, not guineas. The next day, 23rd June 1831, Stephenson himself brought Belvedere to Kirklevington. Bates paid for the bull in Greenwell's presence, and got a receipt as well as a written pedigree signed *John Stephenson*.

BELVEDERE

```
Princess.                                                      
Favourite (252).    } Anne Boleyn.                             
Dam of Robert                         } Angelina.              
  Colling's         } Phenomenon (491).                        
  Wildair.                                                     
Favourite (252).                                } Angelina 2nd,
Cow by                                          }   2nd Feb.   
Favourite (252).    } Princess.                 }    1819.     
Favourite (252).                      } Young                  
Wildair.            } Wellington        Wynyard                
Comet (155).          (680).            (2859).                         } BELVEDERE,
                                                                          formerly
Princess.                                                                 BELVIZER
Favourite (252).    } Anne Boleyn.                                        (1706),
Dam of Robert                         } Angelina.                         calved 6th
  Colling's         } Phenomenon (491).                                   April 1826.
  Wildair.                                                     
Favourite (252).                                } Waterloo     
Cow by                                          }  (2816).     
Favourite (252).    } Princess.                 } 26th Dec.    
Favourite (252).                      } Young      1819.       
Wildair.            } Wellington.       Wynyard                
Comet (155).                            (2859).                
```

No sooner had Bates bought Belvedere (1706) than

---

[1] *Saddle and Sirloin*, p. 153.

he told his acquaintances that by the union of the Duchess and Princess blood, he would produce shorthorns such as the world had never seen. His enemies must have been very incredulous, as in their eyes Belvedere (1706) was 'small and plain, with rather rough shoulders,' although they had to allow that he was 'soft as a mole in his touch.'[1] As a matter of fact Belvedere was a big bull of very great length, with rather large shoulders. Like his sire Waterloo, he had 'a hot bold temper' that made him bad to manage. It took both Stephenson and Thomas Swalwell to lead him from Wolviston to Kirklevington, and Stephenson's son Thomas helped them in the start-off down the Sandy Lane. Bates and Whitaker had a long dispute as to whether Waterloo or Belvedere was the better bull. Whitaker declared for the former although he was old and decrepit at the time. There was, however, little room to doubt that Bates was right.[2]

The Wynyard herd had originated in the purchase of Princess (by Favourite (252), dam by Favourite (252), grandam by Hubback (319)) from Robert Colling by Sir Henry Vane Tempest in 1800, at a price which was never disclosed, but which was popularly believed to be 700 guineas.[3] Her produce is given in the *Herd Book* as

    1803 c. c. Anne Boleyn by Favourite (252).
    1805 c. c. Elvira by Phenomenon (491).
    1806 b. c. Wynyard by Phenomenon (491).
    1809 c. c. Nell Gwynne by Phenomenon (491).
    1810 c. c. Peg Woffington by Wynyard (703).
    1813 b. c. Pilot by Wynyard (703).

Soon after the death of Sir Henry Vane Tempest in 1813, a sale was held at Wynyard. According to what Bates was told, Sir Henry's widow, the Countess of Antrim, instructed her house steward, Mr. Richard Wood of the Close, to buy in Princess and her grand-daughter Angelina by Phenomenon (from Anne Boleyn (491)) calved in 1810,

---

[1] *Saddle and Sirloin*, p. 147.
[2] Memoranda of conversation with Mr. Thomas Stephenson at Wolviston, 21st May 1896.
[3] Bell, p. 279.

and to let her have them again after the sale.¹ Wood bought in Princess himself and got Mr. Wetherell of the Isle to buy in Angelina. The latter was an even roan, three years old at the time, and was remarkable for her fine head, eyes, and horns—in fact she was a very fine-looking heifer.² The two young bulls, Pilot (from Princess by Wynyard) and Wellington (from Alexina by Wynyard), were also bought in. Pilot died the next day; Wellington was kept till the autumn of 1814 and then sold to Mr. Dawson Lambton of Biddick.³

CATALOGUE OF THE IMPROVED SHORT-HORNED CATTLE[4]

Belonging to the late Sir H. V. Tempest, Bart.

WHICH WERE SOLD BY AUCTION AT WYNYARD BY MR. KINGSTON

Tuesday 5th October 1813.

J. APPLETON, PRINTER, STOCKTON.

COWS.

Guineas

1. PRINCESS, 13 years old, by Favorite, her dam by Favorite; dam of Anna Buleyne (Lot 2), Elvira (Lot 3), Wynyard (Lot 24), Nell Gwyn (Lot 6), Peg Woffington (Lot 11), and Pilot (Lot 23) . 36

    Mr. Wood, Close.

2. ANNA BULEYNE, 10 years old, out of Princess, by Favorite; dam of Alexina (Lot 7), Angelina (Lot 12), Artless (Lot 16), and Albion (Lot 22) . 76

    Mr. Alderson, Ferrybridge.

3. ELVIRA, 8 years old, out of Princess, by Phœnomenon; dam of Helen (Lot 17) 96

    Mr. Mills, Ferrybridge.

4. TRINKET, 6 years old, out of Old Tragedy, by Phœnomenon 45

    Mr. Parrington, Middlesbro'.

---

[1] Bell, p. 68. [2] See Appendix F.
[3] Bell, p. 70. [4] The original spelling, etc., has been preserved.

Guineas

5. PAROQUET, 4 years old, got by Wynyard, out of a cow bought of *John Porritt, Esq.*[1] . . . 52
   Mr. R. Wilkinson.

6. NELL GWYN, 4 years old, got by Phœnomenon, out of Princess; dam of Noble (Lot 21) . . 68
   Mr. John Wood, near Durham.

7. ALEXINA, 4 years old, got by Phœnomenon, out of Anna Buleyne, dam of Wellington (Lot 25) . 41
   Mr. Vansittart.

8. TULIP, 4 years old, got by Phœnomenon, out of Tragedy 87
   Mr. Alderson, Ferrybridge.

9. CALISTA, 3 years old, got by Comet out of Cora; dam of Careless (Lot 19) 112
   Mr. Miles, Ferrybridge.

10. TRIFLE, 3 years old, got by Phœnomenon, out of Tragedy 58
    Mr. Vansittart.

11. PEG WOFFINGTON, 3 years old, got by Wynyard, out of Princess 27
    Mr. Vansittart.

12. ANGELINA, 3 years old, by Phœnomenon, out of Anna Buleyne 63
    Mr. Wetherell.

13. RED ROSE, 6 years old, by Phœnomenon, out of a Cow got by *T. Burdon's* Bull . . . 36
    Mr. Dobson.

HEIFERS.

14. YOUNG TRAGEDY, 2 years old, by Wynyard out of Old Tragedy 70
    N.B.—*Old Tragedy died in* 1813.
    Mr. Bower.

---

[1] The same no doubt as 'Mr. Parot of Claxton' and 'Mr. Porritt.'

	Guineas
15. MATCHLESS, 2 years old, by Phœnomenon out of Matron	40

N.B.—Matron was killed for the House, in 1812.
Mr. Hutchinson.

16. ARTLESS, 1 year old, by Wynyard, out of Anna Buleyne	56

Sir B. Graham.

17. HELEN, 1 year old, got by Wynyard, out of Elvira	71

Mr. Cook.

HEIFER-CALVES.

18. PATCH, by Wynyard, out of *J. Porritt's Cow*	11

Mr. R. Wilkinson.

19. CARELESS, by Wynyard, out of Calista	54

Mr. Bower, Welham, Yorkshire.

20. PEERESS, by Wynyard, out of Red Rose	16

Mr. Smith.

BULL-CALVES.

21. NOBLE, by Wynyard, out of Nell Gwyn	51

Mr. Jackson.

22. ALBION, by Wynyard, out of Anna Buleyne	52

Mr. Ord.

23. PILOT, by Wynyard, out of Princess	42

Mr. Beckwith.

BULLS.

24. WYNYARD, 7 years old, by Phœnomenon, out of Princess	210

Mr. Alderson, Ferrybridge.

25. WELLINGTON, 1 year old, by Wynyard, out of Alexina	71

Mr. Wetherell.

Amount of Sale . . £1618 : 1s.

The twenty-five lots averaged £64 : 14 : 5.

On regaining possession of Princess, the Countess of Antrim sent her to Barmpton. The man who took her was told if possible to have her put to the bull Wellington (680), by Comet out of Wildair, and to pay any price Robert Colling chose to charge. Robert Colling, however, told him he never allowed any gentleman's cows to be bulled, and could not comply with Lady Antrim's request. The man was already on the road back to Wynyard, when Colling's servant, like Gehazi, came running after him to say he had told his master Princess was not a gentleman's cow but a lady's, and that, amused with this remark, Colling replied, 'I believe I promised Sir Henry that if ever he was in want of a bull, I would let Princess be served by one of my bulls. She may be put to Wellington at ten guineas.' Lady Antrim's man turned back, and paying the ten guineas, accomplished the object of his mission.[1]

The produce of this episode was Young Wynyard (2850), calved in 1815, and first used when only eight months old. His stock proved their superiority over Old Wynyard's, though the latter's was very good. He was only let out for one year, from October 1818 to March 1820, to Mr. Porritt who had been using Sir H. V. Tempest's bulls from the time he began to breed in 1802, and never used any other's till his own death in 1820. All the calves Mr. Porritt had by Young Wynyard were excellent. Lady Antrim permitted other persons to send their cows to this bull. Bates bought some cows by him, and extraordinarily good shorthorns they were.[2]

In 1815 the Countess of Antrim sent Princess's grand-daughter Angelina to Mr. Baker's Lawnsleeves at Elemore.[3] Mr. John Wood (afterwards of Ricknal Grange, Darlington) happened to see her one day when he was visiting Mr. Baker. Lawnsleeves (365), Mr. Wood described as a large and very lengthy bull with a plain head, good eye, good horn, upright shoulders, somewhat wide shoulder-points, large brisket and deep chest, somewhat defective girth, good back, and good hindquarters. He was roan in colour, and a very good

---

[1] Bell, p. 68.   [2] *Ibid.* p. 69.   [3] *Ibid.* p. 70.

handler. Bred by Mr. Baker, who had purchased his dam of Robert Colling, he seemed to be not less than eight years old in 1815.[1] In due course, Angelina had a heifer-calf, Anna-by-Lawnsleeves, in 1816.

At the last sale of shorthorns held at Wynyard, that of the Countess of Antrim in October 1818, only five animals were sold [2]:—

1. Angelina, calved in 1810; served by Young Wynyard (2850), bought by Mr. Ellerker and resold to Mr. John Stephenson of White House, near Wolviston.[3]
2. A heifer calf from Princess, sire, Young Wynyard (2850), calved in 1817. Bates bought her daughter by Orlando (471) in about 1832, together with her grand-daughter Red Princess and her great-grand-daughters Red Princesses 1st and 2nd from Mr. Fletcher of Blakiston.[4]
3. A White Bull, three years old, from Angelina, sire, Wellington (680).
4. A bull-calf from Angelina, sire, Young Wynyard (2859), calved in 1817.
5. A bull-calf by Young Wynyard (2859), supposed to have been out of a heifer-calf from Princess, bought in at the sale of 1813.

Young Wynyard (2859) and Anna-by-Lawnsleeves were not sold, and were retained by Mr. Porritt till his death in March 1820. Subsequently Young Wynyard went to Ireland.[5]

Up to the time of the Wynyard sale Mr. John Stephenson of Wolviston White House had been principally a 'horse-man.' His family, previously settled at Acklam in Cleveland, came of the same stock as the Stephensons of Ketton. Possessing a certain ambition, he wished to distinguish himself in some line or other, and this led to his purchase of Angelina. In his hands

---

[1] See Appendix F, and pp. 156, 157 above. [2] Bell, p. 70.
[3] *Ibid.* p. 69. [4] *Herd Book*, iii. p. 591. [5] Bell, p. 70.

Angelina produced Angelina 2nd, by Young Wynyard (2850). She was again sent to Young Wynyard, before he left for Ireland, and on 26th December 1819 gave birth to the bull-calf, Waterloo (2816). After Young Wynyard had crossed St. George's Channel, Stephenson fell back on Baron (58), a son of Comet, bred by Colonel Trotter, and an animal of a short and doubtful pedigree. To this bull, Angelina dropped two heifers, one of them being Moss Rose, the origin of Stephenson's Princesses. Belvedere, the produce of Angelina 2nd and Waterloo, was calved 6th April 1826.

Some doubt was thrown on the accuracy of Belvedere's pedigree by the fact that when Mr. Harrison of Streatlam Grange hired the bull Waterloo (2816), then eleven years old, from Stephenson in 1830, a memorandum came with him, not, however, in Stephenson's writing, which stated he was by Young Wellington (Young Wynyard), dam Anna-by-Lawnsleeves. Harrison had some handbills and cards printed giving this pedigree and advertising Waterloo to stand at his house to serve cows. Subsequently to the circulation of these notices Stephenson told Harrison in course of conversation that there was an error in the pedigree they contained, as Waterloo (2816) was not out of Anna-by-Lawnsleeves but out of Angelina by Phenomenon.[1] An application made to Atkinson Greenwell long after Bates's death brought the reply:

I was along with Mr. Bates when he purchased Belvedere, and heard Mr. Stephenson tell him how he was bred. I was also at Mr. Bates's the next day when Mr. Stephenson brought the bull. I saw Mr. Bates pay for him and get a receipt for the money and also receive a written pedigree signed *John Stephenson*, but no bull of the name of Lawnsleeves was in that pedigree, nor was Lawnsleeves ever once named between them, either on that or the preceding day. And I am quite sure of this, that if the bull Lawnsleeves had a right to be in that pedigree, neither Mr. Bates nor Mr. Stephenson were the men to strike it out.[2]

The same year that he bought the Matchem cow and Belvedere, Bates obtained the Waterloo cow and three

[1] See Appendix E.   [2] Bell, p. 284.

other two-year-old heifers[1] from Mr. Thomas Parkin, a large farmer at Thorpe Thewles, about four miles to the north-west of Stockton-on-Tees. Both this cow, calved in 1829, and her dam were by Waterloo (2816). She was a 'short-legged, wide red cow, with the look of a pure shorthorn.'[2] Indeed the idea immediately suggested itself that Bates bought her because she so much resembled the cows in his own herd.[3] At that time it was difficult to distinguish between the Duchesses and Red Roses at Kirklevington, they were nearly all red, the Red Roses rather a lighter red than the Duchesses.[4] On 11th January 1841, Bates made the note: 'I have seen the gentleman who bred the Waterloo cow lately, and he stated to me that he and his father had had the breed for fifty years, and that they were well-descended all that time, having had a son of Comet and other blood before the cross of Waterloo (2816).'[5] Mr. Thomas Parkin's father, William Parkin, had been born in 1757 at a farm at Hemlington near Stainton in Cleveland. After the death of his father he continued farming there till 1793, when he removed to Old Thornaby near Stockton, and thence to Thorpe Thewles in 1808. He appears then to have owned the original cow of the Waterloo family before leaving Hemlington. There exists a tradition that she was one of two cows purchased by him from Mr. Wastell.[6] The son of Comet alluded to was probably Colonel

---

[1] '8th October 1831—Received of Mr. Bates for two two-year-old heifers, £23; for two two-year-old heifers in calf, £28.'—Account Books at Thorpe Thewles. The 'Druid,' *Saddle and Sirloin*, p. 384, inaccurately described Mr. Parkin as 'a small farmer.'

[2] Memoranda of conversation with Mr. Thomas Stephenson at Wolviston, 21st May 1896.   [3] *Ibid.*   [4] *Ibid.*

[5] Bell p. 290. Cf. 'The dam and grandam of Duke of Cambridge (3637) were both by Waterloo Bull, and the blood was good for many generations previously; though I only asked as to the dam, when I bought the first Waterloo cow ten years ago.'—Letter of Thomas Bates to Mr. Bolden of Australia, 1841. 'Mr. Bates set great store on the Waterloo tribe, but he would never say from whom he obtained them.'—*Farmer's Magazine*, 3rd series, xvi. p. 5.

[6] Memoranda of conversation with Mr. Thomas Stephenson. This would explain the reference to the excellence of the descendants of Wastell's heifer, see above, p. 22, which otherwise seems pointless.

Trotter's Baron (58), calved in 1814 (dam by a grandson of Favourite). As Bates did not approve of the use of this bull which served cows at five shillings a head at Wolviston,[1] he may not have cared to carry the pedigree back farther than the Waterloo crosses.

There is no doubt that Waterloo stamped his character very strongly on his daughter-grand-daughter. He was

WATERLOO (2816).[2]

a fine thick dark-red bull with rather an exaggerated neck.[3] His temper was not good, and when a stranger came to look at him, he would thrust his head defiantly over the half-door of his box till calmed by Mrs. Stephenson's 'Cush, Waterloo!'[4]

Put to Belvedere, the Waterloo cow produced a roan calf, Waterloo 2nd, in 1832; but whether from the too close affinities or other causes, the heifer did not breed.

As Mr. Thomas Parkin did a large business in buying

[1] '1st March 1820—Paid Mr. John Stephenson for seven cows bulled at 5s. each.'—Account Books, Thorpe Thewles.
[2] From an oil-painting kindly lent by Mr. Thomas Stephenson of Wolviston.
[3] Memoranda of conversation with Mr. Thomas Stephenson.
[4] Memoranda of conversation with Mr. W. T. Parkin, Thorpe Thewles, 21st May 1896.

and selling cattle, it seems within the bounds of possibility that the three other two-year-old heifers bought from him by Bates with the Waterloo cow, were Lady Barrington, bred at Sedgefield, the Fletcher cow from Blakiston, and Shorthorns by Saladin (1417), all calved in 1829. There is no other clue as to how Bates acquired them.

The Wild Eyes tribe has its origin in a heifer-calf which Bates bought at Mr. Parrington's sale at Middlesbrough on the 24th of April 1832, for £3. Mr. Parrington's house and a few cottages were the only buildings then standing on the south side of the Tees below Stockton, where there is now an important town with a mayor and corporation, a member of parliament, a Latin bishop, and a population of 82,000 souls. Mr. Parrington's house was formed out of the ruins of the old church—the walls a yard thick. When it was pulled down and the ground levelled, bones were discovered under it. Mr. Parrington frequently entertained Mr. Vansittart from Kirkleatham, Sir James Graham, and Lord Althorp when they came to look at his shorthorns. Sir Henry Vane Tempest, who was rather a free liver, often came over and sometimes insisted on taking him back to Wynyard. Shorthorns were not the only animals bred at Middlesbrough. Mr. Leonard Parrington remembers, when a boy of three, seeing a black horse named Lightning groomed down on coming home from the St. Leger; the next year he witnessed a celebrated run of the foxhounds across the estuary of the Tees. He was in his seventeenth year when he acted as clerk of the sale in 1832. The dam of the first Wild Eyes, a roan cow called Wildair, was very wild indeed, and as boys they were always afraid of her very pointed up-turned horns. It was a showery day and there was a poor company. He made out Bates's bill and received a cheque on Backhouse and Co. Bates was a small man in breeches and top-boots with a voice rather shrill in its higher notes. He had almost exactly the same figure as Lord Althorp, only the latter was rather taller and not so well dressed.

Indeed, Lord Althorp affected the get-up of a poor farmer.[1] Mr. Whitaker has preserved the catalogue of the sale; his notes are given in italics :—[2]

		£	s.	d.
Lot 1.	Princess, 12 years old, by Apollo (36), dam by Meteor (431). *A light roan fair old cow.* Mr. Bates	13	10	0
,, 2.	Trinket, 12 years old, by Symmetry (643), dam by Jupiter (345). *Light roan, horns rather large, fair.* Mr. Bates	11	0	0
,, 3.	Countess, 9 years old, by Baron (58), dam Trinket, Lot 2. *Roan neck, like Comet.* J. Whitaker	15	0	0
,, 4.	Jenny, 11 years old, by Baronet (61), dam by Jupiter (345). *Flecked, large, up-horned, fair.* Mr. Hart for Mr. Chaloner	13	0	0
,, 5.	Wildair, 6 years old, by Wonderful, dam by Booth's Cardinal (841) (*no price given*).			
,, 6.	Gaudy, 6 years old, by Wonderful, dam lot 4. *Roan, a useful cow, wide long horns.* Mr. Thompson	20	5	0
,, 7.	Trifle, 5 years old, by Wonderful, dam by Symmetry. *Red and white.* J. Whitaker	20	0	0
,, 8.	Tulip, 6 years old, by Wonderful, dam by Symmetry. *Roan, short poor hind-quarters.* Mr. Hart	16	0	0
,, 9.	Roan bull calf, by Liston, dam lot 2. Mr. Bearshaw	10	5	0
,, 10.	Roan bull calf, by Emperor, dam lot 5. Mr. Goldbro'	9	15	0
,, 11.	Roan bull calf, by Emperor, dam by Charles 2nd. Bought in	20	0	0
,, 12.	Roan bull calf, by Liston, dam lot 3. Mr. Wood	4	0	0
,, 13.	Roan bull calf, by Emperor, dam lot 5. Mr. Farrar	6	0	0
,, 14.	Quey calf, white, by Liston, dam lot 2. Mr. Bates	2	0	0
,, 15.	Quey calf, roan, by Emperor, dam by Wonderful. Mr. Bates	3	0	0
,, 16.	Quey calf. *Sold to butcher*	2	10	0

[1] Memoranda of conversation with Mr. Leonard H. Parrington, 19th July 1895. [2] Holt Beever's *Leading Tribes*, p. 248.

Fortunately, when the roan heifer-calf Wildair by Emperor got to Kirklevington, her name was somehow changed to Wild Eyes. Confusion with Robert Colling's Wildair family was thus avoided. Bates held the Wild Eyes tribe in great esteem; he always spoke of it as giving him the Dobinson strain which was the only good blood that the Colling herds did not contain. It will be remembered that his uncle, William Bates, kept a herd of Dobinson cattle at Chollerton on the North Tyne at the beginning of the century. The steers from this were exhibited for many years at Newcastle as prize beef.

Mr. Hutchinson died in 1833. A few months before his death he invited Mr. Thomas Bell to go to Grassy Nook to see his calf-house, saying that one was much needed at Kirklevington, and it ought to be on the same model as his.[1] Mr. Bell went and made a plan of it, on which Bates at once acted. Hutchinson asked Bell on this occasion to see his shorthorns and tell him which animal he considered the best. Bell pointed out a roan cow and a white heifer but said he preferred the heifer. Hutchinson said the cow's name was Lupin, and the heifer named Blanche was her daughter. Both he said were by Belvedere. 'When Lupin was entered in the Herd Book,' he added, 'the bull Belvedere had not got a name, and was only known as Stephenson's Bull. If I live till Blanche has produce, I shall enter her as got by Belvedere, dam Lupin by Belvedere.' These particulars Bell noted down at the time.[2]

By the time of the sale, which took place on the old man's death, Hutchinson's herd had become, owing to financial and family troubles, 'the wretchedest beasts ever seen, all out of condition and miserably neglected.'[3] Lupin was the only cow in anything like a satisfactory

---

[1] Bell, p. 187.

[2] Bell's *Weekly Messenger*, 10th February 1873, p. 5. Hutchinson had once before made a serious error in the pedigrees of his cattle.—*Sockburn Short-horns*, p. 10.

[3] Memoranda of conversation with Mr. Thomas Stephenson, 21st May 1896.

state.[1] By his purchase of Blanche, Bates rescued the old Sockburn blood from the oblivion that otherwise awaited it. The catalogue, with its genealogical variations, follows in *facsimile:*

THE FOLLOWING SHORTHORNED CATTLE

Late the property of JOHN HUTCHINSON, *Esq., deceased,*

will be Sold by Auction at Grassy Nook, near Stockton-upon-Tees, in the County of Durham, on Thursday the 17th day of October, instant, at 10 o'clock in the forenoon.

(J. APPLETON, Printer, Stockton, 1st October 1833).

COWS.

LOT No.
1. LUPIN, seven years old; got by Mr. Stephenson's Bull (Herd Book, No. 1480), dam Tulip (bred by Major Rudd), by Lancaster (360), gd. by Petrarch (480), gr. gd. by Major (397), gr. gr. gd. Stranger, by Mr. Chapman's Son of Punch (122), gr. gr. gr. gd. Roaned Heifer by Checks (132), gr. gr. gr. gr. gd. Sockburn Sall, by Mr. John Coates's Bull (148). . . . Bulled 13th of July by Reformer. *Mr. Skipsey,* £16:10s.
2. WILDFLOWER, five years old; by Kehama (1150), dam Geraldine, by Sir Leoline (603), gd. Jewel, by Surplice (635), gr. gd. Jessy (bred by Mr. R. Colling), by Wellington (680). . . . Bulled 10th of August by Reformer. *Mr. Walker,* £10:10s.
3. WALLFLOWER, four years old; own sister to No. 2. *Mr. Harding,* £7:2:6.
4. ADELAIDE, three years old; by Pancake (a son of Kehama), dam Violet, by Emperor (243), Dam No. 1. *Mr. Bates,* £11:16s.
5. MARIAN, three years old; by Pancake, dam Jewslip, by Fitz-Marske, gd. Geraldine. . . . Bulled August 17th by Reformer. *Mr. Harding,* £9:12s.
6. JULIA. FAT. *Mr. Harding,* £11:5s.
7. SUNFLOWER, two years old; by Mr. Stephenson's Belvizer, dam Geraldine. *Mr. Harling,*[2] £10.
8. A MILK COW. *Mr. Herd,* £6:5s.
9. A KYLOE.

---

[1] Memoranda of conversation with Mr. Thomas Stephenson.   [2] *Sic.*

HEIFERS.

Lot
No. 10. BLANCHE, two years old; by Belvizer, dam No. 1. Bulled 27th July. *Mr. Bates,* £12 : 15s.
,, 11. WHITELEGS, one year old; by Pancake, dam Breeze, by Fitz-Marske, gd. Princess by Leopold (372), gr. gd. Tulip, by Lancaster. *Mr. Harding,* £6.
,, 12. WOODBINE, one year old; by Sir Cloudesley (a son of Pancake), dam Wallflower, No. 3. *Mr. Hutchinson, Whitton.*

CALVES.

,, 13. MAID OF KENT, six months old; by Tempest (a son of Belvizer), dam No. 4. *Mr. Bates,* £4 : 15s.
,, 14. MISS JEWSBURY, six months old; by Tempest, dam No. 5. *Mr. Cox,* £4.
,, 15. JULIET, six months old; by Sir Cloudesley, dam No. 6. *Mr. Walker,* £4 : 5s.

BULLS.

,, 16. REFORMER, two years old; by Pancake, dam Edith, by Fitz-Marske. *Mr. Wilson,* £15.
,, 17. HAWSER TRUNNION, four months old; by Sir Cloudesley, dam No. 7. *Mr. Eliot,* £4 : 10s.
,, 17. SIR PATRICK SPENCE, four months old; by Sir Cloudesley, dam No. 1. *Mr. Walker,* £4 : 9s.

'A horse,' says the Polish proverb, 'has four legs, yet it sometimes comes down.' Trusting too implicitly to Mr. Thomas Bell's report of his conversation with Hutchinson, Bates identified Blanche's grandsire, Stephenson's Bull (1480), 'got by Waterloo, dam by Wynyard,' with Belvedere himself (1706) instead of with his own brother Blucher (1725). Belvedere was calved in 1826, the same year as Lupin, and therefore could not possibly have been her sire, whereas Blucher was Mr. Stephenson's stock-bull at Wolviston, 1824-1826,[1] and was as an actual fact Lupin's sire.[2] Blucher was four years older than Belvedere; 'though a good bull he was a less beast.'[3] As Blucher and Belvedere were own brothers, the slip is of

---

[1] *Herd Book*, v. p. 165.
[2] Memoranda of conversation with Mr. Thomas Stephenson. [3] *Ibid.*

little genealogical importance; but it gave Bates's enemies an excellent opportunity for cavilling, and it is still suffered to disfigure the *Herd Book*.

In violent contrast with the prices which he gave for well-bred shorthorns at Nunstainton and Grassy Nook, Bates paid 102 guineas for Nonpareil, the dam of Norfolk (2377) at Whitaker's sale in 1833. Nonpareil's grandam, Sally, was bought at the Barmpton sale of 1820, by Mr. William Robinson of St. Helen's, Auckland, for 33 guineas. Sally's dam was really the 'hind's cow' of John Chapman, who had been in Robert Colling's service since a boy. Bred by Mr. Jonathan Dryden and got by Chapman's Son of Punch (122), she had only recently been bought by Colling on the recommendation of Mr. Charles Lenet of Haughton-le-Skerne. She was a good-looking cow and a most capital milker, giving 15 quarts, ale measure (twice a day), after calving. She was in calf when she came to Barmpton, and the catalogue of Colling's sale stated that the calf, Sally, then a two-year-old, was by Alexander (1623).[1]

Bates kept Nonpareil till June 1836; but she never bred again. Twenty weeks before selling her, he put her up to feed with nine other cows and seven steers by Belvedere, all on the same food precisely. He showed them to Mr. Fawkes of Farnley Hall and Whitaker and others with them, and especially called their attention to Nonpareil's rumps, which had no fat on them whatever. It had pleased Whitaker to attribute the want of fat on Norfolk's rumps to his sire Second Hubback, but here was ocular demonstration that this fault came from his dam. Nonpareil did not gain 9 stones in the twenty weeks, while the cow next her in the same house, by Son of Second Hubback (2683), and from a dam which cost only £8 as

---

[1] See Appendix B. Sally's pedigree is given in the *Herd Book*, iii. p. 124, as 'by Alexander, gd. (Sally), by a grandson of Favourite (252), gr. gd. by Punch (531), gr. gr. gd. by Hubback (319).' The Herd Book was, of course, entirely under Whitaker's control. Very true is it that 'whenever the full history of shorthorns is written, the services of Mr. Whitaker to the cause of Shorthorn breeding will be recognised as incalculably valuable.' As a pedigree expander he was indeed unrivalled.

a two-year-old heifer, gained 36 stones, or £20, in twenty weeks, the greatest improvement Bates ever knew an animal to make in the time.

The seven steers by Belvedere gained above 20 stones in the twenty weeks. They had been turned out every day on a very bare pasture, during all the previous autumn, getting only straw at nights, and when they were tied up on 12th January they were not so good as they had been at Yarm Fair (19th October 1835). They had afterwards a few turnips, not 2 stones per day each, and some damaged linseed boiled, the cost of which did not exceed 1½d. per week. At first a little bran was mixed with the linseed to induce them to eat it, but this was soon replaced with chopped straw. The first week in June they averaged £34 for 90 stones, four of them being sold at Leeds for £36 each. Bates had kept these seven, and one he had sold the previous Christmas, eight steers in all by Belvedere (1706), in the company of eight steers by Gambier (2046) from their being calves. At Easter 1835, two graziers from the neighbourhood of Morpeth, who had come to attend Darlington market on Easter Monday, valued those by Gambier at £9 each, and those of Belvedere, all going loose in the same yard, where they had eaten wheat straw all the winter, at 10 guineas each. On the 13th of May the Gambier steers averaged 36 stones, the Belvedere 42 stones. As he had then not sufficient keep to winter the whole sixteen, he sold the eight by Gambier at Yarm Fair, 19th October 1835, for 12 guineas each, to a gentleman's steward, who grazed them near Wakefield, and sold them in the September following at £21 : 10s. a head. The difference between £34 for steers by Belvedere and £21 : 10s. for steers by Gambier was very instructive, especially as the latter had three months' more keep. The four steers by Belvedere were universally admired, and were sold for about £9 a head more than any other cattle of their age that season, either at Leeds, Wakefield, or any markets farther north. Bates intended to have reversed the cows to each bull the following year, but Gambier, as was not unusual with bulls of the Gentle Kitty tribe, got no calves after the first year. He, however, put the cows that had bred to Gambier the next year to Belvedere, and kept the produce in steers in the same manner as in 1836. He sold them in May 1837 at Wakefield, the same place where the Gambier steers had been sold, but more than three months earlier in the year. They brought within five shillings a head of the price the Belvedere steers were sold for in 1836, and the price of beef was nearly the same in both years.

On Easter Monday 1834, Bates was as usual at the Darlington Great Market. Some Americans staying at the King's Head came up and spoke to him. They were, it appeared, the representatives of the Ohio Company for Importing English Cattle—Mr. Felix Renick of Chillicothe and his two assistants, Edwin J. Harness and Josiah Renick. In the course of conversation, Bates soon found that they possessed a great knowledge on the subject of shorthorns, and invited them to Kirklevington. He regretted his house was not more comfortable, but promised he would improve it by the time they came to England again. He gave them full details of his shorthorn experience, telling them, among other things, that Belvedere's sire, Waterloo (2816), then in his sixteenth year, and Norfolk (2377) were the only two bulls, besides Belvedere (1706), that were in his opinion the least likely to get good stock.[1]

He showed them his own cattle, and took them to see the principal herds in the neighbourhood. At Harlsey, Mr. Maynard told them that shorthorns had ruined and would ruin all that had anything to do with them. This was not the way to encourage foreign buyers. It is not therefore surprising that when, on returning through Northallerton, Bates remarked to Renick that he had

[1] Bates and Whitaker had a sharp discussion over the respective merits of Waterloo and Belvedere. Whitaker declared that the former was much the better bull, but Mr. Thomas Stephenson says he was certainly not.—See above, p. 231.

never asked Mr. Maynard the price of any of his animals, the latter replied, 'I would not have one of them as a gift.'

Anxious that America should obtain the best breed of shorthorns, Bates astonished his friends by offering Renick six of his finest cows and heifers. Renick wished to have Jonas Whitaker's advice, so Bates wrote a letter to Greenholme asking him to come over immediately. This Whitaker did, and it was the only favour Bates ever asked of him. Although among the six animals offered were Duchess 33rd, priced at 150 guineas, Duchess 34th (or Brokenleg) at 100 guineas, and the Matchem Cow at 15 guineas, 'surrendering his judgment to Whitaker,' Renick finally settled on 14th April 1834 to take only Red Rose 11th, by Belvedere, at 150 guineas, and Teeswater at 50 guineas, 'the two worst of the lot' in their breeder's opinion.[1] In case Red Rose 11th did not produce a living calf, Bates volunteered to furnish the first sister-in-blood gratis, and if Teeswater proved barren, Renick was to be given the heifer Waterloo 2nd. Bates also sold Renick the two bull calves, Earl of Darlington (1944) and Young Waterloo (2817) for 100 guineas.

A little later, as Bates and the Americans were seated round Whitaker's table at Greenholme, their host burst out into a tirade against Duchess 33rd: 'she was a bad one if ever he saw a bad one.' Bates gave this the lie direct by declaring that she was better than any other animal Whitaker had ever seen, and told the Americans that if they would lay down 300 guineas for her he would not now accept it.[2]

Bates offered Renick a bull (5208), which he had bought for him from Mr. Clarke of Skipton Bridge, but Whitaker, though he had never seen it, abused it so violently that Renick declined to take it at any price.

On the other hand, Whitaker imposed on Renick at

[1] After this it is preposterous to assert, as in *Saddle and Sirloin*, p. 153 n, that Whitaker persuaded Bates not to sell Duchess 34th. He no doubt persuaded the Americans not to buy her—a very different matter.

[2] This and the other similar facts are all recapitulated in Bates's correspondence with Renick, which is sufficient evidence of their truth.

£175 the bull-calf Duke of York (1941), of the bad Hermit strain of the Bright Eyes tribe. This had been sold at his sale a few months before for 13 guineas to Mr. Archbold from Ireland. Bates then told Archbold he had done wrong in buying it, and should have taken The Doctor (2748), which had been sold for 50 guineas. Archbold said that the purchaser of The Doctor (2748) was a friend of his, and if Bates would find him a customer for the Duke of York (1941), he would ask his friend to let him have the better animal. Mr. Paley of Gledhouse, near Leeds, was standing near them at the time, and said that to oblige Bates he would take the Duke of York (1941) off Archbold's hands at cost price, while Archbold's friend gave up The Doctor (2748). In the course of the same day, Mr. Sedgwick of Stone Gap, near Skipton, asked Paley to sell the Duke of York (1941), and Paley, thinking it of little value, taking 3 guineas for his bargain, handed the bull-calf over at 16 guineas. As all these circumstances were perfectly well known to Whitaker, Bates could not conceive how Renick could have been inveigled into giving £175 for this same calf.

The five heifers left in the fold from which Red Rose 11th and Teeswater had been selected were sent by Bates to Mr. Fawkes's, at Farnley, in the company of three older cows, in order that they might be bulled by Norfolk (2377). Before they left Farnley, Whitaker wrote to Bates saying he had been over to see them, and that Duchess 33rd 'had improved beyond any idea he could have formed of an animal's improvement.' 'She has,' he said, 'quite deceived me and must be pronounced a real good one.' With all his poor opinion of Whitaker's knowledge of shorthorns, Bates could not believe that he was 'so destitute of judgment as to be unable to discriminate between the best and the worst.' He was compelled to infer that Whitaker's conduct at Kirklevington and Greenholme had been dictated by a wish to give the Americans a bad opinion of Bates's stock and judgment, in order to gain a monopoly of their trade.

On 6th August 1834 Whitaker and Bates compared the two-year-old Red Rose 11th, just as she was leaving for America, with Duchess 19th, a heifer of about the same age. Whitaker considered Red Rose 11th 'very good; her horns a little wide; head, eyes, crop, back, sides, all good, and bosom extra, but shoulders a little upright.' Bates stood to it that Duchess 19th was the better animal. Whitaker, on the contrary, criticised her crops as deficient and her body as too large, though he was compelled to admit that her head, eyes, and horns were most beautiful.[1]

Renick's purchases arrived safely at Philadelphia, and were driven over the mountains to his farm at Chillicothe, where they were kept as the joint property of the Ohio Company. Red Rose 11th was re-named Rose of Sharon in America.

In November 1834 the Matchem Cow produced the Oxford Premium Cow, so called from her having afterwards taken the first prize at the first meeting of the English Agricultural Society at Oxford in 1839. After her purchase from Mr. Brown at Nunstainton, the Matchem Cow had passed into the hands of the brothers Bell who held some of the Kirklevington farms by a most complicated tenure intended to encourage them in breeding pure shorthorns. Mr. Thomas Bell told Bates that she set all the other cows wrong by leaping the hedges, and that he was determined to part with her, although she was an excellent milker. He took her to Darlington market, hoping to get her prime cost which was £11. After waiting from the morning to the very close of the market, and having only once had an offer of £9 : 15s. for her, he drove her home. On hearing of this Bates said he himself would give the £11, and that he was confident that, put to one of his Duchess bulls, she would breed a calf worth not less than a hundred guineas. Accordingly he had her served by Duke of Cleveland (1937), a bull from Duchess 26th by Bertram (1716), which never weighed more than 40 stones

---

[1] Holt Beever's *Leading Tribes*, p. 194.

(of 14 lbs.) when above three years old. The result in the young Matchem or Oxford Premium Cow surpassed his expectations. Duchess on Princess would seem to have been quite as successful a cross as Princess on Duchess had proved itself to be in the case of Belvedere's stock. The Matchem Cow had previously produced five very inferior calves to Young Monarch, Gambier (2046) and other bulls. She was a capital milker, giving 24 quarts a day when grazing on a poor pasture in summer, and while tied up in an open shed in winter with hay only and water once a day; Bates never allowed his cows to enter a grass field during the six winter months. On being put dry after having had ten calves and having always been in low condition while in milk, this old Matchem Cow, the ancestress of all the Oxfords, made herself fat in a few months and weighed over 60 stones.

Bates was present at the sale held after Major Bower's death, at Welham near Malton in February 1835. 'The stock were large and coarse, fed from being calves; there was nothing worth buying to breed from.'

Later in the year Bates received a letter from Renick of Chillicothe to say that he would take any cattle Bates chose to send him at his own price. Shortly afterwards Whitaker called at Kirklevington and declared that he too had heard from Renick but to the very contrary effect. According to Whitaker all the American trade was to pass through his hands and he was to make the best bargains he could for the Ohio Company on commission. He looked through the Kirklevington herd but never then nor afterwards asked Bates to sell any animal to the Company. One bull particularly struck him: 'I have never seen this one before,' he said, 'where did you get him?' 'If you will give your opinion of him, I will tell you,' was Bates's answer. 'If I am to do that,' continued Whitaker, 'I must say he is a real good one.' 'Well then,' said Bates triumphantly, 'this is the bull I bought of Mr. Clarke at Skipton Bridge (5208)!' Whitaker told Bates to prepare to go with him to Mr. Maynard's at Harlsey. Far from refusing to assist him in procuring cattle for Renick, Bates sent a horse immediately to Yarm, and getting a gig drove Whitaker over. After they had examined the bulls at Harlsey, Whitaker asked Bates his opinion of Comet Halley (1855). Bates replied that if his judgment was worth anything he would not buy him. Whitaker said no more, but went into the house and laid down a hundred guineas on the table before Mr. Maynard for this very bull which Renick had said he would not take as a gift. Soon after this Whitaker's man, Timothy Metcalf, called at Kirklevington and looked through all the stock. He too was struck with the Skipton Bridge Bull (5208) and declared that he was better than any bull he had seen during the many weeks he had been out looking for bulls. He said that as he could not find one so good he would take him, if he thought Renick would not know him again. Bates said that if the bull did go to America, he would take care Renick should know where he came from. Metcalf said that would never do. Determined that the bull should after all be seen and appreciated in the New World, Bates presented him to Kenyon College, Ohio, to serve the college cows.

Whitaker came a second time to Kirklevington that year. Bates asked him if it was true that he had given Mr. John Colling of White House, near Greta Bridge, £100 for the heifer Lady Colling, as all the people he heard speaking of her said she was a bad one, and not likely to breed, and her dam and grandam were cows he himself would not breed from. Whitaker replied that he had given the price, and that he considered Lady Colling a much better heifer than Red Rose 11th. Questioned further if John Colling had agreed to replace her should she prove barren, as Bates had volunteered to do in the case of the Red Rose and Teeswater, Whitaker said he had never so much as asked him. This was quite borne out by Colling, who told Bates that if the heifer had dropped down dead immediately after her purchase, he would not have given Whitaker a farthing.

Very fortunately for Bates, out of the three cows and five heifers sent to Norfolk (2377) only three, Duchess 33rd, Waterloo, and Blanche had calves by him. 'Although the dams were extraordinarily good and were the best adapted to Norfolk of any shorthorns in existence, the calves proved far from good and were not worth one-fourth of the value of their dams. This was especially the case with the calf Duchess 38th, which took entirely after Norfolk's dam, Nonpareil.'[1] On the return of Brokenleg (Duchess 34th) to Kirklevington, Bates put her to her sire Belvedere (1706), and on 15th October 1835 she produced the celebrated Duke of Northumberland (1940). At five months the Duke looked a very delicate calf to one of Bates's critics. Drawing himself proudly up, Bates replied: 'Well, sir, I have the greatest hopes of him.'[2] Nor were these hopes belied, for at ten months the Duke was calculated to weigh 40 stones (of 14 lbs.) the four quarters without offal.

DUKE OF NORTHUMBERLAND

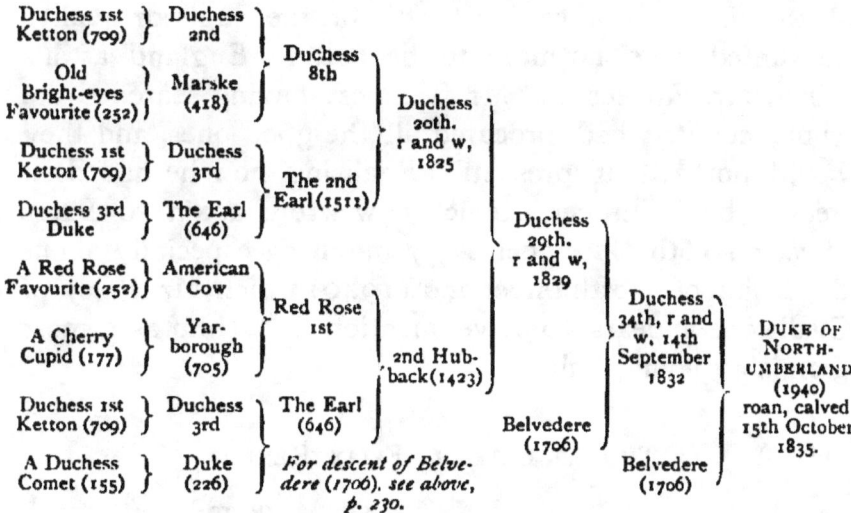

[1] This is borne out by the private testimony of Mr. Thomas Bell: 'The less said about Norfolk the better. It was a great mistake sending any cows to him. Every animal by him was inferior to its dam.'—Letter to Thomas Bates of Heddon, 17th June 1871.

[2] *Saddle and Sirloin*, p. 153 n.

At the great Chillicothe sale, 26th October 1836, Young Waterloo (2817) fetched 1250 dollars from R. D. Lilly, Highland County, Ohio; while Teeswater, bought by John J. Vanmeter, Pike County, headed the cows at 2225 dollars, including her calf Countess. As Bates had foretold, the hundred-guinea cow Lady Colling would not breed, and she brought only 205 dollars.[1] Renick

DUKE OF NORTHUMBERLAND (1940), August 1836.

must have been pleased with the result of his Kirklevington purchases. Whitaker's son, who was in America at the time, told him that 'cattle of the superior quality he wanted were no more to be had in England, as his father, Mr. Fawkes and Mr. Tempest having searched the whole country, had procured all the good ones, and they would not sell at present.' Realising how he had been treated by Whitaker, Renick now wrote again to Bates direct on 16th December 1837 inquiring especially about the Duke of Northumberland (1940) which Mr. Paley of Gledhouse appears to have mentioned to him as a most promising young bull.

THOMAS BATES *to* FELIX RENICK

KIRKLEVINGTON, *April* 1838.

I think it on the whole better not to send you any of my own cattle this season, the exchange being so much against you. Next year, as you say you intend to continue importing, I might

[1] Allen's *History of the Short-horn Cattle*, p. 182.

furnish you with ten young heifers or young cows having had a calf or two, and five or six young bulls, either of the age you got the two last from me or a year older.

The Duke of Northumberland (1940) and Short-tail (2621) are the only bulls I am now using, and their stock is even more promising than that of their sire Belvedere (1706). The four you got of me were all by Belvedere, and all my stock are by him and his sons. After the trials I have now had and seen of short-horns for nearly sixty years, nothing could induce me to use any bull that had not Belvedere's blood. You will find it all money thrown away to buy any bull that has not sprung from him.

Twenty-eight days after the birth of the Duke of Northumberland (1940), Brokenleg (Duchess 34th) whom you will remember was again put to her sire Belvedere and brought 2nd Duke of Northumberland. She has since brought me a heifer to her sire, and is now I expect in calf to Short-tail.

By putting Duke of York (1941) to the heifers you got of me you will bring their produce into disrepute. I will, on no consideration whatever (if you would give me ten times the price I would otherwise have charged you for a heifer) sell you any heifers to put to any bulls but what I have bred, or are of my blood. Nor will I sell you any at any price till you and the Company you act with, under your joint hands, have solemnly promised not to do so. My object has never been to make money by breeding, but to improve the breed of short-horns; and if I know it, I will not sell any to any one who has not the same object in view. On this principle I began breeding, and I am convinced I have a better breed of short-horns in my possession at present than there has been for the last fifty years, even in the best days of the Messrs. Colling.

The bull you ask me about sending you, Duke of Northumberland, is everything I can wish in a bull, and Short-tail has taken after 2nd Hubback, of whom his dam (Duchess 32nd) had two crosses. Short-tail's sister (Duchess 41st), the best animal in my possession, I expect is in calf to the Duke of Northumberland. The six from which your two were taken were good, but the breed of the years 1835-6 were far superior to those six, though very good. Brokenleg (Duchess 34th), I offered you at 100 guineas. If you were to send twenty times that sum for her, and her produce, I would not take it now. You will remember I told you after buying the two heifers, that if either of them died on the passage, or did not breed when you got them home, I would give you the two nearest in blood to them. Now (Red Rose 13th) a sister in blood to your Rose of Sharon (calved since you were here), has produced a heifer (2nd Cambridge Rose) to her sire Belvedere; and for the two I would not take 1000 guineas. These would have been yours now had yours not bred. I will not sell either cow or calf, but I have no objection to sell the bulls I breed from them, or from my Duchess tribe, which are far better animals than the Red Rose tribe. I will not part with the females of these tribes at present.[1]

Bates attended the show of the new Tyneside Agricultural Society held at Hexham in 1837. He has left a gloomy picture of the state of Northumberland:

The short-horned cattle of Tyneside from having been the best shows I ever knew, far exceeding any in the present day, as a whole, had become the worst of any district I know. There was not even the vestige of a good short-horn from Tyneside in Northumberland. The premiums were nearly all carried away by strangers from other districts. With the decline of good short-horns, the agricultural produce of the district fell off to less than one-half of what I had known it on many farms, and probably never again will become equal to what it once was, while in other parts of the kingdom agricultural improvements have greatly advanced. This ought to be a warning to all agricultural societies to prevent the conductors thereof being governed by selfish motives, to advance their own interest, instead of the public advantage, as were the conductors of the old Tyneside Society broken up in 1821. I have held, and ever will hold, that the prosperity of the landed interest, I mean landlord, tenant, and labour conjointly, tends to the prosperity of every other class in the state.[2]

He had used Belvedere (1706) six years (1831-1837), and having his sons Duke of Northumberland, then turned two years old, and Short-tail, 'both superior to their sire, though he was very good,' he thought he had no longer occasion 'to use him and had him slaughtered.'[3]

After the advent of Belvedere, Bates kept more calves for bulls, but very frequently by no means the best-looking or most promising. He often had ten or more bulls in his possession which he never used and had no desire to part with unless he considered they were going into herds which they would improve.

[1] Bell, p. 227. [2] *Ibid.* p. 238. [3] *Ibid.* p. 35.

He used to say, as we have seen, that money-making and not the improvement of shorthorns was the ruling motive that governed the breeders. Had the latter been the main object, shorthorns, in his opinion, might have been of double the value they were in these kingdoms, and have spread over the whole habitable globe. On one occasion, Mr. Wetherell, the 'Nestor of Shorthorn-breeders,' who did an extensive business in buying and selling, and was unsurpassed in his knowledge of stock, paying particular attention to their constitution, came over to Kirklevington and asked Bates if he had any bulls for sale. Bates replied that he had. After viewing the herd Wetherell selected two, and they went into the house and had dinner. Bates then inquired about the herd into which Wetherell proposed to send the bulls. Wetherell asked in reply, 'of what consequence that was as long as he got the money for them?' Bates, however, peremptorily declared that he 'would not sell any man a bull unless he knew the herd to which it was going, for if the cross did not answer, all the blame would be attributed to the bull.' In mounting his horse, Wetherell could not refrain from expressing his opinion of Bates in strong terms for refusing to sell cattle at high prices so long as he got paid for them.[1]

When at home alone, Bates read a great deal, mostly on agricultural subjects and political economy; but he had a most extensive correspondence to keep up. During his leisure, he regularly walked out and visited all his animals, speaking and talking to them as if they understood him. They immediately collected round him, and he generally caressed them by rubbing the under side of their necks. Those that he did not immediately notice poked him with their horns from behind, or pulled at his coat-tails with their teeth. Sometimes he told Robert Thompson the cowman, that he would help him to drive the cows into the field, but this usually ended in Robert becoming irritated and exclaiming, 'I wish you'd keep

[1] Bell, p. 185.

out of t' way, you do far mair ill than good, for they won't leave you, and ther's no driving them.'[1]

Bates carried simplicity of living into all his household arrangements. Although at Halton and Ridley Hall he had good mansions, yet at Kirklevington he never thought of the ordinary comforts of life until his friends remonstrated with him. The dining-room, with its mud floor, was perhaps the smallest room of its sort in the county, but was often crowded with bishops, peers, and agriculturists of all classes. The Duchess of Leeds, who preceded Lady Pigot as an admirer of shorthorns, was a frequent visitor. Although Bates kept only a housekeeper and a housemaid, yet the most fastidious could not object to his style of hospitality.[2]

The same simplicity characterised the accommodation of the stock. The farm buildings at Halton were poor and scattered, in consequence of having been let in separate holdings. At Kirklevington Bates never improved on the original buildings, and the additions he made to them were of the most ordinary and temporary description. The best cattle were kept in cow-sheds, with sides made up with furze, and unpointed roofs of red tile. He had constantly seen the large and superior farm buildings in North Northumberland, but had no idea of imitating them. There was not even a decent calf house at Kirklevington, until Mr. Thomas Bell got him to have the one erected after the model of that at Grassy Nook.

His dress was plain and studiously neat, although it was of the old form and fashion. He was temperate and abstemious in his habits, and generally callous of eating and drinking. The housekeeper had to watch him to see that he took his breakfast, which usually consisted of oatmeal porridge. It frequently happened, however, that he got out without his breakfast, and did not return till mid-day. The good woman would then ask him which he would have, breakfast or dinner. Roused

[1] Bell, p. 186. [2] *Ibid.*

from his reverie, he would exclaim, 'Breakfast? Have I not had any breakfast?' When alone he never took anything except water with his dinner, but with company he always joined in wine. Spirits he never tasted, and he was only once known to touch beer.[1]

He had a great dislike of ostentation, and could never be persuaded to keep a carriage, making it an excuse that neither his father nor grandfather ever kept such a thing. Riding on horseback he said was much healthier. For several years at Halton Castle, he had a very excellent mare, which sometimes carried him a hundred miles a day, in attending markets many miles apart. Mr. Foster of Carlisle afterwards made him a present of a strong pony; though it went blind soon after he got it, he continued to ride it for many years. One day at Kirklevington he hired a dog-cart to take a friend who was visiting him to Mr. Maynard's, at Harlsey. They had not proceeded half a mile when the mare shied at some Irish labourers resting by the roadside. Bates, who was earnestly engaged in talking, suddenly pulled at the reins, which got under the mare's tail; she began to kick, broke the shaft, and turned the vehicle completely over. Bates was much bruised and could proceed no farther; his friend and his luggage were sent on to Harlsey in a cart.[2]

At one time Bates had a fine breed of the old Cleveland road horses, but having lost two fine young ones by their falling into a quarry, he never afterwards took any interest in horse-flesh. He never joined in field sports. When young he shot regularly every season, but meeting with an accident from a large gun bursting in his hands, he never took one up again. The large gun in question was purchased to shoot the ravens, which carried off the lambs, and were, it is said, the last seen on Tyneside.[3]

[1] Bell, p. 187.   [2] *Ibid.* p. 188.   [3] *Ibid.* p. 189.

CHAPTER VIII

KIRKLEVINGTON: THE SHOW-YARDS

1838–1844

IT was an evil day for Bates's comfort and peace of mind when he reluctantly abandoned his determination formed in 1812, and consented to exhibit his herd again for public competition. A stranger who had accidentally viewed it, while buying horses from his tenants, strongly urged him to let the world see what pure shorthorns really were. In order, then, that the breeders of a new generation, especially Americans, should have this opportunity, Bates sent seven animals to the York meeting of the newly-formed Yorkshire Agricultural Society in August 1838.

In all that subsequently happened it is essential to bear in mind that, as far as exhibiting cattle went, Bates was an amateur having to meet and contend with professionals. Believing in an immutable standard of excellence in cattle, fixed by the feeding and milking qualities of themselves and their ancestry, of which their appearance and handling afforded certain indications, he was no votary of fashion, and boldly refused to bow down and worship the golden calves which ignorance or interest set up in the high places of British Agriculture. He took things far too seriously, fondly imagining that the object of agricultural shows was to improve the breeds of live stock and not merely to afford popular hippodromes. With him the neglect or rejection of his cattle was no personal matter, but one of national, or rather international importance.

The shorthorns he sent to York were:—

1. Duke of Northumberland (1940), two crosses by Belvedere.
2. A white three-year-old twin heifer[1] (both breeding), Shorthorns 4th by Belvedere.[2]
3. A three-year-old heifer from Matchem Cow, by Duke of Cleveland (1937).
4. A roan two-year-old heifer, Duchess 41st, by Belvedere.
5. A roan year-old heifer, Duchess 42nd, by Belvedere.
6. A red year-old heifer, Duchess 43rd, two crosses by Belvedere.
7. A roan four-year-old cow, Red Rose 13th, by Belvedere.

Since Duke of Northumberland (1940) was entered for the prize offered for the champion bull as well in the class of two-year-old bulls, Bates might have won eight prizes. As it was he won five, four first prizes and one second:—

CLASS I. *Bulls of any age.* 15 entries. First Premium (£25) to Earl Spencer for Hecatomb (2102). Second Premium (£10) to Mr. Wiley for Carcase (3285).
(Duke of Northumberland passed over unnoticed.)

CLASS II. *Two-year-old bulls.* 9 entries. First Premium (£15) to Mr. Bates for Duke of Northumberland. Second Premium (£5) to Mr. Linton.

CLASS III. *Yearling bulls.* 9 entries. First Premium (£10) to Mr. John Colling. Second Premium (£5) to Mr. Childers, M.P.
(This was the only class in which Bates did not exhibit.)

CLASS IV. *Cows of any age, in calf or in milk.* 9 entries. First Premium (£10) to Mr. John Colling for Rosanne. Second Premium (£5) to Mr. Edwards for Foggathorpe.
(Red Rose 13th passed over unnoticed.)

CLASS V. *Three-year-old cows in calf or in milk.* 6 entries. First Premium (£10) to Mr. Bates for Shorthorns 4th twin heifer by Belvedere. Second Premium (£5) to Mr. Bates for cow out of Matchem Cow.

---

[1] In the parlance of the North of England a heifer does not become a cow until it has had its second calf.   [2] *Herd Book*, v. p. 934.

When the judges went into this last class, they immediately took these two out, but were very long in deciding which to place first. At length they asked the holders of the two cows whose they were, and learnt that they both belonged to me. Seeing me in the next pen examining the aged cows, which they had decided upon, (it being then past the hour that the public were admitted into the yard) they called me and asked which of the two I considered the best. I answered that I thought they need not ask, as the two cows spoke for themselves. The one had calved on the 21st May the bull calf, Omega (4615) which I sold afterwards to Mr. Foster, Springfield, Ireland, and giving a large quantity of milk, could not look as blooming as the other which had gained twenty stones in weight, though the former had paid as much or more in milk as the other had done in fatness. They then placed the first premium on Shorthorns 4th (the twin heifer) and the second premium on the cow off the Matchem cow, by Duke of Cleveland, and did so properly under the circumstances. It is rare, indeed, that judges at exhibitions of stock give premiums to great milkers: they look only at the condition or fatness of the animal.[1] This cow, off the Matchem cow, produced a bull calf, named Locomotive (4242) 5th October 1838, which I sold the March following to J. C. Etches, Esq., then of Liverpool and Barton Park, near Derby, at 100 guineas, and after using him some time he sold him to Mr. J. C. Letton, Piney Grove, Citron Forest, near Millersburgh, Kentucky, U.S.A., at 250 guineas.[2]

CLASS VI. *Two-year-old heifers in calf.* 10 entries. First Premium (£10) to Mr. Bates for Duchess 41st (or York Premium Two-year-old). Second Premium (£5) to Mr. Edwards for White Rose.

CLASS VII. *Yearling heifers.* 8 entries. First Premium (£10) to Mr. Edwards for Malibran. Second Premium (£5) to Mr. Bates for Duchess 42nd.

(Duchess 43rd passed over unnoticed.)

Portraits furnished by the owner of an animal rarely make it look worse than it is. Those of Hecatomb and

---

[1] The meaning is that Bates considered that according to all ordinary rules of judging, the Matchem heifer was much superior to Shorthorns 4th, but if milk was taken into account, as he always contended it should be, the scale was turned in favour of the latter. The judges at York were probably not influenced by the milking properties of Shorthorns 4th, and their decision if based on form and quality only was wrong.

[2] Bell, p. 72.

Duke of Northumberland are given in the *Herd Book*.[1] The former was by Mr. Davies, a faithful animal painter, with whom Lord Spencer found no fault, while Bates was dissatisfied with the latter. Looking, then, on this picture and on that, it must be evident to any one who has the remotest inkling of taste for shorthorns, or indeed for animals of any sort or kind, that the award of the champion prize to Hecatomb, for which John Grey accepted the responsibility, was either proof positive that he did not know the difference between a shorthorn and a buffalo, or that he unconsciously allowed his better judgment to be warped by gratitude to Lord Spencer for obtaining him the place of £1200 a year, as Receiver of the Greenwich Hospital estates.

That the latter was the correct hypothesis appears from the fact that as soon as the judging was concluded, John Grey came up to Bates and said, 'I should be much obliged by your accompanying me to look at three animals which you have exhibited to-day; I consider them the best three animals I ever saw in my life.' He took Bates, to his utter astonishment, first to Duke of Northumberland, then to Red Rose 13th, and from her to the equally rejected Duchess 43rd. These were in fact exactly the three animals which Grey and the two other judges had placed in every case behind two other shorthorns which lacked all the good qualities which he pointed out in them.

Later in the day, Bates had his seven animals placed together for examination by the public. The chief breeders crowded around them, and this impressed him with the idea that at some time during a show, shorthorns ought to be exhibited in family groups.

On the health of the successful candidates being proposed at the dinner by Mr. P. B. Thompson, Bates acknowledged the toast in a very animated speech. He alluded to the condition of cattle-breeding when he first knew it fifty-four years before. The interest of Agricul-

---

[1] *Herd Book*, iv.

ture was, he considered, of permanent national importance. It was a science to which chemistry and all other branches of natural philosophy should be subsidiary. He alluded to the proposal for introducing an agricultural section in the British Association, and regretted that there was no professorship of agriculture in either of the English Universities. His address elicited the warmest applause. Lord Dundas congratulated the meeting on the success likely to attend the Society. He lamented with Bates that there was no professorship of agriculture at either Oxford or Cambridge.[1]

At the general meeting Bates endeavoured to get the Society to obtain full particulars of a reaping-machine that a friend of his had seen in successful operation for three years in Illinois; but Lord Spencer threw cold water on the proposal. John Grey knew Bates was dissatisfied with his awards. In responding to the toast of the judges, he enlarged on the difficulty they had in pleasing every one.[2] The attempt 'to please every one' was exactly the rock on which Grey and his colleagues had allowed their judging to be shipwrecked. They had not sufficient strength of mind, or enough steadfast love of shorthorns, to face what no doubt they imagined would have been been the popular outcry if Bates had swept the decks. Considering more the effect on the public than the merits of the cattle, they probably thought they had gone to dangerous lengths to please Bates as it was. They had given him no prize in Class IV. but they made up for this by giving him both prizes in Class V.; in Classes VI. and VII. the competition was practically restricted to Bates and Edwards; what could be more equitable than to give each a first, and each a second prize? It was thus, as in the case of the championship, that they applied the methods of political opportunism to the distribution of agricultural premiums. On a subsequent occasion, when his judging met with very hostile criticisms, Grey took refuge in the natural partiality of parties for

[1] Bell, p. 246.   [2] *Ibid.* p. 248.

their own stock and the fallibility of human judgment. He very rightly observed: 'Lookers on, or those who examine cattle in pens, have not the same means of comparing the quality or handling that the judges have, when they are brought side by side. That is a point to which much attention should, in my opinion, be given. Form, however correct, without good quality of flesh, does not make a desirable animal.'[1]

At Mr. Henry Edwards's sale at Castle Howard the next year (1839), Bates bought the Foggathorpe cow, which had taken the second prize in the class in which Red Rose the 13th was rejected at York. Mr. Edwards of Market Weighton was son-in-law to Mr. Seaton, who for several seasons had hired Robert Colling's White Bull (151), the only bull bred direct from Hubback to Favourite, and at Seaton's death, Edwards took the best of his stock. Bates gave £113 for Foggathorpe with the auctioneer's fees, though she was ten years old and not expected to breed.[2] Thus, at last, he became possessed of the blood of the White Bull (151) the Collings had refused him the use of.[3] Foggathorpe's character he considered resembled that of Princess, and she might, he thought, be of the same tribe, though of this there was no record. She was, in his opinion, very superior to Mr. John Colling's cow Rosanne, which gained the first prize in the same class at York, but very inferior to Red Rose 13th (Cambridge Premium Rose). He would have given four times the price he did for her if she had been free of Marlborough's (1189) blood.[4] On her arrival at Kirklevington, he had her tied up next to Red Rose 13th. He asked Mr. Thomas Bell what he thought of the new purchase. Bell expressed a very favourable opinion of her shape and points. Bates then told him to put his hands on her and afterwards on Red Rose. Bell immediately felt the difference. In the sequel, Bates took all his visitors to see the two cows and to test them by handling. They were kept side by side, fed precisely

---

[1] Bell, p. 249.   [2] *Ibid.* p. 53.   [3] See above, p. 66.
[4] Letter of Thomas Bates to George Vail, 1st December 1843.

on the same food, and finally sold together. Bell could not help harbouring a suspicion that Bates had bought Foggathorpe on purpose to show his visitors how ignorant judges at the great shows were of what constituted the chief merit of improved shorthorns. Mr. Thomas Stephenson could never understand why Bates had bought this large light-roan cow. Bates told him that she came of an old tribe.[1] There was no doubt in Bell's mind as to the superiority of Red Rose 13th over Foggathorpe, although the latter was in his opinion a good cow. Foggathorpe's produce by Duke of Northumberland (1940) showed, he considered, a very distinct improvement.[2]

Having once begun exhibiting again, and having had what he regarded as his three best cattle—Duke of Northumberland, Red Rose 13th, and Duchess 43rd—pronounced third-rate at the best, Bates resolved that these rejected ones should make their reappearance at the English Agricultural Society's Meeting at Oxford in 1839. When the time came, Red Rose 13th was not in a fit state to travel, so Duke of Northumberland (1940) and Duchess 43rd started in the company of Duchess 42nd and the Young Matchem cow which had been second in her class to Shorthorns 4th at York. Bates went with them in the same steamship from Middlesbrough to London and himself saw to their treatment. In landing at London, Duke of Northumberland (1940) slipped and lay across the gangway; Bates patted him on the head, calling him 'poor boy, poor boy,' and the huge animal remained perfectly passive until he was rescued. Fortunately the Duke received no injury. The four shorthorns proceeded from London in a freight boat by the Aylesbury branch of the Grand Junction Canal. Mr. Fowler of Aylesbury was applied to one evening to make arrangements for their reception. He sent them to the Prebendal farm, alongside the turnpike-road to Oxford. His son never forgot the beauty of the animals, which far exceeded in style and character any that he had ever seen

---

[1] Memoranda of conversation at Wolviston.   [2] Bell, p. 296.

before. They remained for the night at Aylesbury, and the next day were driven on to Thame. Another stage of thirteen miles completed their journey to Oxford, which had occupied nearly three weeks.[1]

At Oxford, Bates was the guest of Mr. John Pinfold of Holywell, a wealthy bachelor on whose farm the show was held. Pinfold supplied most of the colleges with meat, and as he was much in request at this busy time, the task of entertaining Bates devolved on his friend Mr. King of Wealdstone, Middlesex, and his son, with whom Bates spent most of his time.[2]

The judges of the shorthorn classes were Mr. T. Charge, Mr. W. Smith, and Mr. J. Hall. The first prizes were—

1. *Bulls* (£30). 7 entries. To Mr. Bates for 'Duke of Northumberland' (1940), three years and nine months.
2. *Cows in milk* (£15). 4 entries. To Mr. Bates for 'Young Matchem Cow,' four years and eight months.
3. *Heifers in calf, under three years old* (£15). 3 entries. To Mr. Bates for 'Duchess 42nd,' one year and eleven months.
4. *Yearling heifers* (£10). 9 entries. To Mr. Bates for 'Duchess 43rd,' one year and ten months.
5. *Bull calves* (£10). 4 entries. To the Marquess of Exeter for 'Oxford' (4636), seven months.

The show of stock was not so large as had been expected, yet the quality of some was very superior. The animals exhibited by Bates were universally admired as excellent specimens of the shorthorn breed. Mr. George Drewry, with all his long and successful experience as a breeder, could write, more than fifty years afterwards: 'The two things I remember best at Oxford were the Duke of Northumberland and Duchess 43rd. These I still think the two best shorthorns I ever saw!'[3] So that

---

[1] Fowler's *Recollections of Old Country Life*, p. 232.

[2] 'The first two Country Meetings of the Royal Agricultural Society,' by Ernest Clarke in *Royal Agricultural Society's Journal*, 3rd Series, v. p. 209.

[3] *Ibid.* p. 216. Another old judge of shorthorns (Mr. Thomas Stephenson) referred more than fifty years afterwards to the portraits taken of Duchesses 42nd and 43rd as 'the nicest painting he ever saw.'

it is absolutely false to ascribe their honours to the limited competition they encountered. In fact it was only the Kirklevington shorthorns that saved the first great national show of English cattle from a most ignominious collapse. This circumstance called for some show of gratitude from the Royal Agricultural Society, and no one could have supposed that Bates's subsequent communications to them on the subject of cattle-breeding would have been cavalierly ignored.

At the great dinner given in the quadrangle of Queen's College, Daniel Webster, one of the most celebrated orators America has produced, said, in a speech which riveted the attention of the audience:—

In the country to which I belong, societies, like the present, exist on a small scale in many parts, and they have been found to be very highly beneficial and advantageous. They give rewards for specimens of fine animals, and the improvement of implements of husbandry, which may tend to facilitate the art of agriculture and which were not before known. They turn their attention to everything which tends to improve the state of the farmer, and I may add, among other means of advancing his condition, that they have imported largely to America from the best breeds of animals in England, and from the gentleman who has been so fortunate as to take so many prizes to-day. From his stock on the banks of the Ohio and its tributary streams, I have seen fine animals raised which have been supplied from his farms in Yorkshire and Northumberland.

The live weight of Duke of Northumberland (1940), 1st July 1839, two days before starting for Oxford was 180 stones (of 14 lbs.); on arriving home on 29th July, it was reduced to 152 stones. In travelling twenty-six days he had lost 28 stones, or nearly one-sixth of his total weight. During the same period the 'Young Matchem Cow' lost 12 stones, and Duchesses 42nd and 43rd, 6 stones and 2 stones respectively.

Duke of Northumberland (1940) had now secured champion honours, the Young Matchem cow had proved that she was worthy of a first prize, and the Duke's own sister Duchess 43rd, utterly disregarded at York, had also taken a first. In memory of these successes, Bates gave Young Matchem the name of Oxford Premium Cow, or Oxford 1st. She won the first prize of £20 at the Northallerton meeting of the Yorkshire Agricultural Society that same year, with Mr. John Booth's Yorkshire Jenny as her second. If she had been weighed in the summer of 1841, Bates was confident she would have

OXFORD PREMIUM COW.

scaled 100 stones (of 14 lbs.), although she had been milked regularly twice a day, without being put dry for calving, from October 1838 to September 1841, nearly three years, and had had a calf each year regularly. She was a very great milker, but increased all the time her weight of carcase. Bates never saw the same weight of beef on so small a bone. Her own brother, a steer, when killed at two years old in 1838, weighed 72 stones.

In September 1839 Bates received a visit from Mr. Collin, an American, who was making a long agricultural tour in England. Mr. Collin's impressions of four of the celebrated bulls he had seen were—

PRINCE OF NORTHUMBERLAND (4826), fine throat, long body and round barrel, strong horns, tail run into his body and high set on, short hind-quarter.

DAN O'CONNELL (3557), bare between hooks and rumps but long, head long and muzzle good colour, horns long, fine and not wide, good prominent brisket, too high on his legs, handling good.

SIR THOMAS FAIRFAX (5196), an even bull in general appearance, a little want of substance, not a good point, round barrel and good flank, not good in twist.

NORFOLK (2377), bad rumps, small girth and pot-bellied, short legs.

John Grey came and congratulated Bates that autumn on his great success at Oxford. Bates simply replied, 'I sent my cattle rejected at York to Oxford in expectation that those placed before them would have again made their appearance, but there were none of them there.' Grey said nothing, but took great umbrage at this gentle reproof. As a judge he invariably afterwards placed no matter what animal before one that had any trace of Kirklevington blood.[1]

The filial piety of his biographer has enshrined the memory of John Grey of Dilston in a veritable *saga*.[2] As a matter of fact nearly all the agricultural improvements attributed to him in the management of the Greenwich Hospital estates had been effected by the commissioners of the old school under the superintendence of native bailiffs. John Grey merely succeeded to 'a land for which he did not labour and vineyards and oliveyards that he planted not.' His administration was signalised by a dragonade against hedgerow timber, which destroyed some of the finest scenery on Tyneside without any regard for the advantages of shelter; by the pruning of Scots pine and other coniferous trees; by a useless outlay on field-drains four feet deep; by the discouragement of small farms; by the shameless neglect of the homes of the agricultural labourers,[3] and by the imposition on the

---

[1] Bates MS.

[2] *Memoir of John Grey of Dilston*, by his daughter Josephine E. Butler, Edinburgh, 1869.

[3] The cottages at Milfield, in 1810, were amongst the very worst in Northumberland—'dirty thatched hovels, the walls built with mud, and small stones of whin or granite gathered from the fields.'—*Memoir of the Rev. John Hodgson*, by Rev. James Raine, i. p. 71. On the Hospital estates, the hind on the home farm of the Langley Barony had to bring up a family of eleven children in a single dark room 15 feet × 14 feet, and a closet 8 feet × 11 feet. The

turnip-houses of semi-cylindrical roofs, formed with hollow bricks and cement, that would not turn the wet.

A coloured print of Duke of Northumberland (1940) was published by Messrs. Fores of Piccadilly. Bates considered this well executed, and a faithful but not flattering likeness. He forwarded them certain particulars that might, he thought, be pasted on the back of the print if framed:

THOMAS BATES *to* MESSRS. FORES

KIRKLEVINGTON, 25*th October* 1839.

Herewith I send you the pedigree of my shorthorn bull Duke of Northumberland (1940). I can state from measurements I took of the celebrated bull Comet (155) that the Duke was nearly double his weight both at ten months and at two years old.

I have undoubted information from the best authority[1] for saying that the Duchess tribe of shorthorns were in the possession of the ancestors of the present Duke of Northumberland for two centuries; and that Sir Hugh Smithson, the grandfather of the present Duke, kept up the celebrity of this tribe of cattle by paying the utmost attention to their breeding, and that he used regularly to weigh his cattle and the food they ate, so as to *ascertain the improvement made in proportion to the food consumed*—a system that I adopted nearly fifty years ago, not knowing that it had been previously done—and from a minute and close attention to this subject I obtained that knowledge of cattle which enabled me to judge of *their real merits* by their *external character*. From that knowledge I selected this tribe of short-horns as superior to all other cattle, not only as small consumers of food, but as great growers and quick grazers, with the finest quality of beef, and also giving *a great quantity of very rich milk*.

As a proof that this tribe of short-horns are also *good breeders*, the Duke's dam (Duchess 34th) had a calf each of the four years before the Oxford show—two bulls and two heifers—and has since the Oxford show brought *twin bulls*. Also (Duchess 30th) the dam of the yearling in-calver at Oxford has brought three heifer calves since her birth, having had *twin heifer calves* the same week that the Duke's dam brought *twin bulls*.

---

twelve children of the woodman were reared in a living-room 12 feet × 8 feet. Both families of course turned out remarkably well, but this does not justify 'a friend of the labouring classes' with practically unlimited resources.

[1] Lord Prudhoe, afterwards fourth Duke of Northumberland.

The two *yearling heifers* shown at Oxford (Duchesses 42nd and 43rd) are also both near calving, and have greatly improved since their return home after so long a journey.

The Oxford Premium Cow had already produced to Duke of Northumberland (1940) the bull-calf named Locomotive (4242), 5th October 1838, and Bates had sold this the March following for 100 guineas to Mr. J. C. Etches, then of Liverpool. After the Oxford meeting, she dropped another bull-calf, Duke of Wellington (3654), by Short-tail (2621), 24th October 1839. In the spring of 1840 Duke of Wellington was purchased by Mr. Vail, the president of the New York Agricultural Society. He was the first result of the union of the Duchess and Oxford tribes imported into America. With him came the cow Duchess by Duke of Northumberland (1940), dam Nonsuch 2nd by Belvedere.[1]

The object that Bates ever had in view was neither cow-jobbing nor pot-hunting. He merely wished to save the reputation of his herd from the charge of being afraid to exhibit it. The Duchess family had had their honour amply vindicated, and he would not risk the health of his best cows and heifers by exhibiting them. But there was still poor Red Rose the 13th, so summarily rejected at York, and prevented from seeking honours on the banks of the Isis; the males of the Oxford tribe had still to be proved equally valuable as the females, and the last of the first prizes at Oxford had been missed through not entering a bull-calf. Accordingly Red Rose 13th, Cleveland Lad (3407), and a young bull-calf (3407), by Duke of Northumberland 1940 out of Waterloo 2nd, set out for the Royal Agricultural Society's meeting at Cambridge 14th July 1840. Red Rose 13th, then six years old, was very poor in condition compared to the state in which she appeared at York in 1838, in consequence of having had a calf taken from her by a severe operation only a few weeks before.

Bates found that great offence was taken at his again

---

[1] Allen's *History of the Short-horn Cattle*, p. 185. See *Herd Book*, x. p. 141.

being an exhibitor at Cambridge after his brilliant success at Oxford. At the council dinner held at the University Arms before the exhibition opened, he declared that whatever was the verdict of the judges at Cambridge, he would send Cleveland Lad (3407) to the show at Liverpool in the following year. The judges were Mr. John Wright, Mr. Eaton Clarke, and Mr. W. Smith. Red Rose 13th gained the first prize for cows in a class of six, and the bull-calf, eight months old, was equally successful in a class of eight, including one from the Marquis of Exeter, whose bull-calf had won at Oxford. For the future Red Rose 13th was called Cambridge Premium Rose, while the prize Waterloo bull-calf received the title of Duke of Cambridge (3637). Cleveland Lad had not been trained, and his claims were promptly ignored by the judges, much to Bates's annoyance. They proceeded to award the £30 prize for the best bull in his class to Mr. Paul's Hero (4021), bred by Mr. Topham, an animal never heard of before or since.

A gentleman came up to Bates in the show-yard on Parker's Piece, and said, 'Had I been blindfolded, I could have told all your cattle by the feel of my fingers.'[1] 'As the stewards of the yard hear your remarks, I hope in future the judges will be blindfolded,' was Bates's reply.

Bates strongly objected to the fat and trained condition of the whole of the animals, and to the decisions of the judges in the classes in which he was not himself an exhibitor. The bull Locomotive (4242), then a year and nine months old, and a heifer, apparently Anecdote[2] by Duke of Northumberland (1940), dam Craggs, were exhibited by Mr. Etches of Barton Park and Mr. Foreman of Acton Burnell. Neither of these animals was noticed by the judges, who bestowed the premium for young bulls on Mr. Jacques's Clementi (3399), bred by Mr. Parkinson. It happened, however, that Mr. James

[1] Bell, p. 268.
[2] *Herd Book*, v. p. 34. Anecdote was afterwards bought by Mr. Icely, Australia.

C. Letton, of Paris, Bourbon County, Kentucky, had seen Duke of Wellington (3654) land at New York, and was so struck by him that, understanding that he had an elder brother left in England, he crossed the Atlantic on purpose, and immediately gave Mr. Etches the 250 guineas he asked for Locomotive (4242) in spite of his being 'ploughed' at Cambridge. On his return to America, Mr. Letton exhibited Locomotive (4242) ten times, and was invariably successful, though the best bulls in the States were brought out in competition against him.[1]

Messrs. Jobson of Chillingham, 'whose old-established herd had much deteriorated owing to the use of bulls with the Punch and Ben blood,' hired 2nd Duke of Northumberland (3646) from Kirklevington. Bates agreed to take two heifers for his use for three years or £50 a year at Messrs. Jobson's option. They exhibited the bull at the Northumberland Agricultural Show at Alnwick in 1840, and he took the first prize. Bates was present at the dinner. His health, proposed by the chairman, was drunk with enthusiasm. In returning thanks, he said that although he was now living in Yorkshire, his heart was still in Northumberland.[2] He endeavoured to rouse the dormant energies of the breeders of his native county, and to stop the oft-lamented decay of their cattle. His very liberal offers for the further use of the 2nd Duke of Northumberland met, however, with no response. A prophet has no honour in his own country, especially if it happens to be Northumberland.

After the Yorkshire Show at Northallerton in 1840, Mr. Holmes from Ireland, who was staying at Kirklevington, asked to see Mr. Thomas Robinson, formerly of March House near Scorton. As Robinson was living in Yarm next door to old Mr. Jolly, Bates asked them both to come and dine. After dinner, he requested Jolly to give them the history of his famous bull, which the veteran accordingly did.[3]

---

[1] Bell, p. 231. [2] Ibid. p. 239.
[3] Ibid. pp. 16, 17; see above, p. 47.

The honours taken by the Kirklevington cattle at the two Universities led to many inquiries for the blood:

THOMAS BATES *to* RICHARD STRATTON

KIRKLEVINGTON, 18*th December* 1840.

I have just received your letter respecting the twin bulls from Duke of Northumberland's dam.

I have let the younger[1] to a farmer in the county of Durham for two months, to leave here on the 1st of February next, and in the interim I am using him myself. The other twin[2] is under offer to go to Australia with a lot of heifers, daughters of Duke of Northumberland (1940), and I expect the Australian gentleman here to see them.

Now the price of these two bulls would be more than your cows could properly afford; I wish every one to be well repaid who gets stock from me.

My tenants, who got their cows from me and have the use of my bulls, have a bull-calf,[3] calved the beginning of April last, a fine roan colour, and a massive, thick carcase, very promising to make a useful bull, and soon fit to put to cows.

He is got by the bull I showed at Cambridge last year. His dam was by Belvedere, and her pedigree goes back to Mr. James Brown's Old Red Bull. Now you can have this bull-calf for fifty guineas and I will pay the passage to London, so if you have a friend there to see it landed and put it into a railway waggon, you will get it home at little expense. At present no steam vessels go from the Tees to London, but there are packet vessels. If you do not want the calf immediately, I should advise you not to have it sent till February, when I hope the steamers will again commence running from Middlesbrough to London.

I have other bulls, but they are high-priced; there is the bull-calf that got the premium this year at Cambridge,[4] but I asked 300 guineas for him at three months old, and more lately.[5]

From the time of the York Show of 1838, Bates had

[1] 4th Duke of Northumberland (3649).
[2] 3rd Duke of Northumberland (3647).
[3] George (3884) calved 15th April 1840, bred by Messrs. Bell, Kirklevington, got by Cleveland Lad (3407), dam Fletcher 2nd by Belvedere (1706), got by a son of Young Wynyard (2859) great-grandam descended from Mr. James Brown's Red Bull (97). George was afterwards sold to Mr. Etches.
[4] Duke of Cambridge (3637), sold to Mr. Icely, Australia.
[5] *The Stratton Shorthorns*, Newport, p. 22.

threatened that he would continue to exhibit his cattle until public opinion compelled the judges to do them justice. The list of rejected candidates had now been narrowed to one. That one, Cleveland Lad (3407), was sent to the Royal Show at Liverpool in 1841, as Bates had promised at Cambridge. The young bull came off with flying colours. Mr. Philip Pusey, M.P., presided at the council dinner, with Earl Spencer in the vice-chair. The awards of the judges having been read, the president rose and said:

Lord Spencer and Gentlemen—I have now the honour of proposing to you the health of a gentleman who has distinguished himself as a competitor with stock this day. Though not one of the best judges of stock myself, I know enough about it to be aware how much has been done by the Royal Agricultural Society of England for the general benefit of the country. (*Cheers.*) I know, as I daresay all of you do, that there are certain proofs of the excellence of stock, and I know that very great improvements have been made in our breeds, through the efforts of this Society and those of the Smithfield Club. I am unable to state what are all the recent improvements which have been made, but as I was accidentally reading the other day an old history written in the time of Queen Elizabeth, I could not but be struck with one or two curious facts which are worth referring to by way of contrast. The author says that in those days 'England was well known for far surmounting other countries in the breed of cattle, as may be proved with ease, for where are oxen commonly more large of bone?'[1] (*Laughter.*) He then proceeds—'In most places our graziers are now grown to be so cunning, that if they do but see an ox or bullock, and come to the feeling of him, they will give a guess at his weight, and how many score or stone of flesh and tallow he beareth—how the butcher may live by the sale, and what he may have for the skin and tallow—(*laughter*)—which is a point of skill not commonly practised heretofore. Some such graziers also are reported to ride with velvet coats and chains of gold about them.' Gentlemen, I think we have made considerable progress in the weight of cattle, and in judging of their weight since old Holinshed's day, but here follows a point in which I doubt if we have made much progress. He says, 'And in their absence their wives will not let to supply those turns with no less skill than their husbands—(*great laughter*)—which is an hard work for the poor butcher, sith

[1] Holinshed's *Chronicles*, ed. 1807, England, vol. i. p. 369.

he, through this means can seldom be rich or wealthy by his trade.' He proceeds to state that our own cattle have long had the advantage of others: 'Their horns are known to be more fair and large in England than in any other places, except those which are to be seen among the Paeones,[1] which quantity, although it be given to our breed generally by nature, yet it is now and then helped also by art.' So we see in those days science was called in to the assistance of agriculture. (*Laughter.*) I don't, however, think much of their science, because they say, when beasts 'be very young' breeders 'will oftentimes anoint their budding horns or tender tips with honey, which mollifieth the natural hardness of that substance, and thereby maketh them to grow to a notable greatness; certes, it is not strange in England to see oxen whose horns have the length of a yard or three feet between the tips.' (*Renewed laughter.*) Gentlemen, it is not often we see them now. Our first improvement in cattle was in the long-horned breed. I believe they are now little seen. There was one long-horned beast at the last Smithfield Show which was much looked at.

There are few, I presume, to whom we are more deeply indebted for improving the Roast Beef of Old England than the subject of this toast. Gentlemen, I beg to propose to you the health of the successful competitor in Class I, Mr. Bates of Kirklevington. (*Applause.*)

Bates replied:

I beg to return most sincere thanks on behalf of myself and the other successful candidates for the honour you have done us. I may congratulate this assembly on the glorious prospects which are held out to the whole world as the results of our association; for I feel convinced that by the improvement of agriculture we affect the interests of the entire people of the earth. (*Hear, and cheers.*) We have arrived at that glorious era when the welfare of every man is dear to his fellow-creatures. I hope the Royal Agricultural Society will do its utmost to encourage that feeling. (*Hear, hear.*) It is by that alone England has risen so high in the estimation of the world. It is by that alone that she can rise still higher; and I congratulate you on the efforts now being made by the Society to secure that end. I only wish those efforts were increased a thousandfold.

With his experience of the judging at Cambridge, Bates had vowed that after fulfilling his promise to send Cleveland Lad to Liverpool he would never again exhibit

[1] In Macedonia. The classical reference given is Athenaeus, lib. x. cap. 8.

anything at a Royal Show until the whole system of deciding on the merits of breeding stock was radically reformed. His old connection with Scotland induced him to take his Oxford Premium cow to the Highland Society's Meeting at Berwick in 1841. 'The buffoonery of the awards' exceeded that of all agricultural exhibitions he had ever attended: 'an ignorant nobleman dictated to the judges what they were to do, and they obeyed his dictates as apes would have done under similar circumstances.'[1] Fortunately for Bates the portrait of the champion bull, Buchan Hero (3238), is given in the *Herd Book*, and fully corroborates his opinion. The Oxford cow was beaten by Mr. Booth's Necklace; her detractors alleging that she was 'deficient in girth and gaudy behind.'[2] Bates took part, however, in the Yorkshire Society's Show at Hull in 1841. Cleveland Lad (3407), then four years and seven months old, took the champion prize for bulls over Mr. Jacques's Clementi, which had won the premium for young bulls at Cambridge. The yearling Duke of Cambridge (3637) was also first in his class at Hull, one of Mr. Booth's young bulls being 'especially commended.' The honours of the three-year-old cow class were divided between Duchess 42nd and Duchess 43rd. Mr. Booth's Bracelet won the prize for old cows, for which Bates did not compete.

Bates had always been on the most friendly terms with Mr. John Booth of Killerby, who frequently visited Kirklevington and was an annual guest at the Yarm Fair in October.[3] Their views were far from being the same. Bates had only the interest of his herd at heart, and if it only improved in his own estimation, was indifferent, as has been said, to its being a money-making concern; the Booths, on the other hand, knowingly immolated their best females on the altar of the Royal Agricultural Moloch in order to advertise themselves with the public.[4] John Booth

[1] Bates MS. fo. 14.   [2] *Saddle and Sirloin*, p. 154 n.
[3] Bell, p. 258.
[4] *The History of the Rise and Progress of the Killerby, Studley, and Warlaby Herds of Shorthorns*, by William Carr, 1867, p. 94.

took a delight in bantering Bates, especially about his refusing to exhibit any longer at the principal shows. Alluding no doubt to the absence of any old cows from Kirklevington at Hull, and to the discomfiture of the Oxford cow at Berwick, he told Bates that he dared not show a cow at York in 1842, or if he did, he had a rod in pickle for him.[1]

This rod in pickle was none other than 'the all-conquering Necklace,' then four years old; the twin sister of Bracelet, who in her 'victorious career' won sixteen prizes and one gold and three silver medals. Among Booth men, it was a moot question 'to which of the world-renowned twins could be justly awarded the palm of beauty.'[2] Necklace, it appears, had neater fore-quarters, and was better filled up behind the shoulders; Bracelet had fuller, longer, and more level hind-quarters. Through Albion (14) they could boast of direct descent from Colonel O'Callaghan's heaven-sent red Galloway cow, which had so miraculously regenerated Charles Colling's effete herd at Ketton. Their palates were so refined that in the height of summer, on the rich pastures of Swaledale, they never lacked a daily *bonne bouche* of corn and cake.[3] Who dared then challenge the right of Necklace, the best cow, perhaps, that the Booth herds ever produced, to be proclaimed Queen of Beauty in the lists of the Yorkshire tiltyard, as in the heyday of her Royal victory at Bristol she was borne in her triumphal car towards the ancient capital of Imperial Britain?

Now there was in milk at Kirklevington a ten-year-old unregenerate dairy-cow which had never been shown nor had ever been intended to be. When about twelve months old she had broken her leg, and as Bates would not employ a veterinary, Thomas Bell set it with help of the journeyman miller. For some years she had scarce ever tasted a turnip in the winter months.

[1] Bell, p. 258.
[2] Carr's *Killerby, Studley, and Warlaby Herds*, pp. 29, 30.
[3] *Ibid.* p. 29.

Since May Day she had been going in the ordinary cow pasture, and was as ignorant as any northern farmer of what a *bonne bouche* meant. Without any preparatory training at all, old Brokenleg (Duchess 34th) walked by road, about forty miles, to York, in the company of her son, Duke of Northumberland (1940). The judges ordered the fifteen cows entered to parade twice round the ring,

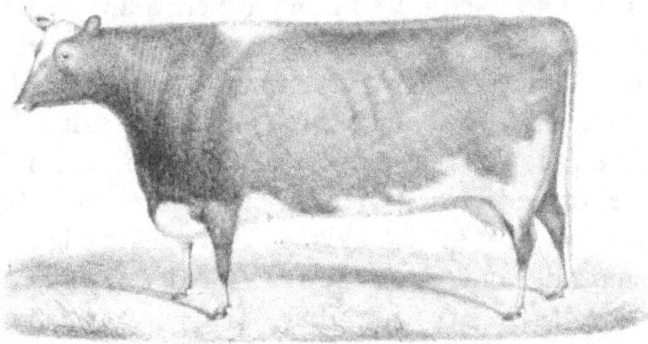

DUCHESS 34th.

and then told old 'Tommy Myers,' the Kirklevington cowman, to stand on one side with Brokenleg.[1] A murmur of indignation broke from the people present, who imagined she was being excluded from the prize-list.[2]

Myers remained for half an hour or so thinking, as he said, 'they were gannin' to use me very badly,' while the judges kept disputing over Necklace and one of Mr. Mason Hopper's cows.[3] 'They could not rightly judge of stars in the presence of the sun.'[4] Myers, who had supposed they were determining which was to be first and which second, was greatly relieved when they sent Brokenleg 'the white rose,' and placed Necklace behind her. 'When the crowning trophy was placed on Duchess 34th's head, there was a burst of applause. She was as like the first Duchess as two animals could be, in colour and in that grandeur of style and appearance such as no animal ever had except a Duchess.'[5]

[1] Bell, p. 261.
[2] Letter of William Charlton of Sutton Farm, Essex, to Thomas Bell, 26th October 1872.   [3] Bell, p. 261.
[4] Letter of William Charlton, 26th October 1872.   [5] *Ibid.*

Bates had good reason to be satisfied with the result of the tug-of-war when Killerby met Kirklevington. It was the only challenge he ever accepted. That the decision was perfectly just was confirmed by Mr. Eastwood, a breeder who had as much admiration for one line of stock as for the other, so long as the animal was a good one, but who thought that a little weight should be allowed to fashion. Mr. John Booth asked him why it was that Brokenleg beat Necklace. 'Well,' he replied, 'I think, Mr. Booth, you are fairly beaten; if I had been one of the judges, I should have done the same.' 'Then,' said Booth, 'I am satisfied.' Bates came up shortly afterwards and asked Eastwood the same question. 'I think you won fairly, Mr. Bates.' 'I am pleased to hear you say that.' 'I told Mr. Booth so.' 'Then,' said Bates, 'I am more pleased still,' and the great rival breeders remained the best of friends.[1]

Duke of Northumberland (1940), who had only received the two-year-old prize from the Yorkshire Society in 1838, was now acknowledged the champion bull in 1842.

La boucherie is the natural end of a policy of *bonne bouche*. Poor Necklace never bred, and went straight from gaining a prize as extra stock at the Royal Show at Newcastle in July 1846, to the Yorkshire Show at Wakefield in August, where she won as the best fat cow or heifer of any age, finishing her career at Smithfield.

HENRY LISTER MAW *to* THOMAS BATES

TETLEY, CROWLE, 23*rd January* 1847.

I was in London at Christmas as a competitor at the Smithfield Show with an ox. He was, however, not fat enough, although he was one of the first beasts, if not the first beast, sold, proving that the butchers did not consider him a bad one. The judges could not give him a premium for fat. The feeding was certainly excessive. Some of the animals were said to have been fed upon sugar; and, although I think feeding is all fair enough at Smithfield, we ought all to feed alike. You are aware that

[1] Thornton's *Shorthorn Circular*, vol. ii. p. 645.

Necklace, after having received premiums at Newcastle and Wakefield,[1] got the gold medal at Smithfield. I do not think she was the best beef, although a good cow; but it proves what the state of the animals is at *Breeding Shows*. They are in fact fed until they will not breed. I understand Bracelet's last calf was not much larger than a cat; she dislocated her stifle joint and was slaughtered. It is a pity to feed animals to such an excess as to destroy them. They cannot go on breeding. I bought the second prize boar at Newcastle and a sow to cross with some pigs I have. Davies made paintings of both of them for some periodical by the editor's directions, but I have not been able to get either of them to breed.

In September 1842 Mr. Henry Colman from America called at Kirklevington on his way back from Scotland. He examined the stock day after day, and Bates, seeing that he was 'earnest to come at the truth,' immediately answered all his questions. On the morning of his departure, Colman said to his host, in the presence of Mr. Phillips, Lord Feversham's agent:

'I came from America much prejudiced against shorthorns, and against yours in particular. I thought them too highly extolled, and nothing but close and repeated examinations of your stock could have removed my prejudices. I had no conception that there could be any such breed of cattle as yours, though I have just come from attending the meetings of the Royal Agricultural Society of England, and of the Yorkshire and the Highland Societies.'

Bates told him to be careful not to say this again in England. Colman then strongly urged him to write a history of shorthorns and to publish it in America. Bates would not promise, but afterwards said he thought of doing so. He wished his intention kept secret so that the work might 'come before the world as a surprise and be unanswerable as a document.'

---

[1] Mr. Carr seems to have been mistaken in representing Birthday, and not Necklace, as the first-prize extra cow at Wakefield, *Killerby, Studley, and Warlaby Herds*, pp. 31, 148. Birthday dropped Lord George to Fitz-Leonard, 2nd June 1846, *Herd Book*, ix. p. 271, so that it is most unlikely that she competed in the fat cow class on the 5th of August.

Cleveland Lad (3407) had been in service with Mr. G. P. Harrison of Low Field, near Piersebridge, and Mr. William Harrison of Mortham. His portrait, which Bates sent to the *New Farmer's Journal* in August 1842 with a letter giving an account of the Oxford family, was followed by those of Duke of Northumberland (1940) and his dam, Duchess 34th, and the history of Duchess's family. Bates tells the editor, 11th November 1842:

I named this bull 'Duke of Northumberland' to perpetuate the commemoration that it is to the judgment and attention of the ancestors of the present Duke of Northumberland that this country and the world are indebted for a tribe of cattle which Mr. Charles Colling repeatedly assured me was the best he ever had or ever saw. As a proof that they have improved under my care, I may mention that 'Duke of Northumberland's' dam consumes one-third less food than my first Duchess, purchased in 1804, and that her milk yields one-third more butter for each quart of milk, while there is also a greater growth of carcase and an increased aptitude to fatten.

It is now above sixty years since I became impressed with the importance of selecting the very best animals to breed from. For twenty-five years afterwards I lost no opportunity of ascertaining the merits of the various tribes of shorthorns. It was only then that this could be done. There is scarce a vestige now remaining of the many excellent cattle then in existence. Since I became possessed of the tribe, I have never used any bull that had not Duchess blood—except Belvedere (1706), and he was the last bull of a long race of well-descended shorthorns—without perceiving immediately the error.

As the post-hour draws near, I must conclude in order to enable you to print this letter in the same paper with the portraits of the 'Duke' and his dam.

I do not expect any artist can do them justice. They must be seen, and the more they are examined, the more their excellence will appear to a true connoisseur; but there are few good judges — a hundred men may be found to make a Prime Minister for one fit to judge of the real merits of shorthorns.

In general Bates claimed that, during thirty-five years of breeding, he had effected an improvement in shorthorns whereby with a third less consumption they gained a third more weight, and that while their milking qualities were unimpaired, the milk yielded a third more butter.

He can hardly be accused of prejudice in the selection of his tribes of cattle, seeing the immense number of experimental purchases he made:

THOMAS BATES *to* GEORGE VAIL

KIRKLEVINGTON, 1*st December* 1843.

I have sent a bull I bought to the College of Ohio,[1] and nine cattle in all I have sold in America of my own breeding. The tribes of really good short-horns are very few. I have tried myself above two hundred varieties; out of these I have kept the ten best, not one in ten. Six of the best tribes I never mean to part with till I decline breeding altogether. This at my time of life must be in ten years at the latest, though I enjoy the best of health. Of the four varieties I have to part with the best is the family of Foggathorpe.[2]

Any bull or cow I have to spare is at your service. You can have good ones better than the one you have got—three or four of the best blood, by Duke of Northumberland and other good bulls, at 100 guineas each; but the three I have offered of the Foggathorpe family are the cheapest. I have ever found the best the cheapest, whatever the cost.

Cleveland Lad 2nd (3408), who had been in service with Mr. Nicholas Burnett at Black Hedley, was placed second to Mr. Forrest's Symmetry (5389), the champion bull at the Yorkshire Show at Doncaster in 1843.[3] He was entered at the Royal Show at Southampton in July 1844, but the first prize in the old bull class there was given to Mr. Cooper's Strelly (7560), an animal known to have hardly any shorthorn blood in its veins, and the second to Mr. Hayter's Standish (7550). A month

---

[1] This was the Skipton Bridge Bull (5208).

[2] The Kirklevington Herd (at the home farm) consisted at this time of ten tribes—the Duchess, Cambridge Rose, Red Princess, Oxford, Wild Eyes, Rosalind, Foggathorpe, Waterloo, Blanche, and White Rose (Secret). It would seem that the first six were those Bates then considered 'his best tribes,' as he sold off his Blanches and White Roses and some of his Waterloos. The Red Princesses and Rosalinds died out.

[3] The judges were John Grey, Dilston, Robert Cattley, Stearsby, and William Torr, junior, Riby. There is a portrait of Symmetry in the *Herd Book*, iv. p. 497.

later Cleveland Lad 2nd was permitted to be the champion bull at the Yorkshire Show at Richmond.[1] Mr. Richard Parkinson of Babworth, near Retford, young Mr. Torr of Riby, and Mr. Anthony Maynard, who were the three judges at Richmond, 'took every possible pains in examining the different animals, and had sufficient judgment and honesty to act uprightly.' Within two months, however, of the show at Richmond the same bulls were exhibited at the Durham Agricultural Society's meeting at Stockton, and one unnoticed at Richmond was placed before the two that had gained the first and second premiums there. Mr. Parkinson had had great reason to be dissatisfied with the decisions of the judges at Durham in 1843, when Sir Thomas Fairfax (5196) himself was placed in the background, behind Harlsonio (6055),[2] a bull that in Bates's opinion was completely destitute of good shorthorn character. Still, the decisions at Durham in 1843 were praiseworthy in comparison to those at Stockton in 1844, and Bates held that after such frequent barefaced unprincipled conduct, agricultural societies ought to take some other mode of ascertaining merit in cattle than the opinions of incompetent men, appointed by some of the exhibitors themselves.[3]

In the autumn of 1844, Mr. Banks Stanhope of Revesby Abbey, near Boston, came to Kirklevington in the company of his neighbour Mr. Dudding of Panton, whom Bates had often had the pleasure of entertaining. Mr. Stanhope wished to purchase two cows, each having a cross of Norfolk (2377). As Bates, it will be remembered, had fed off Nonpareil, Norfolk's dam, and had found her 'the worst grazer he had fed for forty years,' he offered these cows at a lower price than he would have done had

[1] The Colonel (5428), bred by Mr. W. Raine, was placed second. Mr. Thomas Bell's General Sale by Duke of Northumberland, dam by Cleveland Lad, took the first prize in the calf class, while among the three-year-old cows Foggathorpe 2nd was second to Mr. John Booth's Birthday.
[2] Bred and exhibited by Mr. John Beetham of West Harlsey. There is an engraving of it in *Farmer's Magazine*, August 1844.
[3] Letter of Thomas Bates to the *New Farmer's Journal*.

they been free of Norfolk's blood. This Bates did not conceal from Mr. Stanhope, though it was to his own disadvantage. The two cows in question had had a subsequent cross of Duchess blood, and this had greatly restored them to the high merit possessed by their tribes previous to the Norfolk cross. Bates did not wish to see such valuable blood again deteriorated, and asked Stanhope what bull he meant to put to them. Stanhope replied that Cramer (6907), bought of Mr. John Parkinson, of Ley Fields, near Newark, was the bull he used. Not thinking Cramer's blood good either on the sire's side or the dam's, Bates said he would much rather that Stanhope sent the cows if he bought them to any of his own bulls that he chose, and he should have them served gratuitously.[1] Finally, it appears, Stanhope bought Waterloo 5th in calf at the time to 4th Duke of Northumberland (3649). She brought him a bull-calf Duke of Lincoln (7993) in March 1845, and subsequently bred, by the bulls Bates objected to, two bull-calves of no moment.

From Kirklevington, Stanhope went straight to Mr. Parkinson at Ley Fields, and told him of Bates's objection to Cramer (6907). Parkinson roundly accused Bates of libelling his herd and challenged him to exhibit superior animals.[2]

Thomas Bates *to* John Parkinson

Kirklevington, 14*th October* 1844.

You are wrong in stating that I have written to disparage your short-horned cattle generally. I do not know your short-horned cattle generally. In respect to particular blood, I have ever spoken and written of it as my experience convinced me. I did so forty years ago, which I imagine was before you were a short-horn breeder. You are, I presume, a son of Mr. Parkinson the owner of Sir Thomas Fairfax, and I think I gave him my opinion of that bull long before he was the purchaser. What I said to Mr. Stanhope, I said for his good and not to disparage your herd. Further experience may convince him, as it has others, that I was right, though they did not think so when I gave my opinion. It is not the first time that my motives have been misconceived.

[1] Letter of Thomas Bates to the *New Farmer's Journal*.   [2] *Ibid.*

As to accepting challenges, I have always declined to do so, except any great public advantage was to be obtained. After exhibiting successfully for years together, I ceased doing so for twenty-six years. The best cattle at the time I gave up exhibiting in 1812 are now nearly extinct. All I said of them has proved true, and such will be the case again with the stock that has sprung into repute during the last forty years. I have no ill-will to you, but merely express my opinion founded on an experience of above sixty years. This incident may induce me in future both to decline showing as well as speaking of stock. My object in both cases has been to benefit others and not myself.

The only answer Bates received to this temperate letter was an anonymous paragraph communicated to the *Farmer's Journal*, 11th November:—

SHORT-HORNS.—Mr. Bates having lately expressed himself very strongly against the merit of Cramer (by Sir Thomas Fairfax (5196), dam Cassandra by Miracle (2320)) to Mr. Banks Stanhope; his owner offered to show that bull against any of Mr. Bates's bulls under nine years old, at the next meeting of the Yorkshire Society at Beverley; and Mr. Parkinson, the breeder of Cramer, proposed also to show against Mr. Bates thirty short-horned cattle, each on their respective farms, and although it was declared in each instance that the terms of showing might be such as Mr. Bates considered unobjectionable, he has declined exhibiting in either case. CORRESPONDENT.

Justly nettled at this attack, Bates immediately forwarded a copy of his letter to Mr. Parkinson to the paper, and then, at the special request of the editor, gave his reasons for objecting to Cramer's blood:

THOMAS BATES *to the* EDITOR *of the* FARMER'S JOURNAL

KIRKLEVINGTON, 21*st November* 1844.

If the parties, sheltering themselves under the signature of *Correspondent*, wish to have the demerits of the blood of Cramer (6907), I can furnish them to their hearts' content. In the room where I am writing, a breeder from the same blood as on his dam's side, declared to me that he had lost more than ten thousand pounds by breeding short-horns, and he began with one of the best tribes of short-horns then in existence before Robert Colling went to Barmpton. Nor is this a solitary instance, for two near neighbours of mine when I lived at Halton Castle in Northumberland had two very large herds of the hardiest and best constitutioned short-horns, and these were prolific up to their introduction of St. John's blood in 1811. After this one of these breeders lost for two years in succession all his calves except two; the other's short-horns all lived, it is true, but the former always said that he was the greatest gainer of the two, as they became the most delicate and unthriving stock I ever saw. Many breeders who have this St. John's blood have resorted to coarse ill-bred cattle to try and restore their constitutions. This has filled the country with the *improved short-horns* which now generally prevail, and which are in reality the very worst breed of cattle.

I might fill all the columns of your paper by citing instances of the ruin of breeders and the destruction of valuable herds by the use of the same blood as was in Cramer's dam. I shall now state my objections to his sire, Sir Thomas Fairfax (5196). I saw at Mr. Whitaker's his bull called Fairfax (1023), the grandsire of Sir Thomas, and asked him if he had ever used that bull. He clenched his fist, turned round to me, and with great violence said, 'Do you think I would use a bull of a tribe that do not give a drop of milk? I have never used him to my herd of short-horns, nor ever will.' The great-grandam of Fairfax (1023) was, he continued, a cow between a blue and a black colour which he had purchased of a jobber who said he had got her at Stockton. Mr. Whitaker asked me at different times to try and make out her pedigree if she had one. I made inquiries for many years, but no one had ever heard of this blue-black cow. I was repeatedly told that she could not have been a short-horn, for a short-horn was never known of that colour. After all inquiries had proved fruitless, Mr. Whitaker entered the cow in the *Herd Book* as 'supposed by Chapman's son of Punch,' but without any evidence whatever. This blue-black cow was put to Western Comet (689), a bull descended from Colonel O'Callaghan's polled Galloway cow, but with a short-horn cross still worse, for it was from one of the hardest-skinned handlers I ever felt, which gave no milk worth noticing. Now, I will ask seriously whether any one can expect to breed good short-horns from such blood? The dam (Miss Fairfax) of Sir Thomas Fairfax (5196) was a bought cow, not bred by Mr. Whitaker, and where are any good animals of note to be found among her ancestors? On Sir Thomas Fairfax (5196) being knocked down to Mr. Parkinson, senior, at a public sale, he afterwards said to me, 'I know you don't approve of this blood,' and I replied, 'Certainly not.'

When too Mr. Booth told me he had sold his half of Sir Thomas to Mr. Parkinson, senior, I congratulated him, but neither of these gentlemen were offended. It is with great reluctance that I make these disclosures, from my dislike to controversy, and that I do so your *Correspondent* is to be thanked.

By extra feeding the worst of animals may be forced forward to gain the applause of incompetent judges. How the animals have been fed is what we are never told. It may be they have been forced with Indian corn, which I have never used. It is the same with milking trials; a cow that, according to the Rev. Mr. Berry, gave twice a day sixteen quarts of milk at Mr. Whitaker's, never gave more than four quarts when going in my cow-pasture at Kirklevington, and that immediately after calving.

I declined the challenges of Mr. Banks Stanhope and Mr. Parkinson, junior, because I saw no good likely to arise from accepting them. Had I done so, every conceited, purse-proud young coxcomb who began short-horn breeding, without knowing in what a good short-horn consisted, would probably have sent me similar challenges, and surely the public cannot expect that I should accept them? Mr. Henry Lister Maw of Tetley, near Crowle, tells me that he has offered to accept the challenges sent me, and reminds me that more than a year ago I told him that I had made it a rule never to enter into matches and sweepstakes, and advised him to do the same.

CHAPTER IX

KIRKLEVINGTON: LAST YEARS

1844-1849

AGE seemed to sharpen rather than otherwise the keen interest which Bates always took in agricultural politics. He remembered the distress, felt both by owners and occupiers, during the American War which ended in 1785, and the return of prosperity under the administration of Mr. Pitt. He had no doubt that if Pitt had lived to see the conclusion of the great continental wars in 1815, his clear head and sound judgment would have given, as a legislator should, an equal protection to property of every kind. It was in his opinion entirely owing to the flourishing state of British agriculture that the nation had been able to carry those wars to a prosperous issue, and free the world from the military tyranny of Bonaparte. The crucial problem in 1844 was how to properly reconcile the two great interests of Land and Money; the depression of the former for twenty-five years had advanced the interest of the latter in an unprecedented manner, and it was time that a compromise should be effected between them. Lord Brougham, in the House of Commons, had described farmers as 'quiet creatures who came at the call of the Treasury to be shorn,' and even the landlords in both Houses of Parliament showed the same degree of passive submission. Bates endeavoured to rally them to the cause of Agriculture, and especially called on Sir James Graham, now that he held an important post in the Administration, to keep to the texts he had so ably laid down in his work on 'Corn and Currency.' If no one else would undertake the task, and any constituency approving of his views would send him to Parliament, Bates, though nearly seventy years of age, was ready to do his duty.[1]

As a pamphleteer, his style was greatly too diffuse and involved, but he went straight to the root of matters, and his remarks deserve more attention than most of the bi-metallic and pseudo-protectionist literature of modern times:

TO THE MEMBERS OF BOTH HOUSES OF PARLIAMENT AND THE AGRICULTURAL PROTECTION SOCIETIES THROUGHOUT THE UNITED KINGDOM.

I did hope, when the occupiers of land had at length been roused to form themselves into Associations for the Protection of Agriculture, that through their representatives in the Lower House of Parliament, and by their influence with the members of the House of Lords, A GOVERNMENT PAPER CURRENCY would have been adopted. Based upon the property of the country, and payable in bullion at the market price of the day, this issue of paper, for all sums of one pound and upwards, might have been increased or diminished as need required, being ever under the control of the Government, and the greatest advantages would have accrued to all classes. But instead of such a beneficial measure, the Legislature, influenced by Sir Robert Peel, has prohibited all new Banks of Issue being established, and has compelled the Banks in existence to confine their issues in the future to the small sum already in circulation.

The Agricultural Protection Societies, when newly formed, were perhaps scarcely aware of the power they possessed. Now that they begin to feel their strength, I entreat them not to spend that strength in petitioning for the abatement or abolition of the duty on malt. From my own experience of above fifty years, as well as from that of all north-country farmers, malt liquor does not increase the comforts of the agricultural labourer or enable him to do more work, but the very contrary. I am firmly of

[1] Letter, dated Kirklevington, 22nd March 1844, printed in the *New Farmer's Journal*, No. 161, new series, p. 5.

opinion that instead of being a blessing, the removal of the Malt Tax would be a curse, encouraging drinking and dissipation and leading to penury and want. Taking all duties off soap and candles would benefit the working man, and now that an extended intercourse with China is opening out, an abatement of the duty on tea would increase the consumption exceedingly, and the more abundant supply thus required would enable the Chinese to take a larger amount of our manufactured goods in exchange, to the mutual benefit of the inhabitants of both countries.

What, however, I particularly wish to fix upon your undivided attention is the necessity for THE ESTABLISHMENT OF A SINKING FUND to pay off the principal of our National Debt. Till a sinking fund be established, the country can never flourish.

In 1785, when Mr. Pitt brought forward the measure of a sinking fund, I well remember how violently it was at first opposed. The increased taxation was complained of, but the measure soon proved Mr. Pitt's wisdom and foresight. The three per cent stock rose in seven years to ninety-seven, and the increased prosperity of the country, as exhibited by him as Chancellor of the Exchequer in 1791, established for ever his reputation as a statesman. It is only by following the example he set us that universal happiness and contentment can again be produced by providing and ensuring work for all the industrious classes equally, instead of aggrandising the cotton-lords, and driving the rural population into the manufacturing towns. Since the peace of 1815, the manufacturing population has increased threefold, whilst the agricultural has greatly diminished.

I, who well remember the highly flourishing state of our agriculture from 1785 to 1815, deeply deplore the sufferings all persons engaged in agriculture, whether landlords, tenants, or labourers, have had to endure for thirty long years, while the cotton-lords have been reaping golden harvests, almost without interruption. Is such a state of things to be continued? Have the agriculturists for ever to endure a bondage more severe and humiliating than that of the Israelites under their task-masters in Egypt?

The present ministry were placed in office by the agricultural interest in 1841. I freely own I dreaded Sir Robert Peel's coming into power, notwithstanding the high encomiums he paid the suffering tenantry for their patience, prudence, and economy, and that he particularly dwelt upon their self-denial in his after-dinner speech at the Royal English Agricultural Meeting at

Cambridge in 1840. When in 1841, before the general election, I urged the electors not to send party men to Parliament, but those upon whose wisdom and discretion they could depend for the defence of their best interest against improper ministers of the Crown, it was Sir Robert Peel I had in my mind's eye.

Alarming depression and distress are now reiterated wherever you go. The Corn Laws should never have been altered without the Money Laws being changed at the same time. Low prices are sure to follow the fixed diminution of the circulation, and the moneyed interest will eat up the whole landed property.

The head of the present administration was returned to power by the agricultural interest, yet his proposed Corn Laws and New Tariff have brought us into our present situation. His Currency Law of last session renders our escape impossible unless the remedy I proposed be acted on, viz.: *the taxing of the fundholder to the amount of at least one-third of his income as a sinking fund to redeem the National Debt.* This would reduce all interest on dormant securities to two per cent, and enable landowners to lower their rents or, better still, to spend a third of their income in the improvement of their estates.

The annual value of landed property has been decreased by the fall of agricultural produce fully sixty per cent since 1815; while the fundholder's property has increased in value to far more than sixty per cent. For years the fundholder has been purchasing all agricultural produce at less than half the prices he paid when the loans were contracted during the late war, and in future he will purchase agricultural produce at two-thirds less. It requires, indeed, five times the quantity of wheat at present prices to purchase an equal quantity of stock in Three-per-Cent Consols that it did thirty years ago. Well, therefore, may the fundholder allow one-third of his interest to be taken as a sinking fund, and my firm belief is that he would eventually be the greatest gainer of any class in the community; soon after the sinking fund was established the funds would rise more per cent annually than would be lost in interest, as was proved when Mr. Pitt, in 1785, established a sinking fund to a far less amount. The whole National Debt would be paid off before fifty years expired, and then not one-fourth of the revenue would be required to be raised by taxation.

The reason why I urge this measure at present is because the public papers have announced that the Government contemplate taking one quarter per cent from the Three-per-Cents, as was done last year with the Three-and-a-Half-per-Cents, and guaranteeing that interest be not diminished for ten or twenty years. This is one of the most artful measures that could have been thought of to bolster up the interest of money, and if it be adopted, our National Debt will be continued for ever or until the nation can no longer bear the burden. It would be well if the fundholders themselves could see the risk they run of the debt being wiped out once for all with the depression in price of agricultural produce. A more unchristian and inhuman thing than war cannot be imagined. Loans to carry on war are the result of a combination of avarice and bloodthirstiness. Yet what has Britain been doing since loans were first contracted in 1796 but impoverishing herself by borrowing money for the avowed purpose of shedding human blood? Had the nation been obliged to raise the means annually, wars would soon have ceased. What has England gained by the many hundreds of millions she has spent in warfare since 1797? Had a like sum been expended in improving our native soil, the population would by this time have quadrupled, and all had plenty to live upon. No man can say to what extent land may not be improved, more particularly by the improvement of live stock, which is the basis of all real improvement.

It is only by agricultural prosperity that universal prosperity can be attained. Never was there a greater inclination than there is at present to use every effort to promote agricultural improvements. Numerous are the hands wanting employment, and the large capitalists know not how to employ their wealth with safety. A sinking fund of ten millions annually would lower the interest of money so that forty millions might be employed annually in agricultural improvements to the benefit of every class in the United Kingdom. A greater sum was annually so employed during the late war and that principally by the occupiers; but the reduction in the price of all agricultural produce is fully sixty per cent since 1815, so that there are not two-fifths of the means left. It signifies not to the occupier how low the price of corn falls, if rents and the interest of money fall accordingly.

At the conclusion of the war in 1815, and for some years before, the three-per-cent funds stood at fifty-six in a depreciated currency. A guinea then brought in the market 28s., so that in the standard the funds really stood at 42; now they exceed 100, nearly of two and a half times as much. On the other hand wheat, in August 1812, averaged 208s. per quarter, or 26s. per Winchester bushel, and no contracts for land to rent were calculated at less than 108s. per quarter. On this basis, thirty years' purchase was given for land, which in 1823 did not pay two per cent interest and which, with our present prospects, is not likely to pay one per cent.

It is the corn growers and the industrious classes who pay our oppressive taxation, consequent on the enormous National Debt which never could have been contracted in a gold currency. The fundholder can now buy all agricultural produce at one-third of what he did when the debt was contracted, if he lives in this country; but many fundholders live abroad because they can do so at half the cost by avoiding our excessive taxation. Many honest and once opulent farmers, and even labourers have been reduced to beggary; while those who quitted their native land while they had property left and settled in foreign lands or in our colonies, now enjoy prosperity, contentment, and happiness.

Large tracts of land are about to be occupied by Englishmen in the Prussian dominions on the Elbe, because by sending the produce to the English market, they can have ten times the return for their capital that they could make of it by farming in England. And are Englishmen at home to be ruined, one after another, by occupying lands in their native country? Surely not: the remedy is in their own power; they must call on the Legislature to take one-third of the interest of the National Debt as a sinking fund. As the Debt diminishes, our burdens will be lightened, and before seven years elapse, new life and energy will be infused into every transaction. Prices of agricultural produce will, it is true, continue to fall in consequence of the increased supplies resulting from the large expenditure of capital on agricultural improvements, and the price of labour may also fall, but the labourer's wages will enable him to buy agricultural produce in proportion, without diminishing his comforts. The abatement of the rate of interest will enable landowners to lower rents, equivalently to the fall in the price of produce; yet they will be in reality as well off as ever and able to keep up their establishments by reducing the wages of their domestic servants and other expenses equivalently with the fall of rent. And when the Debt becomes extinguished, contemplate the welfare of all classes with not one-tenth of the present taxation. Sir James Graham pointed this out twenty years ago, and had Ministers then acted as men ought to have done and followed his advice, how different would have been the present situation of the country.

Had William Pitt lived to have seen the conclusion of the late war, he would long ere this have freed the nation of its debt by a sinking fund. Pitt had to contend with both parties in the State combined against him, and Sir Robert Peel and Lord John

Russell may again form a coalition to keep themselves in office and thus enrich their adherents who prefer Place and Patronage to honest uprightness of conduct. If Sir Robert Peel stand out and threaten to resign, let him do so; surely out of the admirers of the immortal William Pitt someone of like spirit may be found to tread in his footsteps. THOMAS BATES.

KIRKLEVINGTON, 22nd January 1845.

At the Highland Society's Show at Dumfries in 1845, Walton (6658), a Barrington bull, son of Locomotive

WALTON (6658).

(4242), obtained the highest award,[1] and Mr. Jobson's Bull by 2nd Duke of Northumberland was placed second, Captain Shafto (6833) being nowhere.

As showing the extent to which judges were prejudiced, Bates received a letter from Lincolnshire in 1846, a month before the Royal Show at Newcastle, informing him that Mr. W. Smith, one of the judges, had already settled which animals were to have the prizes, and these eventually received them. Bates was disgusted at the amount of fulsome nonsense written about the 'invincible' Belleville (6778) which won the champion prize, and considered it his duty to warn foreigners against supposing

[1] Walton (6658) also obtained first prizes at the Yorkshire Show at Beverley and from the Liverpool Agricultural Society.

that the decisions at the Royal Shows, given by judges who were indirectly interested in the success of the prize animals, was any guarantee of their usefulness as breeding stock.[1]

It was always Bates's constant endeavour to obtain facts, real solid facts, from other breeders in confirmation or correction of his own knowledge and experience, but he could find no breeder who could supply him with facts of this kind bearing especially on the milking qualities of their cattle, and the ratio between the food they consumed and the good beef they ultimately yielded. One and all relied not on experiments faithfully carried out but on superficial generalities. 'I know hardly any breeder,' he said, 'who, if his livelihood depended on his own skill and knowledge, would not die of want. The little knowledge of shorthorns that most breeders possess is an unintelligible jargon acquired at public exhibitions and gathered from after-dinner speeches.' On one occasion Bates drove a friend over from Kirklevington to see Belleville (6778) at Mr. J. Mason Hopper's at Newham Grange, a few miles off. They met Hopper on the road. Bates greeted him with, 'I am bringing my friend to see your bull. I have told him he is very fat and very quiet.' Hopper, who was rather a rough diamond, replied, 'If that's all you can tell him, gang back; ye need gae no farther.'[2]

Bates was not prone to criticise stock unless it was brought before the public. He then considered it a public duty to use his knowledge and give his opinion. He was often asked to visit herds, and did so, but he never praised, even to please an old friend or kind host, unless praise was really due. Once he happened to be paying a visit to an old friend who had a fine herd and who expected him to appreciate them. He remained, however, silent till, pressed for an opinion, he said they were very fine cattle but not

[1] An engraving of Belleville (6778) in the *Farmer's Magazine*, December 1846, quite justifies Bates's criticisms. Captain Shafto (6833) figures in the January number.
[2] *Live Stock Journal*, 1885.

shorthorns. His friend then explained that they had been bred from a kyloe cow. Another time when, in like manner, a verdict was peremptorily demanded from him, he had the good luck to meet his friend's boys and the nursemaid, and exclaimed, 'I now see some stock I can admire.' He was always very fond of and attentive to children.[1]

Cleveland Lad (3407) proved so successful a sire at Duncombe Park that Lord Feversham applied for the use of his own brother Cleveland Lad 2nd (3408):

Thomas Phillips *to* Thomas Bates

HELMSLEY, 17*th September* 1845.

The bull (2nd Cleveland Lad) arrived safe and well at Griff on Friday night. He had not the slightest appearance of fatigue. Considering his great weight, I was astonished to see him look so well after his journey. Lord Feversham rode two miles or more to meet him the night he came home. He says he is a better animal than our Cleveland Lad, and appears to be highly pleased with him.

Mr. William Jobson, one of the sons of the purchaser of Jolly's bull (337), had a farm sale on his leaving Chillingham New Town, on 7th May 1846. At this sale Bates bought a cow and a heifer for Mr. Topham of Keal Hall, near Old Bolingbroke, in Lincolnshire. They were descended from Jolly's bull (337), and though it was more than seventy years since he had been brought from the Tees to Turveylaws, near Wooler,[2] yet they retained the same character that he had imparted to Mr. Maynard's Favourite tribe.

Mr. George Drewry, then agent to Sir Anthony Buller in Devonshire, had been at Kirklevington in the beginning of September 1845. Bates asked him how the 3rd Duke of Cambridge (5941) had answered for Sir Anthony. 'Very well,' replied Mr. Drewry, 'the stock by him are very promising. There is no fault in the bull, except his temper; but as Sir Anthony has used him for four

---

[1] Bell, p. 298.   [2] See above, p. 49.

years he wishes to change him for another.' Bates offered to exchange him, and arranged that Mr. Beauford of Bletsoe was to see him in London in May 1846, and to take charge of him if he chose to buy. Otherwise he might let him come on to meet Bates's man with the bull[1] he was sending in exchange to Sir Anthony Buller. Mr. Beauford, however, asked to have the 3rd Duke of Cambridge sent on to him, and the bull arrived at Bletsoe after a long and tedious journey. Hearing nothing from

3rd DUKE OF CAMBRIDGE (5941).

Beauford till the 21st of October, Bates concluded that he was quite satisfied. On that day he received a letter to say that Beauford was only willing to take the bull for a year. Bates immediately replied that if he had not liked the bull he ought to have let him come on to Kirklevington, as he had plenty of other customers for him. Beauford said the price was too high, and wished to have settled by arbitration what he ought to pay. To this Bates objected. Letter after letter followed; but Beauford found no fault with the bull or his procreative powers. He even admitted that his temper was better than had been represented to him. Finally Bates gave in and referred the correspondence to open arbitration at Newcastle, on the 25th of November. Bates wrote to Beauford,

[1] Red Duke 2nd (8460).

saying that he had sent his letters to the arbitrators and hoped he had done the same with his. Picture his amazement on receiving the following letter when there was no time left to communicate with the arbitrators :

Henry William Beauford *to* Thomas Bates

Bletsoe, 23rd November 1846.

Before I had received your letter I had forwarded the necessary letters to Mr. Grey. I think that it is quite unnecessary that either of us should attend the meeting. I am only sorry that it should have been requisite to call in any person to decide between us. I fear whoever gets the bull he will get no calves, and I understand from Sir A. Buller's agent (Mr. Drewry) that he had hardly any cows in calf to him.

This was the very first intimation Bates ever had that the 3rd Duke of Cambridge (5941) was not a stockgetter. No hint of this kind had been dropped before the arbitration was agreed upon, and he naturally considered that the point was not included in the reference. It subsequently appeared that Mr. Beauford had all the time been in correspondence with Mr. Drewry, though it was not until the 1st December that he sent the arbitrators declarations as to the inefficiency of the bull from his own cow-herd and some of Sir A. Buller's servants. Sir A. Buller himself refused to answer the inquiries of the arbitrators, and there was no evidence from his cow-herd. The umpire—John Grey, it would seem—chose to consider Sir A. Buller's silence as confirmative of the charge, but allowed a delay in order that Bates might try and induce him to break it :

Thomas Bates *to* Sir Anthony Buller

December 1846.

I find myself obliged to write to you in order to know the truth about the 3rd Duke of Cambridge (5941). Under the circumstances I have no other means of knowing. If it be 'that you had hardly any cows in calf to this bull,' I beg you will candidly write and say so; and if it be not I must request you to have a document to that effect attested before a magistrate, as the arbitrators refuse to accept any other evidence. The question was certainly never referred to them, as I had no intimation of it before the date fixed for their first meeting.

If the 3rd Duke of Cambridge (5941) was really inefficient, and you had let me know at any time, I would have sent you another bull without any charge whatever, rather than you should not have had your cows in calf.

I remember at the Bristol meeting of 1842 a gentleman from Wiltshire, naming his bull, got from my tenants the previous year, not getting all his cows in calf. I immediately told him I would write home to have another sent. He said he did not expect such kindness. I told him that it was nothing more than my usual practice in such a case. He now uses no bulls but those he has got from my tenants, and has been very successful as a breeder and in selling stock to advantage.

It does not surprise me that Mr. Beauford's cows have not held to this bull as they were running a bulling all this last year. I have, too, no means of knowing how the bull has been used, and much depends on that. If he had told me, as he ought to have done if it were so, that the bull was inefficient, I should have written to you, and, had you confirmed it, I should have taken the bull home and there need have been no arbitration at all. As it is, he kept the charge against the bull back, that I might not know of his intentions till it was too late to refute it, if untrue.

There has been a very general complaint in all countries of cows not breeding for some years, particularly the last two years; and the fault has been ascribed not to the bulls but to the cows alone. Having, however, had 3rd Duke of Cambridge (5941) for four years, you should know on which side the fault is if your cows have not bred.

Lord Feversham bought Cleveland Lad of me in 1843 (the same autumn you bought 3rd Duke of Cambridge), and the bull is as vigorous as ever though now turned ten years old. He has taken an aged bull out of Duchess 37th by Duke of Northumberland[1] this autumn to put to Cleveland Lad's heifers (still using Cleveland Lad to his cows), and this makes a good cross. I never had so many applications for bulls that I could not supply as at the end of last September and the beginning of October. Mr. Drewry only got Mr. Bell's bull[2] just in time. I hope he may do you great good.

[1] 2nd Earl of Beverley (5963).   [2] Red Duke 2nd (8460).

The result of the arbitration does not appear:[1]

James Topham *to* Thomas Bates

Keal Hall, 14th January 1848.

I have a calf come by the Duke of Cambridge—a very good one. There can therefore be no mistake about him being a stock-getter. I shall be glad to hear how you and Mr. Beauford go on. I fully expect he will compromise.

Mr. William Parker of Yanwath Hall, near Penrith, and his brother had before this hired the 3rd Duke of Cambridge, 'with a view to receiving support from friends in the neighbourhood to send a few cows, so as to bear one half the expense, as their own herd was not of the extent or value to guarantee such proceedings otherwise.' In this they were not disappointed, but of the six cows sent to the bull not one proved in calf; and the Foggathorpe bull Euclid (9097),[2] whose calves at Keal Hall had 'trained on well,' as Mr. Topham expressed it, though there were not so many of these as he could have wished, was sent as a substitute. Euclid was, however, not thought nearly equal to his predecessor by the neighbourhood, and was finally replaced, in August 1848, by the Hart bull Lord Hardinge (10449)[3] at £100 the first year, and £50 the two following.[4]

Mr. Spraggon told Mr. Trotter of Bishop Middleham in December 1848, very confidently, that the 3rd Duke of Cambridge (5941) was a sure calf-getter; and the end of the long story was that at eight years old the bull was

---

[1] As vol. vii. of the *Herd Book* records the births of six, and vol. viii. of twelve calves by the 3rd Duke of Cambridge at Sir A. Buller's, it is difficult to comprehend how this charge of taurine incapacity ever arose.

[2] Euclid (9097), roan, bred by Mr. T. Bates, got by Cleveland Lad 2nd (3408), dam Foggathorpe 4th by Duke of Northumberland (1949), grandam Foggathorpe, etc. etc.

[3] Lord Hardinge (10449), roan, calved 6th September 1846, bred by Mr. Thomas Bell, Kirklevington, got by 2nd Duke of Oxford (9046), dam Harriet by 4th Duke of Northumberland (3649), grandam Hilpa by Cleveland 2nd (3407), great-grandam Hawkey by Red Rose Bull (2493), great-great-grandam Hart by Rex (1375), great-great-great-grandam bred by Mr. Richardson of Hart.

[4] Letter of Mr. William Parker, Yanwath Hall, 14th August 1848.

exported by Mr. Ambrose Stevens of Batavia, New York. 'After his arrival in America he became the joint property of Colonel J. M. Sherwood of Auburn, and Mr. Stevens, and was kept several years, until he died, on Colonel Sherwood's farm. He did much valuable service as a sire.'[1]

Mr. Glass of Worton, Devizes, appears to have been the Wiltshire breeder to whom Bates alluded in his letter to Sir Anthony Buller:[2]

THOMAS BATES *to* WILLIAM PIPPIT

KIRKLEVINGTON, *9th December* 1847.

The last bull[3] Mr. Glass had from here, and whose stock you have bought, was by Duke of Northumberland (1940), dam by Short-tail (2621), second dam by Belvedere (1706), third dam by son of second Hubback (2683). These are four crosses of blood that cannot be exceeded, and leave only one-sixteenth of the original cow, which, though not of my blood, was a very good one, bought in the market, the seller giving a long pedigree.[4]

I assisted old Mr. Coates in procuring authentic pedigrees of the best short-horns and also of those having inferior crosses. He altered the latter to please as he said 'good friends of his that had that blood.' By this the public are most grossly deceived; since the death of the younger Coates, I never intend sending any more pedigrees to the *Herd Book*. As it was the only livelihood that the Coates had, I would not withdraw my support from them, but I will have nothing further to do with the work. I may some day give the world an authentic history of short-horns, when those who have my blood will receive the benefit.

Mr. Pippit never forgot Bates saying to him on another occasion, 'I shall never live to see the day, you may, when my stock will be duly appreciated.'[5]

As has been said, Bates felt strongly that shorthorns

[1] Allen's *History of the Short-horn Cattle*, pp. 193, 194.
[2] See above, p. 301.
[3] Percy (9472). The *Herd Book* throws no further light on the subject. The reason of Mr. Pippit's application was the fact of his having bought Percy's son, Duke of Somerset (9048), from Mr. Glass. Mr. Pippit, afterwards of Coughton, near Redditch, then resided at Downside College, Bath.
[4] Bell, p. 297.   [5] *Ibid.*

should be exhibited in family groups at some time during an agricultural show, so as to let the public see the results of breeding. With this in view, and at the urgent request of Mr. Milburn, the Secretary of the Yorkshire Society, he sent Oxford 2nd, own sister of the two Cleveland Lads, to the Yorkshire meeting at Scarborough in August 1847, together with her own four youngest calves—two bulls and two heifers—and one of her grandsons. All the six obtained prizes.[1] The 2nd Duke of Oxford (9046) was placed before Captain Shafto (6833), who had been bought by Mr. Parkinson of Ley Fields for 325 guineas, and was the Royal Champion at Northampton a month before.

George Vail *to* Thomas Bates

TROY, U.S.A., *27th September* 1847.

Your two favours I received in due course, the former speaking of Arabella,[2] and giving an account of shorthorns owned by Mr. Stephenson, and the latter giving a detailed account of the fairs [3] which had then recently occurred in England. I am happy to learn the issue of the exhibition of your shorthorns, and those of Mr. Bell, at Scarborough. I cordially congratulate you upon your success. I have delayed writing till after our New York State Fair had been held, which took place on the 14th, 15th, and 16th inst., at Saratoga Springs, about thirty miles from this place. The exhibition was about as usual, though not nearly so good as it would have been had it been held in this city. It was off from the Hudson River, and at a distance from canal navigation, which prevented there being as many articles at the show as there would have been, had it been on the line of our great thoroughfares. The exhibition of stock was, however, creditable to our State, and the attendance was numerous, especially of gentlemen of distinction, from most of the States in the Union, as well as many of the first men in the Canadas.

And now you will allow me to say that while the blood of your stock maintains its deservedly high reputation at home, it is

---

[1] Bell, p. 269.

[2] Arabella, by 4th Duke of Northumberland (3649), dam Annabella by Duke of Cleveland (1937), grandam Acomb by Belvedere (1706), *Herd Book*, ix. p. 256. See also Allen's *History of the Short-horn Cattle*, p. 185.

[3] *Britannice* 'agricultural shows.'

winning not less fame in America. I exhibited Hilpa,[1] and the first premium in the first class of Durhams was awarded to her, and the second to Mr. Prentice's cow, which I suppose is the best he has. I sent my bull Meteor (11811), out of Duchess by Duke of Wellington (3654), up to the show for exhibition only, at the request of some friends, as he had taken the first premium for the best Durham bull in 1844, as well as the first prize for a bull of any breed. The bull Marius,[2] bred by the late Earl Spencer, and owned by Messrs. Bell and Morris of New York, was on the ground and justly took the first premium in Durham bulls. The judges in their report on these, made the following remarks; 'The justly celebrated bull Meteor, belonging to Mr. George Vail of Troy, was on the ground for exhibition only; having taken the first prize at a former fair, he was excluded from competing at the present. *We think he stands unrivalled.*' In the two-year-old bull class, the first premium was awarded to Young Meteor, which I sold to Mr. Wakeman when about three months old. He took the first prize as a bull-calf, and last year, the first prize for the best yearling. He was got by Meteor and out of one of my Durham cows, but not from either of those I had from you. Mr. Wakeman had also his two-year-old heifer, got by Meteor, and her bull-calf on the ground. She has taken the first premium at three successive State Shows as a heifer calf, a yearling, and now as a two-year-old heifer. Her bull-calf, about five months old, took the first premium as a bull-calf, and one I exhibited took the second prize. I had four heifer calves on the ground; one took the first and another the second prize, which were all the prizes offered in that class. I had no yearling bull, nor was there any of my breeding exhibited, and Mr. Prentice took the first prize in this class. My yearling heifer was awarded a third prize. The result is that in every class except one where I had an animal, or there was one of my breeding, they were awarded the first prizes.

I have been particular in detailing the result of this great show, as I doubt not you will feel an interest in it, and that it will be highly gratifying to you. I suppose there were thirty or forty thousand persons present, among them many of the first men in the country, and two Ex-Presidents of the United States.

Hilpa has dropped a white bull-calf, now about six or eight months old, got by Meteor; and Lady Barrington has also dropped a red and white bull-calf, about four weeks ago, also by

[1] Hilpa by Cleveland Lad (3407), dam Hawkey by Red Rose Bull (2493), grandam Hart by Rex, *Herd Book*, ix. p. 397.
[2] See Allen's *History of the Short-horn Cattle*, p. 195.

Meteor. One of these calves I have sold to go to Canada, at the close of navigation, at 300 dollars; the other I think I shall keep for my own herd. I have just got Wellington home, and think I shall put him to Lady Barrington and Hilpa, and hope to get heifer-calves. Lady Barrington's calf last year by Meteor is a heifer, and a promising one. This is the only heifer-calf I yet have had from the cows I received from you. I hope I shall be more successful next year in heifer-calves.

Mr. A. B. Allen of New York, whom you know, is continually urging me to get a young Duchess bull from you. I would much like one, but at present I dare not venture the expense. Meteor is, in some respects, a finer animal than Wellington: he is better in the hind-quarters and across the hips. Wellington has not a broad hip, and is rather thin across the twist, and some of his calves partake of this defect. His fore-end cannot be beat; he is a superior handler, as also is Meteor. They both excel in this valuable quality. I weighed Meteor three days ago: his weight is 2200 lbs., and Wellington, when in order, will weigh about 1800 or 1900 lbs. Meteor makes a splendid show, and, I doubt not, would take a high rank even in your country. Our county show took place in this city last week, and was the best we have had. I was equally successful in winning premiums here as at the State Show. Hilpa took the first prize. A yearling bull which I showed, got by Meteor, but not from your cows, took the first prize for the best bull on the ground; and my two-year-old heifer and my yearling heifer each took the first prize in their respective classes. I only put one animal in each class, and took the first prize in all but one; and I believe this exception had the blood of my herd, and that of yours in part. Arabella, I hope, may make a good milker; she will never make a show cow; she stands entirely too high on her legs; her offspring may be better, as Wellington and Meteor are both right in this particular.

GEORGE VAIL *to* THOMAS BATES

TROY, U.S.A., *June* 1848.

I have recently purchased the entire herd of short-horns belonging to Mr. Prentice. In the present state of demand for short-horns here, it may be considered by some as rather a bold operation. The herd consists of twelve head, being the four reserved cows (which he considered the best in his herd, when three years ago he made his large public sale), and their offspring. Among this number there are a few animals which I conceive will breed well to my bulls Duke of Wellington and Meteor. I propose to retain these, and to sell off the remainder.

I have told Messrs. Allen and Colonel Sherwood, who have had the use of my bulls to their herds, that there is no other way to revive prices, so as to procure renumerative prices for breeding, than by introducing the blood of your herd, and through this medium convincing the public of the superiority of the short-horns. I believe they concur with me in this opinion, though, I fear, they have not entire confidence of success, as I see they are making preparations for a sale of the most of their short-horns, at our New York State show in September, at Buffalo. I do not myself at all like the idea of giving the matter up, as I conceive the short-horns are decidedly preferable, as a combination for the dairy and the shambles, to any other breed of neat cattle. At this crisis, unless some one or other perseveres and keeps up pure bred animals, the loss to the country will be incalculable. What I want to aid me in my endeavours to sustain my stock, is a young red or roan Duchess bull from you, but I cannot, at present, think of incurring this additional expense.

I have to-day received a letter from Mr. John Wittenhall, who (together with the Hon. Adam Ferguson of Nelson Gore, Upper Canada) purchased Lady Barrington's last bull calf, which he calls Halton, after the name of the county he represents in the Canadian House of Assembly. He wishes me to have Halton fully recorded in the forthcoming volume of the British Herd Book. This will be of service to him and Mr. Ferguson. Halton (by Meteor, out of Lady Barrington 3rd) is a red roan, and a fine animal, which will do you much credit in Canada, as descended from your stock. He has recently reached his destination, and the owners are much pleased with him, and intend to keep him as a close bull for their own herds.

Vail gave a letter of introduction to Bates to Mr. Ambrose Stevens, one of the Vice-Presidents of the New York Agricultural Society, who was coming to England in 1848, as a delegate from it to attend the principal British agricultural meetings. Stevens had resided in Buffalo near Mr. L. F. Allen, a brother of Bates's friend, Mr. A. B. Allen, till about 1844. He then went to live in the city of New York, and was introduced by the Allens to Vail. He took an active part in the New York Agricultural Society, and was generally allowed to be a good judge of stock. Vail did not suppose Stevens was a man of much property, and did not know that he intended to purchase any cattle in England, though he thought he might have a commission to purchase one or two shorthorns for Colonel Sherwood of Auburn with whom he was well acquainted.[1]

The Royal Show was to be held that year at York. Remembering the treatment he had formerly received, and knowing that there existed a great jealousy of his cattle, Bates was very unwilling to exhibit any of them for the premiums. He was, however, sincerely anxious for the success of the show, being on the best of terms with Mr. Hudson, the secretary, and only lamenting that like Mr. Milburn, the secretary of the Yorkshire Society, he was 'tied and bound by the directions of his superiors.'

Bates wrote a strongly worded letter to the Council calling their attention to the importance of pedigree, to which the Society had hitherto paid little or no regard. In August 1847 indeed, the Hon. Captain Howard had given notice of the motion, 'That in future in Class I. pedigrees should be required as part of the certificates for shorthorned cattle at the Society's Country Meetings,' but had had to withdraw it in the December following. In order then to open the eyes of the Society to the importance of pedigree, Bates proposed sending four generations of his herd to York. Four of the animals, his best, were to be shown as extra stock in order not to give offence to other exhibitors, and six were to be entered for competition. His three tenants, Messrs. Bell, would also each contribute a shorthorn to the ring. He made it a *sine quâ non* of sending any cattle at all that the judges should examine the four generations from his herd, so that they might form an idea of their value for breeding purposes.

To these overtures came the cut-and-dried answer, dated 31st May, 'The Council return their thanks for your suggestions, and have referred them to the stewards of the yard.' To any one uninitiated in the general policy and administration of the Royal Agricultural Society, this

---

[1] Letter of George Vail to Thomas Bates.

seemed as though his suggestions would be taken into consideration; otherwise, the Council should have said straight out that they would not alter the regulations already issued. Twice again Bates wrote, and twice again he received the same stereotyped response. He vainly expected to hear from Mr. Brandreth Gibbs, the director of the yard, whom he had told that he was perfectly prepared to pay the fines for not sending his cattle, if his conditions were not to be complied with.

Absurd precautions were taken in those days to prevent the judges knowing to whom an animal placed before them belonged or how it was bred. Numbers were sent to each exhibitor to be put on his cattle, and although with such small entries the ownership and lineage of each animal must have been an open secret, the judges were supposed to have nothing but the numbers to go by. When the Society itself thus harboured suspicions of the prejudice and partiality of its judges, criticisms of their decisions might fairly be much more trenchant and pointed than would be right at the present day. On receiving the numbers of his entries from Mr. Gibbs, Bates sent him full particulars of their relationships. They comprised four generations—grandam, dam, son, grandson, grand-daughter, etc. etc. He would have sent a calf to represent a fifth generation but it was just a little too young.[1] He clearly explained to Gibbs the purpose he had in view, and at the same time wrote in robust terms to the secretary to say that as he had received no direct answer from the Society, he assumed that the judges would be instructed to examine the four generations before deciding on the premiums. In Class I. (aged bulls), he entered 2nd Duke of Oxford (9046) and 3rd Duke of Oxford (9047); in Class III. (cows), Oxford 2nd; in Class IV. (in-calving heifers), Wild Eyes 16th and Wild Eyes 17th; and in Class V. (yearling heifers), Oxford 6th. These were all competitive classes, and all the animals except the last were also entered for the local prizes con-

[1] Grand Duke (10284), calved 14th February 1848.

fined to Yorkshire. As extra stock he sent Duchess 34th, then nearly 16 years old, another cow, probably Oxford 4th,[1] and the two heifer-calves Duchess 59th and Oxford 7th.

As far back as 1839, Bates had proposed that on the evening of the day when the judges finished their labours the yard should be opened to the public at 5s. a head, and that all the animals shown by each exhibitor should be grouped together for examination. Lord Portman, the president of 1846, supported these suggestions, and as the result of many discussions, it was agreed, 3rd May 1848, that after the judging, the yard should be opened to the public at 20s., and to governors and members of the Council at 2s. 6d. Mr. Hudson the secretary communicated this decision to Bates in a letter dated 12th May 1848.

At York the judging was not finished on Wednesday, 12th July, till 4 P.M. A lecture had been fixed for that hour, and the Council dinner was to be served at six o'clock. Seeing how limited the attendance in the show-ground was likely to be under these circumstances, Bates vainly urged that the dinner should be postponed to 8 P.M. As it was, the moment the judging was over, Prince Albert arrived in his carriage and entered the yard. In accordance with Lord Portman's resolution, Bates and some sixty others paid down their sovereigns and obtained admission. None of the council or officials were there; there were no catalogues or prize lists, and it is needless to say there was no grouping of animals.

It took a long search before Bates could find his shorthorns. The judges, he learnt, had not only never looked at his four generations but they had never handled the bulls at all. He found that these had been pronounced

[1] As the R.A.S.E. did not then recognise names, to say nothing of pedigrees, it is often difficult to identify animals by their ages, especially as these were in many cases carelessly calculated and inaccurately printed. At the time of the York Show, Oxford 4th's age was '4 years, 11 months, and 5 days,' and not as given in the Royal catalogue, '8 years, 11 months, and 5 days.' She was really ineligible to be shown as extra stock. If she was accepted as such, it must have still further confirmed Bates in the impression that Mr. Gibbs agreed to all his conditions.

upon by Mr. Benjamin Swaffield of Chatsworth, and by Mr. James Walker of Newberries, Hertfordshire, who said he hailed from East Lothian, though none of Bates's many friends there had ever heard of him. A local second prize, adjudged to the 2nd Duke of Oxford, was the only meed Bates received. Even in the local class for bulls both the 2nd and 3rd Dukes of Oxford were beaten by Mr. Ambler's Senator (8548), but Senator had been 'trained' by the celebrated Joe Culshaw. Bates stood looking at his pair with his hat over his brow and could scarcely believe it.[1] The decisions in the bull classes gave universal dissatisfaction.[2] Mr. Ambrose Stevens, the delegate from the New York Agricultural Society, at once said that the prize-winners were not shorthorns at all, so little did they resemble them in colour and character. Mr. Linton's Hudson (9228), the premium bull in Class II., was, he very properly declared, much more like a Devon. The champion bull, Mr. Keevil's Deception (7957), was the colour of no shorthorn bull Bates had ever seen; he was currently reported to be a first cross from a longhorn, and certainly looked like it.[3] Bates considered him the very worst bull in the yard.

Bates had been a regular attendant at agricultural shows ever since 1790, a period of fifty-eight years, but the judging at York was the worst he could remember. If the Royal Agricultural Society gave him a thousand

[1] *Saddle and Sirloin*, p. 330 n. For a portrait of Senator (8548), see *Farmer's Magazine*, April 1849.

[2] Mr. Ellison of Sizergh Castle, the leading farmer in Westmoreland, protested in a letter to the Council, 22nd August 1848, against the prize animals at York being so overfed as to destroy their usefulness. Sir Charles Tempest had given over showing on that account, and Mr. Webb had discontinued exhibiting any rams that were older than shearlings. Some of the finest stock in Yorkshire was kept back for the same reason, and many of the bulls present were too fat for service. As to the pig stock, Mr. Ellison passed by those which could not stand and walk, as he failed to see how these could help *to promote the breeding and improvement of stock*, the end for which the Royal Agricultural Society was instituted.—*Farmer's Magazine*, 1848, ii. p. 273.

[3] Deception (7957), white with a red head and neck and white forehead. See portrait, *Farmer's Magazine*, February 1849. There is nothing to show that many of the early entries in the *Herd Book* were shorthorns at all.

pounds, he would not send ten cows to any of the prize bulls, as the cross would deteriorate the ultimate value of the produce by many thousands. What puzzled him most was the inconsistency of Messrs. Swaffield and Walker in commending his 2nd Duke of Oxford and Mr. Knowles's Tinsley (9748),[1] which were animals the diametrical reverse, whether for good or for bad, of those which carried off the prizes.

The judging of the shorthorn cows and heifers by Mr. Charles Stokes of Kingston, Nottinghamshire, and Mr. William Torr of Aylesby was little better than that of the bulls. Not one of the pretended shorthorns Bates saw, was, he believed, free from an admixture of the Galloway strain.

Mr. J. C. Etches, whom Bates considered the best judge of shorthorns in England, officiated for the Devons, Herefords, and all other breeds except shorthorns, as well as for the extra stock which consisted largely of shorthorns. His decisions Bates believed to be perfectly correct, and they were approved in every instance, even by the unsuccessful exhibitors. Why the shorthorn classes alone should have been excepted from his jurisdiction, and two sets of two judges employed upon them, as was now done for the first time, Bates could not understand.

In the extra stock (not entered for competition) Mr. Etches and his companions Mr. Henry Chamberlain of Desford, and Mr. William James of Hereford, highly commended the cows Duchess 34th and Oxford 4th, and commended Duchess 59th and Oxford 7th, the two heifer calves Bates had sent, there being no calf classes. They also commended Lord Feversham's cow by Cleveland Lad.

A few days later Bates wrote:

With the exception of the horrid decisions on short-horned cattle, the meeting at York went off well in every respect. Prince

---

[1] Tinsley (9748) by General Washington (6036) (a son of Duke of Northumberland (1940)), dam Buttercup by 2nd Duke of Northumberland (3646), a cow purchased by Bates at Mr. William Jobson's sale in 1841 for Mr. Knowles, Tinsley, Sheffield.

Albert's attendance for the first time gave great satisfaction to all. He acquitted himself well on all occasions, and was occupied more and saw more than any man at York from his arrival at half-past three o'clock till eleven that night. The next morning he was again in the show-yard at six. He was at the Council and Pavilion dinners [1] and at the assembly given by the Lady Mayoress of York on the evening of the second day and after the Pavilion dinner. He stayed there till nearly midnight, leaving before supper. He was cheered wherever he went, and during his speeches at the dinners. He left York at 8 A.M. on the Friday.

Bates rightly complained that in the list of commendations printed in the Royal Agricultural Society's Journal there was no sub-heading to show that his cattle, which had been highly commended, were extra stock. Indeed the impression was left that they had competed for the premiums and had been unsuccessful.

On his return home, he addressed a letter of remonstrance to Lord Chichester the President, and Lord Yarborough the past President of the Royal Agricultural Society:

The only true test of excellence in cattle is the comparison of the food they consume with the improvement they make. To this great fundamental point I endeavoured to draw the attention of the Board of Agriculture and all Agricultural Societies in 1807 by means of a pamphlet, a copy of which I sent to the Royal Agricultural Society soon after its formation. This principle has, however, never been acted upon, and until the conductors of Agricultural Societies properly consider in what the merit of cattle consists, all their efforts at improvement will be unavailing, and only tend to increase prejudices and confirm errors.

The proceedings at York prove to me that the council of the Royal Society are determined to resist to the utmost of their power any proper investigation into the merits of shorthorns. I have tested these by the only true test for above fifty years. My judgment has been formed by facts in which there could be no deception. I have never occupied a farm in Northumberland or Yorkshire without being able to double the quantity of cattle

---

[1] During the Council dinner the health of the Royal Family was immediately followed by a set debate 'on rearing cattle from the time that they are dropped to their being a year old.' No wonder that it was impossible to slightly postpone this intellectual feast as Bates had suggested.

hitherto kept. All my cattle shown at York were kept in the ordinary way—the cows and young stock going out at grass night and day, and not kept in the house on hay and linseed-cake to force them forward in condition as was the case with the prize cattle. Indeed, mine were without turnips the previous winter as I had none for them.

One of the stewards of the yard complimented me, on the 14th of July, upon having shown my cattle not made up like the rest but in the proper condition in which cattle should be shown. It is a mere farce when directly contrary to the public declaration of the Royal Society that fat is not to decide merit; the judges appointed hardly ever pay attention to anything else.

If men were good judges of shorthorns—and without handling they cannot be—coarse ill-bred brutes made fat by forced keep would be rejected at once, and quick grazers, however lean when exhibited, would have a chance of gaining prizes; but while size and fat alone are regarded, no right decisions will ever be made.

One tenth of the money prizes would now amply suffice, seeing that the railway companies have consented to convey the stock gratuitously. Animals would then be more likely to be shown without the cost of feeding. Indeed it would be far better if there were no judges and no premiums: good stock would then rest on its own merits and really good cattle be exhibited.

To young breeders in particular the exhibition of my four generations would have been of greater service than a ten years' attendance at ordinary shows. I regret that they were deprived of an opportunity that they may never have again in their life-time at any public exhibition.

Deeply I deplore it, but in all probability the best shorthorns will go abroad and may never return to their native land till the present generation die out, when their successors may be glad to obtain the finest variety of cattle that ever existed.

The American visitors crowded round the Kirklevington cattle at the York Royal. To them Bates, in his white beaver hat, was an object of interest second only to Prince Albert himself; the two conversed earnestly together for some time.[1] With their usual sense and intelligence the Americans adopted Bates's recommendations and began exhibiting their breeding stock in families.

Bates, it is said, introduced Ambrose Stevens to Mr. Etches, who in answer to Stevens's questions replied, 'Yes,

[1] *Ex inf.* the late Mr. Edward Leadbitter.

it was I who purchased some shorthorn cattle of different farmers near the Tees for Mr. Sanders of America in 1817. They were thought by myself and others to be very fine animals.' Asked further if he remembered any of their pedigrees, Etches turned to Bates saying: 'Mr. Bates should know something about the pedigrees of the Shipman heifers.' Bates then told them 'that he recollected very well Mr. Shipman selling a heifer to go to America.' 'She was called,' he continued, 'Mrs. Motte, after a sister of either Mr. or Mrs. Shipman. Mr. Maynard had a cow by a son of Hubback (319) which he called Starling. This cow had three daughters; one of them Mr. Maynard kept, another he sold to me, and the third he sold to Mr. Shipman who called her 'Starling' after her dam. When he bought her, she was in calf to Adam (717); the produce was the heifer "Mrs. Motte" which he afterwards sold to Mr. Etches.'[1]

Ambrose Stevens *to* Thomas Bates

GLASGOW, 15*th August* 1848.

In the many cattle I have seen since I left you, I have found no occasion to alter my opinions, as formed on a visit to you, in regard to your cattle.

The two best cows I saw at the Show of the Highland and Scottish Agricultural Society at Edinburgh, were got by your bulls, and yet they were not noticed. *In quality* they were far superior to anything there, yet this seemed (as everywhere) disregarded—*handling* having to give way to *fat*.

I have to request that you will make no disposition of the young *red* bull (lame) 3rd Duke of York which I saw at Kirklevington, until I see you. You must let me take him to America.

Ambrose Stevens *to* Thomas Bates

LINCOLN, 22nd *September* 1848.

After the sale at Brawith, I came on to York and stayed all night there. I left in the morning and travelled by Hull and New Holland to Great Grimsby. I stopped there and went to

---

[1] Allen's *History of the Short-horn Cattle*, p. 165.

Riby the residence of Mr. Torr. Among his stock one cow struck me as far superior to anything else he has. Mr. Torr seemed reluctant to tell me what she was but did so at length. I knew she had your blood, and so it turned out, as she was bred by the Rev. Mr. Cator and was got by your 4th Duke of Northumberland out of Waterloo by Norfolk. After some conversation Mr. Torr finally admitted that your *quality* was superior to any other in England, and said that all might profit by getting it.

From Mr. Torr's I went on to Revesby Abbey the residence of Banks Stanhope. His whole herd was sold. I bought Waterloo 5th, bred by you. I have a tale to tell about this when we meet. The cow is in fine health; she has bred every year and calved on the 7th of May last.

I have often thought of your kindness of last Tuesday at our parting. I thank you most gratefully for it.

They say favours granted beget requests. May I ask you one? One of the heifers to go to America from Mr. Stephenson, a two-year-old, dropped a calf in August. Will you permit her to be bulled by your lame red bull, 3rd Duke of York? She is quite blood-like and will breed a good calf from him.[1]

I enclose a copy of the certificate said to have been given by Mr. Stephenson to Waistell. I will ferret this whole matter to the bottom. We must be cautious and silent for the present, and not excite their suspicion. It is all aimed at you and you alone. I found out that there was no affidavit made by Waistell. Mr. John Booth told me there was and that he had seen it.

AMBROSE STEVENS *to* THOMAS BATES

LINCOLN, 24*th September* 1848.

I went yesterday to see the Wiseton herd. It wants a bull from you on it. I also called to see the herd of Mr. Watson of Walkeringham. He has one cow by Cleveland Lad and two by 4th Duke of Northumberland, all three bred by Mr. G. P. Harrison of Greta Bridge. These three are all that are of any account in his herd.

AMBROSE STEVENS *to* THOMAS BATES

LONDON, 2*nd October* 1848.

When I return north I shall see Waistell and will learn all the facts for you. If there is anything wrong in the matter, you

---

[1] Bates acceded to this, 13th November 1848.

would not have known and could have had no part in it. It is all a conspiracy to enable your enemies to retort on you for your exposure of their villainy. They will never dare publish their fraud. They know Mr. Stephenson is a quiet retired man, and will probably never hear of their proceedings. They talk of the matter in quarters where they hope to benefit themselves and to injure you, but they will never go farther than talk in private. Publicity would destroy their plot and expose themselves.

Ambrose Stevens *to* Thomas Bates

Babraham (no date).

I have started north. I shall go from here, where I am visiting Mr. Jonas Webb, to Castle Acre to see Mr. Hudson, to the Earl of Leicester's at Holkham, and to Mr. Topham's at Keal. I cannot say just what day I shall be with you, but I shall work my way to Yarm as fast as possible.

In his tour of investigation into the origin of Belvedere it was natural for Ambrose Stevens to address himself in the first place to Mr. Atkinson Greenwell, who had recommended the purchase of the bull to Bates, and had been a witness of the whole transaction. Stevens elicited several new points of detail from Greenwell at Archdeacon Newton, 9th November 1848:[1]—Greenwell was living at Copelaw, about seven miles due north of Darlington, when his cousin John Stephenson of White House near Wolviston came over in October 1818 to see his mother who was then keeping house there. Stephenson offered him at cost price a cow that he had bought a week or ten days before, after the sale of the Countess of Antrim's cattle. He said the cow had lost a teat or two, and as Greenwell was then to some extent a dairy-farmer and not a breeder of blood cattle he declined the offer. Three or four weeks afterwards, Greenwell was at White House and saw the cow. She was a red roan, apparently seven or eight years old and had evidently been milked for three or four years. Greenwell was frequently at White House for years afterwards and

[1] See Appendix D.

recollected Waterloo (2816) and Angelina the 2nd, the first two calves the cow had there and the other stock descended from her. He never saw any other cow at White House that had come from Wynyard,[1] and never heard of any such a cow as Anna-by-Lawnsleeves being there or of Stephenson owning such a one.

Stevens subsequently proceeded to Forcett to see Mr. George P. Harrison, who had hired Waterloo in 1830. Harrison told him that it was true that when Waterloo came to his farm at Streatlam Grange, a memorandum came with him to say that he was got by Young Wellington, dam Anna-by-Lawnsleeves, but that this memorandum was not in Stephenson's writing, and that after he had printed some handbills and cards with this pedigree, Stephenson told him that there was a mistake in it, and that Waterloo was out of Angelina-by-Phenomenon and not out of Anna-by-Lawnsleeves.[2]

George P. Harrison *to* Thomas Bates

Forcett, 23*rd November* 1848.

Your friend Mr. Stevens has just left me. I have been much pleased with him.[3] Enclosed is a certificate regarding the bull Waterloo which I had from Mr. Stephenson. Mr. Stevens requested me to send it to you. It would be desirable to trace out what became of the cow Anna-by-Lawnsleeves, if ever there was such a cow, but I am afraid the parties are all dead.

Bates knew very well that Anna-by-Lawnsleeves had remained with Mr. Porritt after the sale of 1818, but he was no doubt unwilling to produce this in evidence lest it might possibly turn out that his Oxfords were descended from her.

Mr. John Wood of Ricknal Grange, the purchaser of Nell Gwynne at the Wynyard sale of 1813, to whom Stevens applied 25th November 1848, knew nothing

---

[1] 'There was never but one,' Mr. Thomas Stephenson said emphatically, 21st May 1896. [2] See Appendix E.

[3] Mr. Thomas Stephenson describes Ambrose Stevens as 'a pleasant cheery man, rather stouter than Bates. He would talk by the hour on shorthorns till people said he was "shorthorn mad." He had been a barrister, and none ever cross-examined better.'

beyond the facts that he happened to be at Elemore in 1815 when Angelina paid her visit to Lawnsleeves, and that her calf by him was born the following year.[1] Fortunately, George Dobling, who had the entire and sole charge of Sir Henry Vane Tempest's herd from 1802 to 1815, was still alive.[2] During all that period he was very positive that no other bulls but Phenomenon, Wynyard, and Wellington were put to Sir Henry's cows, and that no cow or heifer of Sir Henry's was ever sent to Mr. Baker's Lawnsleeves. Angelina was a heifer, not three years old, at the sale of 1813, and had not then, he knew, had a calf.

After all this explicit and irrefragable testimony, it is amusing now, however exasperating it must have been to Bates, to read the cock-and-bull stories invented by men like Robert Waistell, and eagerly believed in by Mr. John Booth and others. According to Waistell,[3] Angelina was sent to Lawnsleeves at Elemore by Sir Henry Vane Tempest in November or December 1810, when she could not have been more than about a month old. Two witnesses were actually found—each of them ready to swear that he had bought Anna-by-Lawnsleeves for Stephenson at Wynyard in 1818. Edward Hall, then in Sedgefield poor-house, told Ambrose Stevens that he had given twenty or twenty-one guineas for her;[4] 'Handy Smith,' on the contrary, declared that it was he who bought her and that the price was eighteen guineas.[5] According to the circumstantial statements of Waistell, Princess and Angelina were killed at Wynyard before the sale of 1818; according to Hall, Anna was the one shorthorn then offered for sale; while if Smith were to be believed, Princess was sold at that sale, Angelina, then a cow seven or eight years old, was bought in, and Anna-

---

[1] See Appendix F.   [2] See Appendix G.
[3] See Appendix H. Waistell, according to Mr. Thomas Stephenson, was 'a man come completely down in the world, running about and getting a living anyhow.' The two other witnesses suborned to discredit the Kirklevington cattle were men of the same class.
[4] See Appendix I.   [5] See Appendix K.

by-Lawnsleeves was purchased for Stephenson. Again, to compare the descriptions given by these South Durham romancers of the cow bought for Stephenson at Wynyard in 1818, Waistell had the honesty to agree with Greenwell that she was 'an oldish cow, about eight years old, that had lost all of her teats but one'; Hall called her a yellow roan, quite small, probably not so much as four years old, with two of her teats lost; Smith said she was a roan old-fashioned shorthorn, five years old, that had had three calves and had lost one or two of her teats. It was of course absolutely impossible on their own showing that this animal could be Anna-by-Lawnsleeves calved in 1816.

Ambrose Stevens deserves the highest credit for the immense trouble he took to obtain the testimony of these witnesses without giving them an opportunity of laying their heads together. The way in which they flatly contradicted each other, while Bates's account of the Wynyard herd, written two years before from the evidence he had previously collected, remained unshaken, is quite sufficient to disprove any cross by Lawnsleeves in Belvedere's lineage.

The most charitable view that can be taken of the matter is to suppose that it was originally a case of mistaken identity in Robert Waistell's most inaccurate mind. 'Both cows,' he tells us, ' were roans with somewhat long horns which came out level and then turned upwards; Anna was quite like her dam, but not so large.' But even this view will not acquit Waistell of the many falsehoods with which he garnished his old wife's tale, nor other persons of the eager credulity with which they accepted and disseminated it to Bates's prejudice.

GEORGE VAIL *to* THOMAS BATES

TROY, U.S.A., *26th March* 1849.

Mr. Stevens returned about three weeks ago. I have been expecting daily to see him, but he has remained in New York since his arrival.

I am told the heifers he has purchased are from Mr. Stephenson's herd. You may recollect I wrote to you some two years ago to ask what a Princess cow or heifer could be purchased for, and that you replied that Mr. Stephenson's want of judgment in breeding the Princess tribe to bulls of bad blood had so deteriorated them that they were inferior cattle.

Mr. Stevens's heifers will doubtless be exhibited next September at the show at Syracuse in the western part of this State. I have two heifer calves dropt in July last, one got by Duke of Wellington out of Hilpa, and the other got by Meteor out of Lady Barrington 3rd, which I shall probably exhibit there to compete with them. Mine are two most promising heifers, and the new comers must be very extraordinary to beat them. Colonel Johnson, the secretary of our State Society, who has seen the latter speaks well of their appearance. He says the bull, 3rd Duke of Cambridge by Duke of Northumberland, which has come with them, and which is entered in the Herd Book as owned by Sir Anthony Buller, is a good one. I judge from the pedigree that he is equal to my Duke of Wellington and probably better in the hind-quarter and across the hips. Meteor is much superior to the Duke of Wellington in these points, and I think you would say he was a noble animal.

Since last May I have had here the twelve head I purchased from Mr. Prentice. I have now had an opportunity of comparing them with the blood of those I had from you, and am more satisfied than before of the superiority of the blood of your herd. Mr. Prentice made his importations mostly from Whitaker's herd. There are only two animals among the twelve that I shall retain, and these I shall cross with my bulls of your blood.

I have not lost sight of getting by and by a Duchess bull-calf from you, a red or a roan. I want to see the bull and heifers Mr. Stevens has brought over, and if they can fairly compete with those I have, I must try and get from you a young bull from a Duchess cow so as to maintain the ground I now occupy as to shorthorns, and keep the blood of your herd in the ascendency, but I must bespeak your moderation in price.

George Vail *to* Thomas Bates

TROY, U.S.A., *16th May* 1849.

Some days after I wrote my last letter, Mr. Stevens came up from New York and remained one night with me. He delivered the letter you sent me by him. His apology for not sending it up was that he had mislaid it. When he handed it to me it was torn in two so that any one might have read it. He remarked that he had laid it out to bring up and that a little child had torn it in two. He said you had read it to him before sealing it, and as this is now confirmed by you, its being torn is satisfactorily explained.

A week or ten days after the arrival of the cattle in New York, Colonel Sherwood went down there and brought the bull 3rd Duke of Cambridge, and, I believe, the three heifer calves Mr. Stevens had from Mr. Stephenson up to Albany and took them out to his place at Auburn about 170 miles from here. Colonel Sherwood is a high-minded man and all will be right with him, and I hope and trust it will also prove that Mr. Stevens has acted in good faith with you and the gentlemen with whom he had intercourse in England.[1]

Mr. Stevens informed me that he had received intelligence that a cow bred by you, which he had purchased and left her in England,[2] had brought a bull calf. I understood him to say it was a Bates bull, and the conclusion I drew was that the cow having been bred by you and bulled by one of your bulls would make it what he termed a Bates bull. He still speaks in high terms of your herd though he seems to think highly of Mr. Stephenson's *cows*.

The two-year-old Lady Barrington heifer is a good one, and her last August heifer-calf and that of Hilpa of the same age are superior animals. I shall take your advice with regard to them, and if I show them, as I now think I shall do, at our State Show in September I shall not much fear the recent importation from Mr. Stephenson's herd if we have intelligent judges. I have there three heifers from the cows you sent me, and do not intend to sell any heifers till I number as many as I need. If I take the trouble and bear the expense of breeding, I must maintain the ascendency I now have, or I had better sell all out at once. As I have unabated confidence in your stock, it is not my intention that they shall run down here; but you will see from what I have said about the recent importations that I must now depend on your aid in the matter. Having heretofore beaten all competitors, I shall be unwilling to be beaten. You must there-

---

[1] 'Bates showed great kindness and hospitality to one American gentleman who ingratiated himself so much, that he presented him with a bull which was taken to America, and the half share of him sold there for 150 guineas. He was then asked to give a receipt to show that he had been paid 300 guineas for him. He took no notice of the request, and gave orders that if ever the person came again to his house he was to be refused admittance.'—Bell, p. 213.

[2] Waterloo 5th, but there seems to be no record of her breeding after being bought by Mr. Stevens.

fore be prepared to send me at no distant time a young Duchess bull of the first order. As I have great confidence in the superiority of your Duchess tribe, I speak of them in terms of high approbation on all suitable occasions.

The idea you express that Mr. Stevens had purchased Mr. Stephenson's herd must, I think, be incorrect. He would not venture upon such a step unless he is backed by others, and I do not know of any who would now be likely to assume so great a liability.

Bates did not hesitate to recommend other persons' bulls, and these not of his own tribes, when he believed praise was due, and the issue was likely to prove satisfactory:

THOMAS BATES *to* WILLIAM VAUGHAN [1]

KIRKLEVINGTON, 12*th October* 1848.

Since I saw you at the Wiseton Sale last spring, I have been to Helmsley, where I had much pleasure in inspecting Lord Feversham's herd of short-horned cattle, descended from the bull he got from me named Cleveland Lad. They are not kept up by high feeding, and are the better for it to those who buy them.

Lord Feversham has used Cleveland Lad six years, and though six years old when he got him, he is very fresh and likely to be a stock-getter for some years to come. The produce by him are all looking well, and if Lord Harewood is in want of a bull, I would recommend you to go and see this herd. A roan bull, calved 15th February 1847, named Bonny Lad, and out of a cow named Blanche is very promising. He goes back through a cow bought at Mr. Robert Colling's sale at Barmpton in 1820, to Sir James Pennyman's stock.

His opinions found a most able exponent in the North of Scotland:

J. C. GRANT DUFF *to* THOMAS BATES

EDEN, BANFF, 9*th November* 1848.

I think you know that in order to prevent deceptions, or falsifications of pedigrees in the north of Scotland, I introduced annual lists which have, I believe, been of some little use.

[1] Agent to the Earl of Harewood.

Although I may soon retire from the stage as an extensive agriculturist, and my herd of shorthorns now only numbers twenty, I have continued to issue the lists. The space on the sheets enabling me to extend remarks, you will perceive I have taken the liberty of quoting your opinions in regard to the injudicious manner of giving prizes for beef instead of blood. I hope you still enjoy as good health as when I had the pleasure of seeing you at York.

I have always publicly and privately expressed my regret at your not giving the world your opinions of breeding and general management of cattle. No doubt rivalry and the disagreeable necessity of replying to opponents entail obstacles which a wise man is unwilling to encounter, but if you were privately to distribute copies amongst your friends, as I do these lists, the good might be spread without incurring the evil.[2]

In the evening of a life, too, that had encountered so much base ingratitude and malicious misrepresentation, it must have been consoling to receive a kind and appreciatory letter from a veteran breeder nearer home:

T. COWPER HINCKS *to* THOMAS BATES

BRECKENBROUGH, 15*th May* 1849.

I have just returned home after an absence of some months and find that I have several good calves and young things by The Duke (8676) whom I bought of you. He is very fresh and an excellent buller. I should like to buy a young bull from you of another cross which I could use with advantage to the heifers by The Duke which are coming on. I should be glad of your advice. Should you be coming this way, we dine at six o'clock, and if you let me know I will ask Robert Pulleine to meet you. There will always be a bed and a stall for your horse.

With the few really good families in his corner of Yorkshire, Bates always maintained the most friendly relations:

WILLIAM MAULEVERER *to* THOMAS BATES

ARNCLIFFE HALL, 15*th June* 1849.

I beg you to accept my best acknowledgments for the advice contained in your letter relative to the best mode of sending and selling fat stock at Leeds.

[1] Bell, p. 215.

I am sorry to hear your health demands a visit to the sea, where I hope this fine weather will speedily restore it. Farmers yet certainly have no reason to admire the effects of Free Trade. I was always myself of the opinion that the British farmer could not compete with the grower of foreign corn nor with the breeder of foreign stock—more particularly with the Americans, who have no taxes to pay and a boundless extent of fresh land with daily increasing facilities of transit.

Landlords will be forced to attend better than they generally have done to the management of their estates, and to give aid and encouragement to improving tenants, from which a large produce will arise and fuller employment for the agricultural labourer.

The visit to the sea-side at Redcar seemed only to increase the painful renal disease from which Bates was suffering. He returned home to Kirklevington, but it was some time before he could be prevailed on to consult a doctor, and when he did he refused the greater part of the medicine.

His heart was with horn and hoof to the last. Those who strolled with him in his pastures recalled how the cows and even the young heifers would lick his hand, and seem to listen to every gentle word and keen comment, as if they penetrated its import; and even when the last struggle was nigh, and he could wander amongst them no more, he reclined on some straw in the cow-house, that his eye might not lack its solace.[1]

Bates gradually sank, and died on the 25th of July 1849, aged seventy-four. He was buried at Kirklevington on the north-east side of the church. Some years afterwards Mr. Housman and a few other friends who 'appreciated his labours for the improvement of British stock and respected his character' raised a simple memorial to mark his resting-place.[2] In pursuance of the wish of his nephew, Thomas Bates of Heddon and Kirklevington (who died 30th January 1882), a stained-glass window has since been placed to his memory in the new nave.

[1] *Saddle and Sirloin*, p. 150.
[2] The date of his birth is inaccurately given on this monument as '21st June 1776,' and that of his death as '26th July 1849.'

A very painful chord was struck at the Yorkshire Agricultural Meeting of 1849, when hundreds of friends, who expected once more to grasp by the hand 'the old man eloquent' of Kirklevington and to enjoy his half-sportive half-sarcastic lecture on each prize beast, learnt for the first time that Mr. Bates had gone to his rest and that their shorthorn festival was on his funeral day.[1]

The estimate of his character that appeared in the *Farmer's Magazine* is in many particulars an echo of that already given of his mother's first cousin and his own godfather, Arthur Blayney:[2]

As a man, there were few who enjoyed a wider range of popularity. The employment he gave to the poor did not more ingratiate him in their favour, than the unvarying and unmingled kindness he at all times displayed, whether in providing for their cheap and comfortable shelter in his cottages, or ministering to their wants in sickness, infirmity, or age. His kindness as a neighbour was beyond all praise. Scarcely one of the farmers, whose cold barren clay farms surrounded him, but could bear witness to some act of disinterested sympathy. A stranger would have witnessed with surprise the influence his name and opinion had upon them. His word was more relied upon than many men's bonds. In hospitality to all comers he was seldom equalled; his house was open to every one of whatever grade, from the Peer or Member of Parliament, down to the small undistinguished farmer. The longer the visitor stayed, and the more he partook of his liberality, the more welcome he was. On one occasion, two very celebrated short-horn breeders intimated a wish to spend a day at Kirklevington and examine his stock. He immediately wrote in reply, that it was impossible for them to examine them thoroughly in one day, and that they must make up their minds to spend three with him. In fact his house was the home of all who entered it. They had a welcome truly English; no pains nor unostentatious attention were spared to make them comfortable and happy; while his long stories, founded on bygone experience of great breeders of the early part of his life—the dark ages of shorthorns—were so amusing, that the time flew in his society completely unawares, and to no one did it do so more swiftly than himself. In fact but once set him on with his anecdotes and there never seemed to be a

---

[1] *Saddle and Sirloin*, p. 149. [2] See above, p. 17.

termination, nor even a breaking place when the meeting could be separated.

However inconsistent with all this it may appear, he was often in hot water with some of his opulent and influential neighbours, and more than once drove the bench of magistrates from court to court at enormous expense. Amongst many of these he was looked upon as meddling, overbearing, and litigious; certainly the pertinacity with which he opposed many of their measures of a public nature, and the very great expenses he incurred, might seem to justify the opinion. But those who entertained it did not understand Bates. The dispute was not private nor personal; it was of a purely public origin, and with him the course he took was looked upon as a great public duty, and one that he felt he would be false to himself to abandon. His litigiousness was but a discriminating view of public duty; and had the magistracy used a little conciliation to a man so well disposed and time-honoured, they might have achieved any concession they required, consistent with the great sense of public responsibility which Bates held with most scrupulous conscientiousness. On this point, perhaps, his judgment did not always equal his zeal and perception of right. Convince his judgment or appeal to his feelings, and he was gentle and yielding; but once rouse his opposition and he was as untiring in his warfare as he was staunch and unflinching. On one occasion he heard that a living, in a parish in which he felt concerned, was about to be given to a clergyman whom he believed to be unworthy; and though the kindest man living, he determined to prevent it. He wrote to the bishop and the archdeacon, confronted the parties interested, and, by dint of persevering opposition, effected the rescinding of the gift, and obtained an appointment congenial to his wishes. One instance of the general benefit arising from his exertions, was the clause inserted in the Highway Act, removing the power of electing surveyors out of the hands of the magistracy, and placing it in those of the ratepayers.

He was fond of public life, and was not altogether free from a love of excitement. Once he risked the cost and labour of setting afloat a county contest for the representation; and had indeed great delight in addressing the public, using very strong language, and always appearing in earnest. He wrote a vast number of letters to the newspapers, mainly on the politics of agriculture, and was always at his post at a county meeting, or election, where anything agricultural was the subject of investigation or remark. His writing, though neither elegant nor classic, was true and forcible, and he had a remarkable tact in making his facts bear upon his propositions, as well as a wonderful readiness in calculation and mental arithmetic. It was, however, Bates's character as a Christian which gained him the large amount of respect he so generally enjoyed. An undeviating course of moral conduct, absolutely untainted and unimpeachable, gave him a standing, which, though it might for the moment excite the ridicule of the thoughtless, generally created a real respect in their minds. At a period when a profession of religion was by no means so fashionable as it afterwards became, he dared ridicule and despised scorn.[1]

[1] *Farmer's Magazine*, 1850, i. pp. 6, 7.

CHAPTER X

THE DISPERSION

1850–1895

Of Bates's five surviving nephews Edward, the youngest, was the only one brought up to the profession of agriculture. The King of Prussia, Frederick William IV., on his visit to Windsor, had been much struck with English farming, the cattle and hedges in particular. Through the Ritter von Bunsen, then Prussian ambassador at the Court of St. James's, Edward Bates was induced to migrate to Prussia with the promise of a model farm on his own terms. The revolution of 1848 supervening, the new Prussian Ministry refused to be bound by the word of their predecessors. As some compensation for this breach of faith and the expenses he had incurred, Edward Bates received a long lease of the royal domain of Cloeden, situated on the Elbe near Wittenberg, on ordinary conditions of tenure.[1]

---

[1] Edward Bates bought, as will be seen, Wild Eyes 28th and Red Rover (10692) at the Kirklevington sale. During a residence of thirty years in Germany he did all in his power to introduce shorthorns into that country. His own were excellent cattle for the purposes of the dairy and draught. The fine work-oxen he sent over fat to the Newcastle market are still remembered. Germans, however, would not pay for really good animals; one lady insisted on carrying off a young bull and then sent a turkey-cock in exchange for it. The only person in the Province of Saxony who had a real knowledge of shorthorns was the Landrath Freiherr von Kleist at Collochau near Herzberg, who formed an excellent cross-bred herd. Shorthorn crosses were much esteemed by the practical peasantry, but the large landowners and the tenants of the royal domains were influenced by the bad judgment of Herr von Nathusius. The history of the comparative failure of shorthorns in Germany

There was thus no one of the family to interest himself practically in the management of the herd at Kirklevington. Influences were at work that made the task doubly difficult. The sheets with the services of the cows were mysteriously torn out of the private herd-book, and then equally mysteriously replaced. The whole herd was offered to Lord Ducie and declined by him; he also rejected Mr. Harvey Combe's overtures for a joint purchase.[1] It was then determined to sell it by auction. An earlier date was fixed than had been intended, so that Vail and other Americans only reached England when everything was over. All the same a company of at least five thousand persons, representative of the shorthorn interests of the Old and New World, gathered at Kirklevington on Thursday the 9th of May 1850.

The Kirklevington herd, at the time of its dispersion, comprised forty-eight cows, heifers, heifer calves, and twenty bulls and bull calves, and displayed an eminence in every point of excellence which has rarely been attained. It may be asserted with confidence, that it was unequalled by any other of the short-horned variety of cattle in existence. Magnificent size, straight and broad back, arched and well spread ribs, wide bosom, snug shoulder, clean neck, light feet, small head, prominent and bright but placid eyes, were features of usefulness and beauty which distinguished this herd in the very highest degree. Whilst the hide was sufficiently thick to indicate an excellent constitution, its elasticity when felt between the fingers and thumb, and its floating under the hand upon the cellular texture beneath, together with the soft and furry texture of the coat, evinced in an extraordinary degree excellent quality of flesh, and disposition to rapid taking-on fat. In the sixty-eight head of cattle, not one could be characterised as inferior or even as mediocre—all ranking as first-class animals. When any idea of inferiority did arise it was only in reference to a comparison with some others of this

is given in Dr. Georg May's *Die Erfolge der englischen Shorthornzucht in Deutschland*, Berlin 1875. One of the principal shorthorn breeders in Saxony, whose name was well known across the Atlantic—'dessen Name auch jenseits des Oceans einen guten Klang hat'—told Dr. May (p. 53) that half-bred bulls were often shipped as pure shorthorns to Germany, so that it was not strange that frequent complaints were heard of their inability to stamp their image on their offspring.

[1] *Farmer's Magazine*, 1859, i. p. 430.

splendid herd, which, from their most extraordinary excellence, attracted especial attention.[1]

The auction opened with the sale of the eleven-year-old Oxford 2nd to the Marquis of Exeter. Four other old cows followed, and as they were knocked down at about twenty guineas a head, ironical cheers and cries of 'fat prices, butchers' prices' broke from a knot of breeders who had long smarted under Bates's trenchant criticisms of hard handling, lack of quality, and over-training.[2] Their exultation was soon checked by the spirited biddings of Mr. Anthony Maynard.[3] Lord Ducie, who attended the sale in person, though he was suffering from gout and could scarcely walk, secured Duchess 55th for 105 guineas. Mr. Booth and Mr. Torr had a strong notion of going as high as 300 guineas for her jointly,[4] but by some hookey-pookery she was sold as a doubtful breeder, served on All Fools' Day by 2nd Duke of Oxford, whereas she eventually calved on the 25th of October to the 4th Duke of York. Duchess 59th went also to Tortworth at 200 guineas. No sooner had he bought her than Lord Ducie said, 'I'll put her to Usurer and improve her shoulders.' The battle was expected to rage most fiercely round the 4th Duke of York (10167). 'He was the *beau idéal* of bovine excellence. His magnificent size and perfection in every point entitled him to be considered the brightest gem of the herd. If not the very best bull then in existence he certainly could not be surpassed. Bates had valued him at 1000 guineas, and it was expected that he would go for 500 at the least.'[5] Lord Ducie had determined to buy him or make him dear for somebody. The earl was well known and popular among breeders, and it had been dexterously given out that to bid against him was only like taking money out of his pocket. Acting, in racing phrase, as 'a good

[1] *A Review of the Progress of the Improvement and Value of Shorthorn Cattle*, by John Ewart, Newcastle, 1875, p. 36.
[2] Bell, p. 318.
[3] *Ibid.* p. 319.
[4] The Druid in *Farmer's Magazine*, 3rd series, xv. p. 374.
[5] *Farmer's Magazine*, 1850.

beginner,' he promptly put in the sensational bid of 200 guineas. Although one gentleman at least wished to have the bull at 200 more, a sort of stagnation supervened.[1] Bates's representative nephew was only restrained from bidding by the undertaking that the sale was without reserve, and the glass ran out.[2] If Lord Ducie's first bid had been only 100 guineas, at least three competitors would have gone in.[3] It resulted that the highest price of the day, 205 guineas, was given by Mr. Hay of Shethin for the less remarkable bull Grand Duke (10284).

COWS AND HEIFERS.

	Guineas
Oxford 2nd. Marquis of Exeter, Burghley House, Stamford	52
Wild Eyes 5th. Mr. A. Stevens, New York	20
Waterloo 4th. Mr. Singleton, Givendale, Pocklington	21
Foggathorpe, 2nd. Mr. Parker, Yanwath Hall, Penrith	21
Wild Eyes 7th. Mr. Jefferson, Preston Hows, Whitehaven	23
Wild Eyes 8th. Marquis of Exeter, Burghley House, Stamford	40
Duchess 51st. Mr. S. E. Bolden, Red Bank, Lancaster	60
Foggathorpe 4th. Mr. W. Sanday, Holmepierpoint, Nottingham	50
Oxford 4th. Mr. E. James, Wylam Hall, Newcastle-on-Tyne	27
Duchess 54th. Mr. Eastwood, Burnley, Lancashire	90
Duchess 55th. Earl Ducie, Tortworth Court, Wotton-under-Edge	105
Duchess 56th. Mr. H. Ambler, Watkinson Hall, Halifax	52
Oxford 5th. Mr. L. S. Morris, Fordham, Westchester Cy., New York	71
Wild Eyes 14th. Mr. Jonas Webb, Babraham, Cambridge	29
Wild Eyes 15th. Mr. T. Fetherstonhaugh, Kirkoswald, Penrith	31
Wild Eyes 16th. Mr. Higgs, Stamford	22
Wild Eyes 17th. Mr. Faviell, Snydale Hall, Pontefract	41
Wild Eyes 19th. Mr. N. Cartwright, Haugham, Louth	60
Cambridge Rose 5th. Mr. S. E. Bolden, Red Bank, Lancaster	45

[1] *Saddle and Sirloin*, p. 154 n.
[2] Bell, p. 317.   [3] *Saddle and Sirloin*, p. 154 n.

	Guineas
Oxford 6th. Earl Ducie, Tortworth Court, Wotton-under-Edge	125

OXFORD 6th.

	Guineas
Wild Eyes 21st. Mr. A. Morison, Mountblairy House, Turriff, N.B.	47
Waterloo 9th. Mr. R. Ashton, Bury, Lancaster	76
Wild Eyes 22nd. Mr. H. Champion, Ranby House, East Retford	100
Wild Eyes 23rd. Mr. A. L. Maynard, Marton-le-Moor, Ripon	100
Wild Eyes 24th. Mr. Drummond	40
Waterloo 10th. Mr. A. L. Maynard, Marton-le-Moor, Ripon	60
Duchess 59th. Earl Ducie, Tortworth Court, Gloucestershire	200
Wild Eyes 25th. Mr. B. Baxter, Marsden Hall, Colne	71
Waterloo 11th. Mr. Eastwood, Burnley, Lancashire	70
Wild Eyes 26th. Mr. Haigh, Cameron Bridge, Fife	30
Duchess 61st. Earl of Feversham, Duncombe Park, Helmsley	100
Duchess 62nd. Mr. H. Champion, Ranby, East Retford	120
Oxford 9th. Mr. A. L. Maynard, Marton-le-Moor, Ripon	40
Wild Eyes 27th. Mr. N. Cartwright, Haugham, Louth	43
Cambridge Rose 6th. Mr. Harvey Combe, Cobham Park, Surrey	70
Oxford 10th. Mr. L. S. Morris, Fordham, New York	51
Wild Eyes 28th. Mr. E. Bates, Cloeden, Prussia	26
Waterloo 12th. Mr. A. Cruickshank, Sittyton, Aberdeen	42
Wild Eyes 29th. Earl of Feversham, Duncombe Park, Yorkshire	38
Waterloo 13th. Mr. W. Hay, Shethin, Tarves, N.B.	71

	Guineas
Duchess 64th. Earl Ducie, Tortworth Court, Gloucestershire	155
Oxford 11th. Earl Ducie, Tortworth Court, Gloucestershire	125
Oxford 12th. Earl of Feversham, Duncombe Park, Yorkshire	81
Wild Eyes 30th. Mr. S. Townsend, Sapcote Fields, Hinckley	23
Cambridge Rose 7th. Mr. J. H. Downs, Grays, Essex	25
Oxford 13th. Mr. J. Becar, Smith Town, Suffolk Cy., New York	63
Foggathorpe 6th. Mr. Gardiner	30
Oxford 14th. Mr. J. H. Downs, Grays, Essex	20

BULLS.

	Guineas
2nd Duke of Oxford (9046). Earl Howe, Gopsall, Atherstone	105
Duke of Richmond (7996). Mr. A. L. Maynard, Marton-le-Moor	120
Lord George Bentinck (Wild Eyes) (9317). Mr. Annett, Widdrington, Northumberland	28
3rd Duke of Oxford (9047). Mr. J. Robinson, Clifton, Olney	61
3rd Duke of York (10116). Mr. G. D. Trotter, Bishop Middleham, Ferry Hill	71
Euclid (Foggathorpe) (9097). Duke of Sutherland, Trentham, Staffordshire	40
Refiner (Wild Eyes) (10695). *Not returned from hire.*	
4th Duke of York (10167) Earl Ducie, Tortworth Court, Gloucestershire	200
Chevalier (Foggathorpe) (10050). Mr. Pullen, Boroughbridge	41
Parrington (Wild Eyes) (10590). Mr. Fisher	24
Grand Duke (10284). Mr. W. Hay, Shethin, Tarves, N.B.	205
Chieftain (Foggathorpe) (10048). Rev. W. F. Wharton, Barningham, Yorkshire	41
Red Rover (Wild Eyes) (10692). Mr. E. Bates, Cloeden, Prussia	35
Beverley (Oxford) (9964). Mr. S. Townsend, Sapcote Fields, Hinckley	31
Ebor (Foggathorpe) (10184) Earl of Feversham, Duncombe Park, Yorkshire	90
Bates (Wild Eyes) (9918). Earl of Burlington, Holker Hall, Milnthorpe	155

	Guineas
Retriever (Wild Eyes) (10707). Earl of Carlisle, Castle Howard, Yorkshire	50
Duke of Athol (10150). Mr. Parker, Yanwath Hall, Penrith	40
5th Duke of York (10168) Mr. R. Bell, Kirklevington	32
6th Duke of York (10169). *Dead.*	
Crusader (Wild Eyes) (12666). Mr. John Blackstock, Hayton Castle, Maryport, Cumberland	40
Wonderful (Wild Eyes). Mr. H. Smith, The Grove, Bingham, Notts	30

The whole herd brought £4558 : 1s. The six families of which it consisted have since taken rank according to the average prices they realised at the sale—the Duchess (£116 : 5s.); the Oxford (£68 : 16 : 4); the Waterloo (£59 : 10s.); the Cambridge Rose (£49); the Wild Eyes (£48 : 2 : 7); and the Foggathorpe (£46 : 19s.). The entire herd of sixty-eight head averaged £67 : 0 : 7.

Although the prices at the Kirklevington sale fell far short, either individually or on an average, of those obtained at the sale of the Ketton herd in 1810, or at that of the Barmpton herd in 1818, being indeed lower on the average than those at the Wiseton sale in 1846, this depreciation certainly did not arise from any inferiority of the stock. On the contrary, the Kirklevington animals were generally considered the most excellent shorthorns ever seen together, and the comparatively low prices they fetched were to be wholly attributed to the crushing effects of the Free Trade policy on which the country had just embarked. The period was about the blackest of the century; at Christmas 1850, the average prices at Smithfield Great Cattle Market ranged from 3s. to 3s. 10d. per stone of 8 lbs. It should be remembered also that every horned animal on the farm was openly exhibited and honestly sold; there were no stow-aways and there was no pricking.

There can be little doubt that a persistent course of line-breeding and in-breeding, though it had concentrated the physical qualities of the herd, had restricted its numbers. A cross of the prolific Wild Eyes tribe might have produced the same good results in increased fecundity that had been previously gained from the Princess crosses. Bates had no intention himself of keeping to a rigid system of line-breeding, and was perfectly cognisant of the advantages of a large herd for purposes of selection, while the keeping of many bulls was with him a ruinous foible. It was to the union of the three bloods of Duchess, Old Cherry, and Princess that he ascribed the excellence of his cattle; and with these united in the six tribes he kept on his home farm, and in the eight tribes—the Acomb,[1] Darlington, Fletcher,[2] Georgiana,[3] Hart,[4] Hudson,[5] Northallerton,[6] and Place,[7] entrusted to his tenants Messrs. Bell, he considered he had every means not only of preserving their excellent qualities, but of improving them more and more with every fresh cross. Far from advocating a slavish adherence to bulls of the Duchess or Oxford tribes, he specially rejoiced in the fact that he was 'not dependent on any one bull, as every cow in the whole herd had more or less Duchess blood, with all the excellencies cattle so bred possessed.'[8] He realised, too, the dangers of in-breeding.

Instead, however, of being content to preserve and improve their purchases on the safe lines Bates had traced, most of the new owners of the Kirklevington animals rushed at once to the most violent out-crosses. Lord Ducie, as he had threatened, mated Duchess 59th with Usurer (9763), a Mason bull which he had bought for 400

---

[1] In his notice of the Acombs, *Leading Shorthorn Tribes*, p. 2, the Rev. W. Holt Beever erroneously referred to the Messrs. Bell as Bates's 'relatives and assistants.' They were neither one nor the other, but simply his managers and tenants on the understanding that he had control of all shorthorn matters.

[2] See above, p. 240.

[3] 'Bred from the stock of Mr. Thomas, Mount Pleasant, Barnard Castle.' —Catalogue of Brockton House Sale, 1871.

[4] From the stock of Mr. Richardson of Hart.

[5] 'Bred from the stock of Mr. Hudson of Thirsk, who had used Mr. Hincks's bulls from Breckenborough for many years.'—*Ibid*.

[6] From the stock of Mr. Maynard of Eryholme. This tribe afterwards divided into the Kirklevington, Nosegay, and Nettle branches.

[7] 'Bought of Mr. Place of the firm of Hutchinson and Place, bankers at Stockton, who used Mr. Hutchinson's bulls.'—MS. note by Mr. Thomas Bell. *Ibid*. [8] Bell, pp. 81, 82.

guineas at the Wiseton sale of 1848, and which was largely impregnated with the Favourite blood that Bates had striven to eliminate as far as possible from the Duchess tribe. A dead bull-calf was the result of the first essay, a white heifer-calf of the second. This latter, Duchess 67th, was so unpromising that the earl sent for Strafford, the auctioneer and herd-book proprietor, and told him, 'Bates is right and I am wrong. I will never cross the Duchesses and the Oxfords again with anything except themselves.' It is to Lord Ducie's mistaken view of Bates's principles that the system of exclusive breeding in two lines that has so much affected the Kirklevington shorthorns for good or for evil is to be mainly attributed.

Mr. Bolden in his turn put the nine-year-old, Duchess 51st, which he had bought as a doubtful breeder for sixty guineas, to Mr. R. Booth's Leonidas (10414) and was rewarded with a dead heifer-calf. He then tried her with the four-year-old Grand Duke (10284) which Mr. Hay sold him at 205 guineas, his cost price at Kirklevington, and she brought him the renowned Grand Duchesses 1st and 2nd. Grand Duke had been considered an unsatisfactory sire at Shethin, but of his produce with Mr. Bolden, Grand Duchess 2nd, Cherry Duchess 1st, and 2nd Duke of Cambridge, are allowed on all hands to have been animals of very superior personal merit, and the rest, if not equally excellent, to at least take rank in the first class of shorthorns.[1]

The two unalloyed roan cows Grand Duchess and Grand Duchess 2nd, were ushered into the critical presence of 'the Druid' in the yard at Springfield:

A noble pair they were. The elder was a beautiful specimen of a 'toucher,' silky hair on a nice elastic hide, with that peculiarly dainty cellular tissue between the hide and flesh. The head, too, had all the most favourable characteristics of the tribe, slightly dished in the forehead, with a prominent nostril, and a great general sweetness of expression. They were also well down in the twist, and great milkers, combined with heavy

[1] *The Improved Shorthorn*, by William Housman, 1876, p. 52.

flesh. Grand Duchess 2nd bore a strong family likeness to her sister, but she had more substance and gaiety of carriage; and she held up her head as if right conscious of her lineage.[1]

At the time Mr. Bolden bought Grand Duke (10284) he and his father had several cows almost useless, after having been served repeatedly by idle bulls; but with him and successive Duchess bulls, the fertility (which Mr. Bates attributed, in the case of the Duchesses, to the cross with Belvedere) gradually

GRAND DUCHESS 2nd.

returned. The same was observable in other herds where Duchess bulls were introduced, and Earl Ducie did not conceal his opinion that his was saved by the use of them.[2]

After two years' use, Mr. Bolden resold Grand Duke (10284) for 1000 guineas to Mr. Samuel Thorne, U.S.A., and then lapsed to the use of out-crosses, but out-crosses considerably diluted. His 3rd Grand Duke (16182), for example, had only an eighth of Fame blood. That severe critic of unalloyed shorthorns, the Rev. John Storer, described him as 'one of the very best Bates bulls he ever saw. He was of great size, style, and character, and had a beautiful, yet quite masculine head, admirable hair and touch, with abundance of heavy flesh of very fine quality.'[3]

To continue the catalogue of disastrous *mésalliances*, Lord Feversham bred from his 100 guinea Duchess 61st a heifer Desdemona by Ben Nevis (9960), but dam and daughter then both disappear. Mr. Harvey Combe

[1] *Saddle and Sirloin*, p. 386 n.   [2] *Ibid.* p. 387 n.
[3] *Leading Shorthorn Tribes*, by Rev. W. Holt Beever, p. 102.

thought he was going to do wonders by sending Cambridge Rose 6th, which he had purchased for 70 guineas, to be served at Newham Grange by Bates's abhorrence, Belleville (6778), at a fee of 20 guineas; the result was The Beau (12182),[1] a bull 'with a grand forehand but coarse,' like his sire, 'in many other of his points.'[2]

Anxious to obtain the sole control of the Duchess family, Lord Ducie sent his agent to Mr. Dale Trotter's sale, 11th May 1852, and bought the 3rd Duke of York (10166) for 10 guineas, which was then taken to Tortworth and killed. Mr. Fawcett of Childwick described him as the greatest wreck of a pure shorthorn bull that he ever saw.[3] Soon afterwards Lord Ducie met Mr. Tanqueray of Hendon in London and asked him what he was doing in shorthorns. 'I have just got possession of the 5th Duke of York,' was the reply. 'Confound that bull,' said the earl, 'I had lost sight of him.'[4]

In consequence of Lord Ducie's death the Tortworth herd was offered for sale on the 24th of August 1853. This time there was no mistake about the date, and the Americans who had been absent at Kirklevington turned up in full force:

It was a veritable Bunker's Hill removed. England was pitted against America once more—the guineas of the old country against the 'almighty dollar' of the new. Messrs. Becar and Morris bid by their agent; but Mr. Thorne did his own business, in a cool quaker-like style with which it was almost hopeless to cope. But for Mr. Gunter and Mr. Tanqueray, who upset all the wise counsels which had been taken at the Gloucester caucus overnight, the Duchess tribe would have departed bodily across the Atlantic. Previous to the Tortworth sale, Mr. Gunter had only a few Alderneys and ordinary shorthorns. He had not made up his mind as to whether he should buy on that day, but the bitter complaints of some Gloucestershire farmers who shared his waggon, as to the Americans getting Duchess 59th fired him into action at last. He accordingly bid 200 guineas for the twentieth lot, Duchess 64th, but it was hardly taken, and his 400 guineas was soon left in the rear by his

---

[1] *Farmer's Magazine*, 1859, i. p. 430. [2] *Ibid.* p. 431.
[3] *Bell's Weekly Messenger*, 2nd December 1872. [4] Bell, p. 319.

Transatlantic rivals. He did not touch the 700-guinea Duchess 66th, but Duchess 67th, the fifteen months' heifer by Usurer out of Duchess 59th (the highest priced female at the Kirklevington sale), fell to his nod for 350 guineas. Then Duchess 70th by Duke of Glo'ster (11382) out of Duchess 66th followed suit for 310 guineas. She was only a trifle over six weeks. The Americans had no idea of leaving her; but as one of them afterwards said, 'It was the way that other bidder said, "*and* ten guineas" almost before my bidding was out of my mouth that made me falter and give in.'[1]

The Kirklevington animals sold at Tortworth were:—

	Guineas
Duchess 55th, by 4th Duke of Northumberland (3649), Mr. J. S. Tanqueray	50
Oxford 6th, by 2nd Duke of Northumberland (3646), Mr. Tanqueray	205
Duchess 59th, by 2nd Duke of Oxford (9046), Mr. Jonathan Thorne, U.S.A.	350
Duchess 64th, by 2nd Duke of Oxford (9046), Mr. J. Thorne	600
Oxford 11th, by 4th Duke of York (10167), Mr. Tanqueray	250
Duchess 66th, by 4th Duke of York (10167), Messrs. Becar and Morris, U.S.A.	700
Duchess 67th, by Usurer (9763), Mr. Gunter	350
Oxford 15th, by 4th Duke of York (10167), Earl of Burlington	200
Duchess 68th, by Duke of Glo'ster (11382), Mr. J. Thorne	300
Duchess 69th, by 4th Duke of York (10167), Mr. Tanqueray	400
Oxford 16th, by 4th Duke of York (10165), Mr. Tanqueray	180
Duchess 70th, by 4th Duke of Glo'ster (11382), Mr. Gunter	310
Duke of Glo'ster (11382) by Grand Duke (10284), Becar and Morris and Tanqueray	650
4th Duke of York (10167) by 2nd Duke of Oxford (9046), General Cadwallader and Mr. Vail, U.S.A.	500
5th Duke of Oxford (12762) by Duke of Glo'ster (11382), Earl of Feversham	300

It is not surprising that the Bates family felt aggrieved at the result of the Kirklevington sale, when the identical

---

[1] *Saddle and Sirloin*, pp. 155, 156 nn.

six head of cattle which Lord Ducie bought there for £955 : 10s., were sold at Tortworth for £2052 : 15s.

His voyage across the Atlantic proved fatal to the 4th Duke of York, as he broke his neck in a storm. The latter end of the 5th Duke of Oxford, with Lord Feversham at Duncombe Park, is chronicled in *Saddle and Sirloin* :—

The Griff farm to which we were bound, lies about a quarter of a mile from the park, along a field route, lined at intervals with those dark green holly trees peculiar to the North Riding, and which catch a stranger's eye at once from their enormous size. Ear, however, came into play before eye, when we at last neared the box of the 5th Duke of Oxford, and were saluted with a roar quite worthy, in its depth and tone, of a Libyan King of Beasts. He looked the character to the life, with that shaggy, lion-like old head and mane, as he was at last led forth, snorting, in blinkers. The fine length, beautiful touch, and rare union of hip, loin, and rump, take the eye as much as ever; but although he had been reduced some twenty stone since he won the Chester and Northallerton prize ribbons (1858), his day of usefulness like his temper was gone, feeding for show had done its fatal work.[1]

OXFORD 15th.

The Grand Duchesses of Oxford at Holker all descended from Oxford 15th purchased at Tortworth. The Earl of Burlington succeeded his cousin as seventh Duke of Devonshire in 1858.

[1] *Saddle and Sirloin*, p. 217.

Duchess 54th, purchased by Mr. Eastwood at Kirklevington, produced in Mr. Tanqueray's hands Duchess of Athol by 2nd Duke of Oxford (9046) and 2nd Duke of Athol by Lord George (10439), a Toy bull. Both of these crossed the Atlantic to Mr. R. A. Alexander's, Woodburn, Kentucky, and from their union sprang the Duchesses of Airdrie.

The old cow Duchess 55th did not breed again, and Mr. Tanqueray transferred his younger Tortworth venture Duchess 69th at nine months old to Mr. Gunter. She was then sent back to Mr. Tanqueray's Duke of Cambridge (12742), while Duchess 67th went at a 25-guinea fee to the 4th Duke of Oxford (11387). Archduke and Duchess 72nd were the respective produce.

Mr. Gunter's next purchase was the 6th Duke of Oxford at the Hendon sale, 24th April 1855, for 200 guineas, and his dam Oxford 11th for 500 guineas more, when she was just four years old. He originally intended to have bought the Duke of Cambridge (12742); but Mr. Strafford's glass ran out in favour of Sir Charles Knightley, who, strange to say, had his eye rather on the 6th Duke of Oxford (12765).[1]

Next to his favourite maxim that he 'would find a hundred men fit to be Prime Minister for one fit to judge shorthorns,' Bates, with Burley and Farnley in his mind, had another to the effect that there was 'no place for shorthorns like the valley of the Wharfe.' Acting as it were on this, Mr. Gunter moved his herd in August 1857 from Earl's Court near Kensington to his estate of Wetherby Grange. 'The Druid' happened to be at Wetherby on the 24th of January 1859, the day the famous 7th Duke of York (17754) was born, and subsequently recalled the impressions of his visit :—

When we first saw the herd in '59, not long after its removal from Earl's Court, we began with the earliest purchase Duchess

---

[1] *Farmer's Magazine*, 3rd series, xv. p. 374.

67th, and her daughter 72nd, the first calf that Captain Gunter ever bred. Her next daughter the white 75th was third in the array, and the handsomest of the three, and then came 'the twins' 78th and 79th which ran such a splendid career in the show yard. We see the little roan and white through the mist of years once more struggling with the herd boys, and thought the roan rather nicer in her coat, but the white neater, and in after years the bench hardly knew which to take. Having thus exhausted the fruits of the first Tortworth bid, Duchess 70th bore her witness to the second with her calves 73rd and 77th, and we look back to our comment that 'the former had more substance and the latter more elegance of the twain,' and that she was the best, but no one dared predict such a future for her. She rose the Royal ranks step by step, third as a yearling at Warwick, second at Canterbury, and first at Leeds. Duchess 69th had only calved that morning, and though we could not rouse her after the labours of the day we could judge of her fine scale and enjoy the gentle grandeur of the head, which had been specially modelled for Mr. Brandreth Gibbs' testimonial.[1] Sixth Duke of Oxford (12765) was waiting outside to receive us; he was a perfect Esau at his birth, and there could be no doubt whence his stock derived their rich hair.[2]

Mr. Thorne was unfortunate with his lot, for Duchess 68th was killed on board ship by the falling of a mast, and Duchess 59th and Duchess 64th founded no families. The last-named had been originally purchased by Mr. S. E. Bolden, who agreed to make her over to Mr. Thorne on condition of receiving the calf which was due a month after the sale. This, named Grand Duke 2nd (12961), was subsequently purchased as a two-year-old by Mr. Thorne at 1000 guineas in order to replace Grand Duke (12084), who, disabled and useless, was ultimately slaughtered.

On the other hand, Messrs. Becar and Morris were certainly lucky in their purchase of Duchess 66th, 'that brand plucked from the fire,' as Earl Ducie is said to have termed her, when the news of her birth was carried to his

[1] There was a delightful irony in this after the way in which Bates had been treated by the Director of the Yard, previous to the Royal Show at York in 1848.
[2] *Saddle and Sirloin*, p. 276 n.

dressing-room one morning.[1] On Mr. Becar's untimely death in 1854, Colonel Morris purchased his moiety of the herd, and after the sale of many valuable bulls to various breeders, in the different States, transferred the whole fifty-three to Mr. Thorne for £7000. This included among others Duchess 66th and her daughters Duchess 71st and Duchess of Fordham, the cows Oxford 5th, 6th, 13th, 17th, and 20th, Maid of Oxford, Bride of

DUCHESS 66th.

Oxford, Romeo's Oxford, Glo'ster's Oxford, and Beauty of Oxford, together with the Tortworth bull Duke of Glo'ster (11382). Mr. Thorne gave his animals of the Duchess family the titles of Dukes and Duchesses of Thorndale.

On commencing to breed shorthorns Mr. Harvey Combe had been warned that he 'would find them sweal away like a candle.' The crowd of quite fifteen hundred that fourteen years later gathered at the sale at Cobham on the 23rd of March 1859, gave this opinion the lie direct. The first lot fell for 27 guineas, to a gentleman who bore the ominous name of *Butcher*, but all forebodings after so bad a beginning were dispelled the moment the ten-year-old Cambridge Rose 6th—'that rich relic of Kirklevington'—stepped airily into the ring. Her forehand was very gay and beautiful, but rather too light, and her shoulders were a little upright, still she was a very beautiful type of her race. When the glass ran out at 200 guineas, it announced that she was

---

[1] *Ibid.* p. 388 n. The expression seems a strange one on Lord Ducie's part, to hail the birth of a calf on the 25th October, which he had no reason to expect before the 6th of January. See above, p. 331.

destined to leave the Old World for the New.¹ It was the very *beauty* of bidding to see Mr. Downes and Mr. Jonas Webb oppose each other for her heifer (Beauty) by Puritan (9523), a bull which Mr. Combe had bought as a somewhat leggy calf from Lord Ducie, and which did not do much for the herd. There was a sort of calm 'do or die' about the great Southdown King, as he rose in his waggon, catalogue in hand, directly facing Mr. Strafford, and fired off his biddings as regularly as a minute gun. The spectators quite caught the enthusiasm, though they felt assured that the day was Babraham's from the first, and to Babraham The Beauty went at 160 guineas. It was reserved, however, for Cambridge Rose's youngest daughter, Moss Rose, to make the scene of the sale. After much spirited competition, Mr. Hales of North Frith, near Tonbridge, secured her for 260 guineas, and proved that Kent was in earnest in spite of the candle prophet.²

In 1861 the first of all American bulls, 4th Duke of Thorndale (17750), was imported into England and sold to Mr. Hales. At his sale in the following year the fight for this dashing Bates bull was short, sharp, and decisive between Captain Gunter and Lord Exeter's agent. The captain put him up at 200 guineas, and shot his opponent 40 guineas at a time, till 400 guineas was reached. The latter then came again with his '*and* ten' and secured the American roan for Burghley. He earned 250 guineas during his first season in England. In Kent he left at Cobham Park the eleventh and only true-bred heifer of Cambridge Rose 6th. This received the name of Thorndale Rose, and was sold privately in 1864 for 200 guineas to Mr. Betts. Her half sister, The Beauty by Puritan (9523), for which Mr. Jonas Webb gave 160 guineas at the Cobham Park sale, was bought by Lord Braybrooke at the Babraham sale, 24th June 1863, in calf with Heydon Rose by Englishman (19701). With her nine descendants she realised 1253 guineas.

In 1860 Mr. Bolden sold a score of Waterloos by auction at Springfield. Their average was £92 : 13 : 3.

¹ This proved incorrect; she was retained at Cobham by Miss Combe. —*Herd Book*, xvi. p. 370.
² *Farmer's Magazine*, 3rd series, xv. pp. 430, 431.

Sir Curtis Lampson gave 165 guineas for Waterloo 20th, and Mr. E. Bowly 130 guineas for the Waterloo bull Charger (17532). Two years later the whole herd was sold to Mr. Atherton, who soon after parted with the Grand Duchesses—nine cows and four bulls—to Mr. Hegan of Dawpool, Cheshire, by private contract, for £5000. Mr. Hegan died in 1865, and it was arranged to sell his herd at Willis's Rooms without removing it from Dawpool, the auction and the farm being connected by a special telegraph wire. The course was unprecedented. The *Illustrated London News* might well remark:

> Willis's Rooms have been put to many and multifarious uses. Pictures, virtù, and china, have all made fabulous prices in them; but 'the intelligent foreigner,' if he does not happen to be a short-horn man, will be sorely puzzled when he hears that, on 7th June, Mr. Strafford will sell there a number of Grand Duchesses and Grand Dukes. Frenchmen may become puzzled about 'Sir Strafford,' and confound him with the whilom Sir Strafford Canning, or falling back upon their original idea of our wife-selling propensities, conclude that the Grand Dukes, deterred by a more civilised code from getting rid of their spouses on the Continent, have come over in the height of the fashionable season, to do it under the shadow of our English laws and customs. Happily, no foreign States will be bereft of their 'old nobility,' male or female, by the step; but the Dawpool herd will be simply dispersed. The executors reserve to themselves the right of selling the stock in one, two, or more lots. With the exception of Imperial Oxford (18084), the entire herd is descended from the Bates-bred Duchess 51st.[1]

A perfect bridal lunch greeted the congress of about 120 leading short-horn men—peers, M.P.'s, clergymen, and laymen—who attended to see the great battle at Willis's Rooms, over the eighteen Grand Dukes and Duchesses. Lord Feversham was in the chair, supported by General Hood (who came, like several other members of council, direct from Hanover Square), and the Bates men made up a most imposing array; while Mr. Torr and Mr. Thomas Booth were at the head of the great rival house of 'the red, white, and roan.' The noble chairman declared his Kirklevington faith in such unwavering fashion that the Booth men complained he rather ignored Bridecake's share in the Grand

---

[1] *Illustrated London News*, 27th May 1865, p. 514.

Duchess pedigree. Mr. Strafford responded to his own health in the most practical way, by shutting the windows to get rid of the street roar, mounting a table, and proceeding to business, and, we may add, without repeating the new sale creed, which, by the bye, seems most searching, and nearly a foot long. The herd were at Dawpool, where they have had some crowded levees for several days past. It was their late owner's wish that they should be sold in one lot, and it was understood that the Hon. Col. Pennant, M.P., had made an offer of £6000 for the twelve females. Aided by Mr. Drewry, agent to the Duke of Devonshire, who was said to be bidding on behalf of an eminent city merchant somewhat anxious to begin a herd, and Captain Oliver, who was in at 1800, the biddings fairly flew for the first lot (Grand Duchesses 5th, 7th, and 8th); but Mr. Betts's agent was always there, and it fell to him for 1900 guineas. The second lot (Grand Duchesses 9th, 13th, and 18th) followed suit at 1300 guineas; and then the opposition made another rally, but at 1800 guineas Mr. Betts covered them again for the third lot (Grand Duchesses 10th, 15th, and 17th), and when it was seen that he was so determined, no one would bid against him for the fourth (Grand Duchesses 11th, 12th, and 14th), and his 1200 guinea bid remained unchallenged. Their new owner was sent merrily along for Imperial Oxford up to 450 guineas, and then Grand Duke 6th (130 guineas, Mr. Bland), Grand Duke 13th (310 guineas, Mr. T. Walker, of Berkswell Hall, near Coventry), and Grand Duke 10th (600 guineas, Duke of Devonshire). The waiters, whose ideas did not go beyond milch cows standing among ginger-beer bottles in the Mall, stared with amazement again; and the trainer of the Duchesses was so delighted at the last price that he rose and asked the auctioneer, 'How much for the depreciation in shorthorn prices?' an allusion which was caught up and highly relished. Grand Duke 9th and Grand Duke 15th were confessedly rheumatic; but Captain Gunter gave 100 guineas for one, and the other, which is said to be a hopeless case, was left unsold.[1]

The *Times* commented on the sale in a leading article which contained some plain home-truths for the stereotyped run of agricultural societies:

On Wednesday, there was sold at Willis's Rooms a collection of articles, first-rate productions in their way, and the result of extreme industry and skill. Twelve sold for £6510—*i.e.*, for an

---

[1] *Illustrated London News*, 10th June 1865, p. 547.

average of £542 : 10s. each. Five sold for £1699 : 10s.—*i.e.*, for an average of £339 : 18s. each. The average of the whole seventeen was £481 : 3s. each. What could these be? First-class pictures? Works of 'high art'? Mosaics? Manuscripts? They lived and breathed! They were short-horned cattle. The twelve were cows, and the five bulls. They were animals of a noble race. The catalogue is like a bit from the 'Peerage,' giving the pedigree of each Grand Duchess or Grand Duke for a dozen or more generations. Mr. Betts of Preston Hall, Kent, has secured the whole herd of 'Grand Duchesses.' The 'Grand Dukes' are separated; the grandest of all passing, for 600 guineas, to the Duke of Devonshire. The splendour of such an event almost pales the strongest blaze that can be got up by agricultural societies. There is no such test of value, no such triumph of enterprise as that which is obtained, without shows, and judges, and prizes, in the auction room. Here is a plain commercial proof of what can be done, and how far we have advanced upon our forefathers in the matter of kine. But it also proves the difficulty of the work, the necessity of science, and the need there is of agriculturists educated for their profession.

Instead of the out-crosses which Mr. Bolden had tried giving any increased vitality to the Grand Duchess family, the exact contrary was the case, as he should have known from Bates's warnings, Lord Ducie's confession, and his own earliest experiment with Leonidas (10414). He 'corrected the weak places of the Bates type' at the price of a loss of fecundity and constitution. Grand Duchesses 10th, 12th, and 14th all died at Preston Hall, and Grand Duchesses 7th and 13th were slaughtered there. Grand Duke 9th (19876) and Grand Duke 15th, two infirm young bulls, spoilt the sale at Willis's Rooms; an unprejudiced judge described the former, which made 100 guineas, as 'heavy flesh, of fair quality, with a deal of Booth, rheumatic in his fore legs,' and the latter, which had to be withdrawn, as 'a nice young bull but useless from rheumatism or something worse.'[1]

Neither snow nor cold damped the ardour of the 'short-horn

---

[1] *Leading Shorthorn Tribes*, p. 107.

parliament men,' when they met on 14th March 1867, at the Marquis of Exeter's home farm, with Mr. Strafford as 'Mr. Speaker.' The 'Burghley nods' from the Commissioners of Lord Penrhyn and Captain Gunter had a tremendous ten guinea significance, when that 'humble, but meritorious, occupant of our pastures' (as an M.P. recently observed to the House), Fourth Duke of Thorndale (17750), was led into the ring. This rare eight-year-old bull was still very fresh and active, and one of the very best-bred Bates bulls in the world. Fourteen years from the date of the Tortworth sale had not taken the fire out of Captain Gunter's bids, and the bull went to Wetherby Grange at 440 guineas, or a 30 guineas on the late Marquis's purchase money. This is another illustration of the saying that 'really good blood will always pay.' The biddings at Burghley, on behalf of Lord Penrhyn, arose from his wish to have a worthy successor to the celebrated prize bull Duke of Geneva (19614), who died at Penrhyn Castle home farm, on the 26th February 1867, from a tumour, which, for five or six weeks, had been perceptibly closing his gullet. He was just seven years and ten days old, and was by 2nd Grand Duke (12961), from Duchess 71st. His price, on his arrival from America, was 600 guineas.[1]

The breeders of shorthorns and especially the admirers of the Kirklevington blood, enjoyed a rare treat when the magnificent herd of Mr. Edward Ladd Betts of Preston Hall, Aylesford—sixty-two females and ten males—was brought into the auction-ring, on 1st May 1867. It could not be doubted that a large portion of the numerous company were attracted by curiosity to see if the costly animals bought at Willis's Rooms could maintain their price. Many were probably attracted to the sale by a belief, which the result justified, that it would be the last opportunity of seeing the Grand Duchess family intact. Mr. Betts's bad luck pursued him to the very last; his most promising bull calf, Grand Duke 16th, valued at 500 guineas, was perfectly well on the Friday night; on the Saturday morning he was found dead in his stall from hoven. Subtracting the deaths and adding the births, the Grand Duchesses which Mr. Betts had bought for 6200

[1] *Illustrated London News*, 23rd March 1867, p. 291.

guineas in 1865 now brought only 4450 guineas. This was somewhat compensated by the sale of three Grand Dukes for 1025 guineas, while the whole stock sold for £11,187, or the enormous average of £180. The sale was opened by a bid of 100 guineas for Grand Duchess 5th, a truly grand red cow but nine years old. A doubtful breeder, she fell to Mr. Drewry for the Duke of Devonshire at 200

GRAND DUCHESS 17th.

guineas. An on-looker, with no special love for Kirk-levington, described her as 'a trifle leggy perhaps, but with no defect in her general frame, long, lathy, even all over, proud and stylish in her carriage, with long horns as great milkers are apt to have.'[1] Her grand-daughter, the two-and-a-half-year-old Grand Duchess 17th, by Imperial Oxford (18084), a pure Bates bull, made the highest price yet given for a cow or heifer. She was a beautiful cow, red—the Hubback yellow red—and white, with long characteristic hind-quarters, good rib, mossy coat, and splendid touch. When she was put up, 300, 400, 500, 600, and 700 guineas were announced in quick succession, amongst prolonged cheers, the glass then ran, 750 and 800 guineas came more slowly, and finally 850 guineas was debited to Captain Oliver of Sholebroke Lodge, whose pluck in this case proved greater than that of Lord Spencer, the Duke of Devonshire, and others. Grand

[1] *Leading Shorthorn Tribes*, p. 107.

Duchess 18th, calved on the same day, though not so level a cow and of less fashionable colour, fell to the same bidder, after a spirited contest, for 710 guineas. Far advanced in calf, this heifer went amiss when driven home, and never bred again. This loss was quite made up by the good fortune that attended Grand Duchess 17th, who proved well worth the 1560 guineas Captain Oliver

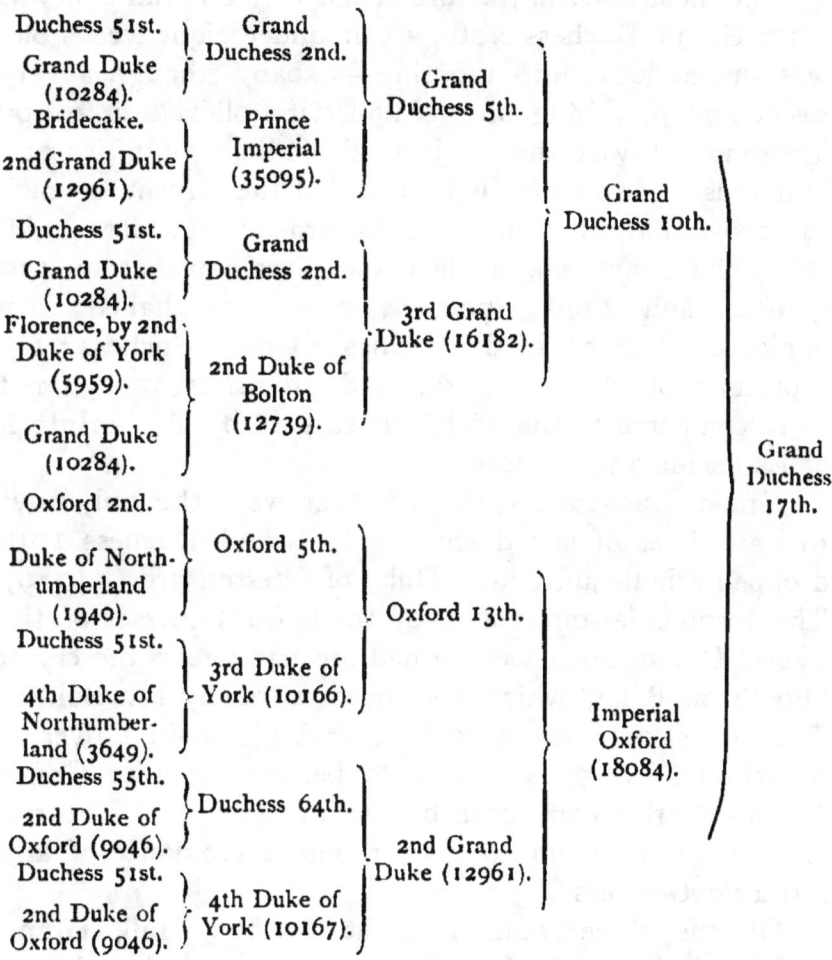

gave for the pair. Ten years afterwards, the Rev. W. Holt Beever described her at Sholebroke as being 'really a marvel of a cow, on such a scale, and yet so refined.

The only possible fault you can point to is a slight droop of the quarter, which may be due to age. She has quite the Duchess head, a perfect front, excels in touch, and through the heart, has a good rib and the levelest of broad backs, although she has bred nine calves. Notwithstanding her great size, she is thoroughly feminine in appearance. She has not in front the breadth of beam a Warlaby prize heifer would exhibit, but she is more like one's idea of a first-class milking cow.'[1]

The most amusing feature of the Preston Hall sale was when Grand Duchess 20th, a calf under eight weeks old, was turned loose into the ring—a baby, though a very sweet and promising one. Whilst it frolicked in happy ignorance, it was the subject of biddings of fifties and hundreds. Roars of laughter hailed the advent of each successive bid, it seemed so absurd to outsiders. All things have an end, at last the glass ran out to 430 guineas, and Lord Spencer's move was hailed with applause. The last of the females, a Grand Duchess 21st, a pretty spotted calf, straight and promising, though not to be compared to the 20th, was secured by Mr. McIntosh of Essex for 330 guineas.

Grand Duchesses 17th and 21st were the only ones to leave lines of she-descendants; Grand Duchess 19th dropped a bull-calf, Grand Duke of Weston 3rd (34079). The terrible lesson taught by the unfruitfulnesses of the hybrid Grand Duchesses hushed for many years the cry of 'Booth on Bates,' which had been raised by Mr. Bolden. The success achieved as a breeder by Captain Oliver at Sholebroke Lodge is greatly to be attributed to Grand Duchess 17th having been by the unalloyed bull Imperial Oxford (18084), and to his scrupulous avoidance of any further out-crosses.

Of the three bulls sold at Preston Hall, Grand Duke 4th (19074), then six years old, was bought by Lord Spencer for 210 guineas. He was the joint property of Mr. Betts and Mr. McIntosh, and had principally been

---

[1] *Leading Shorthorn Tribes*, p. 102.

used by the latter. A right good bull, showing, the Rev. John Storer was kind enough to say, 'almost too much breeding';[1] he was heavy-fleshed all over, low and lengthy, a trifle close behind; his head was kindly but masculine. Grand Duke 16th (24063), a rich red, under two years, with immense length of body, was put up at 300 guineas and rapidly reached 510 guineas, at which sum he was handed over to Mr. Robarts of Lillingstone Dayrell, Bucks. Lastly, Grand Duke 17th (24064), a rich roan with rather low shoulders, long carcase, and great hindquarters, reached by somewhat slow degrees the respectable sum of 305 guineas, bid by Mr. Brogden.

Scarcely less successful at Preston Hall than the Grand Duchesses were the ten descendants of Cambridge Rose 6th.[2] The two older cows, Moss Rose and Red Rose, were by the Gwynne bull Marmaduke (14897); some of the younger by 4th Duke of Thorndale (17750) and Grand Duke 4th (19874)—a combination of the Cambridge Rose and Duchess blood which resulted in some very showy level cows, rich reds and roans, with sweet breedy heads and a deal of substance and good quality. Thorndale Rose fell to Mr. Adams, of blood manure celebrity, but was transferred to Lord Braybrooke.[3]

The sale of Mr. McIntosh's herd at Havering Park, Essex, on the 2nd of May 1867, was quite as great a success as that at Preston Hall on the previous day. It is true there lacked the extraordinary interest attending the distribution of the Grand Duchesses; but a very business-like company assembled, and expressed great satisfaction at the general excellence of the stock, and the fine healthy condition in which they were shown. Mr. McIntosh had been a close follower of Mr. Bates, to whose advice in early life he owed, in some measure, his choice of originals. Looking at the general quality of the herd and the very high average reached, it was evident good judgment had been exercised. The luncheon was unique in its elegance; a Scottish newspaper correspondent was dumbfounded at the sight of 'the boars' heads,

---

[1] *Leading Shorthorn Tribes*, p. 108. [2] See above, p. 345.
[3] Bell, p. 328.

guinea fowls with outspread wings, raised pies, jellies, creams, and towering cakes, decorated with flags, flowers and other ornamentations of festive life.' The most prominent families disposed of were the Oxford and Waterloo; but there had never been a sale with so many lots of Duchess on Knightley. Sir Charles did not like the result when he used Duke of Cambridge (12742), and consigned him and Strafford, who had recommended him, to perdition for having, as he alleged, spoilt the shoulders of his

LADY OXFORD 5th.

cattle. Mr. McIntosh persevered where Sir Charles had failed, and satisfactorily solved the problem. 3rd Duke of Thorndale (17749) and 4th Grand Duke (19874) had been largely used for some years, and both had done good service. The calves by the 4th Grand Duke were a credit to him, and he might have realised a higher figure had he been sold here, in the midst of his offspring, instead of at Preston Hall. There were only four specimens of the Lady Oxford blood, from the American cow Lady Oxford 4th. The first of these, Lady Oxford 5th, a splendid roan, by 3rd Duke of Thorndale, very level, with great substance and quality, and the winner in the calf-class at the Royal Show at Worcester in 1863, was put up at 300 guineas, and attained her 600 guineas in twenty-seven bids. The last bid of 25 guineas was by the Duke of Devonshire. Lady Oxford 6th, her younger sister, was bought in at a reserve of 1000 guineas, Mr. McIntosh having claimed that privilege, and remained with the Grand Duchess calf, purchased at Preston Hall, to commence another herd. Baron Oxford, a two-year-old, by Duke of Geneva (19614), had evidently been fixed upon by more than one, and was bought by Colonel Towneley for 500 guineas. His half-brother by Grand Duke 4th, not a year old, was a very taking animal, and opinion varied as to which was best; he went to

Dumbleton at the same figure. The three animals sold reached the enormous average of £560. It should be mentioned in reference to the Preston Hall sale, that Lord Oxford 2nd, a very grand bull of this blood, imported from America, was kept back on account of lameness, or the bull averages would have been materially increased. The Waterloo tribe was well represented at Havering, and found ready admirers, the eight females and one bull averaging £122 : 12 : 4. Never, perhaps, were such important sales held at a more unfavourable time, in consequence of the cattle plague, yet the result of pluck and judgment was a higher total than had been hitherto reached.[1]

In the autumn of 1867, Mr. J. O. Sheldon of Geneva, Illinois, who had just bought the rest of Mr. Thorne's herd, consisting of forty head, for about £8475, sent over two bulls and a heifer of the Duchess tribe and six Oxford heifers from America. In consequence of the cattle plague regulations they did quarantine on board the *Nestor*, and were then forwarded for sale at the conclusion of a sale of surplus shorthorns at the Shaw Farm, Windsor Park, on the 15th of October. Among the latter there were thirteen good females of the Knightley blood, which averaged about £73, but it was remarked that the prices would probably have been higher if a Bates cross had been used instead of a Booth, as the latter had not 'hit.' After inspecting the American arrivals, the company adjourned for business to the coffee-room of the Castle Hotel. Champagne circulated freely, and for the first time in a long professional career Mr. Strafford sold by candle-light :

Mr. Leney of Kent was the chief bidder, and did it with an emphasis and courage which added great zest to the proceedings. Mr. McIntosh was also highly pugnacious, in the best sense of the word ; he led off with the 3rd Duke of Geneva (550 guineas), and his remorseless 'and ten,' for nearly all the other lots fell like sweet music on every ear but Mr. Leney's and Mr. Culshaw's. Mr. Sheldon's agent had a 600-guinea reserve bid on the white 7th Duchess of Geneva, but he did not exercise it, as a spirited

[1] *Illustrated London News*, April 1867, p. 414.

rally between Kent and Essex made her Mr. Leney's at 700 guineas. 4th Maid of Oxford (Mr. Leney, 300 guineas), 5th Maid of Oxford (Mr. Downing, 200 guineas), and Countess of Oxford (Colonel Kingscote, 250 guineas), followed; and then Mr. Culshaw[1] from his quiet corner answered the Kentish fire, and took 6th Maid of Oxford (400 guineas), and 8th Lady of Oxford (450 guineas) for a new Towneley herd. 7th Maid of Oxford (260 guineas) completed the Leney trio; and then Mr. E. Thorne,

MAID OF OXFORD 7th.

of America, gave 185 guineas for the 12th Duke of Thorndale (a bull bred by his brother, and looking very much out of sorts after his passage), rather than see him go under his value. It was said that 600 guineas had been refused for the 3rd Duke of Geneva in America. The nine averaged £384 : 8 : 4. People differed in opinion as to whether the American lots would have made most under the greenwood or round the mahogany tree, but the sale was unique in character and served to stamp 1867 at the time as an *annus mirabilis* in shorthorn history.[2]

Punch broke out into verse over the sale :—

THE GOLDEN SHORTHORNS

MR. STRAFFORD raised his time-glass, and THORNTON held the pen,
When to a Windsor coffee-room flocked scores of shorthorn men.

---

[1] The Hudsons, Senators, and Deceptions of the Royal Agricultural Society of England had long been forgotten, and Joe Culshaw had come to do homage to the descendants of the Dukes of Oxford, to whose merits his skill in 'training' had completely blinded the 'infallible judges' of the York Show in 1848.

[2] *Illustrated London News*, 19th October 1867, p. 42. 12th Duke of Thorndale was afterwards sold to Mr. D. R. Davies of Mere Old Hall.

They crowded round the table, they fairly blocked the door ;—
He stood Champagne did SHELDON, of Geneva, Illinois.

They talked of Oxford heifers, Duchess bulls, and how the States
Had come into the market with another ' Bit of Bates.'

Their expression is so solemn, and so earnest is their tone,
That nought would seem worth living for but ' Red, and White, and
    Roan.'

All ready for the contest, I view a dauntless three—
The MACINTOSH from Essex, a canny chiel is he.

There's LENEY from the hop yards; 'twill be strange if he knocks
    under,
When once the chords are wakened of that Kentish ' Son of Thunder.'

The Talleyrand of 'trainers' is their 'cute but modest foe,
Him whom the Gods call ' CULSHAW,' and men on earth call ' JOE.'

And sure, it well might puzzle ' the Gentleman in Black,'
When the three nod on ' by fifties,' to know which you should back.

And sure, the laws of Nature must have burst each ancient bound,
When a yearling heifer fetches more than seven hundred pound!

Bulls bring their weight in bullion, and I guess we'll hear of more,
Arriving from the pastures of Geneva, Illinois.[1]

To ' the Druid's ' facile pen we owe two sketches of the herds of Colonel Gunter and the Duke of Devonshire late in the summer of 1868:

Captain Gunter had just been tempted to part with the red yearling heifer Duchess 97th to Mr. M. H. Cochrane of Hillhurst, Canada, for 1000 guineas; the roan calf Wharfdale Rose was sold to go along with her at 100 guineas. The pair held a levee in their barn all the day of the Yorkshire Show, 5th August 1868, and devotees went wandering off through the hot haze into the park to gaze on the four Duchesses, as well as Mild Eyes and her daughter Bright Eyes, and a very fine Waterloo heifer.

The old cows were in the bottom of the park, and took a good deal of finding in the heat. There was the roan Duchess 86th, with the old-fashioned wide-spreading horn; the 87th, of a lighter roan and with a rare loin; the white 88th, which had been amiss; and 91st, one of the same colour and rare substance. The twins (78th and 79th) and the 77th had died or been

---

[1] *Punch*, 2nd November 1867, p. 184. An apocryphal stanza seems to have been added in America in order to bring in ' Barmpton Rose.'

slaughtered, and 96th and 94th were in the home field. Taylor, the herdsman, tells us how once they thought 94th the best, and that the former is the only Duchess which is free from the Usurer cross. The numbers 100th, 99th, 98th, and 97th once roamed together in the home pasture unbroken, but Mr. Cochrane had taken his choice and borne off the last. She is from 92nd, a daughter of 84th, 'which broke down on us as a calf for Leeds.'

DUCHESS 84th.

Her once constant companion 98th from 88th was a white with roan ears. Writers who have to encounter these night-mare numbers may well be among those

> Who dread to speak of '98,
> Who tremble at the name.

The wished-for 100th was reached at last in the shape of a red roan, but a two-days-old roan, half-sister to 'the American lady,' was the latest arrival, and had been entered as Duchess 103rd in Captain Gunter's private herd book. Fourth Duke of Thorndale was the monarch of the yard, and Grand Duchess 8th from Penrhyn Castle was there to share his smiles.

The 2nd Duke of Collingham, Duke of Tregunter, 3rd Duke of Claro, 5th Duke of Wharfdale, and 2nd Duke of Tregunter have since all been sold to English purchasers for 500 guineas each.[1]

At Holker 'the Druid' had only to seek the herd in the field behind Mr. Drewry's:

Eighteenth Duke of Oxford (calved 12th August 1868) was the last hope of the calf-house, and Lady Oxford the 5th (the 600 guinea Royal Worcester calf) was on the eve of calving her

[1] *Saddle and Sirloin*, pp. 276, 277.

Fourth Baron Oxford to that grand old Duchess bull, Seventh Duke of York. Two of her calves were sold for 500 guineas each at Mr. McIntosh's sale, and another for 250 guineas to Lord Kenlis [1] at Killhow.

Countess of Barrington 4th somewhat reminds us of Duchess 77th, and has a son by 10th Grand Duke at her side. The light roan is Blanche 3rd, grand-daughter of old Sylph; and a broken horn marks 7th Grand Duchess of Oxford, who also rejoices in a beautiful-haired daughter, the 12th of that line. Lady Oxford 5th is the queen of the field, fit to found a world of short-horns for substance and true character. Oxford Rose 2nd, by Grand Duke 4th from Rose of Raby, makes a nice pair with Oxford Rose, by Baron Oxford. Old white Dustie [2] has no heifer to perpetuate her line, and Morning Star [3] is the last dying bequest of Lord Oxford. Fifth Grand Duchess of Oxford is a wonderful milk and butter cow. From her we pass through the park to the home of 10th Grand Duke, who is mourning the loss of Mr. Davies's Moss Rose (who bore him Royal Chester) [4] and we bid him be of good cheer, as he puts forth his beautiful head to greet us, and walks most vigorously the whole length of his paddock into his shed for further recognition at Mr. Drewry's hands. [5]

The era of 'long prices' had now fairly set in:

By the outside public the astonishing sums paid for individual animals of fashionable blood was looked upon almost in the light of a mania. These prices were, however, the result of personal enterprise, of increasing demand, and of that abundant wealth and prosperity which the country had been developing during the last few years. Extreme prices may be a source of speculation; but it was difficult to estimate the value of cows when their bull calves sold readily for a thousand and twelve hundred guineas each, and when yearling bulls were let from two to three hundred guineas each for the season. [6]

---

[1] On the death of his grandfather, the 2nd Marquis of Headfort, in December 1870, Lord Kenlis succeeded to his father's courtesy title of Earl of Bective, which he rendered famous in shorthorn history.

[2] A Gwynne cow by the Buck (13836), calved 10th May 1856, and bought from Mr. Harvey Combe for 200 guineas.—*Herd Book*, xiv. p. 437.

[3] A Wild Eyes heifer, calved 22nd February 1866.—*Ibid.* xviii. p. 631.

[4] Royal Chester (29852), roan, calved 25th April 1869, bred by Mr. D. R. Davies, Mere Old Hall, sold to Mr. Taylor, Sydney, N.S.W., by 10th Grand Duke (21848), dam Moss Rose by Marmaduke (14897), gd. Cambridge Rose 6th, etc. etc. [5] *Saddle and Sirloin*, pp. 383, 384.

[6] *Thornton's Circular, a Record of Shorthorn Transactions*, iii. p. 258.

In the spring of 1872 Mr. E. Bowly's draft sale, when 2nd Duke of Tregunter made 900 guineas, gave an average of £153 for thirty lots. It was, however, eclipsed by the astonishing results of the draft sale at Dunmore, when three Oxford heifers realised 3070 guineas, and the dispersion of the Winterfold and Turner's Hill herds, which was a more even sale and increased by the extraordinary price of 1650 guineas given for 8th Duke of Geneva.[1]

It was now clearly shown that Mr. Sheldon's champagne had nothing to do with the high prices obtained at Windsor; souchong and mocca proved quite as stimulating:

The herds at Winterfold and Turner's Hill had a separate existence and treatment, but were one as far as the bulls 3rd Duke of Claro (23729) and 8th Duke of Geneva (28390) were concerned. Mr. Harward's stock was raised at Winterfold, a place both naturally and artificially adapted for well-bred cattle. Mr. Harward bought his first cow, Clear Star of the Wild Eyes strain, for 50 guineas, and how this family has in a few years increased in value remains to be seen. The most successful investment was buying up nearly the whole of Mr. C. W. Harvey's herd at Walton-on-the-Hill, including the Lallys, the Kirklevingtons, and the Wild Eyes. 3rd Duke of Wharfdale was hired from Captain Gunter, and 5th Duke of Wharfdale bought as a calf for 500 guineas. 3rd Duke of Claro was a similar purchase, and with these bulls the herd was fairly launched on the 'Bates and no surrender' principle. On the 18th September 1872, breakfast, in earnest, was laid at eleven o'clock; the company had to do at Rome as the Romans do; tea and coffee took the place of sherry and champagne. Business commenced punctually at noon. Clear Star, one of the best cows in the herd, was of large size, with a great barrel, somewhat thin in the crops, but with the unmistakable Bates neck and style, and darkish up-turned horns. 5th Maid of Oxford, a large fine cow, rather up in her back, but looking well in the ring, created much competition and was secured by Mr. George Moore at 900 guineas. The excellence of the two-year-old and yearling heifers—mostly beautiful roans or reds full of hair—was the talk of the company. They showed in the ring that remarkable elevation of neck and head, that nervous high-bred appearance which seems to sink the back down that is so conspicuous in the highest-bred Bates cattle. Lady Worcester 9th and Lally 15th were two of the finest specimens; the latter, a lovely roan and very full of hair, went finally to the Duke of

[1] *Thornton's Circular, a Record of Shorthorn Transactions*, iii. p. 259.

Devonshire for 500 guineas. The 8th Duke of Geneva, in the very prime of life and vigour, came out as only Bates bulls can, with imposing grandeur of head and crest. He is a capital sire, and nearest to Mr. Bates of anything in the world. 'Who says a thousand?' was answered by 900 from Lord Bective. The biddings then were brisk between Messrs. Drewry, Fisher, and

DUKE OF GENEVA 8th (28390).

Moore from Cumberland and Mr. Leney from Kent, who finally got the bull at 1650 guineas, the highest price that had been reached as yet for a shorthorn.[1]

After the Winterfold sale, a paper bearing the title *Parallel and Parallax*, and in which the average (£356 : 18 : 4) made there was favourably contrasted with that (£157 : 19 : 8) of the late Mr. Pawlett's Booth cattle sold in the spring at Beeston, had an extensive circulation.[2] This was too much for the Rev. John Storer:

John Storer, bold, of Hellidon, by the Nine Gods he swore,
That the great house of Warlaby should suffer wrong no more;
   By the Nine Gods he swore it, and wrote a letter then,
And bade *Bell's Messenger* send forth,
East and west, and south and north,
   This product of his pen.[3]

[1] *Mark Lane Express*, 23rd September 1872.
[2] *Bell's Weekly Messenger*, 28th October 1872.
[3] Letter of the Earl of Dunmore, *Bell's Weekly Messenger*, 4th November 1872.

Instead of being calculated to allay strife between shorthorn breeders, this clerical effusion was most one-sided and prejudiced. It complained that while home-bred Booth bulls had been used at Beeston, high-priced Bates bulls had been used at Winterfold, and alleged that as regards the cows the percentage of profit on the original outlay was much larger in the former case. It proceeded to lay down that prices are no proof whatever of the intrinsic merits of a herd, and then, because the Duke of Richmond was the only noble purchaser at Beeston, launched out into a revolutionary diatribe against peers and men of fortune 'with whom, as a rule, shorthorn breeding was a mere subsidiary pleasure—a fashionable adjunct to the beauties of the demesne.' In purely imaginative contrast to these, the Booths, Torrs, and other heroes were lauded to the skies as 'earning their living by hard work on the land,' and were dogmatically pronounced the only people capable of forming 'a practical judgment as regards the best varieties of shorthorn cattle.' A composition of superlative acerbity and unfairness wound up with a round of mud-throwing at the unfortunate Oxfords:

> Most of the cattle sold at Winterfold were of very ancient descent, but their purity has been largely affected by the blood of those *parvenus*, the Oxfords; and since Mr. Bates's death to a much greater extent than he would probably have approved.
>
> If the cow from which the Oxfords sprung, said to have been by Matchem (2281), dam by Young Wynyard (2859) did not descend from those bulls, and the evidence is at least doubtful, then the Oxford family originated between thirty and forty years since. If, however, she had those two crosses which are claimed for her, still the family does not even profess to go back to the old Teeswaters, and may have been full of every species of alloy. Rave as men will about the 'alloy' introduced by Charles Colling in the last century, it is infinitely worse to fill the veins of Duchesses, Wild Eyes, and others, with such repeated doses of doubtful blood introduced so many years later. Nor will it do to assume, as appears to have been done lately, that nobody knows anything about these things; or that they who do know will be silent under every provocation, even though they are

generally content to pass over with a smile the flattering eulogiums they so often see upon the 'purity' of the Bates cattle, and the 'blue blood of the Oxfords.'[1]

'The wolf'—the Galloway wolf in shorthorn clothing—'indeed, turned round and complained of the lamb for muddying the waters he drank so peacefully below!'

It was easy for Lord Dunmore in reply to point out that it was irrational for Mr. Storer to find fault with Messrs. Harward and Downing for using the best Bates bulls they could get. Respecting the cost of female ancestors the earl pointed to the Lady Worcester family sold at Winterfold, Clear Star, the matron of which, was purchased by Mr. Harward at Mr. Maynard's sale for 50 guineas, while her produce (irrespective of any sold privately) made by public auction over £3200. As to prices being no proof of intrinsic merit, Lord Dunmore rejoined:

By what standard, then, are we to judge of the merits of shorthorns? Surely Mr. Storer is not going to advance the old theory, which never can hold water, that the 'show-ring' is the only place where a good animal is to be seen, for there are more good animals that never enter a show ring than there are that come out of one. And why? For the reason that with a real good animal the honours are not worth the risk, and in consequence the show-yard motto might often be 'plenty of fat covereth a multitude of sins.'

Mutual bovine recriminations can lead to no good result. There is plenty of room in the world for both the Bates and the Booth strains, and long may they both flourish![2]

The editor of *Bell's Weekly Messenger* closed the Great Shorthorn Controversy in the judicial leaderette:

Whether Blucher, Belvedere, or Stephenson's Bull was the sire of Lupin can make little difference. The Sockburns have proved their worth at Grassy Nook, at Kirklevington, at Towneley, and wherever else they have fallen into skilful hands. As to the Duchesses, we should not care to know, except as matter of history, whether they came over with the Conqueror, derive their

---

[1] Letter of Rev. J. Storer, *Bell's Weekly Messenger*, 28th October 1872.
[2] Letter of the Earl of Dunmore, *Ibid.* 4th November 1872.

descent from Hogyn Mogyn, or possess no valid claim of traceable lineage beyond the epoch of Herd Book records. Their bitterest enemies admit that the Duchesses breed true to sort; and the fact of breeding true to sort implies established character and hereditary power. The impression of a good Duchess or Oxford bull upon a well-bred herd is marvellously strong. It is a pity that in the course of controversy any disparaging allusion to Mr. Bates should have been made by one whose usual course is that of generous fair play. Mr. Bates, whatever his errors, however obstinate and unjust his prejudices, was a man of sterling worth, and bore to the grave a reputation unstained by the faintest suspicion of dishonour.[1]

To take up again the main thread of the story of the Kirklevington families:

Twenty-six animals of the Duchess, Oxford, Red Rose, and Princess tribes were imported to this country from the United States and Canada between October 1871 and the close of 1872 by the Earl of Dunmore and Mr. Cheney. These animals were purchased at great cost, which, coupled with the risk and expense of shipping across the Atlantic, was further evidence of the demand for animals of high pedigree. Among them were Duchesses 107th and 108th, and Duchesses of Airdrie 8th, 13th, and 14th, and the bulls 1st Duke of Oneida and 3rd Duke of Hillhurst.

During 1873 the demand for all strains possessing substance was generally good, but no animal publicly realised 1000 guineas in the United Kingdom. Two Duchess calves, at Mr. Cheney's, a bull and a heifer, made 935 guineas and 820 guineas, prices which, though astonishingly high for animals three and seven months old, were below the general expectation. Duke of Oxford 24th (31002) from the Duke of Devonshire's herd at Holker was shipped to New South Wales at 1000 guineas. On the other hand, Lord Dunmore imported ten Bates animals, including the English-born Duchesses 97th and 101st, Lady Worcester and Winsome Eyes 3rd, from Mr. Cochrane's in Canada, and a few Red Rose heifers from Mr. Renick's in Kentucky.[2]

It was impossible to mistake the renewed vibrations felt in this country when the great sale at New York Mills in the autumn of 1873 caused an upheaval of business in America. This new force extended to nearly all classes

[1] *Bell's Weekly Messenger*, 3rd March 1873.
[2] *Thornton's Circular*, iii. p. 624.

of shorthorns, but the influence was by far the strongest on the leading tribes.[1]

Mr. Sheldon of Geneva, Illinois, had disposed of half his herd in 1869 to Messrs. Walcott and Campbell, New York Mills, Oneida, U.S.A., by a process of alternate selection. He led off by choosing 12th Duchess of Geneva, and their representative, Mr. Gibson, chose 6th and 8th Duchesses of Geneva and so on.

DUCHESS OF AIRDRIE 8th.

The six Duchesses made an average of 1100 guineas a-piece, and three Oxfords 560 guineas each. The next year Messrs. Walcott and Campbell purchased the remainder. All the twelve Duchesses sold at the New York Mills 10th September 1873 for the enormous total of 52,570 guineas, and the three Dukes sold for 4720 guineas, were descended from Duchess 66th, bought by Messrs. Becar and Morris at Tortworth for 700 guineas.

The 2nd Duke of Oneida was bought by Mr. T. J. Megibben, Kentucky, for 2400 guineas; Duchess of Geneva 8th reached *the highest figure before or since that a cow ever sold for*—8120 guineas—(for Mr. R. Pavin Davies, Horton, Gloucestershire). Most people thought she went higher than 'the cow that jumped over the moon.' Another Duchess of Geneva, the 10th, went to the Earl of Bective at Underley for 7000 guineas, and Duchess of Oneida to Lord Skelmersdale at Lathom for 6120 guineas.[2]

The year 1874 brought increased averages at the draft sales held at Gaddesby, Holker, Underley, and Wateringbury. The famous 1000-guinea bid for Comet—for more than sixty years unequalled publicly in England—was now doubled for the first

[1] *The Improved Shorthorn*, by William Housman, p. 31.
[2] American Shorthorn Breeders' Association, Circular No. 8, p. 82.

time in public in the purchase of 4th Grand Duchess of Geneva at Mr. Leney's by Mr. Loder of Whittlebury, then a new breeder. The increased demand for all sorts of English bred descendants of Robert Colling's Princess was a prominent feature of the season. Anywhere, with crosses of any kind, the Gwynnes and Princesses met with a ready sale.[1]

The steady upward movement in the value of the females—not now merely of the tribes tracing to the herd on the Kirklevington home farm, but of those which were kept by the tenants—continued throughout 1875. It was conspicuous at the beginning, at Colonel Kingscote's very successful draft sale, and it was to be seen at Messrs. Tunnicliffe's, Robinson's, and Meakin's at the end. The climax for admirers of the Bates blood was reached on the 25th of August, when the Earl of Dunmore's twenty-nine animals made over £672 each, the highest average on record: one Duchess bull, Duke of Connaught, fetching the *price unprecedented and unsurpassed* of 4500 guineas, whilst another, 3rd Duke of Hillhurst, was bought for 3000 guineas.[2]

The shorthorn sale-season of 1876 was looked upon as one of comparative depression, though two animals were sold privately for 3000 guineas each and Duke of Rothesay made 2000 guineas at Weeting Hall.[3] The great feature of 1877 was the importation of thirty-three animals belonging to the Hon. M. H. Cochrane of Hillhurst, Canada (owing to the depression both there and in America) and the extraordinary high prices realised by the two very handsome Duchess heifers at the sale which took place at Millbeckstock, a small farm close by Bowness on Windermere on the 4th of September. One of these, 3rd Duchess of Hillhurst, one year and eight months old, was bought by Mr. Loder for 4100 guineas, Mr. Allsopp keenly competing for her; the other 5th Duchess of Hillhurst, one year and four months old, was purchased by the Earl of Bective in opposition to Mr. Allsopp, Mr. Drewry, and Mr. Loder for 4300 guineas. At Sholebroke Lodge a fortnight later, five of Mr. Oliver's Grand Duchess tribe made over £1733 each. The Earl of Bective bought Grand Duchess 23rd for 2750 guineas and Grand Duke 31st, a very handsome ten months' bull-calf, for 1550 guineas; Mr. Allsopp gave 2450 guineas for Grand Duchess 29th, Sir G. R. Philips 1000 guineas for Grand Duke 29th, and Mr. A. Garfit 505 guineas for Grand Duke 25th.[4]

---

[1] *Thornton's Circular*, iv. p. 293.   [2] *Ibid.* iv. pp. 687, 688.
[3] The Duke of Rothesay (36534) was bought by Mr. H. J. Sheldon, and may be said to have been the foundation stone of the Brailes herd.
[4] *Thornton's Circular*, v. pp. 698, 699.

A general decline in shorthorn prices continued in 1878, though extreme figures were still given for a few animals, especially Oxfords and Wild Eyes. Twenty-three Oxfords were sold at Holker for over £919 each. The International Cattle Show, held at Paris in June, did not lead to much business. An impression prevailed in France that the shorthorn breed of late

DUKE OF CONNAUGHT.[1]

years had run too much to fat and did not carry a proper proportion of lean flesh.[2]

This discrimination on the part of the French brought up again the old, old questions posed by Bates in his memorial of 1807:

What had science done for the shorthorn breeder? Had a greater weight of carcase been gained in a given time, or from a fixed quantity of food? Had a larger supply of milk or butter been yielded? Were the animals hardier in constitution and more impressive to their offspring? Were the younger ones superior to the old?[3]

The dispersion of the two celebrated herds collected during

[1] From a water-colour drawing by the late Mr. A. M. Williams, kindly lent by Mr. John Thornton.
[2] *Thornton's Circular*, vi. pp. 340, 341.   [3] *Ibid.*

ten years at Dunmore and Gaddesby, and both recruited largely from Canada and the United States, gave a peculiar interest to 1879. Lord Dunmore's sale took place under the most unfavourable weather; a heavy rain spoilt the appearance of the animals and damped the spirit of the very large company. Duchess 117th realised 3200 guineas, the highest price of the year; and though many of the animals appeared to sell low in comparison with previous sales, yet the prices, considering the year, could not be looked upon as unsatisfactory. From the time the herd was started in 1868 a little over £29,000 had been expended in the purchase of animals, and three public and several private sales produced considerably over £59,000. The Gaddesby herd was brought out in low condition: most of the early lots were old cows in an unsatisfactory state. For many years no females of the Cambridge Rose tribe had been offered to the public. Consequently Lord Braybrooke's sale at Audley End attracted much attention. Seven of them averaged over £590, Thorndale Rose 7th making 1000 guineas. It was the Lathom sale, however, that realised the highest average of the year; and there was the remarkable feature about it that every animal had been bred by Lord Skelmersdale on the farm. Duchess of Ormskirk having been sent away for service, was in indifferent condition, and did not sell as well as had been anticipated, nor indeed as her high breeding deserved.[1]

The decline in shorthorn prices, which had begun to be serious in America two or three years before, made itself still more felt in England in 1880. Attributed to line-breeding and due no doubt to a blind reliance on pedigree, this led to a revulsion in the contrary direction: a tendency was noted at the ringside to buy animals for their own individual merits. Except at Mr. Lovatt's, rates for cows and heifers were unsatisfactory throughout the spring. During the autumn some prices made at Underley and Wicken Park considerably swelled the average of the season. At the former a Duchess heifer made 2000 guineas; two Oxfords 980 guineas and 515 guineas; and two Princesses, 405 guineas and 435 guineas. At the latter Mr. Holford's Duke of Leicester and Viscount Oxford 3rd, fetched respectively 510 guineas and 300 guineas.[2]

An organised attack on the unalloyed Durhams was now commenced in *The Field* under the title of *A Review of Shorthorn Prices to the Close of 1880*. It is amusing to find the Kirklevington sale denied even a fourth place

---

[1] *Thornton's Circular*, vi. pp. 692, 693.   [2] *Ibid.* vii. p. 246.

in 'the list of shorthorn events of interest.' Notwithstanding its historical errors and its evidently inspired partiality for two particular strains of blood, this article exercised a very important influence 'in setting the fashion' against careful breeding. It did no good to the two strains it was intended to benefit; its prophecies entirely missed fire, but it considerably added to the depression of shorthorn values.

The draft sales of 1881 exhibited a continued decline of prices. Mr. Bowly's at Siddington was largely attended, but the result was unsatisfactory. Sir William Salt's was better, though there was a great difference between the value of the same families sold then and a few years previously. The Havering Park sale did not come up to the anticipations of the large and influential company, though several purchases were made for Belgium. Mr. Longman's sale was brisker and better but the average not so high. At Kimbolton, in July, when a large portion of the Duke of Manchester's herd was sold, business improved. The Oxford tribe was in great demand; Marchioness of Oxford 3rd, a white cow, eight years old, made 750 guineas, and her grand-daughter, Oxford Mary, a beautiful roan two-year-old was the highest public-priced animal of the year. She was keenly competed for by Lord Fitzhardinge, but purchased for Sir Henry Allsopp at 1110 guineas. In marked contrast to these prices were those given for the same family at Killhow. Grand Duchess of Oxford 40th a roan three-year-old cow was bought by Sir Wilfred Lawson for 520 guineas, and her red yearling heifer brought only 320 guineas. This animal had been previously purchased when a calf at Holker by Mr. Foster for 1600 guineas. The export trade was one of the helping features of the year. Mr. Oliver sold privately Grand Duchess 28th and Grand Duchess 35th from his herd at Sholebroke at prices far beyond anything realised by auction. Lord Feversham's fine old bull, 5th Duke of Tregunter, accompanied them to improve Mr. Attrill's herd in Canada.[1]

No entire herd of really first-class cattle came into the market in 1882, and at the draft sales the best animals were kept out of the ring. Another outbreak of foot-and-mouth disease in 1883 caused the spring sales to be postponed. The highest average of the year was made in July by the sale of a portion of Mr. Holford's herd in Dorsetshire, though the weather was very

[1] *Thornton's Circular*, vii. p. 554.

unfavourable and the farm eight miles from the railway in a remote country. Here were some most valuable animals. One four-year-old cow, Duchess of Leicester, was sold for 1505 guineas, the top price of the year; though 1500 guineas was obtained by Mr. Loder privately for the sale of the young bull 2nd Duke of Whittlebury, to Mr. A. J. Alexander of Kentucky, U.S.A. The highest public price paid for a bull, 900 guineas, was given at Mr. Holford's sale by Lord Fitzhardinge for Duke of Leicester 3rd a red two-year-old. The same breeder, in 1875, gave, it will be remembered, 4500 guineas for the Duke of Connaught, a bull which was still living and serviceable, and which had more than paid his way and costs by service fees. The Holker sale, postponed from the spring, took place early in September. In 1878 thirty head averaged £664, now forty-four head of somewhat similar blood brought but £161 each. The condition of the animals on this occasion was less high than heretofore, and the two last sires used had not met with general acceptance.[1]

Most of the spring sales of 1884 were held back until after the dispersion of Mr. Oliver's celebrated herd. This had been in existence about twenty-five years, and the Grand Duchesses, noted for their size, quality, and high breeding, had been a profitable source of income to their owner, until his health broke down and in consequence the entire herd was dispersed. About a thousand people were present, and few who were there can forget the scene beneath the old oak on the stripe at the little hunting-lodge in Northamptonshire, when the old English fox-hunter and breeder was brought in his carriage to the ring-side and received the hearty cheers of the company. Foreign breeders and agents were present from North and South America and the Continent; and out of the fifty lots, thirty left the country. The Grand Duchesses were last sold at the Preston Hall sale, 1867, when thirteen head averaged £443. Mr. Oliver then gave 850 guineas and 710 guineas respectively for two heifers; one proved unfortunate, but the other, Grand Duchess 17th, became the mother of the whole tribe, of which twenty now averaged over £573. In the summer, Messrs. Leney's sale was greatly helped by one young cow, Grand Duchess of Geneva 7th, which was bought to go to South America for 1100 guineas.[2]

Very great expectations were raised by the announcement that Sir Henry Allsopp's herd was coming into the market in May 1885. Not merely had inquiries about them been received from almost all parts of the world, but buyers, from South

[1] *Thornton's Circular*, viii. p. 475. [2] *Ibid.* ix. pp. 255, 256.

America especially, had tendered privately very large sums for the privilege of selecting some of the produce of Sir Henry's high-priced cows. All such offers were declined, and it was decided to retain all and offer all, without reserve, to the public, at one and the same time, giving everybody an equal chance. This was high-minded, but it could scarcely be called politic. The lots were too many and too valuable to be carried off to advantage at any one sale in any one season. For a wet day the company was enormous, but buyers were singularly few. No Australian nor New Zealand orders came to hand; South America kept quite aloof; and a little knot of friendly competitors from Canada and the United States had all the best things at their mercy in their own corner. The consequences were overwhelming. The onlookers saw with dismay some of the very choicest cows and heifers disposed at rates considerably below those of the last ten years.[1]

At Sir C. Lampson's sale at Rowfant, men plucked up a spirit, and there was real competition. Lord Braybrooke's sale proved not only one of the most interesting but one of the most successful sales of the year. The Thorndale Rose tribe, for which Audley End had long enjoyed a world-wide reputation, was a very fine one, and it had been bred upon lines that met public approval. Lord Braybrooke considered that the time had come for some new cross, but preferred to give up the material for others to deal with rather than undertake the responsibility himself. His Knightley experiment was very far less successful from a pecuniary point of view. Mr. Hales at North Frith had made a 'hit' with his Barringtons; they were very good cattle and sold really briskly at good prices. The Lady Worcesters (Wild Eyes) proved attractive at Mr. Barchard's. Although the Horsted farm is in a poor part of Sussex, and there were few strangers in attendance, the sale was creditable. A large draft from Lord Bective's herd was sold at Underley. Both the Waterloos and Darlingtons were in request, showing size, flesh, and constitution.[2]

By 1886 British agriculture had reached a state of depression that had been unparalleled since that brought about by the alteration in the fiscal system and bad seasons preceding 1850. The fine cattle sold by Mr. D. A. Green and Mr. R. H. Crabb in April 1886 fell sadly behind in price what the splendid promise of some of the younger animals would have led one to expect. The last remnant of the Havering Park herd, which had enjoyed much deserved celebrity among the admirers of short-

[1] *Thornton's Circular*, ix. p. 538.   [2] *Ibid.* pp. 539, 540.

horns, was dispersed in the summer in combination with a draft from Sir Hussey Vivian's. This sale illustrated the fact that the very best cattle, shown away from their home, sell at a loss. No better bred Oxfords had been exhibited in the ring for years, nor any which sold worse. To Mr. S. P. Foster is due the credit of having offered the top lot of the year. His Duchess of Killhow (bought in the dam at Hindlip), both in her appearance and her price (for the times), maintained the high prestige of the Stanwick Duchesses. She was a beautiful heifer, and the fact that she was retained in England and made a member of the grand herd at Holker was the theme of frequent congratulation. The Killhow Oxfords also created good competition.[1]

Sales were few in 1887, and the average prices sank lower still. Except for two good old Oxford cows bidding was languid at Lord Penrhyn's. The Waterloos at Mr. Leney's and Mr. Brassey's sale were among the finest animals sold during the whole season. Lord Bective's draft sale in the autumn drew together a large and distinguished company, including several breeders and agents from South America. These took a large number of lots, including Grand Duchess 59th at 500 guineas, the top price of the year.[2]

There was an excellent demand from abroad during 1888. The highest priced animal at Messrs. Evans's sale—Lady Oxford Waterloo 5th, the first-prize heifer at the Royal Show at Newcastle-upon-Tyne in 1887—was bought for exhibition in Canada, and Messrs. Leney's 260-guineas Duchess heifer for Australia. At Lady Camperdown's sale, at Mr. Sheldon's sale, and at Lord Feversham's sale, all the top prices were paid for animals to leave the country. Four figures had not been paid for a shorthorn since 1884, when Lord Feversham, not wishing to lose 9th Duke of York, contrary to the custom of the trade, put a reserve of 990 guineas on him, and a South American agent at once bid 1000 guineas.[3]

Few herds had been formed at greater expense than Sir Robert Loder's at Whittlebury. It was started in about 1872, when prices were at their highest, and contained a number of Duchesses, Gwynnes, and Darlingtons. At its dispersion, in May 1889, it was the remark of the large company assembled that the animals had not improved on the original stock; indeed the young bulls were far from satisfactory. There were, however, a few grand animals in the herd, particularly Duchess of Buckingham (295 guineas), Duchess of Darlington 13th (85

[1] *Thornton's Circular*, x. pp. 219, 221.   [2] *Ibid.* p. 471.
[3] *Ibid.* xi. p. 242.

guineas), and Duchess of Whittlebury 13th (510 guineas). The herd had been closely in-bred until 1887, when the bull Duke of Rosedale 12th (46268) was purchased. The value of this cross was very apparent in the superiority of the stock begotten by this bull; his heifer calves particularly sold remarkably well. Two fine Duchess cows were purchased for France,[1] but there was an absence of South American buyers. A portion of Mr. A. H. Lloyd's highly bred herd in Surrey was sold the following month. Here, too, were some fine specimens. Duchess of Surrey, a handsome roan Grand Duchess cow, was greatly admired. She was purchased by a new breeder to remain in the country at 425 guineas, her heifer going to another new breeder in Hertfordshire at 210 guineas. A yearling Oxford heifer made 105 guineas, and there were several good Waterloos that also made three figures. The bull Duke of Surrey 2nd (52783) went into Sussex at 200 guineas. The old established Kingscote herd attracted a large company, but was not shown at its best. The highest price was paid for the Wild Eyes bull Sir Maurice (56570), bred by Colonel Gunter, who, after competition from a South American agent, finally went to the Underley herd. The sale of the Hon. R. Baillie-Hamilton's herd took place under peculiar circumstances. Bred in an outlying district in the Lothians, it was removed for sale in its entirety to a small farm adjoining the Highland Society's Show at Melrose. The result was the highly satisfactory average of nearly £50 without any competition from foreign buyers. As was anticipated the Duke of Devonshire's sale at Holker resulted in the highest average of the year, without any very extreme prices being paid. Four animals of the Duchess tribe averaged £230, and twenty Oxfords £88. The twin Grand Duchess heifers, which were greatly admired, both went to Sandringham; the Airdrie Duchess heifer, which made the highest price, 285 guineas, was exported to South America.[2]

The June and July of 1890 saw some first-rate herds enter the market. Mr. Herbert Leney had long been known as a spirited buyer of the best animals and a generous keeper. Without making a business of showing, he occasionally sent a few animals for exhibition in the south of England, where they generally held their own even against travelling prize-takers. In consequence of the extension of hop and fruit land, the larger portion of his herd was offered to the public. Eminent home-breeders were present, as well as some distinguished French and other foreign buyers. The thirty-seven head realised nearly

---

[1] *Thornton's Circular*, xi. pp. 519, 520.
[2] See Grollier, *Histoire d'une Étable*, pp. 210-227.

£66, the highest average of the year. Attracted by the historical associations of the place and the popularity of its owner, an enormous company assembled at Tortworth, and Lord Moreton's sale was considered satisfactory. A fine Waterloo bull, belonging to Mr. T. Holford, was sold here at a good price to an Australian breeder. The death of the 7th Duke of Manchester did not interfere with the arrangements made for the sale of about half the herd at Kimbolton. This brought forth several new buyers. Although the young stock by Grand Duke 51st were of singularly good colour and excellence and sold well, it was too apparent that the better half of the herd had been retained. The Oxfords were very good specimens and made the best prices.[1]

1891 was a year in which the lowest prices remembered were obtained for nearly all pure breeds of cattle. The biennial sale at Sandringham was one of the best of the season, and the highest price (200 guineas) was paid there for an individual animal, a promising yearling heifer, Duchess of Lancaster, of the Duchess tribe. The weather was disastrous at the sale of a portion of Sir Hussey Vivian's large herd near Swansea. Notwithstanding this the result was the greatest success of the autumn. M. de Clercq, a large breeder from near Calais, purchased the best heifer, Marchioness of Oxford 2nd, for 130 guineas, and Mr. Mills gave 145 guineas for Lady Worcester 26th, a fine cow of the Worcester branch of the Wild Eyes.[2]

Three sales in June 1892 infused a little more life into the trade. Mr. A. H. Lloyds's entire herd in Surrey attracted breeders from all parts. Mr. Mills gave 195 guineas for Duchess of Surrey 3rd, a young cow of the Grand Duchess tribe, five females of which averaged £125. The herd had been bred on the poor stiff clay on the borders of Surrey and Sussex for fourteen years, and the appearance of the younger animals indicated the difficulty in rearing stock on soil better adapted for woodland than for arable or pasture. Mr. Thomas Holford of Castle Hill brought a good portion of his celebrated herd from Dorsetshire to Warwick for sale during the Royal Show. The auction was much assisted by the presence of several eminent French breeders who bought eight animals. One heifer and a bull were purchased to go to Natal. The selection from the Underley herd in September was not up to the form and quality that had hitherto been associated with that famous stock. The animals had suffered keenly during the preceding winter and spring, roots and hay having been very deficient and poor in the district.

[1] *Thornton's Circular*, xii. pp. 296, 297.   [2] *Ibid.* p. 597.

Grand Duchess 58th and her calf went for 205 guineas; the bulls were not much in demand.[1]

Two of the best sales of 1893 took place in July, both being of Kirklevington blood. Mr. F. Barchard's entire herd in Sussex was one of the finest offered during the season. Two fine young cows, Horsted Rose 4th and Marchioness of Kirklevington 7th, 'the plums' of the sale, were bought for Mr. Potter's herd, near Windermere. Cambridge Duke 26th (58582), a fine red four-year-old bull, and an excellent sire of the Cambridge Rose tribe, went for 70 guineas to Mr. Buckley's herd in Tipperary; twelve of this tribe averaged over £51. The sale of a portion of Mr. P. L. Mill's fine herd at Ruddington attracted a large company, including several eminent French breeders. Cambridge Premium Duke 2nd, one of the best young bulls, went to Sandringham at 100 guineas. A nice lot of heifers, belonging to Mr. Blezard, were sold at the end of August, in North Wales, but did not meet the best market. Mr. Ecroyd's sale at Armathwaite was in a great degree a draft; two good calves by 3rd Duke of Chatsworth (57185), the sire of his prize-winners, being retained, detracted from the sale average. The remainder of Lord Swansea's herd, comprising several old cows, went at moderate figures; a Grand Duchess yearling heifer made 85 guineas, and her sister, a young heifer calf, 45 guineas.[2]

In April 1894 Messrs. Evans sold a large portion of their extensive herd at Sherlowe, Salop. Of the eighty-eight animals catalogued many were local winners, and the whole principally of Kirklevington blood. It attracted a large gathering and resulted in a brisk sale. Mr. P. L. Mills took some of the best lots to Ruddington, including an Oxford heifer at 52 guineas, but Mr. J. A. Preece of Drayton paid the highest price, 65 guineas for Barrington Beauty, a red heifer calf. Several lots went into Kent, Devonshire, and Gloucestershire, whilst Lord Powis, Mr. Humphreys, and Mr. Nevett retained many in the county. Mr. Edmonds's sale in Gloucestershire was, perhaps, one of the most remarkable of the year as indicative of public requirements. It was an old herd of forty years' standing, and had been bred with special regard to the dairy properties, from old Gloucestershire blood, in which the Kirklevington strains prevailed. The cows were of large scale, in good condition with capital udders, and showed milk. No extreme prices were given for females, nor were there any very low ones. Mr. Mills gave 62 guineas for a Blanche cow, and another four-year-old made 51 guineas. A handsome red roan yearling Blanche bull, Earl of Southrop 116th

[1] *Thornton's Circular*, xiii. pp. 521, 522. [2] *Ibid.* pp. 249, 250.

(65431), created keen competition between Mr. C. H. Bassett, Mr. Holt Beever, and Mr. Pinnell, who finally purchased him for 205 guineas to go to South America. It was one of the best sales that has taken place in that part of the country for many years, and was the talk of the local markets for a long time. Mr. Howell's extensive herd, near Cardiff, was dispersed to make more room for a stud of hackneys. Mr. P. L. Mills again paid the highest price, 75 guineas, for a heifer, Grand Duchess 66th, for which Lord Fitzhardinge's agent was a keen competitor. At Castle Hill, Dorsetshire, Mr. Holford was disappointed in his company, and a reserve was put on every animal as it entered the ring. The result was that only four animals out of the thirty-seven catalogued were sold. Two of the best Duchess heifers were bought by Mr. D. R. Scratton at 91 guineas and 95 guineas each; Duchess of Leicester 17th went to the Marquis de Chauvelin in France at 71 guineas, and a bull calf made 26 guineas. Mr. Herbert Leney offered the larger part of his herd in Kent, but competition and prices were somewhat restricted. The roan three-year-old Waterloo Earl Sockburn 3rd (63512) was bought for Australia at 81 guineas. The popularity of the late Earl of Bective, and the reputation of his herd, drew a large gathering to the Underley sale. The absence of foreign buyers and agents, who for some years past had drawn largely from this herd, was remarkable, and competition was left to the admirers of the sort. Duchess of Holker 3rd, a fine roan cow, made the top price, 165 guineas (Mr. J. Harris); Mr. J. D. Fletcher gave 160 guineas for Duchess of Leicester 19th, Mr. E. Potter giving 92 guineas for her roan heifer calf, and Lord Feversham 155 guineas for Duchess 131st. The remainder of the old Scaleby Castle herd was sold the following day. The late Mr. H. Sharpley's large herd in Lincolnshire had suffered much from the drought of 1893. There were a few nice animals of the Thorndale Rose and Waterloo tribes, which made fair prices. Mr. P. L. Mills gave 80 guineas for Thorndale Premium Rose 5th, 51 guineas for the 7th, and Mr. C. R. Lynn 78 guineas for the 6th.[1]

Early in June 1895 the small but very promising herd belonging to Mr. Darcy Taylor was sold near Bath. It was one of the few herds of Kirklevington blood offered, and, though small in number, was excellent in quality. In the presence of a good local company, which included Mr. Hall, a breeder from South Africa, an average of over 31 guineas was obtained for 26 head. Lord Moreton gave the top price, 60 guineas, for Marchioness of Oxford 3rd.[2]

[1] *Thornton's Circular*, xiv. pp. 224-228.   [2] *Ibid.* p. 471.

The fatal result of the policy officially adopted by the Royal Agricultural Society of England, at York in 1848, in spite of Bates's remonstrances, and since financially re-enforced by the prize-scheme of the Shorthorn Society of Great Britain and Ireland, was excellently illustrated at the Darlington Show of 1895. There were no shorthorns of any importance exhibited from the surrounding district; the original home of the Durhams knew them no more. Even the Booth family had been compelled to recognise the justice of Bates's contentions, and the much-admired cattle at Warlaby were sold on the Saturday before the show, being too valuable for breeding purposes to be sacrificed at it. His Highness, Nasrulla Khan, Shahzada of Afghanistan, withdrew in disgust from the parade of the prize bulls; they were, he said, too fat, too fat even for the butcher.

The chief interest in a parade of cattle appears indeed to be whether a given bull will be able to trail round the ring or will expire in the process. The revolt of the London butchers has caused very proper restrictions to be placed on awarding prizes to over-fed animals at Smithfield. It has been the regular custom of professional prize-hunters to 'carry on' for Smithfield the heifers exhibited at the Royal and county shows. We shall now probably witness the crowning absurdity of prize-winners at the breeding shows being disqualified for 'over-fat' when exhibited in the same condition at Christmas. A great deal of trouble and heart-burning might be saved if the Royal Agricultural Society openly followed the practical plan of letting the weighing-machine decide the champion prizes. This was actually done by the Agricultural Society of Upper Hesse in its young bull classes, with the brilliant result that hardly any prize bull ever got a calf.[1] It certainly would have been better for the shorthorn breed if some of the famous show bulls

---

[1] J. v. Kirchbach's *Handbuch für Landwirthe*, 9th ed., Berlin, 1880, ii. p. 355.

of recent years in England had not got any. The final verdict on the reproductive merits of a bull, and the wisdom of the judges who accorded it champion honours at a breeding show, can only be passed some fifty years after. In what pedigrees of the present day do we find the Hecatombs, Heroes, Harlsonios, Strellies, Standishes, Symmetries and Deceptions?

One fatal consequence of the exhibition of breeding stock in a popular circus is that, although the primary value of a shorthorn is its propensity to lay on beef, and 'handling' affords the surest index of that propensity, the ultimate decision of the judges is almost necessarily left to their eyes instead of to their fingers. The public, who pay the gate-money, have to be satisfied, and of the hundred thousand people who see a Royal champion every year, how many ever lay their hands upon him? Under these conditions, if Hubback himself could be resuscitated, what ghost of a chance would he have of being even commended? And yet, what did the Romans mean when they said that cattle should not be hard or rough to the touch—*attactu non asperum ac durum?*[1] Why were the graziers of Elizabeth's reign in their gold chains and velvet coats so eager 'to come to the feeling' of an ox or bullock?[2] What made Charles Colling say on his retirement, 'If I had my eyesight and *the use of my fingers*, I should not despair of forming another herd'?[3] Why did Bates at Cambridge insist that judges ought to be blindfolded?[4] Why in fine does every Irish cattle-dealer bid you, 'Jist touch them, yer honour, and sure yer honour's fingers never felt any to bate them'?

On putting to himself the question, 'What are the distinguishing characteristics of the best short-horns?' Bates, after enumerating the more prominent eyes, the much smaller head, the lighter and finer tail, and the much softer hair and skin, felt his own words inadequate to

---

[1] See above, p. 24.   [2] See above, p. 276.
[3] Bell, p. 200.   [4] See above, p. 273.

convey the full signification of what he meant by 'the touch incomparably more kind and mellow,' and availed himself of a long quotation from that past-master of the art of cattle-feeding, George Culley:

A good judge of cattle, with a slight touch of his fingers upon the hips, rumps, ribs, flank, breast, twist, shoulder score and other fatting points of an animal, knows immediately whether it will make beef or not, and in which part it will lay on most flesh. It is very easy to know where an animal is fattest which is already fat, because we can evidently feel a substance or quantity of flesh upon all the fatty points. The difficulty is to explain how we distinguish among animals in a lean state those that will feed and those that will not, or rather those which will lay on flesh in such and such parts and not in others. This, a person of judgment *in practice* can tell, as it were, instantaneously—I say *in practice*, because I believe that the best of judges *out of practice* are not able to judge with precision, at least I am not. We say a beast touches nicely upon its ribs, hips and other points, when we find a mellow pleasant feel on those parts—a soft feel we do not say, because there are some animals which have a soft loose handle but not that mellow feel. For though two animals may both handle loose and soft, yet perhaps we know that the one will feed and the other will not; and in this lies the difficulty of the explanation. We find a particular kindness or pleasantness in the feel of the one, much superior to that of the other, and immediately conclude that the first will readily make beef and the second not. In this a person of judgment and *in practice* is very seldom mistaken.[1]

As a protest against the disregard of milking properties, the movement for establishing a prize-fund for shorthorns as dairy-cattle is deserving of all encouragement; but there can be as much over-training in milk as in fat, and anything that tends to split shorthorns into two distinct groups, one for the spit and the other for the pail, instead of helping them to regain the proud position they occupied as the good-all-round breed, must have most ruinous consequences. The 'duplex qualification for profit' was the

---

[1] Bell, pp. 64, 65. Mr. William Carr, *Killerby, Studley, and Warlaby Short-horns*, pp. 70-73, seems to have been unable to distinguish between 'the pleasant mellow touch' and the 'soft loose touch'; indeed he preserved a discreet silence on the subject of handling throughout his able encomium.

grand secret of shorthorn supremacy; the disregard of it is the chief reason of shorthorn decadence:

Milk-making and flesh-making capacities are not of themselves in the least inconsistent, but only varying manifestations of the same faculty. One animal gorges himself or herself with herbage, and discharges the same without having assimilated more than merely enough to replace the waste of life. Another (by dint of strong constitutional digestive powers) extracts from its food a superabundance of valuable matter, and converts this from a vegetable to an animal substance. Clearly this latter sort is of the greater value to the owners of the herbage who saw it eaten with complacency, because they intended one day to have the satisfaction of eating it at second-hand. But whether the extra amount absorbed from the food in the case of a cow goes to make meat or milk seems to be a mere accident. In some cows the maternal instinct is so strong that it overcomes the natural tendency to do well to oneself; and this diverts to the calf's benefit the food-extract, the 'chyle,' which was in both instances there, ready to make flesh. Still the bottom of the usefulness of both kinds of cow is the faculty of transmitting the largest proportion of vegetable substance into its animal equivalent. So first of all let us get this admirable digestive apparatus developed as fully as we may. Constitution is the great thing to aim at—vigorous constitution showing itself mainly in a propensity to 'thrive.' This being ensured, at any sacrifice, our result seems tolerably secure. If a heifer in her maiden stage does not lay on flesh, however good a milker she may become, she will not be qualified to hand down through her sons, the beef-making propensity of her race. If, however, she, when no other strain was put on her, did as a heifer and does as a cow, turn her victuals to the natural purposes, then (however poor she may get whilst the milk habit is uppermost) she will be a trustworthy breeder of sires to use in any herd.[1]

Breeders like the late Lord Feversham and Colonel Gunter paid dearly for a knowledge of the evils of prize competition more than twenty-five years ago. Forcing animals to unnatural maturity for the sale-ring is only a less aggravated form of the same ruinous fatuity. The decline of most herds of Kirklevington blood may be attributed, however, to the opposite extreme. In many cases beauti-

[1] Letter signed 'Outsider' in *Bell's Weekly Messenger*, 3rd March 1873, p. 8.

ful cattle purchased at very high prices have been huddled together in low byres with a little fusty hay to eat, and no straw to lie on during winter, and have then been turned fifty or sixty together into large parks for the summer months. With most shorthorn breeders it is never the management that is to blame but the blood. If a herd does not show all the robustness of constitution that can be desired, instead of putting it under a different *régime*, an out-cross is imperatively demanded. Scottish herdsmen have to a great extent kept to the golden mean, and treated their shorthorns in a rational manner. It is submitted that it is to this, and not to any intrinsic merit in concoctions of blood, that Scottish shorthorns owe their value. It seems already doubtful whether they retain this value under altered circumstances. The few herds of Kirklevington origin that have been kept in Scotland have been remarkably successful. It would have been an interesting experiment if some of the high-priced Kirklevington cows and heifers that have been taken to Scotland of late years had been crossed with Kirklevington bulls, so that the offspring might be compared with that got by bulls of the Aberdeen mixture, reared under similar conditions. The great misfortune that has befallen herds of unalloyed Durhams during the last few years has been the fashion of purchasing precisely the best cows and heifers as *corpora vilia* for breeding ventures. In many cases these out-crosses have succeeded admirably in giving the produce 'dour' heads and empty udders. It is a case of 'heads I win, tails you lose'; a good calf is due exclusively to the Sittyton sire, a bad one is all the fault of the Kirklevington dam. Up to the present it has been perfectly easy to obtain among animals of Kirklevington descent plenty of unrelated material to build up any particular type without plunging into what is generally a *terra incognita*. If the English type of shorthorn is to be definitely abandoned in favour of one having for ideal a roan horned Aberdeen-Angus, Kirklevington may yet supply it, if only time be given.

It is very true that it is the inside and not the outside of an animal that is of importance to the consumer, but however much we may admire stolid solidity in an ox, an inert hulk, with a hang-dog look about him, is not the sort of stud-bull we should choose if we have any regard for the constitutional vigour of a herd. As a Northumbrian farmer put it, after a sale in 1872, 'A bull to be a bull should look over a gate, and not through it.'

Some breeders of shorthorns are desirous of having bulls with the heads of heifers and the thighs of bullocks. The offspring of such males is always deficient in quantity, and is of weak constitution. The progeny inherits the paternal effeminacy. When such a blunder has established itself in a herd, it can only be redeemed by recurrence to a male,

> cui turpe caput, cui plurima cervix,
> Et crurum tenus à mento palearia pendent.[1]

These are the true and natural indications of taurility.[2]

It was as a cattle-breeder and not as a shorthorn-breeder that Bates would have preferred to be remembered. He laid clearly down to begin with the tests that must be decisive of the value of a breed of cattle. He found by experience that his Duchesses, especially after the long-desired Princess cross, came nearer his ideal than did any other cattle. Had they not done so he would have sought farther a-field. His affection for them did not lead him to set any limits to further possibilities of improvement:

Every one by trying the cattle they possess by the same tests that I have done, may see what advantage they can gain by my blood or by assimilating it with their own stock. The several varieties, the original cows being different, will enable every one to select the produce he approves. A change to another variety may make a still further improvement in a herd. There are in my view no bounds set to improvement, if breeding be only judiciously carried out. This of course depends upon the judgment exercised. In crossing breeds, if there is any coarseness on

---

[1] Pub. Vergilii Maronis, *Georgic.* lib. iii. 52—'grim-looking with an ugly head, an abundance of neck, and dewlaps hanging down from jaw to leg.'—*The Poems of Virgil*, by Professor Conington, p. 72.

[2] *Quarterly Review*, 1849, lxxxiv. p. 396.

either side, the stock is sure to inherit that coarseness; but when coarseness is absent there is no fear of crossing different varieties. Always, however, let the male animal be the one to improve on the female, that is, that the male be of a superior tribe. I can have no doubt whatever that the best shorthorns will improve all other breeds to which the male animals are put, and when once crossed the owners may increase the shorthorn blood if they think proper, or breed between males and females of their mixed breed, or use those so bred to the best of their original breed, whether Herefords, Devons, Sussex, or what are called home-breds in Norfolk. The breeders of Aberdeenshire cattle have felt the benefit of a cross with what too often were very ordinary shorthorns. But I am sure that with the West Highland heifer and the shorthorn bull the greatest improvement may be made.[1]

His objection to the red Galloway alloy introduced by Charles Colling was that it was a cross with what was at that time a very inferior breed, and that the result of the cross was a neat form, it is true, but hard flesh, bad wiry hair, and no milk. He objected to it not so much because it was alloy, but because it was bad alloy. It cannot be denied that the finest fattening animals are often half-breds, yet to have good half-breds it is necessary to have not merely one but two races of pure-bred cattle with fixed characteristics.

The essential difference between Bates and most other breeders in the latter part of his life, was that he remembered the ancient shorthorns in the early days of the Collings, and was always on the look-out for practical proofs of improvement or deterioration. With the others it was enough to trace back their animals to high-priced lots in the Ketton and Barmpton sale-catalogues, with implicit faith—

> Their's not to make reply,
> Their's not to reason why—

in the judgment of every cross put on by Charles or Robert.[2] Bates was a most determined foe of paper

---

[1] These were Bates's opinions in 1846, and formed part of the final peroration of his intended history.—Bell, pp. 85-87.

[2] Cf. 'The breeders of shorthorns are indifferent as to investigation before the period when Mr. C. Colling first exhibited his stock.'—Berry's *Improved Short-horns*, p. 11.

pedigrees, the authenticity of which might well be called in question unless they were corroborated by the look of the stock.

Nothing can more lucidly express his views on the worth of cattle in general than a passage that appeared in the *Quarterly Review* just after his death:

The real and only question for the farmer is, what breed of cattle will year by year yield me the largest money return per acre, or per given quantity of various sorts of food consumed by them? And this question is not settled by saying, Taken—10 tons of short-horns and 10 tons of Devons; 50 tons of food of equal quality were consumed by each lot; the short-horns gave beef as 21 to 19, or *vice versâ*. First, we must know the respective histories of each 10 tons; we must have a debtor and creditor account of each up to the time of weighing in. The one may have credit for services in the dairy, the other for services in the team; or the creditor side may be blank in the case of either or both. We must *here* consider the breeder and the feeder as one man. Before we can answer the question, so interesting to him, we must know the antenatal cost of each 10 tons, and their respective debits and credits up to the day when they leave the hands of the beef manufacturer for the shambles. Secondly, we must know which fetched the most money—the beef represented by 21 or that by 19. It is easy to say, 'I have bred a beast of rare symmetry, great size, early maturity, first-rate quality.' Equally ready are the inquiries, 'After how many failures?—At what cost?—How stands the balance?' These questions are answered by many brave and contradictory assertions, by many wild and contradictory guesses, but by no statistics on which we can found a safe conclusion. And yet on the answer depends, on average agricultural farms suited to any description of cattle, the whole question of successful breeding and feeding.[1]

It is of course much more important to a true farmer to have cattle of uniform average excellence, giving plenty of rich milk, thriving well, and feeding rapidly, than to have them 'eligible for the Herd Book' and winning occasional prizes. It says little for the science of breeding if the two aims are incompatible.

Yet in the United Kingdom, at any rate, we are no nearer obtaining the statistics indispensable to this end

[1] *Quarterly Review*, 1849, lxxxiv. p. 393.

than we were in 1807 or 1849. It was for the Shorthorn Society of Great Britain and Ireland to institute scientific experiments that should demonstrate the high practical value of the breed under every aspect, and assist in raising it still higher. It has preferred to follow a Mede-and-Persian policy. The result is that a Durham with only five crosses will, *caeteris paribus*, now sell better than one with the longest pedigree. The *Herd Book* may as well be burnt if a shorthorn, descended from only 64 known animals, is worth more for breeding purposes than one derived from 65530 registered ancestors. Yesterday the pedigree was everything, the individual animal next to nothing; to-day the individual animal is everything and the pedigree an encumbrance. Is it too much to hope for a to-morrow when individual merit, based on ancestral merit, scientifically ascertained and authoritatively recorded, will be recognised as the only sure guide in rescuing and improving the most historic race of English cattle? It speaks volumes for the innate qualities of shorthorns, if in face of the facts that most of the best specimens are annually either transported abroad or sacrificed at home for show or sale, their decline has not been more marked and more rapid. We may feel sure that there is plenty of force left in them to do much more than recover any lost ground, if only breeders on their side will contribute a little more scientific study and practical care.

APPENDIX

Note A, p. 35

Alexander Hall's Memoranda

I

Mr. C. Colling bought of me a heifer got by Mr. William Fawcett's Bull, afterwards called Hubback, which heifer bred the bull Mr. George Coates bought of Mr. C. Colling for Mr. Foljambe. This heifer's dam was by a bull of old Mr. C. Colling's, of Ketton, bought by him of the late Mr. John Bamlet, senior, of Newton. Her grandam was by Mr. Harrison's Bull of Barmpton, bred out of Mr. Wastell's old cow that bred Mr. William Robson's Bull (the grandsire of Hubback). Foljambe was got by Mr. Richard Barker's Bull, which was by Mr. Hill's Red Bull out of a cow of his (Mr. Barker's) own. Mr. Hill's Red Bull was by a half-brother of the Dalton Bull, and from a cow of Mr. Hill's own that was got by William Robson's Bull (grandsire of Hubback). This cow's dam was greatly admired by the present Lord Darlington's father.[1]

The late Mr. R. Colling bought a twin heifer of me that was by Mr. Snowdon's Bull (the sire of Hubback), when a heifer, and he bought her sister a year or two after. Their dam was by Mr. James Masterman's old bull (great-grandsire of Hubback), and out of Mr. Thomas Hall's great cow sold to the Duchess of Athol. This last cow was by Mr. Harrison's Bull, that was bred by Mr. Wastell, from the same cow with Mr. Robson's Bull (grandsire of Hubback), and from a famous good cow of Mr. Thomas Hall's that was a great grazer: Mr. Wastell called her Tripes. The dam of the twins, sold to the late Mr. Robert Colling, gave, when getting only grass, with no other food,

---

[1] *I.e.* by Henry, 2nd Earl of Darlington, 1758-1792.

eighteen quarts of milk (ale measure), twice a day, for more than a month together, in June and July; as we sold our milk at Darlington, it was measured regularly. The late Mr. R. Colling called the first twin heifer he bought of me Bright Eyes,—having remarkably bright eyes. The predecessors of this Bright Eyes, and the predecessors of Foljambe's dam, both by Mr. Harrison's Bull, bred by Mr. Wastell, were own sisters, daughters of Tripes.

My great cow, from the dam of Bright Eyes, was put dry in October, and sold to the butcher the first Monday in March, at twenty-five guineas, and weighed 84 stones; the best beef then sold at 4s. 8d. per stone (14 lbs.).

John Hunter, of Hurworth, mason, bred Hubback bull, and his dam, which was by a bull of Mr. Banks's, of Hurworth, which had a great belly, but he was out of a handsome cow of Mr. Banks's own. The grandam of Hubback John Hunter bought of Mr. Stephenson of Ketton.

Mr. Thomas Hall bought the cow which Mr. Wastell called Tripes of Mr. Charles Pickering of Foxton, near Sedgefield, grandfather of Mr. Christopher Mason, of Chilton, but how she was bred I do not know.

The Dalton bull was bred by Mr. Robert Charge, of Low Fields, brother to Mr. John Charge, of Newton, and was by a Red Hazel Bull of Mr. John Charge's, of Newton. This Hazel Bull was bred by Mr. William Dobson, of Croft, and by Mr. Duke Wetherell's old Red Bull.[1]

Mr. Thomas Hall's heifer, sister to Bright Eyes' dam, was sold to go to the south for twenty-five guineas. Another half-sister to Bright Eyes' dam got by old Mr. C. Colling's Bull, was sold to Mr. Hill, of Blackwell, for twenty-five guineas; she bred a heifer that was matched against one of Mr. Hammond's of Hutton Bonville, and won the match. Mr. Hammond's stock in that day were in great repute. A year-old bull, out of the great cow, daughter of Tripes, was sold for twenty-five guineas to Mr. Bell of Halton, in Northumberland, and was also got by old Mr. C. Colling's Bull.

Masterman's Bull got them all with waxy horns, but Hubback's stock were the best of any I ever bred. Their coats were like a moleskin, and their hair stayed on till almost midsummer.

I am quite clear in the whole of the above statements, and am sure they are perfectly correct.

Mr. Snowdon, of Hurworth, told me his stock were from Sir William St. Quintin's herd. His cow that bred Hubback's sire

---

[1] Mr. Wetherell seems to have been nicknamed Duke and Archduke. See below, p. 389.

was a very handsome one, and remarkable for her wide hooks and fine quick eyes; but the bull, sire of Hubback, drooped in his hind-quarters, and was low-sided; but he had a fine fore-end and good crops. These particulars I well remember.—Signed by me in the 66th year of my age,   ALEXANDER HALL.

22nd March 1820.

II

The cow I sold Mr. Robert Colling of Barmpton was by Mr. George Snowdon's Bull. She was a twin calf. Her sister Mr. Colling also bought and fed off, and she was an extraordinary fat one. Mr. Colling put my cow called Bright Eyes to Hubback bull. The dam of Bright Eyes was by James Masterman's Bull, and gave 18 quarts of milk per meal for a month or six weeks after calving, and did this two years, having twin calves each year. The grandam of Bright Eyes was sold to the Duchess of Athol, and was by Mr. Harrison's Bull of Barmpton, bred by Mr. Wastell of Burdon. The great-grandam of Bright Eyes was bought by my brother, Thomas Hall, of Mr. Charles Pickering of Foxton, near Sedgefield (the grandsire of Mr. Christopher Mason of Chilton by his mother's side); she had a wildness and unruliness about her. This cow was called Tripes by Mr. Wastell of Burdon. I had many extraordinary cattle from this tribe of shorthorns and sold them at high prices for those days. The bull of Mr. Harrison's, sire of the Duchess of Athol's cow, had nothing of Masterman's Bull's tribe. The above facts I remember well.   ALEXANDER HALL.

7th May 1832.

NOTE B, p. 245

JOHN CHAPMAN'S RELATION

I lived with Mr. Robert Colling since I was seven years old, and am now nearly fifty, and remember the cow Mr. Robert Colling bought of Mr. Alexander Hall, who lived at Haughton. She was a little red flecked cow, and was called Bright Eyes. This cow was the grandam of Princess, which Mr. Robert Colling sold to Sir Henry V. Tempest when two years old, and Princess's dam was own sister to Mr. Robert Colling's White Bull, and they were both bred direct from Hubback to Favourite Bulls without any other cross intervening.

Mr. Robert Colling's Bright Eyes Cow was from the same original tribe of Mr. Alexander Hall's, but Mr. Alexander Hall's Cow was by Hubback, and was put to Punch Bull and the produce was Bright Eyes's grandam, that was a calf in 1801, the same age as Red Rose in the Catalogue of 1818, sale No. 1.

Old Princess was sent from Wynyard to Barmpton and was bulled by Mr. Robert Colling's Wellington Bull, the sire of Baronet and Lancaster, etc., and they were to pay ten guineas for the bulling of Princess.

Mr. Robert Colling's Wellington Bull was of the Wildair tribe of cows and own brother to Juno No. 3 in catalogue of sale at Barmpton in 1818. He purchased the first of the tribe from Mr. George Snowdon of Hurworth, and she was in calf to Sir James Pennyman's Bull, descended from the stock of Sir William St. Quintin of Scampston.

Sally, lot 28 in the sale at Barmpton, October 3rd, 1820, was bought by Mr. Robinson of St. Helens, Auckland, and was then in calf. She was bulled by Mr. Robert Colling's bull, Old Barmpton. Sally's dam was a hind's cow to Mr. John Chapman, and was a most capital milking and a good-looking cow. She gave 15 quarts of milk after calving, ale measure. She was bought of Mr. Charles Lenet of Haughton, who recommended her to Mr. Robert Colling, and was bred by Mr. Jonathan Dryden and was got by Mr. Chapman's Son of Punch and was in calf when bought. The calf, two years old in 1820, is stated in the Barmpton Catalogue of 1820 to be by Alexander.

The above are correct accounts of the pedigrees as witness my hand this 24th day of October 1831, having read the same over.
JOHN CHAPMAN.

NOTE C, p. 47

MR. MAYNARD'S RECOLLECTIONS

Mr. Maynard's last Favourite (now in the possession of Thomas Bates) is twenty-two years old, spring 1820, being bred by Mr. Maynard, and was got by a son of Mr. Maynard's old yellow cow; her dam by Hubback; grandam (own sister to Young Strawberry the dam of Lord Bolingbroke) by the Dalton Bull (kept by Mr. Maynard and Archduke Wetherell);[1] great-grandam Old Favourite (which Mr. Maynard sold to Mr. C. Colling and was the grandam of Favourite Bull, and great-great-grandam of Comet) by Mr. Ralph Alcock's Bull (bred by Mr. Michael Jackson of Hutton Bonville, near Northallerton—the pedigree of which cannot be traced), remarkable for his handling and lively looks, and all his stock were like himself; great-great-grandam by a bull of Mr. Jacob Smith's of Givendale near Boroughbridge (this bull was of a yellow red

[1] See above, p. 387.

colour, with a white back and white face and white legs to the knee); great-great-great-grandam (called Strawberry) by Mr. Jolly's Bull of Worsall, a dark red with black brindled intermixed; and great-great-great-great-grandam, called Necklace, was a red colour. The dam of Necklace was a black cow with white belly and white legs to the knee, and grandam of Necklace, a grey-coloured cow. The three last were great milkers, and had to be milked before calving. The distemper prevailed among cattle when the last-mentioned cow was young, now above seventy years ago.

Mr. Maynard's old yellow cow was out of the own sister to Young Strawberry (dam of Lord Bolingbroke) and got—either by Mr. Robert Colling's Bull which he bought at Manfield near Piersbridge (the sire of which bull was Mr. James Brown's famous old Red Bull of Aldbrough)—or Mr. George Coates's Bull (the sire of which was the bull Mr. Robert Colling bought at Mansfield), Mr. Maynard not being certain from which of the above bulls his old yellow cow was descended.—The son of Mr. Maynard's old yellow cow (sire of Mr. Maynard's Favourite now twenty-two years of age) was by Mr. Henry Chapman's Grandson of Punch (out of Chapman's famous old cow).—The sister to Strawberry (dam of Lord Bolingbroke) had a thick skin and was a hard handler, her hair as strong as pig bristles, and was plain-shaped and low-sided.—Her daughter by Hubback was the very reverse taking at Hubback, and was a large frame and great feeder, and her handling good.

Starling was, Mr. Maynard believes, got by his bull out of the yellow cow by Chapman's Grandson of Punch, but is not certain —her dam by a son of Hubback (Hubback's son being out of one of the three best cows he had, which he sent to be bulled at Ketton by Hubback in 1787—the year after Mr. Maynard sold Mr. C. Colling Favourite)—her grandam the old yellow cow specified above (either by Mr. G. Coates's Bull or his sire).

JOHN MAYNARD.

13th March 1820.

NOTE D, p. 317

ATKINSON GREENWELL'S NARRATIVE

I

I hereby certify that I am cousin of John Stephenson of White House, near Wolviston, county of Durham; that I recollect well the time at which he bought a cow at a public sale at Wynyard in the month of October in the year 1818. I was intimate with him at the time, living at Copelaw in the parish of Great Aycliffe within ten miles of his residence. Within a fortnight after the sale of the Countess of Antrim's cattle at Wynyard, he came to my house at Copelaw and offered me the cow he had bought a few days before at Wynyard at the price he gave for her. I was then in dairy business in part, and did not wish to have cows that were not in a situation to give milk; and I did not breed blood cattle. Mr. Stephenson stated that the cow had lost one or two teats, and I therefore declined to purchase her. Within three or four weeks after, I visited him at White House, and then for the first time saw the cow. She was a red roan then apparently seven or eight years old. She had evidently been a milk cow, at least three or four years, as I judged from the appearance of her bag. She was thin in condition, apparently from milking. Mr. Stephenson spoke of her high blood to me. I particularly noticed her on account of her high blood and of the low price at which Mr. Stephenson offered her. I was tempted to buy her, and should have done so but for the loss of her teats and her lessened milking capacity. I recollect the cow perfectly well. I knew her from that time on for many years. I was at White House frequently during the years from 1818 to 1831. Mr. Stephenson's mother lived with me as my housekeeper in the year 1818, and I have always continued intimate with Mr. Stephenson up to the present time. I have always known the stock descended from the cow bought by him at Wynyard in the year 1818. I recollect her son Waterloo and her daughter Angelina 2nd which were her two first calves. I never saw any other cow at White House which came from Wynyard than the cow which came first. I never heard of any such cow as Anna by Lawnsleeves being at White House, or that Mr. Stephenson was ever the owner of such a cow as Anna by Lawnsleeves. I always observed that his bulls were superior.

I went to live at Kirklevington in the year 1822 and remained there until the year 1833. While at Kirklevington I made the acquaintance of Mr. Bates, and was frequently at his house. In June 1831 I was there at the time when Mr. Bates's bull Second Hubback came home from Mr. Whitaker, to whom he had been let. Finding him come home alone Mr. Bates expressed regret that Mr. Whitaker had not sent a bull to him as he had many cows and wanted a bull. I advised him to go and see Mr. Stephenson's bull Belvedere and to hire or buy him. Mr. Bates consented to go the next day, if I would accompany him. Accordingly the next day we went to White House, near Wolviston. On seeing Belvedere Mr. Bates remarked that the bull was of the blood which he knew and which he had been seeking for a long time. He determined to buy Belvedere if Mr. Stephenson would sell him. He asked for the particulars of his breeding, and made a memorandum of it from Mr. Stephenson's words at the time. He asked for a price on the bull and Mr. Stephenson named fifty guineas. Mr. Bates agreed to give it.

It was arranged that on the following day Mr. Stephenson should bring the bull over to Kirklevington and receive his money from him. I met Mr. Stephenson with Mr. Bates at Kirklevington. I have this day seen the certificate given to Mr. Bates by Mr. Stephenson. I know it to be the same, and that the signature is in the handwriting of Mr. Stephenson. I have this day made a memorandum on that certificate. ATKINSON GREENWELL.

ARCHDEACON NEWTON, *9th November* 1848.

II

Pedigrees of the following Improved Shorthorned Bulls :—

BELVEDERE, 5 years old last Lady Day, by Waterloo, dam Angelina the Second by Young Wellington, grandam Angelina, calved 1810 by Phenomenon (491), great-grandam Anna Boleyne, calved 1803 by Favourite (252), great-great-grandam Princess by Favourite (252) calved 1800, great-great-great-grandam by Favourite (252) bred by Mr. Robert Colling of Barmpton.

WATERLOO, by Young Wellington, dam Angelina by Phenomenon (491), see above.

YOUNG WELLINGTON, by Mr. Robert Colling's Wellington (680), dam Princess by Favourite (252), see above.

I bought Angelina at Wynyard sale in 1818, and she was in calf of Angelina the Second. I sent Angelina to Young Wellington in 1819, and she bred Waterloo bull, see above. I sold Belvedere to Thomas Bates, June 22nd, 1831.

I never sent any Pedigrees of my cattle to be inserted in the Herd Book.

The above is a true statement, as witness my hand this 23rd June 1831. JOHN STEPHENSON.

WOLVISTON, near STOCKTON.

The above certificate was signed by Mr. Stephenson in my presence. November 9th, 1848, ATKINSON GREENWELL, *Archdeacon Newton.*

Note E, p. 318

Mr. Harrison's Hire of Waterloo

I

In the year 1830 I hired the bull Waterloo of Mr. John Stephenson, of Wolviston, County of Durham. When the bull came to my house at Streatlam Grange, where I resided, there came with him a memorandum of pedigree. That memorandum stated that he was by Young Wellington, dam Anna by Lawnsleeves, gd. Angelina by Phenomenon, g.gd. Princess by Favourite, etc. This memorandum of the pedigree of Waterloo was not in Mr. Stephenson's handwriting; from that memorandum I had some handbills and cards printed stating the pedigree as above, and advertising Waterloo to stand at my house to serve cows.

At a time subsequent to the printing of these handbills, in a conversation with him, Mr. Stephenson informed me that there was an error in the printed pedigree as I had published it, and stated that Waterloo was out of Angelina by Phenomenon, and not out of Anna by Lawnsleeves.   GEORGE P. HARRISON.

FORCETT HOUSE, ALDBROUGH, 23rd *November* 1848.

II

That superior and well-bred Bull Waterloo, the property of Mr. G. P. Harrison, will serve cows at Streatlam Grange at one guinea a cow.

Waterloo was got by Young Wellington, dam Anna by Lawnsleeves, grandam Angelina by Phenomenon, g.gd. Anna Boleyn by Favourite, g.g.gd. Princess by Favourite, g.g.g.gd. by Favourite.

Cows and heifers sent from a distance will be kept on reasonable terms and have every attention paid them.

N.B.—Gentlemen sending only a single cow will be charged three guineas.

STREATLAM GRANGE, *November* 1830.

Note F, p. 319

Mr. John Wood's Statement

I was at the sale of the cattle of the late Sir Henry Vane Tempest, Bart., of Wynyard, in the year 1813, which sale took place in the month of October of that year, after the death of Sir Henry. I bought at that sale the cow Nell Gwynn. I recollect the heifer Angelina put up at that sale and bid in by Mr. Wetherell of the Isle for the Countess of Antrim, the widow of Sir Henry. Angelina was three years old at that sale. I recollect her at that time perfectly well. She was an even roan. She was remarkable for her fine head, eye, and horns, and was a very fine-looking heifer, with particularly good hair.

In the year 1815 I was at Elemore one day to visit George Baker, Esq., of Elemore Hall, and saw there on that day the same heifer Angelina then become a cow. Mr. Baker called my attention to the fact that the Countess of Antrim had sent Angelina to Elemore to be bulled by his bull Lawnsleeves. She was remaining there at that time for that purpose. I have always understood, and am quite certain, that Angelina was at that time bulled by Lawnsleeves. I have always understood that Angelina had in the year 1816 a heifer got by Lawnsleeves called Anna.

I was not at Wynyard after the sale in 1813, while the shorthorn cattle were kept there. I have understood that there was a sale of some of the shorthorn cattle, owned by the Countess of Antrim, at Wynyard in the year 1818; but I was not at that sale and I know nothing of it; nor do I know anything about the cattle at Wynyard from the year 1813 until the year 1818, except what I have stated before.

I knew Lawnsleeves very well. He was owned and used by Mr. Baker. He was bred by Mr. Baker, and Mr. Baker purchased his dam of Robert Colling. Lawnsleeves was a large and very lengthy bull, had a plain head, a good eye, good horn, with shoulders upright, shoulder-points somewhat wide, a large brisket, and deep chest. He was somewhat defective in girth, but had a good back and good hind-quarters, and he was a very good handler. His colour was roan. I think he must have been not less than eight years old in the year 1815, at the time I saw Angelina at Elemore. JOHN WOOD.

RICKNAL GRANGE, DARLINGTON, *November 25th*, 1848.

NOTE G, p. 319

DOBLING'S TESTIMONY

I was herdsman to the late Sir Henry Vane Tempest, Baronet, of Wynyard Park, county of Durham, and lived with him in that capacity for eleven years, from 1802 to 1813, and until the sale of Sir Henry's cattle in the month of October in the year 1813, which sale occurred two months after Sir Henry's death.

I had entire and sole charge during all the period, from 1802 to 1813, of Sir Henry's cattle, and nobody else had any charge of them whatever. During all that period no other bulls were ever put to Sir Henry's cows than Phenomenon, Wynyard, and Wellington.[1] Phenomenon was hired of Robert Colling for three years. After the three years were expired, Sir Henry sent cows to Robert Colling's residence at Barmpton to be bulled by him. I went with those cows in every instance, and no cows ever went from Wynyard to be bulled by any other bull than Phenomenon. No cow or heifer of Sir Henry's was ever put to Mr. Baker's Lawnsleeves during the years from 1802 to 1813.

At the sale of Sir Henry's cattle in October 1813, three or four animals were not put into the catalogue and handbills of the sale. Those three or four animals were all reserved at the sale; they were neither offered nor sold. They were all about two years old. There was a bull and two heifers reserved; that I distinctly now recollect, and I think, perhaps, there was another heifer. These reserved animals were left at Wynyard after the sale, and were there when I left my situation on 13th May 1814.

At the sale of the cattle in October 1813, Mr. Richard Wood of Close was house-steward of Wynyard and directed the sale. He bought in for the Countess of Antrim, the widow of Sir Henry, the cow Princess, and the heifer Angelina, and the one-year-old bull Wellington, and the bull calf Pilot. These four animals were kept at Wynyard by the Countess, as well as those named above, which were not put in the catalogue and were reserved.

Angelina was a heifer at the sale in October 1813, not then three years old, and had then never had a calf. When I left Wynyard all these animals, eight in number, were remaining there except Pilot. After the sale in October 1813, the bull that was used, and the only bull that was used to the cows and heifers at Wynyard up to May Day 1814, was the bull Wellington.[1] Angelina dropped a calf in the spring of 1814, got by Wellington.

I never knew and never saw and never heard of such a heifer or cow as Anna-by-Lawnsleeves. No such heifer as Anna-by-Lawnsleeves was ever at Wynyard during the period that I was there as herdsman from 1802 to 1814; if she had been I must have known it. GEORGE DOBLING.

HAVERTON HILL, *November 30th*, 1848.

[1] Better known as Young Wynyard (2859).

NOTE H, p. 319

ROBERT WAISTELL'S TALES

I

In the year 1810 I was at Mr. Robert Colling's at Barmpton in the month of November or December of that year, on some day in one of those months, when Sir Henry Vane Tempest came to see Mr. Robert Colling, and ask his advice as to what bull Sir Henry should put Angelina, and if he (Colling) had one that he would permit to serve her. Sir Henry stated that she had broken her bulling five times to two bulls. Mr. Colling replied that he had no bull at home old enough to serve Angelina except one, and that one was lame and unable to serve a cow. On ascertaining this from Mr. Colling, Sir Henry asked him what bull he should put her to. Mr. Colling advised him to take Angelina to Mr. Baker's at Elemore, near Durham, and ask Mr. Baker to let her be served by Mr. Baker's bull Lawnsleeves, which Mr. Baker had bought of Charles Colling and was then using.[1] Mr. Robert Colling wrote to Mr. Baker for Sir Henry, requesting him that he would permit Lawnsleeves to serve Angelina. In pursuance of this advice, Angelina was sent to Lawnsleeves and bulled by him. In the year 1811 Angelina dropped a heifer calf, and this heifer was called Anna and was a roan. Anna was about two years old at the time of the sale of Sir Henry Vane Tempest's cattle, which took place just after his death in August 1813. Anna was reserved at that sale of 1813 and was not sold.

Angelina was a roan cow, and had horns somewhat long and pointing upwards. Anna was also a roan, and had horns somewhat long and standing upwards—that is in both cows the horns came out level and then turned upwards. Anna was quite like her dam but not so large.

I was at the sale of Sir Henry's cattle after his death, and Anna was not sold at that sale.

I recollect the time when the cattle owned by the Countess of Antrim were sold at Wynyard in October 1818. The Countess had, through Mr. Richard Wood of Close, purchased at the sale of 1813, the cows Princess and Angelina, and these with Anna (which was reserved) were kept at Wynyard and bred there. The cattle offered for sale by the Countess in 1818 were descendants of these cows, Princess, Angelina, and Anna. Mr. Richard

---

[1] Lawnsleeves was, of course, bred by Mr. Baker, but it is not worth while to point out all the misstatements in this farrago of falsehoods.

Wood of Close was her steward. He circulated some handbills on one day advertising the sale of the Countess's cattle the next day. No pedigrees were given, and but few persons attended the sale. The cattle were sold on the day after the day of printing and circulating the handbills. At the time of this sale Princess and Angelina were dead. At this sale I was not present. I saw afterwards the cow Anna at Mr. Stephenson's, at White House, near Wolviston. I was informed that he had bought her at the sale of the Countess in 1818. Anna was about seven years old in the year 1818, when Mr. Stephenson bought her. I saw Anna at Mr. Stephenson's, and he told me he had bought her at Wynyard in 1818. This cow so bought at Wynyard in 1818 was the dam of Waterloo, and I saw Waterloo sucking this cow. Waterloo was the first calf which Anna dropped after Mr. Stephenson got her; and Waterloo was in her at the time Mr. Stephenson bought her. He sent Anna again to the same bull that got Waterloo, viz., Young Wynyard, and the year after the birth of Waterloo she dropped a heifer calf got by Young Wynyard and called by Mr. John Stephenson Angelina the Second.

Old Princess died at Wynyard and Angelina was killed at Wynyard prior to the sale in 1818. Angelina was in calf and began to make bag. Having bulled while she was in calf, she was not thought to be in calf, and was supposed to have the garget in her bag. A cow doctor was called, and he cut off her teats and bled her excessively, and in consequence of this treatment to cure the garget, Angelina died. On opening her it was discovered that she was in calf, and had been making bag for calving.

I recollect Anna very well when I saw her in 1819, the year after Mr. Stephenson bought her at Wynyard. She was then sucking Waterloo, and was an oldish cow, about eight years old, and had lost all of her teats but one, three of them being ruined at the time.

At the time of the sale of the Countess of Antrim's cattle in 1818, no pedigrees were given and none were printed. Anna was sold merely by the name of Anna, and stated to be in calf by Young Wynyard. Thomas Smith (called Handy Smith, for the reason that he had lost one hand) was employed by Mr. Stephenson to buy the cow at the sale for him; Smith told me this. In 1820 I hired Waterloo of Mr. Stephenson, and this was the first time he was let. When Mr. Stephenson let him to me, I told him how Waterloo was bred, and how his dam was bred, and what she was got by; and at my request Mr. Stephenson gave me a certificate of Waterloo's pedigree, which certificate I now have. That certificate states that Waterloo was got by Young Wynyard, dam Anna-by-Lawnsleeves. As I knew the pedigree of Anna and Mr. Stephenson did not, I requested him to give me the certificate, and he did so.

ROBERT WASTELL.

November 24th, 1848.

II

The within is a copy of a certificate given to me by John Stephenson, when I hired the bull Waterloo of him, and it was written at my residence, Blue House, near Stockton.

ROBERT WASTELL.

November 30th, 1848.

'This is to certify that I have let a bull, rising two years old, to Mr. Robert Wastell for two years, by Young Wynyard, dam Anna-by-Lawnsleeves. JOHN STEPHENSON.'[1]

NOTE I, p. 319

EDWARD HALL'S ALLEGATIONS

I hereby certify that I attended the sale of the cattle of the Countess of Antrim at Wynyard. This sale took place not more than two years after the first sale of cattle at Wynyard, which occurred soon after Sir Henry Vane Tempest's death. I am quite sure that the sale of the cattle of the Countess was not later than two years after the sale of Sir Henry's cattle. At the sale of the Countess's cattle which I attended, there was but one shorthorn offered for sale. All the rest were common cattle, not shorthorns. The one shorthorn so offered was Anna by Lawnsleeves, and so called at the said sale. Mr. John Stephenson of Wolviston was at the said sale; and when Anna was put up, Mr. Stephenson requested me to bid for her on his account, telling me to go as high as twenty guineas. I did bid for her, and she was bought by me for him at twenty or twenty-one guineas. I took the heifer or cow that very day to Mr. Stephenson's place near Wolviston, and left her there. At the time of the sale Mr. Stephenson told me that this heifer Anna was not sold at the first sale because she was a little shabby thing, and was the only shorthorn left at Wynyard at the first sale that occurred just after Sir Henry's death. This heifer so bought by me for Mr. Stephenson at the sale of the Countess was not more than about

---

[1] The original of this does not appear to have been forthcoming.

four years old, probably not so much as four years old. She was quite small and a yellow roan. She had lost two paps or teats, and was said to be in calf at the time of the sale to Young Wellington or Young Wynyard. I saw her often at Mr. Stephenson's after I took her there, and for many years saw her often every year. She grew very much after she went to Mr. Stephenson's, and made a good-sized cow. I saw her at Mr. Stephenson's as many as ten years after she was taken there by me. Her first calf after the sale, and which she had in her when I bought her at Wynyard, was dropped at Mr. Stephenson's. That calf was a bull, a dark red, and at my suggestion Mr. Stephenson called him Waterloo. EDWARD HALL.

SEDGEFIELD, *November 26th,* 1848.

(*Edward Hall, at the time of signing the above certificate, was in the Poor House.*—A. STEVENS.)

NOTE K, p. 319

HANDY SMITH'S STORY

I hereby certify that I attended the sale of the cattle of the Countess of Antrim at Wynyard in the month of October in the year 1818. There were but few cattle offered for sale at that time. They were the remnant of the sale of 1813. I now recollect that Princess and Angelina were among those sold at the sale of October 1818. Angelina was bought in. At that time Angelina appeared to be a cow seven or eight years old. Among others sold was the cow Anna-by-Lawnsleeves. She was then apparently five years old, and had had three calves. She had lost one or two of her teats, and gave no milk out of them. She was a roan—an old-fashioned shorthorn. When she was put up, I was requested by Mr. John Stephenson to bid for her on his account. I did so, and she was struck down to me at eighteen guineas. After the sale was over, Mr. Stephenson gave me the money and I paid for her. She was taken from Wynyard to Mr. Stephenson's residence at Whitehouse, near Wolviston, immediately after the sale. Mr. Stephenson took charge of her at the sale, and from that time I had nothing to do with the cow Anna. I often saw the cow Anna at Mr. Stephenson's, where I frequently went. I was intimate with Mr. Stephenson. I saw Angelina many different times at Wynyard after the sale of 1818. She remained there some years after the sale of 1818. THOMAS SMITH.
BREARTON, near GREATHAM.

STOCKTON, *3rd January* 1839.

Note L

Extracts from the Kirklevington Accounts, 1837-1849

		£	s.	d.
1837.				
June 7th.	Received for 2nd Earl of Darlington Bull[1]	27	0	0
13th.	Received for Red Rose Bull[2]	20	10	0
	Received for 3rd Earl of Darlington Bull	14	0	0
1839.				
Jan. 21st.	Received for a bull off Shorthorns[3]	63	0	0
July 22nd.	Received of an American gentleman for a bull[4] in part	100	0	0
Sept. 22nd.	Received of an Irish gentleman for a bull in full[5]	105	0	0
July 25th.	Thomas Myers, expenses to Leeds with Blanche bull[4]		15	6
Sept. 17th.	Paid Mr. Edwards for a Cow[6]	110	5	0
Nov. 23rd.	Paid for conveyance of bull and 3 females from London 29th July last, £8 : 8s., and charges for wharfage in London, £1 : 10s.	9	18	0
	Paid for a bull going to London, £2 : 2s., and 7s. 6d. wharfage, London	2	9	6
Dec. 18th.	Received of Mr. John Backhouse for a bull[7]	105	0	0
1840.				
May 2nd.	Received of Mr. J. Grant Duff for a bull[8]	157	10	0
4th.	Paid James Grange taking a bull to Hull—five days, to York and a Sunday's rest, and the first day to Selby and next to Hull, and had three days rest before being shipped to Aberdeen	3	11	1

[1] (1945) of the Tragedy tribe, got by Belvedere (1706).
[2] (2493) by Belvedere (1706), d. Red Rose 10th by Bertram (1716).
[3] Van Amburg (5543), sold to Mr. Barrett, Stratton Park, and Mr. H. Beauford, Bletsoe.
[4] The Blanche bull Yorkshireman (5700), sold to Mr. Cope, West Chester, Philadelphia, U.S.A.
[5] Omega (4615), sold to Mr. Foster, Springfield, Ireland.
[6] Foggathorpe by Malbro' (1189).
[7] Prince Albert (4781).   [8] Holkar (4041).

			£	s.	d.
June 3rd.	Received of Mr. Etches for a bull-calf, and one year-old quey for America[1]		200	0	0
	Received of Mr. Bolden for a cow[2]		80	0	0
	Received of Mr. Bolden for Messrs. Bell[3]		80	0	0
July 29th.	Received of Messrs. Foreman and Etches for two heifers at Cambridge,[4] £210, and returned as they got no premiums, £20		190	0	0
	Received of Mr. Robinson, Bletsoe, for Nonsuch Bull[5]		105	0	0
11th.	Paid expenses Middlesbro', shipping cattle			9	0
	Paid conveyance of cattle to London		9	12	0
	Paid two men for journey with them (£4 and £5)		9	0	0
16th.	Paid expenses self for lodging and eating at Cambridge		3	0	0
29th.	Paid conveyance of cattle to and again in 3 waggons from Middlesbro' to Yarm		1	10	0
	Paid conveyance in steamer from London to Middlesbro' for 3 cattle		4	13	6
	Paid, 3 cattle, grass at Cambridge, 1s. 6d. per night		1	1	0
Sept. 16th.	Received of Messrs. Etches for two heifers[6]		168	0	0
Oct. 1st.	Received of Messrs. Jobson for the use of 2nd Duke[7] three seasons		105	0	0
21st.	Received of Mr. Wilson, Cumledge, near Coldstream or Dunse, for a bull-calf[8]		52	0	0
1841.					
Jan. 13th.	Remitted Messrs. Fores for 5 prints of Duke of Northumberland for the American Ambassador		3	0	0

[1] Apparently Duke of Wellington (3654) and Duchess by Duke of Northumberland (1940), d. Nonsuch 2nd—*Herd Book*, v. p. 742—both bought by Mr. George Vail, but this heifer seems to have been two years old at the time.

[2] Brown 2nd by Duke of Cleveland (1937), sold to Messrs. Bolden, Port Phillip, Australia—*Herd Book*, v. p. 117.

[3] Lord Percy (4266) by Duke of Northumberland (1940), d. Nancy, sold to Messrs. Bolden, Port Phillip.

[4] Probably Ancedote (*sic*) by Duke of Northumberland (1940), d. Craggs—*Herd Book*, v. p. 34, and Kirklevington by Short Tail (2621), d. Shorthorns 2nd—*Ibid.* v. p. 934. [5] Nonsuch Bull (4581).

[6] Perhaps Annabella 2nd and Belle 2nd—*Herd Book*, v. pp. 39, 79.

[7] 2nd Duke of Northumberland (3646).

[8] Dunse (3661) by Duke of Northumberland (1940), d. Blossom.

		£	s.	d.
	Remitted Mr. Reginald Jennings, Bishop Stortford, for conveying cattle to Cambridge.	10	5	0
April 13th.	Received of Mr. Etches for bull calf, Hill Farm [1]	31	0	0
May 29th.	Received for a heifer off Craggs by Duke of Northumberland, sold to Mr. Etches [2]	84	0	0
July 26th.	Received of Mr. Unsworth for 2 cows, Hill Farm,[3] £130 and £80, for bull-calf, Town Farm [4]	210	0	0
Aug. 7th.	Received of Mr. Icely for Duke of Cambridge [5]	157	10	0
12th.	Received of Mr. Etches, Liverpool, 3 queys, two-year-olds, Hill Farm	140	0	0
Sept. 4th.	Received of Mr. C. W. Harvey, near Liverpool, for a bull-calf, Duke of Cambridge [6]	105	0	0
Oct. 4th.	Received of Messrs. Jobson for a year's use of 2nd Duke	105	0	0
	Received of Mr. Lowndes for bull-calf off Wild Eyes [7]	90	0	0
1842.				
Jan. 17th.	Received of Hayem Worms, Esq., Metz, France, for a year-old bull, Edward [8]	42	0	0
Feb. 2nd.	Received of Mr. Anthony Ridley for C. H. Leigh, Esq., Pontypool Park, Monmouthshire, for John (bull), calved Sept. 3rd, 1840 [9]	100	0	0
Mar. 2nd.	Received of Mr. Thomas Wilson, Shotley, for Queen's bull-calf [10]	52	10	0

[1] George (3884) by Cleveland Lad (3407), d. Fletcher 2nd.

[2] This heifer does not seem to have been Anecdote, which had already brought Messrs. Foreman and Etches the b. c. Earl of Liverpool in October 1840—*Herd Book*, v. p. 34.

[3] One of these cows was Belle by Belvedere (1706), d. Craggs—*Ibid.* v. p. 79.

[4] Crofter (5898) by Duke of Northumberland (1940), d. White Rose.

[5] Duke of Cambridge (3637), sold to Mr. Icely, Australia.

[6] 2nd Duke of Cambridge (3638) died 24th July 1842.

[7] Apparently General Washington (6036), afterwards the property of Mr. Joshua Knowles, Attercliffe, Sheffield.

[8] Edward (3697) by Duke of Northumberland (1940), d. Dinah.

[9] John (4108) by Duke of Northumberland (1940), d. Greenwell.

[10] Earl of Durham (9059), afterwards the property of the Rev. J. Vane, Wrington, Somerset.

		£	s.	d.
July 14th.	Received of Mr. J. C. Etches for a bull-calf off Oxford[1]	49	10	0
	Received of Mr. John Clifford Etches for three heifers	148	10	0
25th.	Paid Thomas Myers taking bull to —— for Mr. James Glass	8	0	0
Aug. 18th.	Paid Mr. Strafford for 4 prints of Duke of Northumberland for Colonel Trotter, Mr. Bellerby, Mr. Knowles, and Mr. Thomas Booth	2	8	0
22nd.	Received of Sir A. Buller for a bull-calf[2]	107	6	6
Feb. 1st.	Received of Mr. Cator for two cows[3]	147	0	0
Mar. 15th.	Received for a bull-calf to Mr. McIntosh[4]	157	10	0
April 21st.	Received of Messrs. Harrison for the use of Cleveland Lad	84	0	0
July 15th.	Received of Mr. Robinson's son for a bull-calf getting Bletsoe, Bedfordshire	10	10	0
Sept. 23rd.	Paid Female Manager of Cattle, Duncombe Park[5]	1	0	0
24th.	Received of Mr. J. Grant Duff for 2nd Duke of Northumberland (hire)	50	0	0
	Received of William Peel, Esq., Taliaris, near Llandilo, for 3 heifers	210	0	0
	Received of William Peel, Esq., for 2 cows and 5 heifers for Mr. Burnett, £98, and expenses of same to Gloucester	105	0	0
	Received of William Peel, Esq., for a bull-calf off Rosalind 4th[6]	105	0	0
27th.	Received of Thomas Phillips, Esq., for Lord Feversham, for Cleveland Lad	300	0	0

[1] Locomotive 2nd (6139) by Duke of Northumberland (1940), d. Oxford Premium Cow.

[2] 3rd Duke of Cambridge (5941).

[3] Blanche 2nd by Norfolk (2377), and Waterloo 3rd by Norfolk (2377), sold to the Rev. Thomas Cator, Skelbrook Park.—*Herd Book*, v. pp. 91, 1084.

[4] The Lord of Hainault (6588), by 4th Duke of Northumberland (3649), d. Wild Eyes 6th, see illustration, *Herd Book*, iv. p. 756.

[5] 'Old Anna,' yet the ungrateful old woman told 'the Druid' years after, 'As for Bates and Booth, she might have heard their names.'—*Saddle and Sirloin*, p. 215.

[6] Bearl (5781) by 4th Duke of Northumberland (3649), d. Rosalind 4th.

		£	s.	d.
Oct. 13th.	Received for hire of 2nd Duke, of Captain Barclay Allardice, per J. Grant Duff	60	0	0
	Received for Sockburn Bull[1] of Mr. Archbold, Ireland, and carriage	80	0	0
1844.				
Jan. 5th.	Received of Mr. Henry Lister Maw, Tetley, Crowle by Bantry, for a two-year-old roan heifer off Blanche[2]	101	10	0
8th.	Received of S. O. Priestley, Esq., Trefan near Criccieth, Carnarvonshire, for bull from Hill Farm,[3] off Craggs by Son of 2nd Hubback, sire Duke of Northumberland	62	0	0
Sept. 11th.	Received of Mr. Mitchell, Arthington Hall, near Otley, for 2nd Duke of York Bull[4] for the season	78	15	0
Oct. 16th.	Received of Mr. C. W. Harvey for Robert Bell's heifer, Lady Barrington[5]	85	0	0
Nov. 22nd.	Received of Mr. C. W. Harvey for White Rose Cow[6]	20	0	0
1845.				
Jan. 22nd.	Paid Thomas Myers' Expenses with T. Bell's bull sold to Mr. Glass[7]	5	0	0
May 3rd.	Received from the Dowager Duchess of Leeds for keeping a bull-calf by 4th Duke	21	0	0
June 3rd.	Received from Henry Lister Maw, Esq. of Tetley, Crowle, for a bull-calf off Blanche[8]	105	0	0
July 7th.	Received for John Bell's cow, Darlington (Mr. Squires) Romanby	18	18	6
Aug. 19th.	Paid for the last print of Duke of Northumberland given to Mrs. Appleton, Crathorne		12	0

[1] Sockburn (6509) by 4th Duke of Northumberland (3649), d. Blanche 3rd.
[2] Blanche 5th by Duke of Northumberland (1940).   [3] Percy (7326).
[4] 2nd Duke of York (5959) by Duke of Northumberland (1940), d. Duchess 41st.
[5] Lady Barrington 4th by Cleveland Lad (3407)—*Herd Book*, vii. p. 412.
[6] Probably Secret by Short Tail (2621), d. White Rose—*Ibid.* v. p. 1101.
[7] Percy (9472), see above, p. 303.
[8] Roan Duke (8486) by 4th Duke of Northumberland (3649), d. Blanche 3rd.

		£	s.	d.
Oct. 15th.	Received of Mr. Charles Whitfield Harvey for Blanche Cow[1]	105	0	0
Nov. 20th.	Received of Mr. Vail for Mr. Bell's cow (of Lady Barrington tribe), got last September twelvemonth[2]	67	10	0
1846.				
April 29th.	Received of Godfrey Wentworth, Esq., for Foggathorpe 2nd bull named Chevy Chace[3]	150	0	0
Sept. 30th.	Received of Mr. Grant Duff, bull, Duke of Richmond's[4] hire, due 17th Oct. next	52	10	0
Oct. 6th.	Received of Mr. Harvey for two Blanche cows (105 guineas and 50 guineas)[5]	162	15	0
10th.	Received of Mr. James Topham for Foggathorpe bull-calf ten months old (Euclid)[6]	140	0	0
Oct. 17th.	Paid Thomas Myers taking bull off Wild Eyes 2nd[7] to Mr. Burnett, and bringing back 2nd Earl of Beverley[8] from Black Hedley	1	10	7
Nov. 3rd.	Received of the Honourable Octavius Duncombe for a bull-calf[9] off 7th Wild Eyes by 2nd Cleveland Lad	105	0	0
1847.				
March 1st.	Received of Mr. Harrison, Greta Bridge, for his share of Duke of Norfolk,[10] one year	26	5	0
	Received of Mr. ——, Middlesbro', for old Wild Eyes	5	0	0

[1] Blanche 4th by Duke of Northumberland (1940), 'died of distemper caught on the road.'   [2] Lady Barrington 3rd, pp. 305, 321.

[3] Chevy Chace (7897) by 4th Duke of Northumberland (3649), d. Foggathorpe 2nd.

[4] Duke of Richmond (7996) by Cleveland Lad 2nd (3408), d. Duchess 50th.

[5] Blanche 3rd by Short Tail (2621), and Blanche 6th by 4th Duke of Northumberland (3649). The *Herd Book*, vii. p. 270, is wrong in describing Blanche 6th as bred by Mr. C. W. Harvey; she was calved at Kirklevington 4th Jan. 1844.

[6] Euclid (9097) by Cleveland Lad 2nd (3408), d. Foggathorpe 4th.

[7] Lord George Bentinck (9317) by 2nd Duke of Northumberland (3646), d. Wild Eyes 2nd.

[8] 2nd Earl of Beverley (5963) by Duke of Northumberland (1940), d. Duchess 37th.   [9] Wizard.

[10] Duke of Norfolk (5952) by 4th Duke of Northumberland (3549), d. Duchess 38th.

		£	s.	d.
Nov. 1st.	Paid Robert Bell for a bull off Blossom [1] in exchange for 3rd Duke of Cambridge	73	10	0
1848.				
Jan. 8th.	Received of Mr. Carrington for hire of bull Refiner [2] . . .	63	0	0
Aug. 23rd.	Received for Euclid Bull's use for a year of Messrs. Parker [3] . .	50	0	0
Oct. 6th.	Received of Mr. Phillips for hire of 2nd Earl of Beverley [4] . . .	50	0	0
1849.				
Jan. 17th.	Received of Mr. Timothy Featherstonhaugh, Kirkoswald, for Harmless Cow [5]	80	0	0
20th.	Received of Mr. Carrington for hire of a bull to October last . .	63	0	0
April 16th.	Received for a year-old bull off Foggathorpe 4th [6] of Messrs. Sanday and Smith	84	0	0
May 4th.	Received of Mr. Joseph Roberts, Mickleton, near Campden, Gloucestershire, for Thomas Bell's yearling bull [7]	50	0	0

NOTE M

A Selection of Prices paid for Kirklevington Cattle at Public Auctions

The late T. Bates's Sale, Kirklevington, 9th May 1850
See above, p. 332.

The late Earl Ducie's Sale, Tortworth, 24th August 1853
See above, p. 340.

Mr. J. S. Tanqueray's Sale, Hendon, 24th April 1855

	Guineas
Olive Leaf, by Earl of Liverpool (9061), *Captain Blathwayt*	66

---

[1] Red Duke 2nd (8460) by Duke of Northumberland (1940), d. Blossom.
[2] Refiner (10695) by Cleveland Lad 2nd (3408), d. Wild Eyes 8th.
[3] See above, p. 302.   [4] At Duncombe Park.
[5] Harmless by Cleveland Lad (3407), d. Hawkey—*Herd Book*, x. p. 392.
[6] Baron Foggathorpe (9931) by Cleveland Lad 2nd (3408), d. Foggathorpe 4th.   [7] Willie (11050) by Cleveland Lad 2nd (3408), d. Nosegay.

	Guineas
Lady Barrington 8th, by 2nd Duke of Oxford (9046), *Lord Burlington*	170
Oxford 11th, by 4th Duke of York (10167), *R. Gunter*	500
Oxford 16th, by 4th Duke of York (10167), *Becar and Morris*, U.S.A.	480
Lady Bates, by Duke of Glo'ster (11382), *Harvey Combe*	105
Lady Blanche, by 4th Duke of York (10167), *N. G. Barthropp*	100
Silence, by Earl of Derby (10177), *N. G. Barthropp*	94
Surprise, by Gilliver (11529), *Becar and Morris*, U.S.A.	80
Surmise, by Duke of Glo'ster (11382), *F. Sartoris*	45
Duke of Cambridge (12742), by Grand Duke (10284), *Sir Charles Knightley*	280
6th Duke of Oxford (12765), by Duke of Glo'ster (11382), *R. Gunter*	200
Barrington (12447), by Duke of Glo'ster (11382), *Fisher*, Australia	200

Mr. HARVEY COMBE's Sale, COBHAM, 23rd March 1859[1]

Cambridge Rose 6th, by 3rd Duke of York (10166), *Fisher*, Australia	200
Darlington 5th, by 4th Duke of Oxford (11387), *J. Simpson*	71
The Beauty, by Puritan (9523), *Jonas Webb*	160
Darlington 8th, by 4th Duke of Oxford (11387), *Hon. Col. Pennant*	120
Lady Bates, by Duke of Glo'ster (11382), *H. E. Surtees*	73
Asia, by 2nd Grand Duke (12961), *Lord Dacre*	55
Barbara, by The Buck (13836), *D. McIntosh*	100
Blush, by Marmaduke (14897), *E. Hales*	110
Ayah, by Marmaduke (14897), *J. S. Crawley*	110
Moss Rose, by Marmaduke (14897), *E. Hales*	260
Duchess (Darlington), by Marmaduke (14897), *Marquis of Exeter*	82
Diadem (Darlington), by Marmaduke (14897), *S. Marjoribanks*	40
The Beau (12182), by Belleville (6778), *Walesby*	75
Marmaduke (14897), by Duke of Glo'ster (11382), *Hon. Col. Pennant*	350
The Briar (15376), by Puritan (9523), *Marquis of Exeter*	100

[1] See above, p. 344.

The late Mr. HEGAN's Sale, WILLIS'S ROOMS, 7th June 1865
See above, pp. 346, 347.

Mr. E. L. BETTS's Sale, PRESTON HALL, 1st May 1867 [1]

	Guineas
Grand Duchess 5th, by Prince Imperial (15095), *Duke of Devonshire*	200
Grand Duchess 8th, by Prince Imperial (15095), *Lord Penrhyn*	550
Grand Duchess 9th, by Grand Duke 3rd (16182), *F. Leney*	210
Grand Duchess 11th, by Grand Duke 3rd (16182), *Lord Spencer*	400
Grand Duchess 17th, by Imperial Oxford (18084), *Captain Oliver*	850
Grand Duchess 18th, by Imperial Oxford (18084), *Captain Oliver*	710
Grand Duchess 19th, by Imperial Oxford (18084), *C. H. Dawson*	700
Grand Duchess 20th, by Imperial Oxford (18084), *Lord Spencer*	430
Grand Duchess 21st, by Baron Oxford (23375), *D. McIntosh*	330
Moss Rose, by Marmaduke (14897), *J. P. Foster*	220
Red Rose, by Marmaduke (14897), *F. Leney*	330
Thorndale Rose, by 4th Duke of Thorndale (17750), *Adams*	355
Red Rose 3rd, by Englishman (19701), *F. Leney*	150
Red Rose 4th, by Grand Duke 4th (19874), *A. Brogden*	240
Moss Rose 2nd, by 4th Duke of Thorndale (17750), *D. R. Davies*	160
Thorndale Rose 2nd, by Grand Duke 4th (19874), *Lord Braybrooke*	200
Grand Duke 4th (19874), by Grand Duke 3rd (16182), *Earl Spencer*	210
Grand Duke 16th (24063), by Imperial Oxford (18084), *Roberts*	510
Grand Duke 17th (24064), by 2nd Duke of Wharfdale (19649), *A. Brogden*	305
Cambridge Duke 3rd (23503), by Grand Duke 4th (19874), *F. Leney*	210

[1] See above, p. 349.

Mr. D. McIntosh's Sale, Havering Park, 2nd May 1867[1]

	Guineas
Lady Oxford 5th, by 3rd Duke of Thorndale (17749), *Duke of Devonshire*	600
Baron Oxford, by Duke of Geneva (19614), *Colonel Towneley*	500
Baron Oxford 2nd, by Grand Duke 4th (19874), *E. Holland, M.P.*	500
Waterloo 24th, by Grand Duke 3rd (16182), *Z. Walker*	105
Wellingtonia, by 3rd Duke of Thorndale (17749), *D. R. Davies*	120
Wellingtonia 2nd, by 3rd Duke of Thorndale (17749), *Duke of Devonshire*	160
Wellingtonia 4th, by Grand Duke 4th (19874), *F. Sartoris*	110
Wellingtonia 5th, by Grand Duke 4th (19874), *J. Whitworth*	140
Wellingtonia 6th, by Grand Duke 4th (19874), *J. Whitworth*	125
Wellingtonia 7th, by Grand Duke 4th (19874), *Lord Penrhyn*	120
Wellington 4th (25427), *Tetley*	115

Mr. J. O. Sheldon's Sale, Windsor, 15th Oct. 1867[2]

7th Duchess of Geneva, by 3rd Lord Oxford (22200), *F. Leney*	700
3rd Duke of Geneva (2373), by Imperial Oxford (24185), *D. McIntosh*	55
12th Duke of Thorndale (26020), by Baron Oxford (23371), *E. Thorne*[2]	185
4th Maid of Oxford, by Baron Oxford (23371), *F. Leney*	300
5th Maid of Oxford, by 7th Duke of Airdrie (23718), *I. Downing*	200
Countess of Oxford, by 7th Duke of Airdrie (23718), *Colonel Kingscote*	250
6th Maid of Oxford, by Imperial Oxford (24185), *Colonel Towneley*	400
8th Lady of Oxford, by Imperial Oxford (24185), *Colonel Towneley*	450

[1] See above, p. 353.   [2] See above, p. 355.

	Guineas
7th Maid of Oxford, by 7th Duke of Airdrie (23718), *F. Leney*	260

Mr. J. P. Foster's Sale, Killhow, 23rd Sept. 1868

Moss Rose, by Marmaduke (14897), *D. R. Davies*	400
Royal Cambridge (25009), by Grand Duke 4th (19874), *Sir W. Lawson*	240
Royal Cumberland (27358), by Grand Duke 4th (19874), *J. Fawcett*	160

The Duke of Devonshire's

Baron Oxford 3rd (25579), by Grand Duke of Essex 2nd (21860), *Lord Kenlis*	250
Windermere (27814), by Grand Duke 10th (21848), *G. Hunt*	105
17th Duke of Oxford (25994), by Grand Duke 17th (24064), *J. P. Foster*	100

Mr. S. Rich's Sale, Didmarton, 7th Oct. 1868

Waterloo 28th, by 3rd Grand Duke (16182), *F. Leney and Son*	270
Kirklevington 12th, by 4th Duke of Oxford (11387), *F. Leney and Son*	380

Mr. S. Rich's Sale, Didmarton, 23rd March 1869

Waterloo 32nd, by 7th Duke of York (17754), *Lord Fitzhardinge*	360
2nd Duke of Collingham (23730), by 3rd Duke of Wharfdale (21619), *Earl of Dunmore*	650
3rd Duke of Waterloo (23801), by 7th Duke of York (17754), *J. W. Larking*	140
Duke of Kirklevington (25982), by 7th Duke of York (17754), *J. W. Larking*	105

Mr. E. Bowly's Sale, Siddington, 22nd April 1869

Siddington, by 4th Duke of Oxford (11387), *Earl of Dunmore*	260
Siddington 4th, by 7th Duke of York (17745), *Earl of Dunmore*	400

	Guineas
Siddington 7th, by 7th Duke of York (17745), *Earl of Dunmore*	370

Mr. C. W. HARVEY's

Lord Wild Eyes 5th (26762), by 3rd Duke of Oxford (22200), *Lord Fitzhardinge*	110

Lord PENRHYN's Sale, WICKEN PARK, 7th May 1869

Darlington 12th, by Duke of Geneva (19614), *H. J. Sheldon*	105
Grand Duchess 3rd (Wild Eyes), by Duke of Geneva (19614), *Hon. C. W. Fitzwilliam*	105
Grand Duchess 5th (Wild Eyes), by Duke of Geneva (19614), *F. Leney and Son*	105
Grand Duchess 6th (Wild Eyes), by Duke of Geneva (19614), *J. P. Foster*	145
Darlington 15th, by Grand Duke 11th (21849), *Lord Kenlis*	120
Grand Duchess of Wales (Wild Eyes), by Grand Duke 11th (21849), *Earl of Dunmore*	200
Darlington 17th, by Grand Duke 11th (21849), *Lord Kenlis*	120
Darlington 18th, by 3rd Duke of Wharfdale (21619), *Lord Kenlis*	195

Mr. H. J. SHELDON's Sale, BRAILES, 17th September 1869

Lady Ellen Barrington, by Lord Stanley (24467), *Duke of Devonshire*	155

Mr. W. W. SLYE's Sale, BEAUMONT GRANGE, 25th February 1870

Lady Thorndale Bates 2nd, by 4th Duke of Thorndale (17750), *Earl of Dunmore*	300

Mr. D. R. DAVIES's Sale, MERE OLD HALL, 13th July 1870

Moss Rose, by Marmaduke (14897), *E. H. Cheney*	350
Wellingtonia, by 3rd Duke of Thorndale (17749), *W. W. Slye*	160
Moss Rose 2nd, by 4th Duke of Thorndale (17750), *E. H. Cheney*	800
Wellingtonia 2nd, by 12th Duke of Thorndale (26020), *F. Leney*	200

	Guineas
Royal Chester (29852), by 10th Grand Duke (21848), *Barnes*, Australia	200
Royal Lancaster (29870), by 10th Grand Duke (21848), *J. Knowles*	130

Mr. C. R. Saunders's Sale, Nunwick, 23rd September 1870

	Guineas
Wild Eyes Duchess, by 9th Grand Duke (19879), *M. H. Cochrane*, Canada	275
Waterloo 36th, by Earl of Eglinton (23832), *Lord Kenlis*	475
Waterloo 37th, by Royal Cambridge (25009), *R. E. Oliver*	500
Waterloo 38th, by Earl of Eglinton (23832), *M. H. Cochrane*, Canada	300
Waterloo 39th, by Earl of Eglinton (23832), *Lord Skelmersdale*	150
Wild Eyes Duchess 2nd, by Earl of Eglinton (23832), *Earl of Dunmore*	120

Mr. William Butler's Sale, Badminton, 5th October 1870

	Guineas
Darlington 13th, by Earl of Glo'ster (21644), *W. Pierse*, California	105
Darlington 15th, by Grand Duke of York (24071), *R. P. Davies*	100
Darlington 17th, by Grand Duke of York (24071), *R. P. Davies*	155
Darlington 18th, by Grand Duke of York (24071), *W. Pierse*	125
Lord Collingham (Fidget) (29089), by 2nd Duke of Collingham (23730), *Earl of Aylesford*	115

Col. Kingscote's Sale, Kingscote, 8th March 1871

	Guineas
Dora, by 2nd Duke of Airdrie (19600), *Sir John Rolt*	200
Doralice, by 2nd Duke of Wetherby (21618), *S. P. Savage*	165
Duke of Fussbox (Fidget) (28389), by 3rd Duke of Clarence (23727), *Sir John Rolt*	200
Oxford Beau (29485), by 3rd Duke of Clarence (23727), *Lord Penrhyn*	330

Mr. A. J. Robarts's Sale, Lillingstone Dayrell, 30th March 1871

	Guineas
Lady Barrington 8th, by the Duke of York (23032), *D. McIntosh*	100
Wild Duchess, by 3rd Grand Duke (16182), *Capt. Webb*	100
Barringtonia, by Duke of Tregunter (26021), *R. E. Oliver*	270
Bridesmaid, by Duke of Tregunter (26021), *D. McIntosh*	200
Duke of Tregunter (26021), by 3rd Duke of Wharfdale (21619), *Earl of Dunmore*	165

Mr. E. H. Cheney's Sale, Gaddesby, 5th April 1871

Lady Waterloo 16th, by 3rd Viscount Waterloo (25387), *Lord Skelmersdale*	200
Bright Eyes 5th, by Grand Duke 6th (19876), *Duke of Devonshire*	170
Cherry Countess, by Grand Duke 6th (19876), *H. J. Sheldon*	410

Mr. D. McIntosh's Sale, Havering Park, 3rd May 1871

Lady Bates 7th, by 3rd Duke of Geneva (23753), *Earl of Bective*	815
Ladybird 6th, by 3rd Duke of Geneva (23753), *Earl of Dunmore*	105

Messrs. J. Harward and I. Downing's Sale, Winterfold, 20th July 1871

Kirklevington 16th, by Duke of Wetherby (17753), *Earl of Bective*	355
Kirklevington 19th, by 7th Duke of York (17754), *R. P. Davies*	175
Kirklevington 22nd, by 5th Lord Wild Eyes (26762), *Earl of Bective*	300
Kirklevington 24th, by 5th Duke of Wharfdale (26033), *W. Ashburner*	100

The Duke of Devonshire's Sale, Holker, 6th September 1871

Grand Duchess of Oxford 7th, by Lord Oxford (20214), *Lord Penrhyn*	915
Winsome 2nd, by Lord Oxford (20214), *Earl of Bective*	355

	Guineas
Winsome 7th, by Grand Duke 10th (21848), *A. Brogden*	300
Grand Duchess of Oxford 16th, by Grand Duke 17th (24064), *Rev. P. Graham*	610
Winsome 8th, by Grand Duke 17th (24064), *Earl of Bective*	320
Winsome 9th, by Grand Duke 17th (24064), *Earl of Bective*	405
Winsome 10th, by Duke of Oxford 18th (25995), *Rev. P. Graham*	370
Grand Duchess of Oxford 18th, by Baron Oxford 4th (25580), *Earl of Bective*	100
Lady Laura Barrington, by Baron Oxford 4th (25580), *Earl of Bective*	355
Winsome 11th, by Baron Oxford 4th (25580), *Earl of Feversham*	350
Duke of Oxford 19th (27909), by Grand Duke 10th (21848), *R. Botterill*	335
Barden (Barrington), by Grand Duke 10th (21848), *B. Marshall*, Australia	220
Duke of Oxford 20th (28432), by Baron Oxford 4th (25580), *Earl of Feversham*	1000
Duke of Oxford 21st (30999), by Grand Duke 10th (21848), *Right Hon. J. E. Denison*	155
Duke of Oxford 22nd (31000), by Baron Oxford, 4th (25580), *J. P. Foster*	305
Duke of Oxford 23rd (31001), by Baron Oxford 4th (25580), *A. Brogden*	155

Mr. W. W. SLYE's Sale, BEAUMONT GRANGE, 7th Sept. 1871

Lady Tregunter Bates, by 2nd Duke of Tregunter (26022), *J. Fawcett*	500

Mr. J. P. FOSTER's Sale, KILLHOW, 8th September 1871

Carolina 5th, by 7th Duke of York (17754), *D. McIntosh*	215
Grand Duchess 6th (Wild Eyes), by Duke of Geneva (19614), *Sir C. M. Lampson*	325
Fantail 5th, by Royal Cambridge (25009), *W. Angerstein*	360
Grand Duchess Surmise, by Grand Duke 10th (21848), *F. Sartoris*	210
Grand Duchess Carolina, by Grand Duke 10th (21848), *W. Angerstein*	325

Mr. Thomas Bell's Sale, Brockton, 12th September 1871

	Guineas
Princess Victoria 7th (Place), by 13th Duke of Oxford (21604), *Lord Skelmersdale*	100
Duke of York 8th (28480), by 4th Duke of Thorndale (17750), *J. Fawcett and others*	1065

Sir John Rolt's Sale, Ouzleworth, 19th September 1871

Dora, by 2nd Duke of Airdrie (19600), *Lord Fitzhardinge*	240
Duke of Fussbox (28389), by 3rd Duke of Clarence (23727), *Sir W. C. Trevelyan*	105

Mr. H. J. Sheldon's Sale, Brailes, 21st September 1871

Countess of Barrington 2nd, by 9th Duke of Oxford (17738), *W. W. Slye*	170
Darlington 12th, by Duke of Geneva (19614), *E. H. Cheney*	150
Antoinette, by 4th Duke of Thorndale (17750), *E. Leney*	255
Grand Duchess of Barrington, by Grand Duke 7th (19877), *R. E. Oliver*	415
Antonia, by Duke of Darlington (21586), *Lord Skelmersdale*	180
Lord Barrington (31616), by Duke of Brailes (23724), *J. Knowles*	150

Mr. E. Bowly's Sale, Siddington, 25th April 1872

Wild Eyes 28th, by 3rd Lord Lally (24408), *H. de Vitre*	120
Siddington 5th, by 7th Duke of York (17754), *W. W. Slye*	570
Siddington 9th, by 2nd Duke of Tregunter (26022), *G. Moore*	330
Siddington 10th, by 2nd Duke of Tregunter (26022), *H. de Vitre*	505
2nd Duke of Tregunter (26022), by 4th Duke of Thorndale (17750), *Earl of Bective*	900
Wildfire (32866), by 2nd Duke of Tregunter (26022), *J. d'A. Samuda*	105

Messrs. Walcott and Campbell's, sold at Burghley, 9th May 1872

5th Lord of Oxford, by 4th Duke of Geneva (30958), *W. Angerstein*	500

Mr. W. W. Slye's Sale, Beaumont Grange, 21st August 1872

	Guineas
Lord Thorndale (31756), by Grand Duke of Kent 2nd (28759), *H. de Vitre*	180

The Earl of Dunmore's Sale, Dunmore, 5th September 1872

	Guineas
Siddington 1st, by 4th Duke of Oxford (11387), *J. W. Larking*	260
Siddington 7th, by 7th Duke of York (17754), *Earl of Bective*	500
Lady Thorndale Bates 2nd, by 4th Duke of Thorndale (17750), *Earl of Bective*	805
Ladybird 6th, by 3rd Duke of Geneva (23753), *Sir W. G. Armstrong*	150
Wild Eyes Duchess 2nd, by Earl of Eglinton (23822), *Lord Fitzhardinge*	265
Grand Duchess of Athol (Wild Eyes), by 2nd Duke of Collingham (23730), *Major Stapylton*	255
Marchioness 2nd (Kirklevington), by 2nd Duke of Collingham (23730), *Earl of Bective*	455
Marchioness 3rd (Kirklevington), by 2nd Duke of Collingham (23730), *Earl of Bective*	535
Marchioness 5th (Kirklevington), by 2nd Duke of Collingham (23730), *Rev. P. Graham*	270
Marchioness of Oxford, by 4th Duke of Geneva (30958), *R. P. Davies*	1010
Lady Bright Eyes 3rd, by 7th Duke of York (17754), *Duke of Devonshire*	375
Oxford Duchess, by 6th Duke of Geneva (30959), *R. P. Davies*	1200
Marchioness of Oxford 2nd, by 6th Duke of Geneva (30959), *W. Angerstein*	860
Lady Mary (Ladybird), by Baron Oxford 5th (27958), *T. Gow*	110
Baron Oxford 5th (27958), by 2nd Duke of Claro (21576), *Duke of Devonshire*	400
Marquis 3rd (Kirklevington) (31826), by 2nd Duke of Collingham (23730), *E. J. Coleman*	255

Messrs. J. Harward and I. Downing's Sale, Winterfold, 18th September 1872[1]

	Guineas
Wild Eyes 24th, by 4th Duke of Oxford (11387), *R. H. Crabb*	180

[1] See above, p. 360.

IL LIBRO D'ORO

	Guineas
Kirklevington 17th, by Lord Lally (22161), *J. W. Larking*	248
5th Maid of Oxford, by 7th Duke of Airdrie (23718), *G. Moore*	900
Lally 7th, by 3rd Lord Oxford (22200), *J. W. Larking*	305
Wild Eyes 29th, by Earl of Glo'ster (21644), *T. Barber*	160
Wild Eyes 30th, by 7th Duke of York (17754), *W. Angerstein*	200
Lally 8th, by 7th Duke of York (17754), *W. Angerstein*	600
Lally 9th, by 7th Duke of York (17754), *Mrs. Fawcett*	300
Lady Worcester 3rd, by 3rd Duke of Wharfdale (21619), *Earl of Dunmore*	505
Kirklevington 21st, by 4th Duke of Thorndale (17750), *E. Bowly*	210
Lally 10th, by 5th Lord Wild Eyes (26762), *Lord Skelmersdale*	405
Lady Waterloo, by 3rd Duke of Claro (23729), *F. Leney and Son*	265
Lady Worcester 5th, by 3rd Duke of Claro (23729), *Earl of Dunmore*	510
Lally 13th, by 3rd Duke of Claro (23729), *H. J. Sheldon*	620
Lady Worcester 6th, by 3rd Duke of Claro (23729), *J. W. Larking*	360
Lady Worcester 7th, by 3rd Duke of Claro (23729), *A. Garfit*	215
Lally 14th, by 8th Duke of Geneva (28390), *Lord Skelmersdale*	305
Wild Eyes Lassie, by 3rd Duke of Claro (23729), *J. P. Foster*	420
Lady Worcester 9th, by 3rd Duke of Claro (23729), *Earl of Dunmore*	425
Kirklevington 25th, by 3rd Duke of Claro (23729), *A. Brogden, M.P.*	220
Lady Wild Eyes, by 8th Duke of Geneva (28390), *A. Brogden, M.P.*	290
Lady Worcester 10th, by 8th Duke of Geneva (28390), *G. Moore*	240
Lally 15th, by 8th Duke of Geneva (28390), *Duke of Devonshire*	500
Lally 16th, by 3rd Duke of Claro (23729), *G. Moore*	205
3rd Duke of Claro (23729), by 2nd Duke of Wharfdale (19649), *W. Angerstein*	620

	Guineas
8th Duke of Geneva (28390), by Baron of Oxford (23371), *F. Leney and Son*	1650
Earl of Strafford (Lally) (31084), by 8th Duke of Geneva (28390), *J. Blundell*	155
Wild Eyes Lad (37680), by 8th Duke of Geneva (28390), *Major Fanning*, Australia	170
Lord Claro 2nd (Wild Eyes) (31640), by 3rd Duke of Claro (23729), *J. Pulley*	170

Mr. Alexander Brogden's Sale, Lightburne Park, 8th April 1873

Grand Duke of Lightburne 3rd (Cambridge Rose) (28761), by Grand Duke 10th (21848), *G. Ashburner*	115
Baron Lightburne (Wild Eyes) (30468), by Baron Oxford 4th (25580), *Gordon*	220

Col. Towneley's Sale, Towneley Park, 1st May 1873

6th Maid of Oxford, by Imperial Oxford (24185), *H. Strafford*	800
Baron Oxford (23375), by Duke of Geneva (19614), *Rev. W. Sneyd*	250

Mr. A. Dugdale's Sale, Rose Hill, 1st May 1873

Annette 2nd, by 4th Duke of Oxford (11387), *W. Ashburner*	250
Oxford's Waterloo 4th, by 13th Duke of Oxford (21604), *Whitehead*	100
Wild Flower Duchess 5th, by 13th Duke of Oxford (21604), *R. Botterill*	105

Lord Penrhyn's Sale, Wicken Park, 8th May 1873

Waterloo 26th, by Duke of Geneva (19614), *A. Brogden, M.P.*	450
Waterloo 30th, by 3rd Duke of Wharfdale (21619), *F. Leney*	500
Waterloo 33rd, by 11th Grand Duke (21849), *Lord Skelmersdale*	550
Waterloo 36th, by 11th Grand Duke (21849), *R. H. Crabb*	305
Waterloo 37th, by Oxford Beau (29485), *C. A. Barnes*	415
4th Duke of Wellington (31030), by 11th Grand Duke (21849), *Major Fanning*, Australia	340

Mr. E. H. Cheney's Sale, Gaddesby, 10th July 1873

	Guineas
Lady Waterloo 14th, by 2nd Lord of Waterloo (22198), *R. Botterill*	150
Lady Waterloo 15th, by 3rd Viscount Waterloo (25387), *G. Fox*	165
Lady Sale of Putney, by 9th Duke of Thorndale, *Earl of Bective*	470
Wild Duchess of Geneva, by 9th Duke of Geneva (28391), *Sir W. G. Armstrong*	335
Water Lass 3rd, by 7th Duke of York (17754), *Sir John Swinburne*	305
Geneva's Minstrel, by 9th Duke of Geneva (28391), *J. P. Foster*	600
Lady Waterloo 25th, by 9th Duke of Geneva (28391), *H. A. Brassey, M.P.*	305
Wild Princess 2nd, by 9th Duke of Geneva (28391), *Sir W. Lawson*	460
14th Lady of Oxford, by 9th Duke of Geneva (28391), *Earl of Bective*	905
Wild Duchess of Geneva 2nd, by 9th Duke of Geneva (28391), *Sir W. G. Armstrong*	355
Fantail's Duchess, by 9th Duke of Geneva (28391), *Sir John Swinburne*	115
12th Duchess of Geneva, by 9th Duke of Geneva (28391), *Sir W. Lawson*	935
Wild Princess 3rd, by 9th Duke of Geneva (28391), *Sir John Swinburne*	205
Lady Waterloo 26th, by 9th Duke of Geneva (28391), *W. Ashburner*	105
Lady Waterloo 27th, by Duke of Oxford 20th (28432), *Sir W. C. Trevelyan*	110
3rd Duke of Glo'ster (33653), by 10th Duke of Thorndale (28458), *Earl of Bective*	820

Mr. H. J. Sheldon's Sale, Brailes, 8th August 1873

Lady Emily Darlington, by Duke of Darlington (21586), *J. W. Wilson*	155
Arethusa, by Duke of Brailes (23724), *F. Leney and Son*	205
Arethusa 2nd, by 3rd Earl of Fawsley (28506), *Y. R. Graham*	260
Lally 18th, by 8th Duke of Geneva (28390), *Lord Skelmersdale*	630

	Guineas
18th Duke of Oxford (25995), by Grand Duke 10th (21848), *Col. Gunter*	150
Duke of Barrington 4th (30924), by 9th Duke of Geneva (28391), *Hon. G. Brown*, Canada	110

Mr. SAMUEL CAMPBELL'S Sale, NEW YORK MILLS, U.S.A., 10th September 1873

	Guineas
2nd Maid of Oxford, by Grand Duke of Oxford (16184), *A. W. Griswold*	1200
3rd Maid of Oxford, by Grand Duke of Oxford (16184), *Warnock and Megibben*	200
12th Duchess of Thorndale, by 6th Duke of Thorndale (23794), *A. B. Conger*	1140
8th Duchess of Geneva, by 3rd Lord Oxford (22200), *R. Pavin Davies*	8120
2nd Countess of Oxford, by 2nd Duke of Geneva (23752), *A. W. Griswold*	420
13th Duchess of Thorndale, by 10th Duke of Thorndale (28458), *A. B. Conger*	3000
10th Duchess of Geneva, by 2nd Duke of Geneva (23752), *Earl of Bective*	7000
Lady Bates 5th (Fletcher), by 11th Duke of Thorndale (31024), *G. M. Bedford*	220
Lady Bates 4th (Fletcher), by 11th Duke of Thorndale (31024), *E. G. Bedford*	650
Lady Bates 8th (Fletcher), by 2nd Duke of Oneida, *Col. King*	320
Lady Worcester 4th, by 2nd Duke of Wetherby (21618), *T. Holford*	600
12th Lady of Oxford, by 10th Duke of Thorndale (28458), *T. Holford*	1400
1st Duchess of Oneida, by 10th Duke of Thorndale (28458), *Lord Skelmersdale*	6120
3rd Duchess of Oneida, by 4th Duke of Geneva (30958), *T. Holford*	3120
Lady Worcester 5th, by 4th Duke of Geneva (30958), *T. Holford*	400
3rd Countess of Oxford, by Baron of Oxford (23371), *A. B. Conger*	1820
4th Duchess of Oneida, by 4th Duke of Geneva (30958), *E. G. Bedford and Megibben*	5000
Lady Bates 6th (Filbert), by 4th Duke of Geneva (30958), *G. M. Bedford*	460

	Guineas
Lady Bates 7th (Filbert), by 4th Duke of Geneva (30958), *A. B. Cornell*	320
7th Duchess of Oneida, by 2nd Duke of Oneida, *A. J. Alexander*	3800
12th Maid of Oxford, by 4th Duke of Geneva (30958), *Col. Morris*	1200
8th Duchess of Oneida, by 4th Duke of Geneva (30958), *Earl of Bective*	3060
9th Duchess of Oneida, by 2nd Duke of Oneida, *Earl of Bective*	2000
10th Duchess of Oneida, by 2nd Duke of Oneida, *A. J. Alexander*	5400
2nd Duke of Oneida, by 4th Duke of Geneva (30958), *T. G. Megibben*	2400
4th Duke of Oneida, by Baron of Oxford (23371), *A. B. Cornell*	1520
7th Duke of Oneida, by 4th Duke of Geneva (30958), *A. W. Griswold*	800
10th Earl of Oxford, by 2nd Duke of Oneida, *A. B. Cornell*	500
6th Lord of Oxford, by 2nd Duke of Oneida, *M. H. Cochrane*	260

Mr. W. W. Slye's Sale, Beaumont Grange, 22nd April 1874

Lady Oxford Bates, by 17th Grand Duke (24064), *J. Fawcett*	120
Lady Clarence Bates, by Grand Duke of Thorndale (31297), *J. Fawcett*	250

Messrs. F. Leney and Sons' Sale, Orpines, 2nd July 1874

Lily, by Lord Liverpool (22168), *Duke of Devonshire*	170
Thorndale Duchess (Waterloo), by 12th Duke of Thorndale (26020), *Jonathan Rigg*	100
4th Grand Duchess of Geneva, by 8th Duke of Geneva (28390), *R. Loder*	2000
Surprise 3rd, by 8th Duke of Geneva (28390), *Lord Feversham*	200
Wellingtonia 4th, by 6th Duke of Oneida (30997), *Rev. W. Sneyd*	260
Lady Bates 2nd (Hawkey), by 6th Duke of Oneida (30997), *Duke of Devonshire*	120

	Guineas
Thorndale Duchess 2nd (Waterloo), by 8th Duke of Geneva (28390), *Rev. W. Sneyd*	105
Surprise 4th, by 6th Duke of Oneida (30997), *A. S. Hill, M.P.*	100

The Duke of Devonshire's Sale, Holker, 9th September 1874

	Guineas
Grand Duchess of Oxford 6th, by Imperial Oxford (18084), *T. Allen*	805
Countess of Barrington, by Lord Oxford (20214), *H. J. Sheldon*	300
Winsome 18th, by Baron Oxford 4th (25580), *T. Wilson*	310
Grand Duchess of Oxford 11th, by Grand Duke 10th (21848), *G. Moore*	1000
Grand Duchess of Oxford 12th, by 2nd Duke of Wetherby (21618), *A. Brogden, M.P.*	1010
Bright Eyes 5th, by Grand Duke 6th (19876), *Lord Skelmersdale*	500
Winsome 5th, by Grand Duke 10th (21848), *Earl of Feversham*	350
Winsome 12th, by Baron Oxford 4th (25580), *T. Holford*	360
Countess of Barrington 6th, by Baron Oxford 4th (25580), *F. Leney and Sons*	560
Carry 2nd, by Baron Oxford 4th (25580), *Sir J. Whitworth*	205
Winsome 14th, by Baron Oxford 4th (25580), *Sir W. Lawson, M.P.*	605
Grand Duchess of Oxford 25th, by Baron Oxford 4th (25580), *W. Ashburner*	760
Winsome 16th, by Baron Oxford 4th (25580), *G. Fox*	700
Winsome 17th, by Duke of Oxford 24th (31002), *Lord Skelmersdale*	310
Baroness Oxford 3rd, by Duke of Hillhurst (28401), *T. Holford*	1100
Grand Duchess of Oxford 28th, by Duke of Oxford 24th (31002), *W. Ashburner*	675
Baron Oxford 5th (27958), by 2nd Duke of Claro (21576), *Duke of Roxburghe*	250
Baron Winsome 3rd (33108), by Baron Oxford 4th (25580), *E. J. Coleman*	250
Baron Barrington 4th (33006), by Baron Oxford (25580), *W. H. Wakefield*	210
Baron Barrington 5th (33007), by Baron Oxford 4th (25580), *Lord Skelmersdale*	300

	Guineas
Duke of Oxford 26th (33708), by Baron Oxford 4th (25580), *J. Robinson*	420
Baron Tregunter (33085), by 3rd Duke of Tregunter (31026), *J. Postlethwaite*	250
Baron Barrington 6th (33008), by Baron Oxford 4th (25580), *J. Harward*	185
Duke of Oxford 28th (33710), by Duke of Oxford 24th (31002), *Lord Chesham*	550
Duke of Oxford 29th (33711), by Baron Oxford 4th (25580), *Sir J. Whitworth*	400
Duke of Oxford 30th (33712), by 5th Duke of Wetherby (31033), *J. Grant Morris*	275

The EARL OF BECTIVE'S Sale, UNDERLEY, 10th September 1874

	Guineas
Duchess Gwynne, by Duke of Wetherby (17753), *W. H. Salt*	430
Kirklevington 16th, by Duke of Wetherby (17753), *Sir J. Swinburne*	420
Winsome 2nd, by Lord Oxford (20214), *E. J. Coleman*	305
Siddington 4th, by 7th Duke of York (17754), *Earl of Ellesmere*	600
Carolina 7th, by 13th Grand Duke (21850), *E. J. Coleman*	170
Darlington 17th, by 11th Grand Duke (21849), *R. Loder*	500
Siddington 7th, by 7th Duke of York (17754), *Lord Fitzhardinge*	750
Darlington 19th, by 3rd Duke of Wharfdale (21619), *R. Loder*	659
Winsome 9th, by Grand Duke 17th (24064), *T. Wilson*	505
Marchioness 3rd (Kirklevington), by 2nd Duke of Collingham (23730), *Earl of Ellesmere*	600
Dentsdale (Darlington), by 2nd Duke of Collingham (23730), *R. Loder*	550
Lady Laura Barrington, by Baron Oxford 4th (25580), *Duke of Devonshire*	305
Duchess Gwynne 4th, by Baron Oxford 5th (27958), *R. Loder*	580
Winsomedale, by Baron Oxford 4th (25580), *Lord Fitzhardinge*	650
Princess Gwynne 2nd, by Royal Cambridge (25009), *Rev. P. Graham*	250
Duchess Gwynne 5th, by Grand Duke of Kent 2nd (28759), *T. Holford*	320

	Guineas.
Empress of Oxford, by Grand Duke of Kent 2nd (28759), *T. Holford*	770
Deepdale (Darlington), by 2nd Duke of Tregunter (26022), *G. Fox*	315
Dentsdale (Darlington), by Grand Duke of Kent 2nd (28759), *Lord Penrhyn*	420
Princess 6th, by Grand Duke of Kent 2nd (28759), *Marquis of Headfort*	450
Princess Sale, by 2nd Duke of Tregunter (26022), *Sir J. Swinburne*	370
Winsomedale 2nd, by 2nd Duke of Tregunter (26022), *J. W. Larking*	330
3rd Duke of Gloucester (33653), by 10th Duke of Thorndale (28458), *E. J. Coleman*	900
Grand Duke of Kent 2nd (28759), by Lord Oxford 2nd (20215), *H. D. de Vitre*	750
Marquis 3rd (Kirklevington) (31826), by 2nd Duke of Collingham (23730), *Captain Sandy*	175
Duke of Dentsdale 2nd (33617) (Darlington), by 2nd Duke of Tregunter (26022), *W. Hutchinson*	125
Duke of Kirklevington (33683), by 2nd Duke of Tregunter (26022), *Sir R. C. Musgrave*	305
Lord Lunesdale Bates (34592), by Baron Oxford 5th (27958), *J. H. Casswell*	570
Duke of Tosca (Gwynne) (33742), by Grand Duke of Tregunter (28759), *M. Kennedy*	300
Marquis 4th (Kirklevington) (34776), by Grand Duke of Kent 2nd (28759), *E. Williamson*	220
Marquis 6th (Kirklevington) (34777), by Grand Duke of Kent 2nd (28759), *Major C. J. Webb*	330
Lord of Garsdale (Gwynne) (34617), by Grand Duke of Kent 2nd (28759), *W. H. Wakefield*	115
Visigoth (Winsome) (35907), by Grand Duke of Kent 2nd (28759), *F. W. Low*	120
Duke of Dentsdale 3rd (Darlington) (33618), by 2nd Duke of Tregunter (26022), *Marquis of Headfort*	110
Sir Lawrence Barrington (35582), by 2nd Duke of Tregunter (26022), *C. A. Barnes*	100

Mr. E. H. CHENEY'S Sale, GADDESBY, 23rd September 1874

Duchess of Airdrie 8th, by Royal Oxford (18774), *D. McIntosh*	1700

	Guineas
Lady Waterloo 22nd, by 7th Duke of York (17754), *G. Fox*	225
Peach Blossom 8th, by 8th Duke of York (28480), *Col. Kingscote, M.P.*	185
Princess of Geneva, by 9th Duke of Geneva (28391), *D. McIntosh*	860
Princess of Geneva 2nd, by 9th Duke of Geneva (28391), *H. J. Sheldon*	800
Lady Oxford 15th, by 9th Duke of Geneva (28391), *R. P. Davies*	310
Duchess of Glo'ster, by 9th Duke of Geneva (28391), *Sir C. M. Lampson*	1785
Lady Elizabeth (Princess), by 9th Duke of Geneva (28391), *F. Leney*	430
Lady Waterloo 28th, by 9th Duke of Geneva (28391), *G. Fox*	530
Wild Duchess of Geneva 3rd, by 9th Duke of Geneva (28391), *T. Gow*	555
Rosalie (Princess), by Saladin (35461), *G. Fox*	700
Fantail's Duchess 2nd, by 9th Duke of Geneva (28391), *Lord Stamford*	135
Lady Wellesley (Princess), by 9th Duke of Geneva (28391), *R. P. Davies*	360
16th Lady of Oxford, by 9th Duke of Geneva (28391), *R. H. Crabb*	605
Lady Waterloo 29th, by 9th Duke of Geneva (28391), *T. Gow*	110

Mr. R. Pavin Davies's

Kirklevington 18th, by 3rd Lord Oxford (22200), *W. H. Salt*	385
Kirklevington Duchess 2nd, by Duke of Athelstane (Place) (21562), *J. H. Casswell*	410
Kirklevington Duchess 6th, by 2nd Duke of Claro (21576), *J. Snodim*	280
Kirklevington Duchess 12th, by 2nd Duke of Glo'ster (28392), *J. H. Casswell*	250

Col. Kingscote's Sale, Kingscote, 21st April 1875

Marchioness of Bickerstaffe, by 3rd Duke of Clarence (23727), *Duke of Manchester*	170
Ariel Marchioness, by 3rd Duke of Clarence (23727), *Earl of Bective*	450

	Guineas
Papaver, by 3rd Duke of Clarence (23727), *H. W. Beauford*	190
Ariel Countess, by 3rd Duke of Clarence (23727), *Lord Penrhyn*	350
Georgie Hillhurst, by Duke of Hillhurst (28401), *Richardson and Boswell*, U.S.A.	155
Oxford Ida, by 3rd Duke of Clarence (23727), *R. Loder*	380
Georgie Clarence, by 3rd Duke of Clarence (23727), *Richardson and Boswell*, U.S.A.	170
Lady Bickerstaffe, by Duke of Hillhurst (28401), *Richardson and Boswell*, U.S.A.	135
Water Girl, by Grand Duke of Waterloo (28766), *Richardson and Boswell*, U.S.A.	265
Georgie Hillhurst 2nd, by Duke of Hillhurst (28401), *Richardson and Boswell*, U.S.A.	165
Papaver 2nd, by Duke of Hillhurst (28401), *R. Pavin Davies*	180
Carrie Craggs, by Grand Duke of Clarence (28750), *R. Loder*	300
Velvet Eyes, by Duke of Hillhurst (28401), *W. Angerstein*	760
Lady Secret, by Duke of Hillhurst (28401), *Sir C. M. Lampson*	320
Georgie Hillhurst 4th, by Duke of Hillhurst (28401), *H. A. Brassey, M.P.*	160

Mr. Edward Bowly's Sale, Siddington, 22nd April 1875

Siddington 13th, by 3rd Duke of Clarence (23727), *D. McIntosh*	650
Siddington 15th, by 3rd Duke of Clarence (23727), *Earl of Bective*	600
Siddington 16th, by 3rd Duke of Clarence (23727), *J. W. Larking*	505
Duke of Siddington 2nd (33732), by 3rd Duke of Clarence (23727), *Lord Fitzhardinge*	400

Lord Penrhyn's Sale, Wicken Park, 4th May 1875

Waterloo 32nd, by 11th Grand Duke (21849), *Lord Feversham*	500
Cherry Duchess 24th, by 20th Grand Duke (31281), *Rev. P. Graham*	900
Waterloo 38th, by 11th Grand Duke (21849), *D. McIntosh*	560

	Guineas
Cherry Duchess 22nd, by 11th Grand Duke (21849), *Earl of Bective*	900
5th Belle of Oxford, by Oxford Beau (29485), *Rev. P. Graham*	1050
Dowager Duchess 4th, by 11th Grand Duke (21849), *D. McIntosh*	230
Dowager Duchess 5th, by Oxford Beau (29485), *G. Fox*	300
Cherry Duchess 25th, by 20th Grand Duke (31281), *G. Fox*	720
Waterloo 40th, by 20th Grand Duke (31281), *T. Holford*	210
2nd Beau of Oxford (33129), by Oxford Beau (29485), *E. Bowly*	105
9th Duke of Wellington (33754), by Oxford Beau (29485), *F. Nichol*	105

Mr. D. McIntosh's Sale, Havering, 6th May 1875

	Guineas
Duchess Carolina, by 17th Duke of Oxford (25994), *Lord Penrhyn*	430
Lady Barrington 9th, by 3rd Duke of Geneva (23753), *Lord Penrhyn*	200
Duke of Carolina 2nd (33590), by 3rd Duke of Geneva (23753), *F. N. Smith*	150

Mr. J. W. Philip's Sale, Heybridge, 13th May 1875

	Guineas
Fuchsia 9th, by Grand Duke of York (24071), *J. W. Larking*	960
Winsome 6th, by 10th Grand Duke (21848), *R. Loder*	610
Winsome Wild Eyes 1st, by Bolton (25650) (Lally), *G. Ashburner*	175
Winsome Wild Eyes 2nd, by Bolton (25650) (Lally), *Sir R. C. Musgrave*	300
Lady Fuchsia, by Bolton (25650) (Lally), *J. Hope*, Canada	140
Lady Fuchsia 2nd, by Bolton (25650), *D. McIntosh*	205
Lady Fuchsia 3rd, by Lord Tregunter (31758) (Fletcher), *D. McIntosh*	430
Winsome Wild Eyes 3rd, by Lord Tregunter (31758), (Fletcher), *R. Loder*	405
Airdrie Geneva (Red Rose) (32920), by Airdrie 3rd, *J. Wilson*	200

Messrs. F. Leney and Sons' Sale, Orpines, 1st July 1875

	Guineas
7th Maid of Oxford, by 7th Duke of Airdrie (23718), *B. B. Groom*, U.S.A.	365

	Guineas
Kirklevington 20th, by 5th Lord Wild Eyes (26762), T. Lister	580
Wellingtonia 3rd, by Grand Duke of Kent (26289), *Duke of Manchester*	200
Baroness Fawsley 3rd, by Grand Duke 15th (21852), *D. McIntosh*	300
Surprise, by 8th Duke of Geneva (28390), *J. H. Blundell*	215
Wild Princess 2nd, by 8th Duke of Geneva (28390), *Lord Penrhyn*	400
Baroness Fawsley 6th, by 8th Duke of Geneva (28390), *Sir G. R. Philips*	200
Wild Duchess 4th, by 6th Duke of Oneida (30997), *Lord Penrhyn*	430
Surprise 5th, by 8th Duke of Geneva (28390), *Duke of Manchester*	140
8th Duke of Geneva (28390), by Baron of Oxford (23371), *B. B. Groom, U.S.A.*	2000
Earl of Oxford, by 8th Duke of Geneva (28390), *B. B. Groom, U.S.A.*	305

Mr. W. ANGERSTEIN'S

3rd Duke of Claro (23729), by 2nd Duke of Wharfdale (19649), *Col. Kingscote, M.P.*	500

The EARL OF DUNMORE'S Sale, DUNMORE, 25th August 1875

Wild Eyes Duchess, by 9th Grand Duke (19879), *T. Wilson*	480
Winsome Eyes 3rd, by 5th Duke of Wharfdale (26033), *Earl of Feversham*	329
Red Rose of the Isles, by Airdrie (30365), *Earl of Bective, M.P.*	1950
Lady Worcester 5th, by 3rd Duke of Claro (23729), *H. A. Brassey, M.P.*	620
Lady Worcester 9th, by 3rd Duke of Claro (23729), *A. Brogden, M.P.*	440
Water Flower, by 6th Duke of Geneva (30959), *T. Holford*	620
Wild Rose, by 6th Duke of Geneva (30959), *Col. Kingscote, M.P.*	350
Fuchsia 12th, by Duke of Albany (25931) (Fletcher), *T. Lister*	900
Oxford Duchess 2nd, by 2nd Duke of Collingham (23730), *Lord Fitzhardinge*	1000

	Guineas
Fuchsia 13th, by Duke of Albany (25931), *J. W. Larking*	650
Lady Worcester 11th, by 3rd Duke of Claro (23729), *Duke of Manchester*	550
Wild Eyebright, by 6th Duke of Geneva (30959), *T. Wilson*	455
Lady Worcester 12th, by 8th Duke of Geneva (28390), *Earl of Bective, M.P.*	555
Marchioness of Oxford 3rd, by 2nd Duke of Collingham (23730), *Duke of Manchester*	1810
Red Rose of Balmoral, by 3rd Duke of Hillhurst (30975), *Earl of Bective, M.P.*	1280
Sparkling Eyes, by 6th Duke of Geneva (30959), *Earl of Feversham*	350
Lady Worcester 13th, by 3rd Duke of Hillhurst (30975), *G. Fox*	450
Fuchsia 14th, by Duke of Albany (25931), *T. Lister*	360
Water Lily, by 3rd Duke of Hillhurst (30975), *T. Holford*	520
Hazel Eyes, by 3rd Duke of Hillhurst (30975), *Earl of Bective, M.P.*	400
Blythesome Eyes, by 3rd Duke of Hillhurst (30975), *Earl of Bective, M.P.*	605
Lady Worcester 15th, by 3rd Duke of Hillhurst (30959), *R. Loder*	360
Lady Worcester 14th, by 6th Duke of Geneva (30975), *Earl of Bective, M.P.*	550
Duke of Connaught (33604), by Duke of Hillhurst (28401), *Lord Fitzhardinge*	4500
3rd Duke of Hillhurst (30975), by 6th Duke of Geneva (30959), *J. W. Larking*	3000
2nd Marquis of Worcester (34789), by 3rd Duke of Hillhurst (30975), *J. H. Kissinger and Co., U.S.A.*	150
Marquis of Oxford (34786), by 3rd Duke of Hillhurst (39075), *J. d'A. Samuda, M.P.*	300
Scots Fusilier (35483), by 3rd Duke of Hillhurst (30975), *Earl of Zetland*	155

The late Mr. JAMES FAWCETT's Sale, SCALEBY CASTLE, 27th August 1875

Butterfly Princess 24th, by 8th Duke of York (28480), *G. Fox*	120
Peach Blossom 10th, by 8th Duke of York (28480), *Sir C. M. Lampson*	200

	Guineas
Kirklevington Duchess 15th, by 2nd Duke of Glo'ster (28392), *A. Crane*, U.S.A.	750
8th Duke of York (28480), by 4th Duke of Thorndale (17750), *T. Holford*	1000

Lord SKELMERSDALE's Sale, LATHOM HOUSE, 7th September 1875

	Guineas
Princess Victoria 5th, by Lord Oxford 2nd (20215), *C. Magniac*	200
Lady Geneva Waterloo, by 9th Duke of Geneva (28391), *T. Holford*	500
Fluffy Gwynne, by 1st Duke of Oneida (30996), *Earl of Bective, M.P.*	520
Princess Victoria 10th, by 1st Duke of Oneida (30966), *J. Hope*, Canada	385
Waterloo Bienvenue, by Oxford Beau (29485), *R. E. Oliver*	520
Princess Victoria 11th, by 1st Duke of Oneida (30996), *Cochrane and Beattie*, Canada	375
Princess Victoria 12th, by Baron Oxford 4th (25580), *H. W. Beauford*	225
Baron Barrington 5th (33007), by Baron Oxford 4th (25580), *A. Dalzell*	165

Mr. GEORGE MOORE's Sale, WHITEHALL, 9th September 1875

	Guineas
Grand Duchess of Oxford 11th, by Grand Duke 10th (21848), *Sir C. M. Lampson*	2000
Flighty Gwynne, by Grand Duke of Lightburne (26290), *Jonathan Rigg*	400
Wild Maid, by 17th Duke of Oxford (25994), *Sir R. C. Musgrave*	450
Fantail 6th, by 17th Duke of Oxford (25994), *G. Ashburner*	310
Siddington 9th, by 2nd Duke of Tregunter (26022), *Rev. P. Graham*	850
Graceful Duchess, by Baron Oxford 4th (25580), *Sir Wilfrid Lawson, M.P.*	500
Lady Worcester 10th, by 8th Duke of Geneva (28390), *Duke of Manchester*	760
Lally 16th, by 3rd Duke of Claro (23729), *Duke of Manchester*	470

	Guineas
Lily Gwynne, by 17th Duke of Oxford (25994), *Earl of Bective, M.P.*	515
Baroness Lally, by 6th Baron Oxford, *W. Ashburner*	315
Grand Duchess of Oxford 31st, by 5th Duke of Wetherby (31033), *Duke of Devonshire*	1000
Lady Mild Eyes, by 8th Duke of York (28480), *T. Holford*	285

Mr. ALEXANDER BROGDEN, M.P.'s Sale, CONISHEAD PRIORY, 14th September 1875

Lady Wild Eyes, by 8th Duke of Geneva (28390), *Lord Skelmersdale*	820
Princess 4th, by Baron Oxford 3rd (25579), *Duke of Manchester*	300
Kirklevington 25th, by 3rd Duke of Claro (23729), *T. Lister*	510
Princess of Lightburne, by Duke of Oxford 23rd (31001), *W. W. Slye*	300
Lightburne Winsome, by Duke of Oxford 23rd (31001), *Jonathan Rigg*	410
Princess of Lightburne 2nd, by 2nd Grand Duke of Kent (28759), *Lord Skelmersdale*	500

Mr. WILLIAM ASHBURNER'S

Lady Waterloo 18th, by 2nd Lord of Waterloo (22698), *Jonathan Rigg*	150
Lady Barrington 9th, by Wild Duke (27808), *E. Holden*	360
Kirklevington 24th, by 5th Duke of Wharfdale (26033), *T. Lister*	420
Kirklevington Duchess 7th, by Duke of Kirklevington (25982), *E. Holden*	660
Lady Waterloo 26th, by 9th Duke of Geneva (28391), *J. Martin*	280
Lady Barrington 11th, by Grand Duke of Oxford (28764), *Cochrane and Beattie*, Canada	254
Lally Duchess 2nd, by Grand Prince of Claro (28718), *Cochrane and Beattie*, Canada	490
Wild Eyes 33rd, by 2nd Grand Duke of Kent (28759), *Cochrane and Beattie*, Canada	160
Kirklevington 26th, by 2nd Grand Duke of Kent (28759), *Hon. H. M. Cochrane*, Canada	390

	Guineas
Kirklevington 27th, by 2nd Grand Duke of Kent (28759), *G. Ashburner*	360
Master Wild Eyes (34882), by 2nd Duke of Glo'ster (28392), *E. Fell*	135

Mr. H. J. Sheldon's Sale, Brailes, 22nd September 1875

	Guineas
Grand Duchess of Barrington 2nd, by Duke of Brailes (23724), *Lord Skelmersdale*	655
Grand Duchess of Barrington 3rd, by 2nd Duke of Collingham (23730), *Hon. H. M. Cochrane*, Canada	600
Lally 19th, by 2nd Duke of Collingham (23720), *H. A. Brassey, M.P.*	500
Duke of Barrington 6th (33576), by Duke of Hillhurst (28401), *G. C. Spencer*	150

Mr. Thos. Barber's Sale, Sproatley Rise, 6th April 1876

	Guineas
Bright Eyes 4th, by Beau of Oxford (21254), *R. Blezard*	265
Duchess of Clarence 3rd, by Grand Duke 6th (19876), *D. R. Scratton*	170
Water Duchess 3rd, by Oxford's Baronet (29499), *R. Blezard*	215
Water Duchess 4th, by Oxford's Baronet (29499), *J. I. D. Jefferson*	150
Duchess of Clarence 10th, by Oxford's Baronet (29499), *D. R. Scratton*	280
Duchess of Clarence 12th, by Oxford's Baronet (29499), *J. P. Foster*	200
Bright Eyes 9th, by Oxford's Baronet (29499), *Earl of Bective, M.P.*	300
Oxford's Baronet (29499), by Baron Oxford (23375), *C. Miller*	140

Lord Fitzhardinge's Sale, Berkeley, 19th April 1876

	Guineas
Lady Wild Eyes 7th, by Grand Duke of Waterloo (28766), *R. Blezard*	555
Grand Duke of Waterloo (28766), by 3rd Duke of Wharfdale (21619), *Duke of Manchester*	155
Wild Duke of Geneva (36004), by 6th Duke of Geneva (30959), *Sir W. Miles*	115

####### Mr. H. D. DE VITRE'S

	Guineas
C.C. from Alexandra (Acomb), by Lord Thorndale (31756), *J. Richardson*	115

####### Sir C. M. LAMPSON and J. W. LARKING'S

Grand Duke of Geneva (28756), by 15th Grand Duke (21852), *Hill*	380

####### Mr. W. H. SALT'S Sale, BIRD'S NEST, 11th May 1876

Duchess Gwynne, by Duke of Wetherby (17753), *R. Spencer*	265
Flossy Gwynne, by 12th Duke of Oxford (19633), *G. Fox*	100
Flavia Gwynne, by Oxford Gwynne (24711), *D. R. Scratton*	225
Blanche 4th, by Marquis of York (34791), *Major Webb*	130
Flossy Gwynne 2nd, by Marquis of York (34791), *D. A. Green*	278
Silky Gwynne, by Duke of Waterloo (28464), *S. Shaw*	105
Dolly Gwynne 3rd, by Marquis of York (34791), *J. Martin*	130
Blanche 5th, by Marquis of York (34791), *A. and R. Mann*	100
Geneva's Minstrel 2nd, by 9th Duke of Geneva (28391), *W. Ashburner*	500
Flossy Gwynne 3rd, by Marquis of York (34791), *J. Martin*	175
Princess Gwynne, by 5th Duke of Oxford (31738), *W. Ashburner*	505
5th Lord Oxford (31738), by 4th Duke of Geneva (30958), *J. Watts*	115

####### Mr. T. LISTER'S

Fantail's Duchess 2nd, by 9th Duke of Geneva (28391), *Sir C. M. Lampson*	250
Lord Oxford 7th (38645), by 9th Duke of Geneva (28391), *D. A. Green*	300

####### Mr. E. J. COLEMAN'S Sale, STOKE PARK, 17th May 1876

Winsome 2nd, by Lord Oxford (20214), *Jonathan Rigg*	110
Wild Eyes Gwynne, by Baron Wild Eyes (19290), *J. W. Larking*	100

	Guineas
Winsome 3rd, by 13th Duke of Oxford (21604), *Lord Skelmersdale*	450
Princess Gwynne, by Royal Cambridge (25009), *Earl of Bective*	410
Winsome Duchess, by Airdrie Geneva (32920), *Earl of Bective*	700
3rd Duke of Glo'ster (33653), by 10th Duke of Thorndale (28458), *E. H. Cheney*	1250

Lord CHESHAM's Sale, RAYANS FARM, 18th August 1876

Waterloo 34th, by Wallace (23166) (Wild Eyes), *W. Ashburner*	130
Duke of Oxford 28th (32710), by Duke of Oxford 24th (31002), *J. I. D. Jefferson*	330
Baron Wastwater (30492), by Baron Oxford 4th (25580), *T. Statter*	105

Mr. JOSEPH ROBINSON's Sale, NORTHCHURCH, 19th May 1876

Lady Hudson's Duchess 4th, by 6th Duke of Oneida (30997), *W. Ashburner*	160
Duke of Oxford 26th (33708), by Baron Oxford 4th (25580), *Jonathan Rigg*	700

Mr. W. W. SLYE's Sale, BEAUMONT GRANGE, 5th Sept. 1876

Lady Walton 2nd, by Earl of Glo'ster (21644), *Colonel Kingscote*	155
Lady Thorndale Bates, by 4th Duke of Thorndale (17750), *J. d'A. Samuda*	255
Tregunter Gwynne, by Barrington Duke (27985), *J. H. Casswell*	290
Siddington 7th, by Grand Duke of Thorndale (31297), *M. Fawcett*	430
Princess of Lightburne, by 23rd Duke of Oxford (31001), *W. Ashburner*	360
Siddington Duchess, by Grand Duke of Thorndale 2nd (31298), *Earl of Ellesmere*	800
Duchess of Thorndale, by Grand Duke of Thorndale 2nd (31298), *Colonel Kingscote*	145
Duchess of Thorndale 2nd, by Grand Duke of Thorndale 2nd (31298), *J. Croudson*	175

IL LIBRO D'ORO

	Guineas
Duchess Gwynne, by Grand Duke of Thorndale 2nd (31298), *G. Fox*	250
Duchess of Thorndale 3rd, by Grand Duke of Thorndale 2nd (31298), *J. Hope*	110
Siddington Grand Duchess, by Grand Duke of Thorndale 2nd (31298), *Earl of Ellesmere*	400
Thorndale Grand Duke 2nd (39218), by Grand Duke of Thorndale 2nd (31298), *Lord Leigh*	100

Mr. J. P. Foster's Sale, Killhow, 6th September 1876

Fame Gwynne, by Grand Duke of Lightburne (26290), *C. Magniac*	305
Ross Gwynne, by Royal Cambridge (25009), *C. Williams*	165
Siddington 12th, by 2nd Duke of Tregunter (26022), *Sir C. M. Lampson*	920
Fancy Gwynne 2nd, by 22nd Duke of Oxford (31000), *Lord Penrhyn*	505
Rival Gwynne, by 22nd Duke of Oxford (31000), *H. J. Sheldon*	205
Minstrel Maid, by 6th Baron Oxford (33075), *Lord Skelmersdale*	550
Duchess of Clarence 7th, by Oxford's Baronet (29499), *J. Hope*, Canada	210
Waterloo 24th, by 22nd Duke of Oxford (31000), *J. d'A. Samuda*	205

Lord Dunmore's

2nd Lord of the Forth (38639), by 6th Duke of Geneva (30959), *R. Thompson*	150

Sir Wilfrid Lawson's Sale, Brayton, 7th September 1876

Waterloo 36th, by Royal Cambridge (25009), *J. Hope*, Canada	120
Polly Gwynne 10th, by Grand Duke of Lightburne 2nd (26291), *J. Hetherington*	105
Minstrel 5th, by 9th Duke of Geneva (28391), *Lord Skelmersdale*	410
Wetherby Duchess, by 5th Duke of Wetherby (31033), *A. H. Longman*	225

Mr. William Angerstein's Sale, Weeting Hall, 20th September 1876

	Guineas
Fantail 5th, by Royal Cambridge (25009), *Sir John Swinburne*	130
Lady Bates (Hawkey), by 15th Grand Duke (21852), *J. Peter*	210
Blanche 10th, by Grand Duke 10th (21848), *R. Blezard*	135
Lady Waterloo 23rd, by 7th Duke of York (17754), *Sir John Swinburne*	210
Fantail 8th, by 17th Duke of Oxford (25994), *Jon. Rigg*	135
Musical 16th, by 17th Duke of Oxford (25994), *Lord Moreton*	120
Grand Duchess Carolina 2nd, by 8th Duke of Geneva (28390), *Jonathan Rigg*	390
Fantail 9th, by 5th Lord of Oxford (31738), *Sir J. Swinburne*	140
Velvet Eyes, by Duke of Hillhurst (28401), *R. Loder*	900
Carolina 4th, by 3rd Duke of Claro (23729), *G. Graham*	100
Fantail 10th, by 3rd Duke of Claro (23729), *G. Graham*	200
Wild Eyes 32nd, by 3rd Duke of Claro (23729), *Lord Penrhyn*	800
Lady Bates 2nd (Hawkey), by 3rd Duke of Claro (23729), *Colonel Kingscote, M.P.*	165
Fantail 11th, by 3rd Duke of Claro (23729), *Sir J. Swinburne*	125
Musical 18th, by 3rd Duke of Claro (23729), *Lord Moreton*	100
Lady Waterloo 24th, by 3rd Duke of Claro (23729), *Sir J. Swinburne*	140
Musical 19th, by 3rd Duke of Hillhurst (30975), *Sir J. Swinburne*	110
Grand Duchess Carolina 3rd, by Duke Lally, *W. H. Salt*	150
Duke of Rothesay (36534), by 6th Duke of Geneva (30959), *H. J. Sheldon*	2000
Duke Lally (36457), by 3rd Duke of Claro (23729), *G. Graham*	135

Mr. George Fox's Sale, Elmhurst Hall, 5th July 1877

2nd Cambridge Lady (Red Rose), by 4th Duke of Geneva (30950), *H. Allsopp*	1100
Christmas Gwynne, by Christmas Duke (33378), *Earl of Bective*	305

	Guineas
Harefield Gwynne, by Grand Duke of Weston 3rd (34079), *J. Knowles*	360
Harefield Darlington, by Grand Duke of Weston 3rd (34079), *Earl of Bective*	185
Weston's Gwynne, by Grand Duke of Weston 3rd (34079), *H. Lovatt*	250
Geneva's Kirklevington Duchess, by 9th Duke of Geneva (28391), *H. Allsopp*	700
Elmhurst Princess, by Grand Duke of Weston 3rd (34079), *Earl of Bective*	275
Lady Worcester Wild Eyes, by Grand Duke of Weston 3rd (34079), *R. Lodge*	200
Red Rose of Severn, by 24th Duke of Airdrie (36460), *H. Allsopp*	255
Water Belle, by 24th Duke of Airdrie (36460), *H. Lovatt*	100

Hon. M. H. Cochrane's (Canada) Sale, Millbeckstock, 4th September 1877

	Guineas
Marchioness of Barrington, by Grand Duke 22nd (34062), *Sir W. H. Salt*	800
3rd Duchess of Hillhurst, by 2nd Duke of Hillhurst, *R. Loder*	4100
5th Duchess of Hillhurst, by 2nd Duke of Hillhurst, *Earl of Bective*	4300
Lady Surmise, by 2nd Duke of Hillhurst, *Sir W. H. Salt*	400
2nd Duke of Hillhurst (39748), by 6th Duke of Geneva (30959), *A. H. Longman*	800

Mr. William Ashburner's Sale, Conishead Grange, 6th September 1877

	Guineas
Fuchsia 12th, by Duke of Albany (25931), *H. Allsopp*	110
Mild Eyes 4th, by Royal Lancaster (29870), *H. Allsopp*	610
Royal Gwynne, by 22nd Duke of Oxford (31000), *H. Allsopp*	430
Rose of Lightburne 3rd (Elvira), by Grand Duke of Kent 2nd (28759), *Hon. C. Duncombe*	105
Bright Eyes 6th, by Baron Oxford 4th (25580), *H. Allsopp*	780
Wild Eyes 37th, by Lord Darlington 8th (34520), *H. Lovatt*	120
Conishead Fuschia, by 24th Duke of Airdrie (36460), *H. Allsopp*	170

	Guineas
Conishead Wild Eyes, by 24th Duke of Airdrie (36460), *H. Allsopp*	610

Mr. E. H. Cheney's Sale, Gaddesby, 18th September 1877

	Guineas
13th Duchess of Airdrie, by 10th Duke of Thorndale (28458), *R. Loder*	2200
10th Maid of Oxford, by 4th Duke of Geneva (30958), *Earl of Bective*	1605
13th Lady of Oxford, by Baron of Oxford (23371), *H. Allsopp*	1900
Wild Duchess of York, by 7th Duke of York (17754), *J. I. D. Jefferson*	470
Lady Wellesley 2nd, by 9th Duke of Geneva (28391), *L. Rawstorn*	425
Princess Alexandra 2nd, by 9th Duke of Geneva (28391), *Lord Moreton*	110
Lady Angelina, by 9th Duke of Geneva (28391), *Earl of Bective*	415
11th Maid of Oxford, by 9th Duke of Geneva (28391), *H. Lovatt*	1400
Wild Duchess of Geneva 5th, by 9th Duke of Geneva (28391), *Major Chaffey*	200
Lady Angelina 2nd, by 3rd Duke of Glo'ster (33653), *T. Holford*	230
Wild Duchess of Glo'ster, by 3rd Duke of Glo'ster (33653), *Sir W. H. Salt*	340
3rd Duke of Glo'ster (33653), by 10th Duke of Thorndale (28458), *J. Lynn*	550
7th Duke of Glo'ster (39735), by 9th Duke of Geneva (28391), *Duke of Devonshire*	1850
Lord Wild Eyes, by 9th Duke of Geneva (28391), *Sir J. Whitworth*	110

Captain R. E. Oliver's Sale, Sholebroke Lodge, 19th September 1877

	Guineas
Grand Duchess 23rd, by Grand Duke 7th (19877), *Earl of Bective*	2750
Cherry Grand Duchess 4th, by 9th Duke of Geneva (28391), *Lord Skelmersdale*	1800
Grand Duchess 29th, by Grand Duke 22nd (34062), *H. Allsopp*	2450

	Guineas
Cherry Grand Duchess 8th, by Grand Duke 21st (34061), *H. Allsopp*	900
Grand Duchess of Barringtonia 4th, by Grand Duke 25th (34065), *H. Lovatt*	375
Grand Duke 31st (38374), by Grand Duke of Waterloo (34077), *Earl of Bective*	1550
Grand Duke 29th (38372), by Grand Duke 21st (34061), *Sir G. R. Philips*	1000
Grand Duke 25th (34065), by Grand Duke 21st (34061), *A. Garfit*	505

The late Mr. JAMES FAWCETT's Sale, SCALEBY CASTLE, 27th September 1877

Peach Blossom 13th, by 8th Duke of York (28480), *J. Graham*	100
Peach Blossom 14th, by Grand Duke of Kirklevington (34071), *J. Turner*	100
Lady Thorndale Glo'ster Bates, by Grand Duke of Kirklevington (34071), *W. Ashburner*	550
Grand Duke of Kirklevington (34071), by 8th Duke of York (28480), *J. Turner*	100

Mr. E. BOWLY's Sale, SIDDINGTON, 25th April 1878

Siddington 10th, by 2nd Duke of Tregunter (26022), *Lord Moreton*	500
Gazelle 37th, by Beau of Oxford 2nd (33129), *Lord Fitzhardinge*	180

Lord PENRHYN's Sale, WICKEN PARK, 14th May 1878

Lady Barrington 9th, by 3rd Duke of Geneva (23753), *J. I. D. Jefferson*	120
Duchess Carolina, by 17th Duke of Oxford (25994), *W. Ashburner*	150
Waterloo 42nd, by Grand Duke 20th (31281), *W. Ashburner*	210
Ariel Countess 2nd, by Grand Duke 20th (31281), *H. Allsopp*	145
Waterloo 43rd, by Grand Duke of Oxford (31293), *R. E. Oliver*	215
6th Belle of Oxford, by Grand Duke of Oxford (31293), *H. Allsopp*	920

	Guineas
7th Belle of Oxford, by Grand Duke of Oxford (31293), *H. Allsopp*	920
Waterloo 44th, by Beau of Oxford 3rd (33130), *H. Allsopp*	160
Lady Barrington 10th, by Oxford Waterloo (35002), *W. M. Christison*	105

Mr. J. W. LARKING'S Sale, CANSIRON, 15th May 1878

	Guineas
Siddington, by 4th Duke of Oxford (11387), *H. Allsopp*	250
Fuchsia 10th, by Grand Duke of York (24071), *W. Ashburner*	240
Lady Worcester 6th, by 3rd Duke of Claro (23729), *R. Loder*	275
Gazelle 26th, by 2nd Duke of Tregunter (26022), *W. McCulloch*	425
Kirklevington Duchess 9th, by Grand Duke of Clarence (28750), *Lord Moreton*	460
Fuchsia 13th, by Duke of Albany (25931), *W. Ashburner*	310
Cherry Queen, by Oxford 5th (27598), *L. H. Wraith*	680
Gazelle 29th, by 2nd Duke of Tregunter (26022), *Earl of Bective*	455
Marchioness of Worcester, by 8th Duke of Geneva (28390), *Duke of Devonshire*	850
Kirklevington Princess, by Grand Duke of Geneva (28756), *Lady Fitzhardinge*	340
Winsomedale 2nd, by 2nd Duke of Tregunter (26022), *Duke of Devonshire*	550
Fuchsia's Duchess, by Lord Tregunter (31758), *W. McCulloch*	400
Siddington 15th ⎫ twins by 3rd Duke of ⎧ *Lord Moreton*	500
Siddington 16th ⎭ Clarence (23727) ⎩ *J. H. Blundell*	340
Kirklevington Princess 3rd, by Grand Duke of Geneva (28756), *Lord Feversham*	460
Fuchsia's Duchess 2nd, by Airdrie Geneva (32920), *W. McCulloch*	400
Kirklevington Princess 4th, by 3rd Duke of Glo'ster (33653), *H. Allsopp*	600
Countess of Worcester, by 3rd Duke of Glo'ster (33653), *H. Allsopp*	580
Kirklevington Princess 5th, by 3rd Duke of Hillhurst (30975), *R. Lodge*	450

	Guineas
Fuchsia's Duchess 4th, by 3rd Duke of Hillhurst (30975), *W. Ashburner*	265
Cherry Duchess of Hillhurst, by 3rd Duke of Hillhurst (30975), *Duke of Devonshire*	905
Kirklevington Princess 6th, by 3rd Duke of Hillhurst (30975), *Earl of Bective*	520
Belle of Worcester, by 3rd Duke of Hillhurst (30975), *H. Allsopp*	600
Winsome Lass, by 3rd Duke of Hillhurst (30975), *Rev. P. Graham*	340
3rd Duke of Hillhurst (30975), by 6th Duke of Geneva (30959), *Sir C. M. Lampson*	1530
Fusilier (39905), by 3rd Duke of Hillhurst (30975), *G. H. Watts*	100
Viscount Worcester (40881), by 3rd Duke of Hillhurst (30975), *Webb and Chalk*	110
Siddington Kirklevington (42379), by 3rd Duke of Hillhurst (30975), *Jonathan Rigg*	210

Mr. Joseph d'A. Samuda, M.P.'s Sale, Chillies, 17th May 1878

Cleopatra, by 3rd Grand Duke (16182), *Prince of Wales*	135
Lady Thorndale Bates, by 4th Duke of Thorndale (17750), *Rev. P. Graham*	305
Gazelle 25th, by 2nd Duke of Tregunter (26022), *Prince of Wales*	185
Waterloo 37th, by Oxford Beau (29485), *H. Allsopp*	315
Cherry Duchess 20th, by Grand Duke 11th (21849), *Jonathan Rigg*	410
Gazelle 32nd, by Wildfire (32866), *J. P. Clark*	135
Wild Musical, by Wildfire (32866), *Prince of Wales*	125
Cherry Duchess of Oneida, by 6th Duke of Oneida (30997), *H. Allsopp*	820
Gazelle of Chamonix, by 8th Duke of Geneva (28390), *Earl of Bective*	100
Waterloo of Oneida, by 6th Duke of Oneida (30997), *Jonathan Rigg*	330
Lady Tregunter 2nd, by Baron Wastwater (30492), *J. I. D. Jefferson*	110
Gazelle of Isis, by 26th Duke of Oxford (33708), *C. Samuda*	140
Waterloo 24th, by 22nd Duke of Oxford (31000), *H. Allsopp*	155

442 APPENDIX

	Guineas
Gazelle of Oxford, by 26th Duke of Oxford (33708), *Lord Fitzhardinge*	120
Lady Hillhurst Bates, by 3rd Duke of Hillhurst (30975), *H. Allsopp*	410
Fuchsia of Hillhurst, by 3rd Duke of Hillhurst (30975), *Prince of Wales*	125
Marquis of Oxford (34786), by 3rd Duke of Hillhurst (30975), *H. W. Beauford*	150
Carlisle (Barrington) (41190), by 22nd Duke of Oxford (31000), *Earl of Annesley*	100
Marquis of Hillhurst (40309), by 3rd Duke of Hillhurst (30975), *J. Robinson*	130

Mr. F. Sartoris's Sale, Rushden, 31st May 1878

	Guineas
Duchess of Waterloo, by Duke of Kingscote (25981), *Hon. C. Duncombe*	195
Fleur de Lys Gwynne } twins by 12th Duke of Oxford (19633) { *Earl of Bective*	185
Fragrant Gwynne } { *C. Howard*	180
Duchess of Waterloo 2nd, by Baron Oxford 5th (27958), *A. H. Lloyd*	465
Lady Wild Eyes Surmise, by Wild Eyes Duke (36007), *Sir C. M. Lampson*	200
Flippant Gwynne, by Earl of Leicester 3rd (33804), *S. P. Foster*	280
Lady Waterloo Surmise, by Grand Duke of Waterloo (28766), *Duke of Manchester*	160
Violet Gwynne, by Earl of Leicester 7th (33806), *R. Loder*	280

Mr. John H. Blundell's Sale, Woodside, 2nd July 1878

	Guineas
Duke of Oxford 32nd (36527), by 5th Duke of Wetherby (31033), *J. A. M. Cope*	235

Mr. Thomas Holford's Sale, Papillon Hall, 4th July 1878

	Guineas
Winsome 12th, by Baron Oxford 4th (25580), *A. H. Lloyd*	810
Lady Geneva Waterloo, by 9th Duke of Geneva (28391), *Sir G. R. Philips*	110
Countess of Clarence, by 9th Duke of Geneva (28391), *R. Botterill*	110

	Guineas
Water Lily, by 3rd Duke of Hillhurst (30975), *W. McCulloch*	455
Charming Duchess Echter, by 8th Duke of York (28480), *H. J. Sheldon*	210
Lady Mild Eyes, by 8th Duke of York (28480), *W. Ashburner*	405
Violet Eyes, by Grand Duke 21st (34061), *H. Allsopp*	560
Elmhurst Princess, by Grand Duke of Weston 3rd (34079), *H. Allsopp*	410
Countess of Clarence 2nd, by Grand Duke 23rd (34063), *H. Allsopp*	230
Lady Angelina 2nd, by 3rd Duke of Glo'ster (33653), *H. Allsopp*	410
Venusta, by Grand Duke 23rd (24063), *H. J. Sheldon*	180
Princess Morwydd, by 2nd Duke of Hillhurst, *H. Allsopp*	360
Violet Eyes 2nd, by Grand Duke 23rd (34063), *H. Allsopp*	600
Water Lily 2nd, by Grand Duke 23rd (34063), *H. Allsopp*	380
Viscount Oxford (40876), by Grand Duke 23rd (34063), *D. McIntosh*	800

Mr. GEORGE FOX'S Sale, ELMHURST, 5th July 1878

Kirklevington 25th, by 3rd Duke of Claro (23729), *Jonathan Rigg*	305
Lady Waterloo 28th, by 9th Duke of Geneva (28391), *W. McCulloch*	225
Duchess Gwynne, by Grand Duke of Thorndale 2nd (31298), *W. Ashburner*	250
Duchess 20th (Red Rose), by 4th Duke of Geneva (30958), *W. McCulloch*	300
Red Rose of Avon, by 24th Duke of Airdrie (36460), *W. McCulloch*	225
Duchess 26th (Red Rose), by 4th Duke of Geneva (30958), *W. McCulloch*	365
Geneva's Rose of Sharon, by 8th Duke of Geneva (28390), *A. H. Lloyd*	145
Pearlie Gwynne, by 24th Duke of Airdrie (36460), *A. H. Lloyd*	300

Mr. GEORGE ALLEN'S Sale, KNIGHTLEY HALL, 6th Aug. 1878

Kirklevington Lady 3rd, by Duke of Wetherby 6th (33756), *G. Garne*	203

	Guineas
Duchess 8th (Acomb), by Duke of Clarence 5th (36479), *G. Garne*	120
Kirklevington Lady 7th, by Duke of Clarence 5th (36479), *F. Welsh*	140

Mr. Thomas Wilson's Sale, Shotley Hall, 4th Sept. 1878

Winsome 9th, by Grand Duke 17th (24064), *S. P. Foster*	320
Wild Eyebright, by 6th Duke of Geneva (30959), *W. McCulloch*	195
Winsome 18th, by Baron Oxford 4th (25580), *Lord Skelmersdale*	750
Winsome Colleen, by Duke of Connaught (33604), *Earl of Lonsdale*	715
Waterloo Belle, by Oxford Beau 3rd (32013), *Earl of Lonsdale*	160
Wild Erin, by Duke of Connaught (33604), *Lord Skelmersdale*	500
Waterloo Rose, by Duke of Oxford 31st (33713), *Captain Gandy*	110
Winsome Isis, by Duke of Oxford 31st (33713), *Sir J. Swinburne*	585
Winsome Oxonia, by Duke of Oxford 31st (33713), *Lord Moreton*	505
Duke of Oxford 31st (33713), by Baron Oxford 4th (25580), *W. McCulloch*	435

The Earl of Feversham's Sale, Duncombe Park, 6th September 1878

Lady Bright Eyes 2nd, by General Napier (24023), *Duke of Devonshire*	240
Waterloo 32nd, by Grand Duke 11th (21849), *Sir Harcourt Johnstone*	170
Secrecy, by Duke of Underley (33745), *Earl of Lonsdale*	155
Piercing Eyes, by 20th Duke of Oxford (28432), *Earl of Bective*	175
Wild Winsome 3rd, by Duke of Underley (33745), *Earl of Lonsdale*	455
Winsome Winnie, by Duke of Tregunter 5th (33743), *Col. Gunter*	200
Fair Kirklevington, by Fugleman (36670), *Lord Moreton*	275

	Guineas
Duke of Oxford 20th (28432), by Baron Oxford 4th (25580), *Rev. W. H. Beever*	105

The DUKE OF DEVONSHIRE's Sale, HOLKER, 18th Sept. 1878

	Guineas
Lady Ellen Barrington, by Lord Stanley (24467), *G. Ashburner*	155
Grand Duchess of Oxford 19th, by Grand Duke 10th (21848), *Major Chaffey*	855
Oxford Rose 5th, by Baron Oxford 4th (25580), *Earl of Dunmore*	205
Lady Bright Eyes 3rd, by 7th Duke of York (17754), *Lord Skelmersdale*	305
Grand Duchess of Oxford 21st, by Baron Oxford 4th (25580), *Lord Penhryn*	1550
Musical 2nd, by Baron Oxford 4th (25580), *Sir J. H. Greville Smyth*	160
Grand Duchess of Oxford 22nd, by Baron Oxford 4th (25580), *W. McCulloch*	2100
Oxford Rose 10th, by Baron Oxford 4th (25580), *A. H. Longman*	150
Countess of Barrington 7th, by Baron Barrington 4th (33006), *Earl of Dunmore*	505
Oxford Rose 12th, by 5th Duke of Wetherby (31033), *Earl of Dunmore*	350
Baroness Oxford 5th, by 5th Duke of Wetherby (31033), *D. McIntosh*	2660
Blanche 15th, by 5th Duke of Wetherby (31033), *J. I. D. Jefferson*	110
Dainty 2nd (Nettle), by Duke of Hillhurst (28401), *S. P. Foster*	240
Countess of Barrington 9th, by 5th Duke of Wetherby (31033), *W. Ashburner*	360
Winsome 20th, by 5th Duke of Wetherby (31033), *Sir J. H. G. Smyth*	805
Grand Duchess of Oxford 38th, by 24th Duke of Airdrie (36460), *Sir J. Swinburne*	1450
Grand Duchess of Oxford 40th, by 5th Duke of Wetherby (31033), *S. P. Foster*	1600
5th Duke of Wetherby (31033), by 3rd Duke of Wharfdale (21619), *J. Fox*	705
Duke of Oxford 44th (39774), by 5th Duke of Wetherby (31033), *H. A. Brassey*	1650

	Guineas
Duke of Oxford 45th (39775), by 5th Duke of Wetherby (31033), *Lord Fitzhardinge*	1500
Duke of Oxford 46th (41413), by Baron Oxford 7th (36199), *Earl of Ellesmere*	660
Duke of Barrington 4th (39712), by Duke of Ormskirk (36526), *Col. Webb*	160
Duke of Barrington 6th (39714), by Baron Oxford 7th (36199), *Sir Wilfrid Lawson*	105
Baron Winsome 5th (39450), by Baron Oxford 7th (36199), *Earl of Bective*	130
Duke of Barrington 7th (39715), by 5th Duke of Wetherby (31033), *R. Blezard*	185

Mr. ALEXANDER BROGDEN, M.P.'s Sale, STONE CROSS, 19th September 1878

Grand Duchess of Oxford 12th, by 2nd Duke of Wetherby (21618), *E. Baillie*	510
Princess 4th, by Royal Cambridge (25009), *Earl of Bective*	300
Water Lass 4th, by 9th Duke of Geneva (28391), *Earl of Bective*	300
Grand Princess of Lightburne 3rd, by 2nd Duke of Glo'ster (28392), *Earl of Bective*	780
Lightburne's Duke of Oxford, by Duchess of Oxford, 34th (36529), *A. H. Lloyd*	955
Princess of Lightburne 3rd, by Lightburne's Duke of Oxford 2nd (38564), *Earl of Bective*	590
Lightburne's Duke of Oxford 2nd, by 5th Duke of Wetherby (31033), *J. J. Hetherington*	330

Mr. JOHN MARTIN'S

Gazelle 8th, by Baron Barrington 7th (33009), *W. Ashburner*	100
Flora Gwynne, by Duke of Oxford 34th (36529), *Earl of Bective*	275
Minstrel Princess, by Duke of Oxford 34th (36529), *R. Lodge*	125

Mr. A. CRANE'S (KANSAS, U.S.A.)

Kirklevington Duchess 15th, by 2nd Duke of Glo'ster (28392), *A. H. Lloyd*	330

	Guineas
Grand Duchess of Kirklevington, by Royal Lancaster (29870), *A. H. Lloyd*	360
Duke of Airdrie 27th (41351), by 14th Duke of Thorndale (28459), *A. H. Lloyd*	505

Messrs. AVERY and MURPHY's (MICHIGAN, U.S.A.)

Fordham Duke of Oxford 4th (41569), by Beau of Oxford, *R. Botterill*	150

Lord FITZHARDINGE's Sale, BERKELEY, 26th March 1879

Lady Wild Eyes 6th, by 2nd Duke of Tregunter (26021), *D. R. Scratton*	200
Dorothea (Darlington), by Grand Duke of Waterloo (28766), *W. McCulloch*	265
Lady Wild Eyes 9th, by Grand Duke of Waterloo (28766), *D. R. Scratton*	210
Lady Wild Eyes 10th, by Duke of Siddington 2nd (33732), *D. R. Scratton*	300
Kirklevington Empress 2nd, by Duke of Connaught (33604), *J. A. Rolls*	770
Musical 18th, by 3rd Duke of Glo'ster (33653), *J. H. Angas*, Australia	125
Minstrel 7th, by Duke of Connaught (33604), *R. Blezard*	125
Lady Wild Eyes 12th, by Duke of Connaught (33604), *J. H. Angas*	400
Minstrel 8th, by Duke of Connaught (33604), *J. H. Angas*	105
Lady Gracious (Gazelle), by Duke of Connaught (33604), *J. H. Angas*	205
Lady Wild Eyes 14th, by Duke of Connaught (33604), *S. P. Foster*	250
Baron Graham (41030), by Duke of Connaught (33604), *J. H. Angas*, Australia	155
Duke of Woodford (Winsome) (41453), by Duke of Connaught (33604), *W. J. Marsh*	110

Colonel KINGSCOTE's Sale, KINGSCOTE, 27th March 1879

Carolina Craggs, by Duke of Hillhurst (28401), *W. Ashburner*	100
Honey 62nd, by Duke of Hillhurst (28401), *Prince of Wales*	165

	Guineas
Honey 64th, by Duke of Hillhurst (28401), *Prince of Wales*	165
Ariel Viscountess, by Grand Duke of Glo'ster (36721), *Col. Luttrell*	110
Oxford Belle 5th, by Grand Duke of Glo'ster (36721), *Lord Fitzhardinge*	1100
Oxford Beau 7th (42082), by Duke of Hillhurst (28401), *J. H. Angas*, Australia	675

Mr. HENRY LOVATT'S Sale, BUSHBURY, 29th April 1879

	Guineas
Lady Mild Eyes, by 8th Duke of York (28480), *G. Ashburner*	350
Kirklevington Duchess 16th, by 2nd Duke of Glo'ster (28392), *G. Ashburner*	395
Gwynne Princess 8th, by Oxford le Grand (29496), *W. Bliss*	125
Lady York and Underley Bates, by 2nd Duke of Underley (26022), *Rev. P. Graham*	280
Grand Duchess of Barringtonia 4th, by Grand Duke 25th (34065), *J. Martin*	315
Duchess Lally 3rd, by Grand Duke of Kirklevington 2nd (38379), *Rev. P. Graham*	200
Gwynne Duchess, by Grand Duke of Kirklevington 2nd (38379), *W. Bliss*	115
Countess of Kirklevington, by Lord of the Isles (34631), *J. H. Angas*, Australia	260
Countess of Kirklevington 2nd, by Baron Turncroft Oxford 4th (37822), *J. Martin*	155

The late Sir W. C. TREVELYAN'S Sale, WALLINGTON, 14th May 1879

	Guineas
Lady Waterloo 30th, by Oxford Beau 4th (34964), *R. Botterill*	110

Messrs. LENEY and SONS' Sale, ORPINES, 24th June 1879

	Guineas
Countess 6th, by 6th Duke of Oneida (30997), *H. J. Sheldon*	110
Oneida Gwynne, by 6th Duke of Oneida (30997), *S. P. Foster*	190
Acomb 2nd, by 6th Duke of Oneida (30997), *H. J. Sheldon*	100

	Guineas
Wild Princess 4th, by 6th Duke of Oneida (30997), *H. A. Brassey*	100
Princess Oneida 2nd, by 6th Duke of Oneida (30997), *Earl of Bective*	515

Mr. D. McIntosh's Sale, Havering, 26th June 1879

Dowager Duchess 4th, by Grand Duke 11th (21849), *R. Morton*	115
Dowager Duchess 5th, by 3rd Duke of Geneva (23753), *R. Morton*	110

Lord Braybrooke's Sale, Audley End, 27th June 1879

Heydon Rose 2nd, by 3rd Duke of Geneva (23753), *F. Barchard*	120
Heydon Rose 3rd, by Grand Duke 17th (24064), *R. Botterill*	300
Grand Duchess of Oxford 5th, by Heydon Duke 2nd (31370), *H. C. Smith*, Australia	165
Thorndale Rose 7th, by 6th Duke of Oneida (30997), *H. Allsopp*	1000
Christmas Rose 3rd, by Duke of Rosedale 3rd (33723), *F. Leney and Sons*	205
Heydon Rose 5th, by Duke of Rosedale 3rd (33723), *H. C. Smith*	240
Grand Duchess of Oxford 8th, by Duke of Rosedale (33721), *J. H. Angas*, Australia	165
Thorndale Rose 9th, by Duke of Underley (33745), *Earl of Bective*	900
Thorndale Rose 12th, by 6th Duke of Oneida (30997), *Sir C. M. Lampson*	600
Thorndale Rose 13th, by 6th Duke of Oneida (30997), *H. Allsopp*	565
Thorndale Rose 14th, by 6th Duke of Oneida (30997), *A. H. Lloyd*	535
Duke of Rosedale 8th (39780), by Duke of Underley (33745), *Major Chaffey*	210
Duke of Rosedale 9th (41419), by Duke of Underley 3rd (38196), *Col. Kingscote*	125

The Earl of Dunmore's Sale, Dunmore, 27th August 1879

Oxford Rose 5th, by Baron Oxford 4th (25580), *R. Botterill*	100

	Guineas
Red Rose of Killigrey, by 4th Duke of Geneva (30958), *F. Barchard*	250
Blythesome Eyes, by 3rd Duke of Hillhurst (30975), *Prince of Wales*	270
Countess of Barrington 7th, by Baron Barrington 4th (33006), *H. Lovatt*	170
Red Rose of Strathearne, by 6th Duke of Geneva (30959), *Earl of Bective*	150
Lady Worcester 17th, by 3rd Duke of Hillhurst (30975), *H. Lovatt*	150
Red Rose of Virginia, by 6th Duke of Geneva (30959), *H. Allsopp*	250
Oxford Rose 12th, by 5th Duke of Wetherby (31033), *R. Botterill*	100
Lady Worcester 18th, by Duke of Connaught (33604), *H. Allsopp*	410
Fuchsia 15th, by Duke of Connaught (33604), *F. Barchard*	150
Duchess 114th, by 6th Duke of Geneva (30959), *H. Allsopp*	2700
Lady Worcester 20th, by 6th Duke of Geneva (30959), *H. A. Brassey*	390
Red Rose of Benledi, by 6th Duke of Geneva (30959), *G. Fox*	180
Lady Worcester 21st, by 6th Duke of Geneva (30959), *Prince of Wales*	350
Red Rose of Strathspey, by Duke of Underley (33745), *Sir C. M. Lampson*	210
Duchess 117th, by 2nd Marquis of Oxford (37055), *H. Allsopp*	3200
Duke of Cornwall 2nd (43082), by 2nd Marquis of Oxford (37055), *Sir C. M. Lampson*	1250
Marquis of Oxford 2nd (37055), by 3rd Duke of Hillhurst (30975), *Sir C. M. Lampson*	220
Sir Glo'ster Barrington (44034), by 7th Duke of Glo'ster (39735), *J. P. Clark*	100
Marquis of Worcester 8th (43623), by 2nd Marquis of Oxford (37055), *J. Darling*	110

Lord Skelmersdale's Sale, Lathom House, 3rd September 1879

Maud Waterloo, by 4th Baron Oxford (25580), *D. Reynolds Davies*	150

	Guineas
Viscountess Barrington, by 6th Duke of Barrington (33576), *H. J. Sheldon*	130
Frosty Gwynne, by 4th Baron Oxford (25580), *Lord Penrhyn*	310
Winsome Beauty, by 4th Baron Oxford (25580), *S. P. Foster*	350
Duchess of Ormskirk, by 5th Duke of Tregunter (33743), *R. Loder*	2000
Winsome Beauty 4th, by 4th Baron Oxford (25580), *Duke of Devonshire*	400
Princess of Blythe, by Duke of Ormskirk (36526), *Sir W. H. Salt*	585
Viscountess Barrington 2nd, by Baron Oxford (25580), *W. Ashburner*	130
Cherry Grand Duchess 11th, by Grand Duke 25th (34065), *F. Leney and Sons*	325
Florence Waterloo, by Lord Bright Eyes (36918), *D. Reynolds Davies*	310
Lally Barrington 2nd, by 4th Baron Oxford (25580), *R. Blezard*	105
Faith Gwynne, by 4th Baron Oxford (25580), *P. L. Mills*	115
Fortune Gwynne, by 4th Baron Oxford (25580), *H. Rawcliffe*	120
Minstrel Gwynne, by 4th Baron Oxford (25580), *H. J. Sheldon*	110
Winsome Beauty 6th, by 6th Duke of Rosedale (38176), *Major Chaffey*	165
Bertha Waterloo, by 4th Baron Oxford (25580), *D. Reynolds Davies*	300
Wild Prince 7th (44260), by 4th Baron Oxford (25580), *H. Rawcliffe*	120

Mr. WILLIAM ASHBURNER'S Sale, CONISHEAD GRANGE, 5th September 1879

	Guineas
Kirklevington Duchess 21st, by 2nd Duke of Glo'ster (28392), *H. Lovatt*	190
Countess of Kirklevington 2nd, by Baron Turncroft Oxford 4th (37822), *Earl of Bective*	200
Conishead Lally 3rd, by 2nd Lightburne's Duke of Oxford (38564), *P. L. Mills*	125
Conishead Wild Eyes 2nd, by 2nd Duke of Glo'ster (38392), *H. Lovatt*	170

	Guineas
Earl of Oxford (41481), by Baron Turncroft Oxford 4th (37822), *G. Fox*	200

Mr. R. Lodge's

	Guineas
Duchess of Wellington 6th, by Baron Barrington 5th (33007), *Captain Liddon*	235
Wild Boy of the Valley (42609), by Lightburne's Duke of Oxford (36895), *Major Conwy*	150

Lord Lonsdale's

	Guineas
Waterloo Belle, by Oxford Beau 3rd (32013), *Major Conwy*	160

Mr. H. J. Sheldon's Sale, Brailes, 17th September 1879

	Guineas
Charming Duchess 6th, by 2nd Duke of Collingham (23730), *E. Leney*	110
Cherry Duchess of Brailes 3rd, by Duke of Rothesay (36534), *Duke of Devonshire*	205
Prince of Brailes (42193), by Duke of Rothesay (36534), *J. A. Rolls*	130
Duke of Barrington 9th (44650), by Duke of Rothesay (36534), *D. Reynolds Davies*	135

Mr. E. H. Cheney's Sale, Gaddesby, 19th September 1879

	Guineas
Lady Wellington, by Duke of Putney (33717), *H. J. Sheldon*	100
Wild Duchess of Geneva 4th, by 9th Duke of Geneva (28391), *C. Magniac*	195
Lady Wellesley 3rd, by 9th Duke of Geneva (28391), *L. Rawstorne*	300
Lady Angelina 3rd, by 3rd Duke of Glo'ster (33653), *H. J. Sheldon*	160
Lady Wellesley 4th, by 3rd Duke of Glo'ster (33653), *Earl of Bective*	185
Wild Duchess of Glo'ster 2nd, by Duke of Glo'ster 6th (39734), *Major Chaffey*	310
3rd Duke of Glo'ster (33653), by 10th Duke of Thorndale (28458), *J. Lynn*	150
Duke of Glo'ster 6th (39734), by 9th Duke of Geneva (28391), *F. Leney and Sons*	280
Lord Oxford 13th (40230), by 9th Duke of Geneva (28391), *F. H. Jennings*	110

IL LIBRO D'ORO

Col. KINGSCOTE'S, sold at WINTERFOLD, 12th May 1880

	Guineas
Duke of Hillhurst (28401), by 14th Duke of Thorndale (28459), *Lord Fitzhardinge*	200

Mr. HENRY LOVATT'S Sale, LOW HILL, 21st May 1880

Lady Wellesley, by 9th Duke of Geneva (28391), *Earl of Bective*	155
Conishead Wild Eyes 2nd, by 2nd Duke of Glo'ster (28392), *G. Fox*	155
Countess of Oxford, by Baron Turncroft Oxford 4th (37822), *Earl of Bective*	525
Lady Oxford Barrington, by Baron Turncroft Oxford 4th (37822), *R. Gibson*, U.S.A.	150
Lady Worcester 23rd, by Baron Turncroft Oxford 4th (37822), *F. Barchard*	150

The EARL OF BECTIVE'S Sale, UNDERLEY, 8th September 1880

Gazelle 29th, by 2nd Duke of Tregunter (26022), *Prince of Wales*	160
Winsome Duchess, by Airdrie Geneva (32920), *S. P. Foster*	400
Princess 8th, by Duke of Underley (33745), *A. H. Lloyd*	405
Kirklevington Princess 6th, by 3rd Duke of Hillhurst (30975), *H. M. Vail*, U.S.A.	170
Lady Edith Bates, by Lord of the Isles (34631), *G. Fox*	300
Duchess of Underley 3rd, by Grand Duke 31st (38374), *Earl of Feversham*	2000
Lady Ann 2nd (Red Rose), by 3rd Duke of Glo'ster (33653), *G. Fox*	310
Marchioness 12th (Kirklevington), by Duke of Underley (33745), *Earl of Lathom*	310
Wayward Eyes, by Lord of the Isles (34631), *G. Ashburner*	145
12th Maid of Oxford, by Duke of Underley (33745), *Sir C. M. Lampson*	980
Countess of Kirklevington 2nd, by Baron Turncroft Oxford 4th (37822), *Sir W. Armstrong*	405
Marchioness 13th, by Baron Turncroft Oxford 4th (37822), *Earl of Lathom*	350
Underley Princess, by Lightburne's Duke of Oxford 2nd (38564), *A. H. Lloyd*	435

	Guineas
Windermere, by Duke of Underley (33745), *Lady Howard de Walden*	120
Countess of Oxford, by Baron Turncroft Oxford 4th (37822), *G. Fox*	515
Fair Gwynne, by Grand Duke 34th (41642), *S. P. Foster*	100
Duchess Gwynne 5th, by Grand Duke 31st (38374), *Captain Gandy*	105
Marquis 10th (Kirklevington) (46749), by Grand Duke 34th (41642), *J. Farmer* (New Zealand)	130
Turcoman 4th (Underley Darling), by Grand Duke 31st (38374), *Lady Howard de Walden*	105

Sir R. C. MUSGRAVE'S

Winsome Wild Eyes 2nd, by Bolton (25650), *Earl of Bective*	195
Winsome Wild Eyes 5th, by Royal Cambridge 4th (40624), *G. Fox*	225
Wild Maid 2nd, by Duke of Underley (33745), *Sir C. M. Lampson*	160

Mr. S. P. FOSTER'S

Oxford Duke of Killhow 2nd (43720), by Duke of Ormskirk (36526), *A. P. Heywood Lonsdale*	450

Sir H. ALLSOPP'S

Lord of the Tweed (Red Rose) (41902), by Duke of Underley (33745), *Earl of Bective*	220

Mr. HETHERINGTON'S Sale, BRAMPTON, 8th October 1880

Red Rose of Rodil, by 4th Duke of Geneva (30958), *G. Fox*	100
Polly Gwynne 13th, by Duke of Underley (33745), *S. P. Foster*	120
Polly Gwynne 14th, by Duke of Underley (33745), *S. P. Foster*	180

Lord PENRHYN'S Sale, WICKEN PARK, 28th October 1880

8th Belle of Oxford, by Grand Duke of Oxford (31293), *T. and C. Horsfall*	340
Waterloo 46th, by Beau of Oxford 4th (36231), *Earl of Bective*	180

	Guineas
Waterloo 47th, by Beau of Oxford 3rd (36230), *Earl of Bective*	120
Cherry Duchess 28th, by Grand Duke of Oxford (31293), *Earl of Lathom*	150
Wild Eyes 33rd, by Beau of Oxford 3rd (33130), *Lord Fitzhardinge*	140
Wild Eyes 34th, by Grand Duke of Oxford (31293), *Lord Fitzhardinge*	115
Cherry Duchess 30th, by Grand Duke of Oxford (31293), *Earl of Lathom*	130

Mr. T. HOLFORD'S

Duke of Leicester (43112), by Viscount Oxford (40876), *Earl of Bective*	510
Viscount Oxford 3rd (44208), by Grand Duke 23rd (34063), *F. Barchard*	300

Sir W. H. SALT's Sale, BIRD'S NEST, 26th April 1881

Maplewell 2nd (Kirklevington), by 5th Lord Oxford (31738), and c.c. by Duke of Glo'ster 5th (36494), *Earl of Stamford*	126
Lady Worcester 19th, by 6th Duke of Geneva (30959), *A. H. Lloyd*	165
Maplewell Darling, by 5th Duke of Glo'ster (36494), *Earl of Bective*	110
Duchess of Barrington, by 5th Duke of Glo'ster (36494), *G. Fox*	155
Maplewell 4th, by 5th Duke of Glo'ster (36494), *Earl of Stamford*	120
Duchess Wild Eyes, by Duke of Oxford 47th (41414), *R. Gibson*	130
5th Duke of Glo'ster (36494), by 9th Duke of Geneva (28391), *J. A. Rolls*	290

Mr. D. McINTOSH's Sale, HAVERING, 19th May 1881

Havering Waterloo 3rd, by Duke of Havering (33664), *R. Gibson*, Canada	100
Baron Oxford 3rd (42737), by Duke of Havering (33664), *C. H. Bassett*	310

DUKE OF MANCHESTER's Sale, KIMBOLTON, 7th July 1881

Marchioness of Oxford 3rd, by 2nd Duke of Collingham (23730), *Earl of Feversham*	750

	Guineas
Sequoiah (Waterloo), by 3rd Duke of Geneva (23753), *H. Hussey Vivian*	165
Lally of Ellington, by Duke of Hillhurst (28401), *W. Murray*, Canada	140
Oxford Mary, by Duke of Underley 3rd (38196). *Sir H. Allsopp*	1110
Oxford Augusta, by Duke of Underley 3rd (38196), *C. Magniac*	270
Lally of Littlehurst, by Duke of Underley 3rd (38196), *G. Fox*	150
Lady Montagu 3rd (Wild Eyes), by Duke of Underley 3rd (38196), *H. A. Brassey*	200
Lally of Lymage, by Duke of Underley 3rd (38196), *Lord Fitzhardinge*	160

Mr. George Fox's Sale, Elmhurst, 12th July 1881

	Guineas
Cherry Duchess of Elmhurst 2nd, by Duke of Airdrie 24th (36460), *Lord Fitzhardinge*	305
Princess Airdrie, by Duke of Airdrie 24th, *Earl of Bective*	105
Deepdale 3rd, by Duke of Airdrie 24th (36460), *H. Hussey Vivian*	105
Cherry Duchess of Elmhurst 4th, by Duke of Oxford 39th (38173), *J. I. D. Jefferson*	130
Duke of Oxford 39th (38173), by Duke of Wetherby 5th (31033), *J. Darling*	110

Mr. H. Lovatt's

	Guineas
Lady Oxford Bates, by Lightburne's Duke of Oxford 2nd (38654), *G. Fox*	155

Mr. S. P. Foster's Sale, Killhow, 7th September 1881

	Guineas
Flippant Gwynne, by Earl of Leicester 3rd (33804), *T. D. Grissell*	105
Royal Gwynne 2nd, by Duke of Oxford 22nd (31000), *Sir W. Lawson*	100
Wild Eyes Lassie 3rd, by Duke of Ormskirk (36526), *R. G. Richardson*	175
Lady Cumberland (Wild Eyes), by Duke of Underley (33745), *J. I. D. Jefferson*	150
Grand Duchess of Oxford 40th, by 5th Duke of Wetherby (31033), *Sir W. Lawson*	520

	Guineas
Wild Eyes Lassie 4th, by Duke of Ormskirk (36526), *H. Lovatt*	130
Lady Cumberland 2nd (Wild Eyes), by Duke of Ormskirk (36526), *R. Blezard*	120
Oxford Duchess of Killhow 4th, by Oxford Duke of Killhow (42091), *Rev. P. Graham*	320
Lady Warner (Winsome), by Duke of Ormskirk (36526), *Duke of Devonshire*	160

Lord FITZHARDINGE's Sale, BERKELEY, 7th July 1882

Kirklevington Princess, by Grand Duke of Geneva (28756), *H. Lovatt*	155
Lady Wild Eyes 11th, by Duke of Connaught (33604), *F. Barchard*	100
Dowager 3rd (Darlington), by Duke of Connaught (33604), *Earl of Bective*	160
Charming Daisy, by Duke of Connaught (33604), *C. H. Bassett*	105
Wisdom 3rd, by Duke of Connaught (33604), *A. H. Lloyd*	250
Wisdom 4th, by Duke of Connaught (33604), *A. H. Lloyd*	210

Mr. S. J. MURPHY's (Detroit, U.S.A.), sold at SPARROW FARM, 13th July 1882

Airdrie Duchess 9th, by Duke of Airdrie 23rd (41350), *Duke of Devonshire*	700
Airdrie Duchess 10th, by Knight of Oxford (40082), *Duke of Devonshire*	730

The EARL OF FEVERSHAM's Sale, DUNCOMBE PARK, 5th Oct. 1882

Wild Winsome 4th, by Duke of Tregunter 5th (33743), *Rev. P. Graham*	220
Wild Winsome 8th, by Duke of Tregunter 5th (33743), *S. White*, Canada	135
Wild Winsome 9th, by Baron Oxford 9th (42738), *Duke of Devonshire*	255

Mr. S. P. FOSTER's

Oxford Duke of Killhow 4th (48399), by Duke of Ormskirk (36526), *Hon. Baillie Hamilton*	250

Mr. Robert Lodge's Sale, Dringhouses, 6th October 1882

	Guineas
Kirklevington Princess 5th, by Duke of Hillhurst 3rd (30975), *H. Lovatt*	195
Princess of the Valley, by Grand Duke of Thorndale 2nd (31298), *S. P. Foster*	210
Duchess of Wellington 7th, by Duke of Clarence 4th (33597), *R. Briggs*	105
Kirklevington Duchess 25th, by Oxford's King (34997), *Earl of Bective*	135
Kirklevington Duchess 26th, by Grand Duke of Glo'ster (36271), *J. I. D. Jefferson*	125

The late Earl of Lonsdale's

Winsome Colleen, by Duke of Connaught (33604), *H. Lovatt*	210
Wild Winsome 3rd, by Duke of Underley (33745), *R. Botterill*	130
Blanche Winsome, by Grand Duke 31st (38374), *T. Horsfall*	140
Winsome Care, by Duke of Underley (33745), *A. H. Lloyd*	155

Mr. D. A. Green's Sale, East Donyland, 4th May 1883

Duchess of Oxford, by Duke of Underley 3rd (38196), *A. J. Alexander*, U.S.A.	435
Kirklevington Lady, by Lord Oxford 7th (38645), *Sir C. M. Lampson*	150
Duchess of Oxford 2nd, by Duke of Oxford (39770), *R. Briggs*	275
Kirklevington Lady 2nd, by Duke of Oxford (39770), *Earl of Bective*	165
Kirklevington Lady 3rd, by Duke of Oxford (39770), *Sir C. M. Lampson*	130

Mr. Alfred H. Lloyd's Sale, Harewoods, 31st May 1883

Wisdom 4th, by Duke of Connaught (33604), *R. Briggs*	140
Countess of Worcester, by Duke of Glo'ster 5th (36494), *D. Hume*, for U.S.A.	230
Kirklevington Countess, by Duke of Airdrie 27th (41351), *G. Fox*	115

	Guineas
Queen of Oxford 3rd, by Grand Duke 41st (46439), *R. Briggs*	300
Winsome Duke (48968), by Grand Duke 41st (46439), *E. Ellis*	180

Sir H. ALLSOPP'S

Knight of Oxford 6th (46574), by Grand Duke of Airdrie (43310), *H. Lovatt*	150

Mr. THOMAS HOLFORD'S Sale, CASTLE HILL, 11th July 1883

Airdrie Duchess 7th, by Duke of Hillhurst 2nd (39748), *Earl of Bective*	500
Beaming Eyes 2nd, by Grand Duke 23rd (34063), *Sir C. M. Lampson*	130
Lady Bright Eyes 3rd, by Grand Duke 24th (34064), *C. Ramsey*, U.S.A.	100
Beaming Eyes 3rd, by Grand Duke 23rd (34063), *Sir C. M. Lampson*	125
Duchess of Leicester, by Viscount Oxford (40876), *Earl of Bective*	1505
Duchess of Leicester 3rd, by Viscount Oxford 4th (44209), *Lord Fitzhardinge*	1150
Viscountess Oxford 4th, by Grand Duke 23rd (34063), *A. Leney*	420
Duchess of Vittoria 2nd, by Grand Duke 23rd (34063), *Sir C. M. Lampson*	220
Viscountess Oxford 5th, by Duke of Underley (33745), *Sir C. M. Lampson*	320
Duchess of Leicester 6th, by Viscount Oxford 2nd (42558), *B. C. Ramsey*, U.S.A.	355
Duke of Leicester 3rd (46256), by Grand Duke 23rd (34063), *Lord Fitzhardinge*	900
Duke of Leicester 4th (47774), by Viscount Oxford 6th (47216), *R. H. Crabb*	205
Duke of Leicester 6th (49461), by Viscount Oxford 2nd (42558), *S. P. Foster*	355

The DUKE OF DEVONSHIRE'S Sale, HOLKER, 6th Sept. 1883

Grand Duchess of Oxford 33rd, by Duke of Wetherby 5th (31033), *Earl of Bective*	155

460   APPENDIX

	Guineas
Grand Duchess of Oxford 39th, by Baron Oxford 7th (36199), *J. Harris*	310
Cherry Duchess of Brailes 3rd, by Duke of Rothesay (36534), *C. Leney*	130
Baroness Oxford 4th, by Duke of Hillhurst 3rd (30975), *G. Fox*	365
Baroness Oxford 9th, by Duke of Glo'ster 7th (39735), *A. J. Alexander*, U.S.A.	520
Countess of Barrington 10th, by Duke of Glo'ster 7th (39735), *H. Goodriche*	105
Grand Duchess of Oxford 48th, by Duke of Glo'ster 7th (39735), *L. H. Wraith*	240
Winsome 25th, by Duke of Rosedale 6th (38176), *Sir C. M. Lampson*	130
Grand Duchess of Oxford 49th, by Duke of Glo'ster 7th (39735), *J. Harris*	565
Baroness Oxford 12th, by Duke of Glo'ster 7th (39735), *H. Goodriche*	275
Grand Duchess of Oxford 67th, by Baron Oxford 12th (45926), *Hon. R. Baillie Hamilton*	200
Grand Duchess of Oxford 59th, by Baron Oxford 8th (41057), *Sir C. M. Lampson*	205
Grand Duchess of Oxford 60th, by Baron Oxford 8th (41057), *H. W. B. Berwick*	305
Grand Duchess of Oxford 63rd, by Baron Oxford 8th (41057), *H. W. B. Berwick*	340
Winsome 28th, by Baron Oxford 8th (41057), *Prince of Wales*	200
Countess of Barrington 13th, by Baron Oxford 8th (41057), *Prince of Wales*	210
Winsome 29th, by Duke of Leicester (43112), *Earl of Bective*	115
Countess of Barrington 14th, by Baron Oxford 8th (41057), *Hon. R. Baillie Hamilton*	165
Grand Duchess of Oxford 69th, by Baron Oxford 12th (45926), *W. Murray*, Canada	150
Duke of Oxford 60th (46265), by Duke of Glo'ster 7th (39735), *R. Gibson*, Canada	105
Duke of Oxford 61st (46266), by Duke of Glo'ster 7th (39735), *Sir J. Whitworth*	105
Duke of Oxford 63rd (49470), by Baron Oxford 8th (41057), *Sir W. Lawson*	180
Duke of Holker 7th (49456), by Baron Oxford 8th (41057), *J. Harris*	165

	Guineas
Baron Oxford 16th (49090), by Grand Duke 41st (46439), *Lord Moreton*	200
Duke of Oxford 65th (49472), by Baron Oxford 12th (45926), *Lady Howard de Walden*	170
Baron Oxford 17th (49091), by Duke of Oxford 56th (46263), *Viscountess Ossington*	150

Mrs. STARKIE's Sale, ASHTON HALL, 7th September 1883

Lady Ashton Wild Eyes 4th, by Duke of Wetherby 5th (31033), *G. Fox*	215
Lady Ashton Wild Eyes 5th, by Baron Oxford 8th (41057), *Hon. R. Baillie Hamilton*	160
Lady Ashton Wild Eyes 7th, by Grand Duke 39th (43308), *Hon. R. Baillie Hamilton*	160

Rev. P. GRAHAM's

Wild Winsome 4th, by Duke of Tregunter 5th (33743), *R. Gibson*, Canada	125
Turncroft Belle of Oxford, by Duke of Tregunter 7th (38194), *J. I. D. Jefferson*	150
Lady Turncroft Bates 8th, by Grand Duke 39th (43308), *R. Briggs*	100
Turncroft Duchess of Oxford, by Grand Duke 39th (43308), *J. Harris*	100

Mr. WRAITH's

Cherry Queen 2nd, by Duke of Tregunter 7th (38194), *Millar*	215

Captain R. E. OLIVER's Sale, SHOLEBROKE LODGE, 8th May 1884

Grand Duchess 32nd, by Grand Duke 25th (34065), *Sir C. M. Lampson*	400
Waterloo 43rd, by Grand Duke of Oxford (31293), *Hon. R. Baillie Hamilton*	205
Grand Duchess 37th, by Grand Duke 25th (34065), *C. T. Lucas*	155
Grand Duchess 39th, by Grand Duke 30th (38373), *Sir C. M. Lampson*	1060
Grand Duchess 40th, by Duke of Underley 3rd (38196), *A. H. Lloyd*	675

	Guineas
Grand Duchess 41st, by Duke of Underley (33745), *Duke of Devonshire*	1005
Grand Duchess 42nd, by Grand Duke 30th (38373), *Earl of Bective*	1120
Grand Duchess 43rd, by Grand Duke 30th (38373), *J. J. Hill*, U.S.A.	500
Grand Duchess 44th, by Grand Duke 30th (38373), *Earl of Bective*	705
Waterloo Bienvenue 2nd, by Duke of Connaught (33604), *H. A. Brassey*	305
Grand Duchess 46th, by Duke of Connaught (33604), *Sir H. Hussey Vivian*	300
Grand Duchess 47th, by Grand Duke 30th (38373), *J. J. Hill*, U.S.A.	600
Grand Duchess of Waterloo 2nd, by Grand Duke 33rd (39946), *C. T. Getting*	340
Grand Duchess 48th, by Grand Duke 33rd (39946), *A. H. Lloyd*	850
Grand Duchess 49th, by Grand Duke 41st (46439), *Earl of Bective*	910
Grand Duchess 51st, by Duke of Connaught (33604), *P. L. Mills*	430
Grand Duchess 52nd, by Duke of Connaught (33604), *Lord Fitzhardinge*	500
Grand Duchess 53rd, by Cherry Grand Duke 8th (39575), *Earl of Bective*	280
Grand Duke 41st (46439), by Grand Duke 30th (38373), *A. H. Lloyd*	375
Grand Duke 44th (46440), by Grand Duke 30th (38373), *Mrs. McIntosh*	200
Grand Duke 46th (49671), by Duke of Connaught (33604), *Duke of Devonshire*	750
Grand Duke of Waterloo 7th (49679), by Grand Duke 41st (46439), *C. T. Getting*	150

DUKE OF MANCHESTER'S

Oxford de Vere 5th (48397), by Duke of Underley 3rd (38196), *Mrs. Perry Herrick*	160

Mr. ROBERT LODER'S Sale, WHITTLEBURY, 9th May 1884

Duchess Craggs 3rd, by Grand Duke 22nd (34062), *Earl of Lathom*	135

	Guineas
Lovely Eyes, by Grand Duke 25th (34065), *R. Gibson*	105
Grand Duchess Darlington 2nd, by Grand Duke 25th (34065), *Lord Fitzhardinge*	105
Duchess Darlington 9th, by Grand Duke 25th (34065), *P. L. Mills*	105
Duchess of Wappenham, by Grand Duke 25th (34065), *J. J. Hill*, U.S.A.	650
Grand Duchess Acomb 4th, by Grand Duke of Oxford 3rd (39953), *P. L. Mills*	110
Duchess Fawsley 9th, by Grand Duke 25th (34065), *C. T. Getting*	100
Winsome Wild Eyes 8th, by Grand Duke 25th (34065), *Lord Moreton*	105

Sir H. ALLSOPP'S

Duke of Cornwall 4th (47726), by Knight of Oxford 3rd (43441), *R. Loder*	230
Knight of Oxford 8th (48118), by Grand Duke 40th (43309), *Earl of Lathom*	275

Mr. S. P. FOSTER'S

Oxford Duke of Killhow 6th (50130), by Duke of Ormskirk (36526), *Duke of Manchester*	290
Oxford Duke of Killhow 7th (50131), by Duke of Oxford 50th (43121), *R. H. Allen*, U.S.A.	280

Messrs. F. LENEY and SONS' Sale, ORPINES, 2nd July 1884

Cherry Grand Duchess 11th, by Grand Duke 25th (34065), *Dr. Frias*, Buenos Ayres	185
Wild Princess 6th, by Duke of Oxford 44th (39774), *E. Ellis*	120
Grand Duchess of Geneva 7th, by Duke of Glo'ster 6th (39734), *Dr. Frias*, Buenos Ayres	1100
Countess 14th, by Duke of Glo'ster 6th (39734), *Dr. Frias*, Buenos Ayres	100
Wateringbury Rose 3rd, by Duke of Oneida 6th (30997), *H. A. Brassey*	225
Kirklevington Lady 10th, by Duke of Glo'ster 6th (39734), *R. Rostrepo*, South America	130
Grand Duke of Geneva 3rd (49677), by Rowfant Duke of Glo'ster 2nd (48610), *Earl of Bective*	355

Mrs. McIntosh's, sold at Preston Hall, 3rd July 1884

	Guineas
Baroness Oxford 5th, by Duke of Wetherby 5th (31033), *Sir H. Hussey Vivian*	210
Lady Oxford 9th, by Duke of Havering (33664), *H. A. Brassey*	410

Mr. Henry A. Brassey, M.P.,'s

Lady Worcester 20th, by Duke of Geneva 6th (30959), *H. Leney*	105
Duke of Oxford 44th (39774), by Duke of Wetherby 5th (31033), *T. Holford*	105

The Earl of Lathom's Sale, Lathom House, 3rd September 1884

Princess of Blythe 4th, by Duke of Oxford 51st (43122), *Earl of Bective*	165
Duchess of Ormskirk 6th, by Grand Duke of Worcester 2nd (43323), *R. Briggs*	305
Duchess of Ormskirk 7th, by Grand Duke of Worcester 2nd (43323), *R. Briggs*	720
Princess of Blythe 5th, by Grand Duke of Worcester 2nd (43323), *R. Briggs*	105

Lord Bective's

Grand Duke 47th (51354), by Grand Duke 31st (38374), *T. Horsfall*	100

Sir Henry Allsopp's Sale, Hindlip, 18th May 1885

Airdrie Duchess 3rd, by Duke of Geneva 11th (41385), *T. Tempest Radford*	150
Water Lily 2nd, by Grand Duke 23rd (34063), *J. J. Hill*, U.S.A.	200
Duchess 117th, by Marquis of Oxford 2nd (37055), *T. Nelson and Sons*, Canada	430
Duchess 127th, by Knight of Worcester 3rd (46581), *Earl of Bective*	175
Thorndale Rose 13th, by Duke of Oneida 6th (30997), *A. H. Lloyd*	140
Darlington Belle 3rd, by Grand Duke of Airdrie (43310), *T. E. Walker*	100
Duchess of Hindlip 3rd, by Duke of Hillhurst 3rd (30975), *T. Nelson and Sons*	360

	Guineas
Duchess of Hindlip 4th, by Knight of Oxford 3rd (43441), *A. Leney*	260
Duchess of Hindlip 4th's c.c., by Duke of Somerset (49485), *Sir J. Swinburne*	120
Duchess 122nd, by Knight of Oxford 3rd (43441), *Sir H. H. Vivian*	130
Oxford 26th, by Grand Duke of Worcester 2nd (43323), *E. Ellis*	175
Kirklevington Queen 5th, by Duke of Connaught (33604), *J. J. Hill, U.S.A.*	150
Waterloo Belle 7th, by Duke of Glo'ster 7th (39735), *J. J. Hill, U.S.A.*	250
Cambridge Rose 10th, by Knight of Oxford (40082), *T. Nelson and Sons*	200
Duchess 123rd, by Grand Duke 41st (46439), *S. P. Foster*	480
Belle of Worcester 2nd, by Knight of Oxford (40082), *Sir H. H. Vivian*	110
Oxford 29th } twins, by Knight of Oxford (40082) *J. J. Hill, U.S.A.*	195
Oxford 30th }	160
Worcester Rose, by Duke of Glo'ster 7th (39735), *T. Nelson and Sons*	170
Duchess of Hindlip 7th, by Duke of Cornwall 4th (47726), *E. Ellis*	200
Belle of Worcester 3rd, by Duke of Glo'ster 7th (39735), *T. Nelson and Sons*	125
Duchess 124th, by Duke of Connaught (33604), *T. Nelson and Sons*	660
Duchess 125th, by Duke of Connaught (33604), *J. J. Hill, U.S.A.*	320
Worcester Rose 2nd, by Knight of Oxford (40082), *A. H. Lloyd*	110
Duke of Somerset (49485), by Grand Duke 30th (38373), *Sir W. G. Armstrong*	150
Duke of Cumberland (49439), by Grand Duke 41st (46439), *T. Nelson and Sons*	140
Lord Rosebery (51644), by Knight of Oxford (40082), *R. Blezard*	170
Duke of Brunswick (51109), by Duke of Cornwall 4th (47726), *Mrs. Bubb*	100
Duke of Saxony (51149), by Duke of Cornwall 4th (47726), *Hon. R. Baillie Hamilton*	120
Knight of Oxford 12th (51500), by Grand Duke of Worcester 2nd (43323), *Lord Middleton*	210

Sir Curtis M. Lampson's Sale, Rowfant, 1885

	Guineas
Rowfant Oxford, by Duke of Hillhurst 3rd (30975), *Herbert Leney*	200
Duchess of Rowfant, by Duke of Underley 2nd (36551), *J. J. Hill*, U.S.A.	500
Duchess of Rowfant c.c., by Rowfant Duke of Oxford 4th (47011), *J. J. Hill*, U.S.A.	200
Duchess of Leicester, by Viscount Oxford (40876), *J. J. Hill*, U.S.A.	550
Grand Duchess 39th, by Grand Duke 30th (38373), *Herbert Leney*	610
Winsome 29th, by Duke of Rosedale 6th (38176), *Sir H. H. Vivian*	105
Thorndale Rose 23rd, by Duke of Cornwall 2nd (43082), *J. J. Hill*, U.S.A.	320
Kirklevington Lady 3rd, by Duke of Oxford (39970), *J. J. Hill*, U.S.A.	190
Viscountess Oxford 5th, by Duke of Underley (33745), *P. L. Mills*	255
Rowfant Thorndale Rose, by Grand Duke 37th (43307), *J. J. Hill*, U.S.A.	210
Rowfant Grand Duchess, by Grand Duke 33rd (33946), *Earl of Bective*	520
Grand Duke 37th (43307), by Duke of Underley (33745), *A. Irvine Fortescue*	145
Rowfant Duke of Leicester (52013), by Grand Duke 37th (43307), *Earl of Bective*	270

Lord Braybrooke's Sale, Audley End, 26th June 1885

Thorndale Rose 6th, by Duke of Geneva 8th (28390), *B. Langdale Barrow*	125
Heydon Rose 11th, by Duke of Connaught (33604), *F. Barchard*	110
Thorndale Rose 22nd, by Duke of Connaught (33604), *P. L. Mills*	405
Thorndale Rose 24th, by Duke of Connaught (33604), *H. Y. Attrill*, U.S.A.	500
Thorndale Rose 25th, by Grand Duke 30th (38373), *Charles F. Leney*	185
Thorndale Rose 26th, by Grand Duke 30th (38373), *Earl of Lathom*	160

	Guineas
Thorndale Rose 28th, by Grand Duke 41st (46439), *Earl of Feversham*	195
Thorndale Rose 29th, by Grand Duke 41st (46439), *T. A. Titley*	300
Thorndale Rose 30th, by Duke Oneida (43151), *Earl of Feversham*	195
Duke of Rosedale 19th (47782), by Duke Oneida (43151), *P. L. Mills*	225
Duke of Rosedale 20th (51148), by Grand Duke 37th (43307), *J. Pulley, M.P.*	125

Mr. LLOYD'S

Thorndale Duke 3rd (53754), by Duke of Airdrie 27th (41351), *C. Hobbs*	100

Mr. EDWARD HALES's Sale, NORTH FRITH, 1st July 1885

Lady Rosedale Barrington, by Duke of Rosedale 6th (38176), *T. Nelson and Sons*, Canada	195
Lady Rosedale Barrington 2nd, by Duke of Rosedale 6th (38176), *T. Nelson and Sons*, Canada	230
Lady Rosedale Barrington 3rd, by Duke of Rosedale 6th (38176), *J. J. Hill*, U.S.A.	215
Lady Underley Barrington 2nd, by Duke of Underley 7th (46273), *T. Nelson and Sons*	100
Lady Huntsland Barrington, by Duke of Huntsland 4th (47769), *J. J. Hill*, U.S.A.	110

Mr. F. BARCHARD's Sale, HORSTED, 2nd July 1885

Lady Worcester 23rd, by Baron Turncroft Oxford 4th (37822), *H. Leney*	120
Lady Sussex (Wild Eyes), by Viscount Oxford 3rd (44208), *H. Leney*	140
Lady Sussex 2nd, by Viscount Oxford 3rd (44208), *J. Pulley*	105

EARL OF BECTIVE'S SALE, UNDERLEY, 3rd September 1885

Grand Duchess of Oxford 33rd, by Duke of Wetherby 5th (31033), *A. Leney*	105
Waterloo 46th, by Beau of Oxford 4th (36231), *Earl of Lathom*	100

	Guineas
Dowager 3rd (Darlington), by Duke of Connaught (33604), *P. L. Mills*	170
Waterloo 47th, by Beau of Oxford 3rd (33130), *A. Leney*	160
Kirklevington Lady 2nd, by Duke of Oxford (39770), *G. Fox*	110
Windermere 3rd, by Grand Duke 31st (38374), *C. Hills*, U.S.A.	150
Underley Oxford, by Grand Duke 31st (38374), *J. Harris*	230

Mr. D. A. GREEN's Sale, EAST DONYLAND, 29th April 1886

King of Oxford (53066), by Knight of Oxford 3rd (43441), *D. R. Scratton*	120

Mr. CHARLES F. LENEY's Sale, HILDEN, 21st May 1886

Cambridge Rose 8th, by Knight of Oxford 3rd (43441), *B. L. Barrow*	155

H.R.H. The PRINCE OF WALES' Sale, SANDRINGHAM, 15th July 1886

Lady Blanche Rose, by Duke of Norfolk (44699), *Lord Fitzhardinge*	230

Mrs. McINTOSH's, sold at LANGLEYBURY, 23rd September 1886

Lady Oxford 8th, by Duke of Havering (33664), *F. Torromé* for South America	160
Lady Oxford 10th, by Grand Duke 41st (46439), *Sir J. Whitworth*	125
Lady Oxford 12th, by Grand Duke 44th (46440), *Earl of Bective*	175

Mr. ROBERT BLEZARD's Sale, POOL PARK, 4th August 1886

Lady Surmise, by Duke of Hillhurst 2nd (39748), *W. Lynd*	105
Lady Blithfield 4th (Waterloo), by Wild Eyes Marquis (47278), *S. P. Foster*	100
Lady Surmise 7th, by Grand Duke of Barringtonia 7th (46450), *Lord Middleton*	100
Lady Blanche 22nd, by Grand Duke of Barringtonia 7th (46450), *Sir W. H. Salt*	105

IL LIBRO D'ORO

Mr. S. P. Foster's Sale, Killhow, 9th September 1886

	Guineas
Oxford Duchess of Killhow 3rd, by Duke of Ormskirk (36526), *T. A. Titley*	105
Duchess of Killhow, by Lord Rosebery (51644), *Duke of Devonshire*	505
Oxford Duchess of Killhow 7th, by Duke of Underley 3rd (38196), *C. T. Getting*	300
Oxford Duchess of Killhow 8th, by Duke of Ormskirk (36526), *J. Harris*	250
Oxford Duchess of Killhow 9th, by Duke of Leicester 6th (49461), *R. Blezard*	125

Earl of Lathom's

Duke of Ormskirk 5th (52768), by Knight of Oxford 8th (48818), *T. Horsfall*	120

Earl of Bective's

Grand Duke 48th (52961), by Grand Duke 31st (38374), *C. T. Getting*	200

Mr. J. Harris's

Oxford Duke of Calthwaite 1st (53367), by Grand Duke 31st (38374), *J. Coux*	105

The late Lord Penrhyn's Sale, Wicken Park, 12th May 1887

Archduchess of Oxford 2nd, by Beau of Oxford 3rd (33130), *R. Blezard*	100
Archduchess of Oxford 3rd, by Duke of Oxford 42nd (39772), *R. Blezard*	190

Mr. Herbert Leney's Sale, Barming, 3rd June 1887

Waterloo Victor (56728), by Duke of Leicester 4th (47774), *P. L. Mills*	225

Mr. H. A. Brassey's Sale, Preston Hall, 30th June 1887

Waterloo Bienvenue 2nd, by Duke of Connaught (33604), *P. L. Mills*	205
Waterloo Bienvenue 3rd, by Rowfant Duke of Glo'ster (48610), *P. L. Mills*	165

The EARL OF BECTIVE's Sale, UNDERLEY, 7th September 1887

	Guineas
Dentsdale 8th (Darlington), by Grand Duke 31st (38374), C. T. Getting, Buenos Ayres	100
Grand Duchess 59th, by Grand Duke of Geneva 3rd (49677), C. T. Getting, Buenos Ayres	500
Princess Sale 4th, by Duke of Oxford 69th (49475), J. W. Schwager, Chili	130
Grand Duchess 60th, by Duke of York 9th (51159), C. T. Getting	400
Turcoman 12th, by Grand Duke of Geneva 3rd (49677), T. Cockbain, South America	105

Mr. HARRIS's

Oxford Duke of Calthwaite 2nd, by Duke of Leicester 6th (49461), C. T. Getting	105

DUKE OF MANCHESTER's, sold at BIDDENHAM, 28th September 1887

Oxford de Vere 9th (54768), by Oxford Duke of Killhow 6th (50130), G. Fox	100

The late Mr. F. SARTORIS's Sale, RUSHDEN HALL, 3rd May 1888

Coral Waterloo, by Lord Oxford Charmer 3rd (48236), Lord Middleton	105

The late Mr. T. A. TITLEY's, sold at DRINGHOUSES, 17th May 1888

Thorndale Rose 29th, by Grand Duke 41st (46439), J. Harris	210

Mr. AUGUSTUS LENEY's Sale, ORPINES, 5th July 1888

Duchess of Hindlip 4th, by Knight of Oxford 3rd (43441), W. McCulloch, Australia	185
Orpines Grand Duchess of Oxford, by Rowfant Duke of Glo'ster 2nd (48610), P. L. Mills	225
Duchess of Orpines, by Duke of Oxford 65th (49472) W. McCulloch	260

	Guineas
Grand Duke of Kent 3rd (54394), by Duke of Rosedale 12th (46268), *S. Browning*	110
Waterloo Beau, by Rowfant Duke of Glo'ster 2nd (48610), *S. Browning*	125

Mr. EDWARD LENEY'S

	Guineas
Waterloo 59th, by North Pole (Gwynne) (46807), *F. Birnley*	120
Grand Duke of Connaught (Waterloo) (57412), by Lord of the Manor (Waterloo) (56078), *Jonas Webb*	135

Mr. T. HOLFORD'S, sold at WESTON PARK, 17th July 1888

	Guineas
Duke of Vittoria 10th (55636), by Duke of Barrington 15th (52745), *J. H. Yeomans*	160

Sir JOHN SWINBURNE'S, sold at DRINGHOUSES, 6th Sept. 1888

	Guineas
Prince Waterloo 5th (56356), by Grand Earl of Waterloo 3rd (52967), *J. Pulley*	120

The DUKE OF MANCHESTER'S

	Guineas
Oxford de Vere 10th (56260), by Oxford Duke of Killhow 6th (50130), *H. F. Smith*, New South Wales	105

The EARL OF FEVERSHAM'S Sale, DUNCOMBE PARK, 7th September 1888

	Guineas
Oxford Helen, by Duke of York 9th (51159), *Lord Middleton*	210
Duchess of York 5th, by Knight of Oxford 12th (51500), *Prince of Wales*	300
Duchess of York 6th, by Earl of Oxford 3rd (51186), *W. McCulloch*, Australia	290
Duke of York 9th (51159), by Duke of Connaught (33604), *C. T. Getting and Sons*, South America	1000
Duke of York 10th (54231), by Grand Duke 46th (49671), *Evan Baillie*	210
Earl of Oxford 3rd (51186), by Baron Oxford 9th (42738), *H. Sharpley*	120

The EARL OF BECTIVE'S

	Guineas
Grand Duke 51st (57409), by Rowfant Duke of Leicester (43112), *Duke of Manchester*	190

Mr. H. J. SHELDON's Sale, CROUCH FARM, 19th Sept. 1888

	Guineas
Duchess of Barrington 11th, by Duke of Barrington 10th (43067), *Herbert Leney*	190
Duke of Charmingland 54th (55608), by Duke of Barrington 10th (43067), *H. Hamkens*, Germany	215
Duke of Barrington 24th (57173), by Duke of Barrington 10th (43067), *J. Livesey*	130

Mr. S. P. FOSTER's, sold at CALTHWAITE, 28th September 1888

Duchess of Killhow 2nd, by Duke of Leicester 6th (49461), *Earl of Bective*	405

The late Sir ROBERT LODER's SALE, WHITTLEBURY, 15th May 1889

Duchess of Buckingham, by Duke of Underley 3rd (38196), *P. L. Mills*	295
Duchess of Whittlebury 6th, by Grand Duke 41st (46439), *H. Leney*	150
Duchess of Whittlebury 9th, by Grand Duke 41st (46439), *Mons. L. Grollier*, France	165
Duchess of Whittlebury 10th, by Duke of Whittlebury (47788), *Prince of Wales*	160
Duchess of Whittlebury 11th, by Duke of Cornwall 4th (47726), *Madame L. Grollier*	270
Duchess of Whittlebury 13th, by Grand Duke 25th (34065), *A. H. Lloyd*	510
Duchess of Whittlebury 14th, by Duke of Whittlebury 4th (54228), *J. Harris*	175
Duchess of Whittlebury 15th, by Duke of Whittlebury (47788), *Sir W. H. Salt*	390
Duchess of Whittlebury 17th, by Grand Duke of Oxford 3rd (39953), *Prince of Wales*	305
Duchess of Whittlebury 18th, by Duke of Rosedale 12th (46268), *Earl of Bective*	375
Duchess of Whittlebury 19th, by Duke of Rosedale 12th (46268), *J. Harris*	330
Duchess of Whittlebury 20th, by Duke of Rosedale 12th (46268), *Earl of Feversham*	140
Duke of Rosedale 12th (46268), by Duke of Connaught (33604), *F. S. Stanley*	130

Mr. A. H. Lloyd's Sale, Harewoods, 13th June 1889

	Guineas
Countess of Waterloo 4th, by Duke of Airdrie 27th (41351), *J. Livesey*	100
Countess of Waterloo 5th, by Grand Duke 41st (46439), *Earl of Bective*	105
Duchess of Surrey, by Duke of Connaught (33604), *W. Farnell Watson*	425
Duchess of Surrey 4th, by Duke of Gloucester 7th (39735), *A. Deacon*	210
Queen of Oxford 7th, by Duke of Surrey 2nd (52783), *Earl of Bective*	105
Duke of Surrey 2nd (52783), by Duke of Airdrie 27th (41351), *G. M. Alexander*	200

H.R.H. The Prince of Wales's Sale, Sandringham, 2nd July 1889

Barrington Prince (53989), by Baron Oxford 18th (50830), *D. Maclennan*, South America	100
Rowfant Prince (56645), by Baron Oxford 18th (50830), *D. Maclennan*, South America	130

Sir Nigel Kingscote's Sale, Kingscote, 5th July 1889

Sir Maurice (56570), by Prince Airdrie (48472), *Earl of Bective*	170

Hon. R. Baillie Hamilton's Sale, Melrose, 31st July 1889

Oxford Grand Duchess 2nd, by Duke of Saxony (51149), *J. Harris*	155
Oxford Grand Duchess 3rd, by Duke of Saxony (51149), *J. Harris*	110
Duke of Waterloo 11th (57231), by Duke of Somerset (49485), *A. H. Lloyd*	105

The Duke of Devonshire's Sale, Holker, 5th Sept. 1889

Baroness Oxford 19th, by Grand Duke 46th (49671), *T. Horsfall*	100
Grand Duchess of Oxford 79th, by Grand Duke 46th (49671), *E. Ecroyd*	110

	Guineas.
Grand Duchess of Oxford 80th, by Grand Duke 46th (49671), *J. Lindow*	110
Grand Duchess of Oxford 82nd, by Grand Duke 46th (49671), *C. Magniac*	100
Grand Duchess of Oxford 84th, by Duke of Oxford 68th (49474), *Prince of Wales*	110
Baroness Oxford 23rd, by Grand Duke 46th (49671), *P. L. Mills*	140
Grand Duchess of Holker } twins, by Grand Duke 46th (49671) *Prince of Wales*	160
Grand Duchess of Holker 2nd	260
Grand Duchess of Oxford 89th, by Grand Duke 46th (49671), *Captain Duncombe*	115
Grand Duchess of Oxford 90th, by Grand Duke 46th (49671), *T. Horsfall*	105
Baroness Oxford 24th, by Grand Duke 46th (49671), *C. R. Lynn*	185
Baroness Oxford 27th, by Duke of Barrington 13th (46191), *P. L. Mills*	205
Duchess of Holker 3rd, by Grand Duke of Holker (52963), *C. T. Getting and Sons*	285
Grand Duchess of Oxford 102nd, by Knight of Rosedale (54517), *T. Horsfall*	105
Grand Duchess of Oxford 107th, by Duke of Barrington 13th (46191), *Earl of Bective*	100
Duke of Chatsworth 3rd (57185), by Grand Duke of Holker (52963), *E. Ecroyd*	170
Duke of Oxford 90th (57212), by Grand Duke of Holker (52963), *Captain Duncombe*	130
Baron Winsome 22nd (56921), by Lord Rosebery (51644), *Lord Armstrong*	200

The Earl of Bective's

Grand Duke 52nd (57410), by Waterloo Victor (56728), *T. Cockbain*, Chili	175

Mr. Herbert Leney's Sale, Barming, 20th June 1890

Barming Barrington 2nd, by Rowfant Grand Duke (52014), *Earl of Bective*	205
Barming Cherry Duchess, by Rowfant Grand Duke (52014), *Madame de Clercq*	105

	Guineas
Duchess of Whittlebury 22nd, by Duke of Rosedale 12th (46268), *J. Livesey*	160
Barming Foggathorpe, by Lord of the Manor (Waterloo) (56078), *Earl of Bective*	110
Barming Waterloo Emperor (58411), by Rowfant Grand Duke (52014), *Rose Innes, Cox and Co.*, Chili	140
Barming Barrington Duke (60290), by Duke of Charmingland 51st (55607), *P. L. Mills*	175

Mr. T. Holford's, sold at Tortworth, 26th June 1890

Duke of Leicester 11th (57209), by Duke of Barrington 15th (52745), *C. P. Lancaster*	110
Duke of Vittoria 13th (58834), by Duke of Barrington 15th (52745), *T. Hawkins Smith*, Australia	160

The late Duke of Manchester's Sale, Kimbolton, 3rd July 1890

Oxford Helene, by Duke of Underley 3rd (38196), *Prince of Wales*	160
Oxford Kate, by Grand Duke 51st (57409), *Earl of Bective*	160
Grand Duke 51st (57409), by Rowfant Duke of Leicester (52013), *J. Linton*	115

Mr. T. Holford's

Duke of Leicester 13th (58819), by Duke of Vittoria 10th (55636), *F. H. Jennings*	140

H.R.H. The Prince of Wales's Sale, Sandringham, 3rd July 1891

Georgie Sandgrove, by Lord Sandgrove 15th (56092), *Lord Fitzhardinge*	100
Duchess of Lancaster, by Duke of Leicester 9th (58818), *Baron F. de Rothschild*	200

Sir H. Hussey Vivian's Sale, Singleton, 25th August 1891

Lady Worcester 26th, by Waterloo de Breos (53819), *P. L. Mills*	145
Marchioness of Oxford 2nd, by Grand Duke of Geneva 3rd (49677), *Madame de Clercq*	130

Mr. J. H. Caswell's Sale, Laughton, 21st April 1892

Guineas

Duchess of Laughton 23rd (Kirklevington), by Duke of
Stroxton 2nd (55635), *Prince of Wales* . . 100

Mr. A. H. Lloyd's Sale, Harewoods, 10th June 1892

Duchess of Surrey 3rd, by Duke of Glo'ster 7th (39735),
P. L. Mills 195
Duchess of Surrey 5th, by Waterloo Count 7th (56723),
A. P. Clear 100
Duchess of Surrey 6th, by Duke of Waterloo 11th
(57231), *J. Harris* . . . 150

Mr. T. Holford's, sold at Warwick, 22nd June 1892

Duchess of Leicester 19th, by Prince of Airdrie 4th
(57916), *Earl of Bective* 185
Duchess of Leicester 23rd, by Duke of Vittoria 13th
(58834), *Lord Middleton* . . . 150
Duke of Leicester 18th (63908), by Duke of Vittoria 13th
(58834), *Earl of Bective* . . . 200

The Earl of Bective's Sale, Underley, 14th Sept. 1892

Grand Duchess 58th, by Baronet (47396), *H. Leney* . 140

Mr. F. Barchard's Sale, Horsted, 9th June 1893

Horsted Rose 4th, by Duke of Whittlebury 4th (54228),
E. Potter 155
Marchioness of Kirklevington 7th, by Rowfant Grand
Duke (52014), *E. Potter* . . . 155

Mr. Philo L. Mills' Sale, Ruddington, 16th June 1893

Dowager of Barrington, by Barming Barrington Duke
(60290), *F. Platt* 115
Viscountess Oxford of Ruddington 4th, by Waterloo
Victor (56728), *Lord Middleton* . . 110
Cambridge Premium Duke 2nd (63751), by Bienvenue
Duke of Rosedale (58515), *Prince of Wales* . 100

H.R.H. The PRINCE OF WALES's Sale, SANDRINGHAM,
13th July 1893

Guineas

Grand Duchess of Babingley, by Duke of Oxford 88th
(55624), *T. Holford* 110

Mr. W. J. EDMONDS's Sale, SOUTHROP, 2nd May 1894

Earl of Southrop 116th (Blanche) (65431), by Barrington
Surmise 2nd (60343), *C. Pinnell*, South America . 205

The late EARL OF BECTIVE's Sale, UNDERLEY, 12th
July 1894

Duchess 127th, by Knight of Worcester 3rd (46581), *J.
Harris* 100
Duchess of Holker 3rd, by Grand Duke of Holker (52963),
J. Harris 165
Duchess of Leicester 19th, by Prince of Airdrie 4th
(57916), *J. D. Fletcher* 160
Duchess 131st, by Rowfant Duke of Leicester (52103),
Earl of Feversham 155
Duchess of Whittlebury 22nd, by Yosemite 6th (56787),
J. D. Fletcher 100
Duchess of Holker 4th, by Duke of Underley 8th (57229),
J. Harris 105

GENERAL INDEX

ABYSSINIAN cattle at Halton and Scaleby, 140
Acklam, 150, 201, 229, 236
Acton Burnell, 273
Afghanistan, H.H. Nasrulla Khan, Shahzada of, 377
Agricultural Societies, *The English*, show at Oxford, 1839, 266; *The Royal, of England*, shows at Cambridge, 1840, 272-4; at Liverpool, 1841, 276; at Bristol, 1842, 279, 301; at Southampton, 1844, 284; at Newcastle, 1846, 281, 282; at Northampton, 1847, 304; at York, 1848, 308-15, 377; at Newcastle, 1887, 372; at Darlington, 1895, 377; *The Highland*, show at Berwick, 1841, 278; at Dumfries, 1845, 296; at Edinburgh, 1848, 315; *The Durham*, shows at Durham, 1843, and Stockton, 1844, 285; *The Northumberland*, show at Alnwick, 1840, 274; *The Tyneside*, show at Hexham, 1837, 256; *The Yorkshire*, show at York, 1838, 260-4; at Northallerton, 1839, 269; at Hull, 1841, 278; at York, 1842, 279; at Doncaster, 1843, 284; at Richmond, 1844, 285; at Beverley, 1845, 296 n.; at Wakefield, 1846, 281, 282; at Scarborough, 1847, 304; at Wetherby, 1868, 357
Agriculture, the Board of, Bates's Address to, 71; the receipt of it unacknowledged, 84, 86; applies to Bates for the result of his experiments, 138; awards premium to Sir Leoline, 177

Aislabies, Herd of the, at Studley, 29, 36
Albert, H.R.H. Prince, at the York Show, 1848, 310, 313
Alderneys, classed with Dutch cattle, 28; Sir W. St. Quintin crosses his shorthorns with, 32; Ayrshire cattle crossed with, 96
Alexander, R. A., Kentucky, 342; W. J., 370
Allen, scenery of the, 174, 183
Allen, A. B., U.S.A., 306, 307; L. F., 307
Allen, George, his sale at Knightley, 1878, 443
Alloy, Galloway, 55, 288, 383
Allsopp, Mr., 366; Sir Henry, 369; his sale, 1835, 370, 371, 464
Alnwick, Northumberland show at, 1840, 274
Althorp, Lord, 149, 150; letters from, 151, 153, 155, 161, 162-8, 188, 203; his opinion of Bates, 164; his cattle, 165, 178, 179, 190, 191, 202, 213, 214; his appearance, 240, 241: *see* Spencer, 3rd Earl
Ambler, Mr., 311
America, Bates intends emigrating to, 215; the idea abandoned, 217
American Herd Book, drawing of the Dun Cow in the, 45 n.
Amherst, Lady, buys northern moiety of Kirklevington, 183
Angerstein, Wm., his sale at Weeting Hill, 1876, 436
Angus, Mr., 218
Anick Grange, 79
Anna, Old, 403 n.

Antrim, the Countess of, her sale at Wynyard, 1818, 225, 231, 234, 235, 236, 317, 318
Appleby, Thos., 43, 117
Applecross: *see* Mackenzie of
Appleton, Mrs., 150; of Crathorne, 405
Archbold, Mr., 249, 404
Archdeacon Newton, 317
Ardnave, Islay, 124 n.
Argyle, the Duke of, his Highland heifers, 70; his steers, 124
Armstrong, Wm., 173
Arrowsmith of Ferryhill, his twin steers, 20, 60; his losses through St. John's blood, 133, 287
Ashburner, Wm., his sales at Conishead, 1875, 431; 1877, 437; 1879, 451
Ashton Hall, Mrs. Starkie's sale at, 1883, 461
Atherton, Mr., 346
Athol, the Duchess of, 35, 388
Atkinson, George, of Staingills, 85; the Misses, 82; Matthew, Bates's letter to, 88
Attrill, Mr., his herd in Canada, 369
Audley End, Lord Braybrooke's sales at, 1879, 368, 449; 1885, 371, 466
Australia, 238 n., 275 n., 359 n., 364, 372, 376
Aydon, John Cook's house at, 5; Castle, 1, 6, 61-3, 98, 109; White House, 5, 6; Bates tenant of, 13
Aylesbury, Duke of Northumberland at, 266
Ayrshire, Bates visits, 68, 93; cattle there improved by Teeswater and Alderney crosses, 96

BABRAHAM, Mr. Webb's sale at, 1863, 345
Backhouse, John, 400
Badminton, Mr. Butler's sale at, 1870, 411
Bailey, John, his *Survey of Durham*, 110; Bates's letters to, 111, 113, 115
Baillie-Hamilton, Hon. R., his sale at Melrose, 1889, 373, 473
Baker, George, of Elemore, his Devons and French bull, 110 n.; his shorthorns, 156, 157; libels Sir Leoline, 177; Mr. Wood's visit to, 235
Bakewell, Robert, Charge's visit to, 20; recommends West Highland foundation, 20; imports horses from Holland, 30; Colling's visit to, 39; his management of cattle, 39, 95, 135
Bamlet, John, 35, 386
Banks, Mr., of Hurworth, 36
Barber, Thos., his sale at Sproatley Rise, 1876, 432
Barchard, Mr., of Horsted, his sales, 1885, 371, 467; 1893, 375, 476
Barclay-Allardice, Captain, 404; of Ury, 213
Barforth, 37 n.
Barker, Richard, of Oxenfield, 51; Wm., of Kipling, 52
Barming, Mr. H. Leney's sales at, 1887, 469; 1890, 474
Barmpton, R. Colling takes, 22, 64; sales at, 1782, 22; 1818, 149, 155, 169, 245
Barnard Castle, ox killed at, 33
Barningham, 32, 37 n.
Barrett, Mr., of Stratton Park, 400 n.
Barrington, Hon. Shute, bishop of Durham, 170, 172, 212; Lady, 171
Basnett, purchaser of Hubback, 38
Bassett, Mr. C. H. 376
Bates, Catherine (Mrs. Donkin), 19, 107; Diana, 2, 3, 67, 83, 98; Edward (Oberamtmann), 329; Elizabeth (Mrs. Culley), 19; George, at Aydon Castle, 2; his marriage, 3; at Buttermere, 4, 6; buys Wark Eals, 18, 109; John, buys Dobinson cattle, 6, 31; John Moore, 61, 62, 100, 213; Thomas, of Halton, Brunton, and Coupland Castle, 21; Thomas, of Heddon, 325; Thomas, of Kirklevington, *passim*; Thomas, of Prudhoe, 4; Wm., of Clarewood, 32, 242; Wm., of Easby, 104
Bath, Agricultural meeting at, 1807, 70, 71
Baxter of Acklam, 156, 169
Bearl, Mr. Wm. Charlton's sale at, 202
Beauford, H. W., 299; letter from, 300, 400 n.

Beaumont Grange, Mr. W. W. Slye's sales at, 1870, 411; 1871, 414; 1872, 416; 1874, 421; 1876, 434
Becar and Morris, Messrs., U.S.A., 339, 343, 344, 365
Bective, Earl of, 359 n., 361, 365, 366; his sales, 1874, 423; 1880, 368, 453; 1885, 371, 467; 1887, 372, 470; 1892, 374, 476; 1894, 376, 477
Bedford, the Duke of, 77, 79
Beeston, Mr. Pawlett's sale at, 361, 362
Beever, Rev. W. Holt, 28 n., 29 n., 336 n., 351
Bell, family, at Halton, 21, 35, 387; at Kirklevington, 250, 275 n., 308; Mr. Holt Beever's error respecting them, 336 n.; Robert, steward at Halton, 98, 109, 110, 113, 145, 159; Robert jun., 406; Thomas, 219, 242, 244, 250, 258, 265, 266, 279, 285 n., 302 n., 406; his love of Halton, 147 n.; his sale at Brockton, 415
Bell, Matthew, M.P., 206 n.
Bell and Morris, Messrs., U.S.A., 305
Berkeley, Lord Fitzhardinge's sales at, 1876, 432; 1879, 447; 1882, 457
Berry, Rev. Henry, 197; letter from, 198, 200; his 'production,' 201, 289; his indifference to early history, 383 n.; Bates's copy of his *Improved Short-horns*, 11
Berwick-on-Tweed, the 'Hen and Chickens,' at, 135; Highland Show at, 1841, 278
Best, George, of Manfield, 52, 144
Betts, Edward Ladd, 347-9; his sale at Preston Hall, 408
Bewick, Thos., his woodcuts of Dutch cattle, 30, 31; of Improved Dutch cattle, 56; his opinions on in-breeding, 57
Bible Society, the, 171, 207
Bird's Nest, W. H. Salt's sales at, 1876, 433; 1881, 455
Black, original colour of Yorkshire cattle, 24; restricted to Welsh and Scotch runts, 26, 96
Black cattle, generic name, 26
Black Hedley, 171, 284
Blackmail, original meaning of, 26

Black-nosed calf, Mr. Rhodes's, 177-181
Blackett, Alderman, 64; Sir Edward, of Newby, 29; his shorthorns, 33
Blackwood's Magazine, review of *Fleurs* in, 209
Blakiston, 240
Blayney, family, 7; name, 8 n.; peerage, 7 n.; Arthur, 7; Bates's godfather, 8; his death, 15; his eccentricities, 16; his portrait, 16; his character, 17, 326; Diana, 83; Mary, her portrait, 8
Bletsoe, 299, 401, 403
Blezard, Mr., his sales, 1886, 468; 1893, 375
Block test, 12
Blundell, Mr. J. H., his sale at Woodside, 1878, 442
Board of Agriculture: see Agriculture, Board of
Bolam, the vicar of, 208
Bolden, Mr., Australia, 401; Bates's letter to, 238 n.; S. E., 337, 343; his sale, 1860, 345; sells his herd to Mr. Atherton, 346
Boldon, rectory of, 212
Bolton Abbey, 191
Bonaparte, a puppet, 62
Bone-mill at Ridley Hall, 207
Booth, Mr., 199; John, 269, 278, 281, 289, 316, 331, 346; Thos., 50; Thos. jun., 403
Bourne, Rt. Hon. Sturges, 187
Bower, Major, 233, 234; his sale at Welham, 251
Bowly, Mr. E., 346; his sales, 1869, 410; 1872, 360, 415; 1875, 426; 1877, 439; 1881, 369
Bowmont, the Marquis of, 208, 209
Brafferton, 46
Brailes, 366 n.; Mr. H. J. Sheldon's sales at, 1869, 411; 1871, 415; 1873, 419; 1875, 432; 1879, 452
Brampton, Mr. Hetherington's sale at, 1880, 454
Brassey, Mr., his sale at Preston Hall, 1887, 372, 469
Brawith, sale at, 315
Braybrooke, Lord, 345, 353; his sales at Audley End, 1879, 368, 449; 1885, 371, 466

Brayton, Sir Wilfrid Lawson's sale at, 1876, 435
Breckenbrough, 324, 336 n.
Briggs, Rawdon jun., 169, 170, 182, 183, 194, 198, 201; letters from, 191, 199
Bristol, the R.A.S.E. at, 1842, 279, 301
Broadpool Common, 60, 80
Brogden, Mr. A., his sales, 1873, 418; 1875, 431; 1878, 446
Brough Hill Fair, 125
Brougham, Lord, 290
Brown, Mr., of Tallentire, 92; John, of Nunstainton, 219, 220
Browne, Col., of Mellington, 61
Bruce, Mr., offers 600 gs. for Young Star, 151; Dr. J. Collingwood, 206 n., 207
Buller, Sir Anthony, 298, 403: Bates's letter to, 300
Bulls, importance of constitutional vigour in, 382
Bunsen, Ritter von, 329
Burdon, Great, 47
Burghley, sale at, 1867, 349; 1872, 415
Burlington, Earl of: *see* Devonshire, Duke of
Burnett, Nicholas, of Black Hedley, 171, 206, 284, 403, 405
Bushbury, Mr. Lovatt's sale at, 1879, 448
Butler, Mr. Wm., his sale at Badminton, 1870, 412
Butter, from Duchess-by-Daisy-Bull, 70; at Halton, 111-3, 132; at Workington, 118; at Kirklevington, 284
Buttermere, George Bates at, 4
Butterworth, Jos., M.P., 186

CADZOW, wild cattle at, 96
Cambridge, Royal Show at, 1840, 272-4, 401, 402; Parker's Piece, 273; 'University Arms,' 273
Campbell, Duncan, of Ardnave, Bates's letters to, 124-9; Samuel, his sale at New York Mills, 365, 420
Camperdown, Countess of, her sale, 1888, 372
Cansiron, Mr. J. W. Larking's sale at, 1878, 440
Capital for farming should be five times the rent, 103

Carlisle, floods at, 1809, 92
Carr, Mr. Wm., of Stackhouse, 226, 379 n.
Carrington, Mr., 406
Castle Hill, Mr. Holford's sales at, 1883, 369, 459; 1894, 326
Castle Howard, Mr. Edwards's sale at, 265
Casswell, Mr. J. H., his sale at Laughton, 1892, 476
Cator, Rev. T., 316, 403
Chaloner, Col., 201
Chamberlain, Mr. Henry of Desford, 312
Chapman, Mr., banker, Newcastle, 206; John, 245; his account of shorthorn pedigrees, 388
Charge, John, his steers, 20; his visit to Bakewell, 20; his sale at Newton Morrell, 229; Robert, of Low Fields, 387; T., 267
Charlton, John, cow-doctor, 133; Mr. Wm., of Bearl, 202; of Sutton, 59 n., 120 n., 139, 162 n., 195, 280 n.
Chase, Philander, bishop of Ohio, 215
Chauvelin, Marquis de, 31 n., 35 n., 376
Cheese, sent to Mr. Curwen, 84; to the Mayor of Newcastle and Lord Collingwood, 84
Cheney, Mr. E. H., of Gaddesby; his importations from America, 364: his sales, 1871, 413; 1873, 419; 1874, 424; 1877, 438; 1879, 368, 452
Chesham, Lord, his sale at Rayans Farm, 1876, 434
Chesterholme, 207
Chichester and Yarborough, Lords, Bates's letter to, 313
Chillicothe, 250; sale at, 254
Chillies, Mr. Samuda's sale at, 1878, 441
Chillingham, wild cattle at, 29 n., 96; New Town, Wm. Jobson's sale at, 298
Chilton, Sharter's cattle at, 33; Mason's sale at, 213, 219
Chollerton, Wm. Bates's herd at, 32, 242
Clarewood, Wm. Bates's herd at, 32

Claridge, Mr., of Jervaux, 200
Clarke, Mr., of Skipton Bridge, 248; Mr. Eaton, 273
Clercq, M. de, 374
Cleveland, state of, 104; rotations in, 105
Cleveland Bays kept by Bates, 259
Cloeden-on-the-Elbe, 324
Coates, George, his origin, 160; editor of the *Herd Book*, 160, 169, 184, 185, 190-3, 198, 228, 303; his criticisms, 50, 51, 54, 168 n., 192; his carelessness, 35 n., 68
Coatham Mandeville, 53
Cobham, Mr. Harvey Combe's sale at, 1859, 344, 407
Cochrane, M. H. of Hillhurst, Canada, 357, 358, 366; his sale at Millbeckstock, 1877, 437
Coke of Holkham, 175
Coleman, Mr. E. J., his sale at Stoke Park, 1876, 433
Collin, Mr., U.S.A., 269
Colling, Charles, senior, 38, 194; Charles, 21, 35, 38; at Dishley, 39, 42; buys Duchess, 43, 283; at Eryholme, 46; owns the four best cows, 49; his opinions of Foljambe, 52; his error in using Lame Bull, 52; did not use Bolingbroke, 53; in-bred accidentally, 54; introduced alloy accidentally, 55; declined showing, 55; his first 100-guinea heifer, 58; his handling, 65, 378; wishes to retain Duchess-by-Daisy-Bull, 67; his breach of faith as to Duke and his apology, 67; his kyloe cow, 116; his letters to Bates, 117, 161; at Croft, 148, 160; his opinion of Acklam Red Rose, 150, 168, 179, 190, 193; John, of White House, Greta Bridge, 252; Mary, 42, 54; retains Duchess heifer, 64; her shorthorn-kyloe heifer, 112, 114, 117, 154; Robert, 21; takes Barmpton, 22, 35, 38, 42, 43, 63; refuses Mr. Colling the use of Ben, 54; his 53-qrt. cow, 65; refuses Bates the use of White Bull, 66; his high sense of honour, 67; his fat heifer, 86; his kyloe crosses, 111, 113, 115; his triplets, 147; his opinion of Acklam Red Rose, 150; his sale, 1818, 149, 155; his death, 161; his conduct to Lady Antrim, 235; his hind, John Chapman, 245, 388
Collingwood, Lord, his aunts, 3; Bates dines with, 62; and sends a cheese to, 85
Colman, Henry, U.S.A., 282
Colpitts, Mary: *see* Colling, Mary
Combe, Mr. Harvey, 330, 338; his sale at Cobham, 1859, 344, 407
Compton, Mr., of New Learmonth, 150, 158
Conishead, sales at, 1875, 431; 1877, 437; 1879, 451
Cook, John, of Aydon, 5
Cope, Mr., U.S.A., 400 n.
Coventry, Dr., his lectures on agriculture, 94-6, 131
Crabb, Mr. R. H., his sale, 1886, 371
Craven longhorns, 96
Croft of Barforth, family of, 37 n.
Crouch Farm, Mr. H. J. Sheldon's sale at, 1888, 472
Culley family, 5; their shorthorns, 19, 41; their Leicester sheep, 74; George, 13, 19, 39, 41, 110; his sale, 138; on handling, 379; Matthew, 19
Culshaw, Joe, 311, 355, 357
Cunningham, Col., 71
Curry of Brandon, 181
Curwen, J. C., his son, 71; his experiments at Workington, 76, 84; Bates's opinion of, 87; and letters to, 85, 90, 91, 101, 119, 122; letters from, 118, 120, 121, 123, 124; his *Report*, 121; his shorthorns, 169
Cusworth, 176
Cuthbert, St., 25, 45
Cuthbertson, Mr., his farm at Linton Mains, 137

Darlington, great markets at, 19, 20, 245, 246; Henry, 2nd Earl of, 386
Davidson, John, buys Ridley Hall, 206
Davies, Mr., animal painter, 263, 282; D. R., of Mere Old Hall, 359; his sale, 1870, 411; R. Pavin, of Horton, Gloucestershire, 365,

Dawpool, Mr. Hegan's herd at, 346
Day and Martin, 175
Demidov, Prince, 79
Denton, Culley's shorthorns at, 41
Devonshire, Duke of, 341, 348, 350; his sales: *see* Holker
Didmarton, Mr. Rich's sales at, 1868, 410
Dishley, 39
Dixon, General, 39; Henry Hall: *see* 'Druid, the'
Dobinson, of Great Whittington, 98; Michael, 6, 31, 242
Dobling, George, 319, 394, 395
Doncaster, agricultural meetings at, 165, 191; Mr. Rhodes's purchases at, 175, 177, 180; Yorkshire Show at, 1843, 284
Donkin, Wm., of Sandhoe, 98, 107-109, 147; his letter to Maynard, 158; his hallucinations, 159; his sale, 201, 203
Drewry, Mr. George, 267, 298, 300, 301, 347, 350, 358, 359, 361, 366
Dringhouses, sales at, Mr. R. Lodge's 1882, 458; Mr. T. A. Titley's, 1888, 470
'Druid, The' (H. H. Dixon), at Springfield, 337; at Duncombe Park, 341; at Wetherby, 1859, 342; 1868, 357; at Holker, 358
Dryden, Jonathan, 245
'Duchess Spot, the,' 148
Ducie, Earl, 330-2; his out-crosses and line-breeding, 337, 338; his sale at Tortworth, 1853, 339, 343
Dudding, Mr., of Panton, 285
Duff, J. C. Grant, 400, 403, 405; letter from, 323
Dugdale, Mr. A., his sale at Rose Hill, 1873, 418
Dun Cow at Durham, the, 25; the new, 45
Duncombe, Hon. Octavius, 405
Duncombe Park, 298, 341, 403; sales at, 1878, 444; 1888, 471
Dundas, Lord, 264
Dunmore, Earl of, his controversy with Rev. J. Storer, 361-3; his importations from America, 364; his sales at Dunmore, 1872, 360, 416; 1875, 360, 428; 1879, 368, 449

Durham, Bailey's Survey of, 110; the Dun Cow at, 25; the new, 45; bishops of: *see* Barrington and Van Mildert
Dutch, or Flemish, cows, white, 27; classed with Alderneys, 28; ancient and modern races of, 31, 96

EALS, WARK, 18, 172
Easby in Cleveland, 104
Eastwood, Mr., 281
Ecroyd, Mr., his sale at Armathwaite, 1893, 375
Edinburgh, Bates's motive in going to, 93; course of studies at, 94; their excellence, 97, 111; his lodgings at, 99; his return to, 109; a fore-chine sent to, 128
Edmonds, Mr., his sale, 1894, 375, 477
Edward IV. permits export of cattle to Holland, 29 n.
Edwardes, Rev. Sir Thos., 8, 16
Edwards, Mr. Henry, his sale at Castle Howard, 265, 400
Elbe, English settlers on the, 295
Elemore, 156, 157, 235, 319
Ellfoot, 217
Ellis, Mr., 179; Wm., 27
Ellison of Sizergh, his protest to the Council of the R.A.S.E., 311 n.
Elly Hill, 42
Elmhurst, Mr. G. Fox's sales at, 1877, 436; 1878, 443; 1881, 456
Eryholme, Mr. Maynard's cattle at, 47; Colling's visit to, 46
Etches, John Clifford, 262, 272, 273, 275 n., 314, 315, 401-3
Evans, Messrs., their sales, 1888, 372; at Sherlowe, 1894, 375
Exeter, Marquis of, his prize bull-calf at Oxford, 267, 331; his sale at Burghley, 1867, 349

FAIRFIELD, of Coatham, 53
Farnley, 245, 249
Fawcett, Rowland, of Scaleby, 140, 205, 277; James, 139; his reminiscences of Halton, 146; at Pitcorthie, 151, 195 n.; his sales, 1875, 429; 1878, 439; Mrs. and Major, their sale, 1894, 376; Wm., of Haughton Hill, 38; Mr., of Childwick, 339

GENERAL INDEX

Fawkes, Mr., of Farnley, 245, 249, 254
Featherstonhaugh, Timothy, 406
Fenwick, John, 207
Ferguson, Hon. Adam, of Nelson Gore, Canada, 307
Ferrybridge, sales at, 175, 191
Feversham, Lord, 298, 301, 323, 338, 341, 346, 380, 403; Earl of, 369, 376; his sales, 1878, 444; 1888, 372, 471
Fitzhardinge, Lord, 369, 370; his sales, 1876, 432; 1879, 447; 1882, 457
Fitzwilliam, Earl, 196
Fletcher, of Blakiston, 240; Mr. J. D., 376
Fleurs, a poem, 208; Christopher North's review of, 209-11
Floors Castle, 208, 211
Foreman, Mr., of Acton Burnell, 273, 401
Fores, Messrs., 401; Bates's letter to, 271
Forrest, Mr., 284
Foster, Mr., of Carlisle, 259; of Springfield, 262; J. P., his sales at Killhow, 1868, 410; 1871, 414; 1876, 435; S. P., his sales, 1881, 456; 1886, 372, 469
Fowler, Mr., of Aylesbury, 266
Fox, Mr. George, his sales at Elmhurst, 1877, 436; 1878, 443; 1881, 456
Foxton, 35

GADDESBY, Mr. Cheney's sales at, 1871, 413; 1873, 419; 1874, 424; 1877, 438; 1879, 368, 452
Garbutt, Mr., 199
Garfit, Mr. A., 366
Garrow, models by, 95
Germany, shorthorns in, 329 n.
Gibbs, Mr. Brandreth, 309, 343
Gibson, Mr., of Stagshow, 66, 67
Giles, Barbara, 98, 132
Gilsland, 83, 208
Givendale, 48
Glass, Mr., of Worton, Devizes, 303, 403
Gledhouse, near Leeds, 249
Goliath, the Romish, slain with his own weapon, 171
Graham, Sir James, 240, 291, 295; letters from, 142, 144; Bates's letters to, 143, 186; Mr., of Edmond Castle, 206
Grange, James, 400
Grassy Nook, 176, 177, 242; sale at, 243
Gray, Charles Gordon, 70, 81
Green, Mr. D. A., his sales, 1883, 458; 1886, 371, 468
Greenholme, near Burley, 169, 228, 248
Greenwell, Atkinson, 227, 229, 237, 317, 390
Greenwood, Mr., 183
Gregynog, 7, 8, 15
Greta Bridge, 252
Grey, Sir Henry, 9, 33; John, 151, 263, 265, 284 n., 300; letters from, 152, 153, 181; *Memoir* of, 270
Griff Farm, Duncombe Park, 298, 341
Grollier, *Histoire d'une Étable*, 373 n.
Guisbrough, Col. Chaloner's sale at, 1817, 201
Gunter, Mr., 339, 342; Captain, 343, 345, 347, 349, 360; Col., 380
Guzerat cattle, 140

HAGGERSTON, large shorthorns at, 101
Hales, Mr., of North Frith, 345; his sale, 1886, 371, 417
Hall, Alexander, of Haughton-le-Skerne, 46, 386, 388; Thomas, 34, 386-8; Edward, 319, 398; Mr., of Wiseton, 164, 201; Mr., of Natal, 376
Halton, farms at, 21; capital employed on, 103; Castle, 58; chapel, 98; auction, 70; fire, 88; feeding experiments at, 128, 129; Mr. Fawcett's reminiscences of, 146; Bates leaves, 170
Halton County, Canada, 307
Haltwhistle, 170-3, 208; tithes in, to parish of, 204
Hamburg, colour of cattle from, 30 n.
Hammond, Mr., of Hutton Bonville, 387
'Handling,' 12, 24 n., 65, 110 n., 330, 379
Harbottle, George, 79; Bates's letter to, 87
Harewoods, Mr. Lloyd's sales at,

1883, 458; 1889, 473; 1892, 476
Harlsey, 155, 158, 247, 252, 259
Harness, Edwin J., 247
Harris, Mr. J., 376
Harrison, of Barmpton, 22, 65; G. P., of Streatlam, 237; of Lowfield, 283, 316, 318, 392, 393, 403, 404; Rev. Thos., 165, 198; Wm., of Mortham, 283
Harrowgate, near Darlington, 38
Hart, 336 n.
Harvey, Mr. C. W., 296 n., 402, 404, 405
Havering Park, Mr. McIntosh's sales at, 1867, 353, 409; 1871, 413; 1875, 427; 1879, 449; 1881, 369, 455; 1886, 371
Hay, Mr., of Shethin, 332, 337
Haydon Bridge, 9
Hedley, Rev. Anthony, 207
Hegan, Mr., of Dawpool, his sale, 346
Heighington, 53
Hemlington, 238
Hendon, Mr. Tanqueray's sale at, 342, 406
Hepburn, Baron, Bates's visit to, 90, 91; letters to, 90, 92; letters from, 136, 146
Herbert of Cherbury, Lord, 14 n.
Herd Book, Coates's, origin of, 160, 184, 185; printing of, 190; illustrations, 191; red ink notes in, 192; merely a copy of pedigrees furnished, 193; errors in, 35 n., 221 n., 245, 303, 405 n.; disregard of, 385
Herefords, Bolingbroke's stock resembled, 56; bad milkers, 110 n.; Whitaker's opinion of, 185; Rev. H. Berry's disparagement of, 198, 200
Hesse, Upper, Agricultural Society of, 377
Hetherington, Mr., his sale at Brampton, 1880, 454
Hexham, presentation of colours at, 61; Tyneside Show at, 1837, 256
Heybridge, Mr. J. W. Philips's sale at, 1875, 427
Highlands, the Scottish, Bates's tour in, 68
Hilden, Mr. C. F. Leney's sale at, 1886, 468

Hill of Blackwell, family of, 35; Christopher, 44
Hillhurst, Canada, 357
Hilton Castle, 86
Hincks, T. Cowper, of Breckenbrough, letter from, 324
Hindlip, Sir Henry Allsopp's sale at, 1885, 464
Hobhouse, Mr., president of the Bath Society, 70, 81
Holderness, 27 n.; cattle, 27
Holford, Mr. T., 368; his sales at Papillon Hall, 1878, 442; at Castle Hill, 1883, 369, 459; at Warwick, 1892, 374, 476; at Castle Hill, 1894, 376
Holinshed, his description of cattle, *temp.* Elizabeth, 276
Holker, 341, 364; 'the Druid's' visit to, 1868, 358; the Duke of Devonshire's sales at, 1871, 413; 1874, 422; 1878, 367, 445; 1883, 370, 459; 1889, 373, 473
Holland, exportation of cattle to, 29 n.; importation of bulls from, 30, 31, 35 n.
Holland, in Lincolnshire, 30
Hollingsworth, Rev. N. J., 170-3, 204-9
Holme House, Mr. Wetherell's sale at, 1828, 227
Holmes, Mr., from Ireland, 219, 220, 274
Holstein or Dutch cattle, 30, 31; Improved, 56, 57
'Hoodle' in calves, 218
Hopkins, Mr. and Miss, 208, 209
Hopper, J., of Witton Castle, 110 n.; J. Mason, of Newham Grange, 280, 297
Horace, on the virtues of *sires*, 33
Hornby, 38
Horns of cattle, 277
Horsted, Mr. Barchard's sales at, 1885, 371, 467; 1893, 375, 476
Housing, Bakewell's idea of, 40
Housman, Mr., 325
Howard, Hon. Captain, 308
Howell, Mr., his sale, 1894, 376
Howitt, Wm., his rebuff at Seaton Delavel, 220
'Hubback, Mr.', 50
Hudson, Mr., of Thirsk, 336 n.; Mr., secretary of the R.A.S.E., 308, 310

Huggup, Mr. William, of North Seaton, 50 n.
Hughes, Professor M'Kenny, 29 n.
Hull, Yorkshire Show at, 1841, 278
Humphreys, Mr., 375
HUNNUM, 58
Hunt of Thornington, 181
Hunter, John, of Hurworth, 36-8, 387
Hurworth, 36, 38
Hustler, Mr., of Acklam, 150, 229
Hutchinson, Mr., 44; John, 36, 175, 199, 201; his mistake about Favourite, 55 n.; letter from, 164; his *Sockburn Short-horns*, 176, 180; his failure, 202; his death, etc., 242; catalogue of his sale, 243; Thos., 45
Hutton Bonville, 387, 389
'Hyanstriking,' the murrain called, 28, 47

IBBETSON, Sir Henry, 190, 191; his death, 202
Icely, Mr., Australia, 273 n., 275 n., 402
Importation of cattle prohibited, 28
In-breeding, alleged early, 35 n.; avoided by Colling, 51; and then accidentally adopted, 54; success of, in Northumberland, 57
Islay, stots from, 127
Isle, the, near Sedgefield, 6, 31

JACKSON, Michael, 389
Jacqueline of Luxemburg, 104
Jacques, Mr., 273, 278
James II., his alleged present of cattle to William of Orange, 29
James, Wm., of Hereford, 312
Jameson, Professor, 94, 130, 131
Jennings, Mr. Reginald, 402
Jervaux Abbey, 200
Jobling, Wm., of Newton Hall, 59, 98; buys Lord Bolingbroke, 56; his sale, 99; Robert, 99; Thos., of Styford, 63
Jobson, Wm., buys Jolly's bull, 49; his sale at Chillingham, 298; Messrs., hire 2nd Duke of Northumberland, 274, 401, 402
Johnson, Col., U.S.A., 321
Jolly, Mr., of Worsall, 47, 49, 150; his sale, 191; dines at Kirklevington, 274
Joshua, Robert Colling's servant, 65

KEAL HALL, 298, 302
Keevil, Mr., 311
Kenlis, Lord, 359: *see* Bective, Earl of
Kenyon College, Ohio, Bates presents bull to, 252
Ketton, Stephenson at, 35; J. Walker at, 36; the Collings at, 38; Bates's visit to, 1804, 64-6; Charles Colling's sale at, 1810, 108, 117
Killhow, Messrs. Foster's sales at, 1868, 410; 1871, 414; 1876, 435; 1879, 447; 1881, 369, 456; 1886, 372, 469
Kimbolton, the Duke of Manchester's sales at, 1881, 369, 456; 1890, 475
King, Mr., of Wealdstone, 267
Kingscote, Col. (Sir Nigel), 356; his sales at Kingscote, 1871, 412; 1875, 366, 425; 1879, 447; 1889, 373, 473
Kipling, near Scorton, 52
Kirklevington, Bates buys a moiety of the manor of, 103, 107, 120, 125; history of, 104; removal of the herd to, 218; calf-house at, 242, 258; shorthorn tribes at, 284, 336; sale, 1850, 330; disappointing results of, 335, 340
Kleist, Freiherr von, 329 n.
Knightley, Sir Chas., 342, 354; the Duchess cross on his cattle, 354, 355
Knightley, Mr. George Allen's sale at, 1878, 443
Knowles, Mr., of Tinsley, 312; of Attercliffe, 402 n., 403

LAMBERT, Mr., 218
Lambton, Dawson, of Biddick, 232
Lampson, Sir Curtis, 346; his sale at Rowfant, 1885, 371, 466
Lancashire longhorns, 40
Langleybury, Mr. McIntosh's sale at, 1886, 468
Larking, Mr. J. W., his sale at Cansiron, 1878, 440
Lartington, Mr. Witham's sale at, 1819, 149
Lathom, 365; Lord Skelmersdale's sales at, 1875, 430; 1879, 368, 450; (Earl of Lathom's), 1884, 464

Laughton, Mr. J. H. Casswell's sale at, 1892, 476
Lawrence on Live Stock, 95
Lawson, Sir Wilfrid, 369; his sale at Brayton, 1876, 435
Leeds, the Duchess of, 257, 404
Lenet, Chas., 245
Leney, Mr. F., 355-7; his sales at Orpines, 1874, 366, 421; 1875, 427; 1884, 370, 463; Mr. Augustus, his sale at Orpines, 1888, 372, 470; Mr. C. F., his sale at Hilden, 1886, 466; Mr. Herbert, his sales at Barming, 1887, 372, 469; 1890, 373, 474; 1894, 376
Lethbridge, Sir T. B., letter from, 189
Letton, James C., Kentucky, 262, 274
Lightburne, Mr. A. Brogden's sale at, 1783, 418
Lilburn, 130
Lillingstone Dayrell, Mr. Robarts's sale at, 1871, 413
Lime, operation of, 106
Lincolnshire, cattle in, originally 'pied,' 26; Blackwell shorthorns in, 44; large variety of shorthorns in, 101; Bates's visit to, 101
Lindisfarne Gospels, figure of ox in the, 24, 25
Line-breeding, initiated by Lord Ducie, 337
Linton, Mr., 311
Linton Mains, Mr. Cuthbertson's farm at, 137
Liverpool, Royal Show at, 1841, 276
Lloyd, Mr. A. H., his sales at Harewoods, 1883, 458; 1889, 373, 473; 1892, 374, 476
Loder, Mr. Robert, 366; his sales at Whittlebury, 1884, 462; (the late Sir Robert), 1889, 372, 472
Lodge, Mr. Robert, his sale at Dringhouses, 1882, 458
London, Duke of Northumberland lands at, 266
Londonderry, Lord, his Corn Law legislation, 186, 188, 189
Longhorns, origin of, 96; formerly great milkers, 96; Curwen's antipathy to, 121; Mr. Pusey on, 277
Longman, Mr. A. H., his sale at Shendish, 1881, 369

Lothian, East, its prosperity, 137
Low Field, near Piersebridge, 283
Low Fields, near Richmond, 387
Low Hill, Mr. Lovatt's sale at, 1880, 453
Lowndes, Mr., 402
Lowes, Miss, of Ridley Hall, 171; John, of Allen's Green, 213 n.
Lovatt, Mr. Henry, his sales, 1879, 448; 1880, 368, 453
Luke, St., his ox in the Lindisfarne Gospels, 24, 25
Lynn, Mr. C. R., 376

M'Dougall of Lorne, 60; of Kerrera, 70
M'Gibbon, 70
McIntosh, D., 352, 355, 357, 403; his sales at Havering Park, 1867, 353, 354, 409; 1871, 413; 1875, 427; 1879, 449; Mrs., her sale at Langleybury, 1886, 468
Mackenzie, Thos., of Applecross, 130, 131; his opinion of Northumbrian society, 133; sends his bailiff to Halton, 145; M.P. for Ross-shire, 145; letters from, 167, 187, 189
M'Neill of Oransay, 70, 86 n.
Magistrates at Stockton, 195; at Yarm, 216, 327
Manchester, Duke of; his sale, 1881, 369, 456; death of, 374; (the late), sale, 374, 475
Manfield, 52, 191
Markham, Gervase, his *Cheap and Good Husbandry*, 25
Marton exhibition (Major Rudd's), 150
Marton-le-Moor, 201
Mason, Christopher, 59, 68, 86, 199; his visit to Halton, 88, 132; puffs off his stock, 120; his feeding experiments, 90, 129; his ruinous system of breeding, 133; his blood, 133, 151, 198, 204, 214, 287, 288; his sale at Chilton, 1829, 213, 219
Matfen, 100
Mauleverer, Mr., 216; letter from, 324
Maw, Henry Lister, 289, 404; letter from, 281
May, Dr. Georg, his *Erfolge der Shorthornzucht*, 330 n.
Maynard, John, of Eryholme, 20, 68,

336 n.; his use of Laird, 89, 119, 122, 201; his account of the Lady Maynard family, 161, 389; his reason for parting with it, 122; John Charles, of Harlsey, 119; buys Marske, 149; his letters to Bates and Donkin, 158; his *dictum* on shorthorns, 247; sells Comet Halley, 251, 252; Anthony, of Marton-le-Moor, 201, 285, 331

Measuring cattle, 12

Megibben, Mr. T. J., Kentucky, 365

Melmerby, 81, 89 n.

Melrose, Hon. R. Baillie-Hamilton's sale at, 1889, 373

Mere Old Hall, Mr. D. R. Davies's sale at, 1870, 411

Metcalf, Timothy, 252

Metz, France, 402

Middleham, Bishop, 302

Middlesbrough, 240; Mr. Parrington's sale at, 240, 241; cattle shipped at, 266; steamers from, 275

Milbank of Barningham, 32, 33, 37 n.

Milburn, Mr., 304, 308

Milk, 48 qrts. *per diem* from Stamfordham cow, 8; 36 qrts. from Bright Eyes, 35; 53 qrts. from Barmpton cow, 65; 28 qrts. from Duchess-by-Daisy-Bull, 70, 114, 229; 32 qrts. from Old Daisy, 150, 229; 30 qrts. from R. Colling's Bright Eyes, 229; 12 quarts sufficient, 95; quantity and quality of, 112, 113; hereditary, 113; good yield from all 2nd Hubback cows, 195, 228; unimpaired at Kirklevington, 284; forced trials of, 289, 379; danger of prize-fund for, 379

Mills, Mr. Philo L., 374, 376; his sale at Ruddington, 1893, 375, 476

Mirehouse, near Keswick, 86 n.

Mitchell, Mr., of Arthington, 404

Moore of the Moore, family of, 6, 7; Royal Standard Bearers in Wales, 7; Diana, *see* Bates, Diana; Joyous, 14, 15; letters from, 22, 61, 62; her death, 63; legacy from, 64

Moore, Mr. George, 360; his sale at Whitehall, 1875, 430

Moorhouse, Mr., of Skipton, 60

Moreton, Lord, his sale at Tortworth, 1890, 374

Morgan, Sir Chas., 176

Morris, Col., U.S.A., 344

Mortham, 283

Motte, Mrs., 315

Murphy, Mr. S. J., U.S.A., his sale at Sparrow Farm, 1882, 457

Myers, Thos., cowman at Kirklevington, 280, 400, 403-4

Myton, Mr. Stapylton's sale at, 1824, 201

NASRULLA KHAN, H.H., Shahzada of Afghanistan, 377

Natal, 374

Nathusius, Herr von, 329 n.

Natrass, purchaser of Hubback, 38

Nelson, Wm., of Lilburn, 130, 137

Nelson Gore, Canada, 307

Netherwitton, 100

Nevett, Mr., 375

New York Mills, Oneida, sale at, 1873, 365, 420

Newby, Blackett shorthorns at, 29, 33

Newcastle races, 1808, 82; Royal Shows at, 1846, 281, 282; 1887, 372

Newham Grange, 297, 339

Newton Hall, near Corbridge, 56; the late Wm. Jobling's sale at, 1810, 99

Newton Morrell, 20; Mr. Charge's sale at, 1828, 229

'North, Christopher,' his review of *Fleurs*, 209-11

North Frith, near Tonbridge, Mr. Hales's sale at, 1885, 371, 467

Northallerton, 247; Yorkshire Show at, 1839, 269

Northampton, Royal Show at, 1847, 304

Northchurch, Mr. Joseph Robinson's sale at, 1876, 434

Northumberland, backbiting in, 134; blood feuds in, 206; industry and intelligence of the country folk, 215; magistrates in, 216; Bates's affection for, 173, 274; Agricultural Show at Alnwick, 1840, 274

Northumberland, Duke of, 43 n.: *see* Smithson, Sir Hugh

Nunstainton, Mr. Brown's sale at, 1831, 219-24; bad results of most purchases at, 226

Nunwick, Mr. C. R. Saunders's sale at, 1870, 412

OAKLEY, Mr., of Oakley, Salop, 83
O'Callaghan, Col., 55, 279, 288
Ohio, bishop of, *see* Chase, Philander; Company for importing English cattle, 247, 250, 251; Kenyon College, 252, 284; Kirklevington cattle on the, 268
Oliver, Captain, 347, 350, 352, 369; his sales, 1877, 366, 438; 1884, 370, 461
Ormesby, 32, 36
Orpines, Messrs. Leney's sales at, 1874, 366, 421; 1875, 427; 1884, 370, 463; 1888, 372, 470
Ostler, Mr., of Audleby, 36
Ouzleworth, Sir J. Rolt's sale at, 1871, 415
Oxford, meeting of the English Agricultural Society at, 1839, 266; Bates, 267
Ovingham, Tyneside Shows at, 69, 87, 91, 132

PAEONES, in Macedon, their longhorns, 277
Paley, Mr., of Gledhouse, 249, 254
Parallel and Parallax, 361
Paris, international cattle show at, 1878, 367
Parker, Mr. Wm., of Yanwath, 302, 406
Parkin, Mr. Thos., 238, 239; Wm., 238; W. T., 239 n.
Parkinson, Richard, of Babworth, 273, 285, 288; John, of Leyfields, 286, 287, 304; letter to, 286
Parot, Mr.: *see* Porritt
Parrington, Mr., his sale at Middlesbrough, 1832, 240, 241; Mr. L. II., his reminiscences, 240
Pattinson, Rev. J., Bates's letter to, 89
Paul, Mr., 273
Pedigrees, kept first in the male line, 33; in the female, 2, 34; Bates offers to print those of the Ketton stock, 115-8; little attention paid to, 161; 'cooked,' 193, 303, 384; not recognised by the R.A.S.E., 308, 310 n.; Grant Duff tries to stop their falsification, 323
Peel, Mr. Wm., of Taliaris, 403; Sir Robert, 291-3
Pelham, Mr., introduces Blackwell shorthorns into Lincolnshire, 44

Pennant, on wild cattle, 96; Hon. Col., 347
Pennyman, Sir James, 32, 36, 323
Penrhyn, Lord, 349; his sales at Wicken Park, 1869, 411; 1873, 418; 1875, 426; 1878, 439; 1880, 368, 454; (the late) 1887, 372, 469
Percy family, owners of Kirklevington, 104; John, of Haram, 23
Phantassie, 96, 97, 100, 137
Philips, Sir G. R., 366; J. W., his sale at Heybridge, 1875, 427
Phillips, Mr. Thos., 282, 406, 407; letter from, 298
Pickering, Chas., of Foxton, 35, 387, 388; Mr., of Ellfoot, 217
Pilley, Mr., his cattle, 175 n., 178
Pinnell, Mr., 376
Pippit, Mr., 303
Pitcorthie, General Simson's sale at, 1818, 151
Pitt, William, 290, 292, 295, 296
Place, Mr., banker at Stockton, 202, 336 n.; Thos., of Spennithorne, 153
Poland, Suffolk duns alleged to have come from, 96
Pollard, Robert, 80
Pontypool, 402
Pool Park, Mr. R. Blezard's sale at, 1886, 468
Porritt, John, of Claxton, 224, 225, 233 n., 235, 236, 318; Henry, of Togston, 224 n.
Portman, Lord, 310
Portraits of cattle, importance of accurate, 80, 125, 155, 283
Potter, Mr. E., 375, 376
Powel, Col., of Pennsylvania, 228
Powis, Earl of, 375
Preece, Mr. J. A., of Drayton, 375
Prentice, Mr., U.S.A., 305, 306, 321
Preston Hall, near Aylesford, 348; Mr. Betts's sale at, 1867, 349, 352, 353, 408; Mr. Brassey's sale at, 1887, 469
Priestley, Mr. S. O., of Trefan, 404
Prudhoe, Lord, 271 n.
Pulleine, Mr. Robert, 324
Pusey, Mr. Philip, 276

QUARTER-ILL, 120, 182, 202

RADCLIFFE, Mr. Thos., 12, 214
Raine, Mr., of Gainford, 133
Ramsay, Mrs., Edinburgh, 99
Rayans Farm, Lord Chesham's sale at, 1876, 434
Red colour of cattle, preferred by the Romans, 24; prevalent in Somerset and Gloucestershire, 26; distinctive of English cows, 27
Redcar, 179, 325
Remenham, 79
Renick, Felix, of Chillicothe, Ohio, 247-51; Bates's letter to, 254; importations of his Red Roses, 364; Josiah, 247
Rennie, George, of Phantassie, 96, 136, 137; letters to, 97, 101; letter from, 100
Reuss, princes of, their numerical series, 149
Revesby Abbey, Banks Stanhope's sale at, 1848, 316
Rhodes, Rev. J. A., 165, 177-9; letters from, 174, 180, 191, 194
Riby, Ambrose Stevens at, 316
Rich, Mr. S., of Didmarton, his sales, 1868, 410
Richardson, Mr., of Hart, 336 n.
Richley, Mr., butcher at Corbridge, 11
Richmond, Yorkshire Show at, 1844, 285
Ridley, Mr. Anthony, 402
Ridley Hall, purchased by Bates, 170; his removal to, 173, 174; its situation, 174; bridge at, 204; sold to Hollingsworth, 206; Bates leaves, 218
Robarts, Mr. A. J., of Lillingstone Dayrell, 353; his sale, 1871, 413
Roberts, Mr. Joseph, of Mickleton, 406
Robertson of Ladykirk, buys Lame Bull, 52
Robinson, Wm., of St. Helen's, Auckland, 245; Thos., of March House near Scorton, 274; Mr., of Bletsoe, 401, 402; Mr. Joseph, his sale at Northchurch, 1876, 434
Robson, Mr., butcher, 162, 163; Wm., in charge of R. Colling's papers, 161; Wm., of Colmhill, 171, 213 n.
Rogers of the Home, Salop, 83

Rohde, Dr. O., his *Rindviehzucht*, 31 n.
Rolt, Sir John, his sale at Ouzleworth, 1871, 415
Romans, the, preferred 'good touchers' of red colour, 24
Ross-shire election, 145
Rotations of crops in Cleveland, 105
Rotterdam, colour of cattle from, 30 n.
Rowfant, Sir C. Lampson's sale at, 1885, 371, 466
Roxburghe, the Duke and Duchess of, 208
Rubbing-posts, Bakewell's, 41
Rudchester, Dutch cattle at, 30 n.
Rudd, Major, 32 n., 180, 201
Ruddington, Mr. Philo L. Mill's sale at, 1893, 375, 476
'Runts,' black Welsh and Scotch, 26
Rushden, Mr. F. Sartoris's sales at, 1878, 422; (the late) 1888, 470
Russell, Lord John, 296
Ryle, Mr., of Biddick, 69

SADDLE AND SIRLOIN, errors in, 45 n., 228, 238 n., 248 n.
St. Quintin, Sir Wm., 32, 36
Salt, Mr. W. H., his sales at Bird's Nest, 1876, 433; (Sir), 1881, 369, 455
Salvin, Mr., of Croxdale, his Herefords, 110 n.
Samuda, Mr. d'A., his sale at Chillies, 1878, 441
Sanday and Smith, Messrs., 406
Sanders, Mr., U.S.A., 315
Sandhoe, 98, 107, 109, 110, 158, 161; Donkin's sale at, 1825, 203
Sandringham, H.R.H. the Prince of Wales's sales at, 1886, 468; 1889, 473; 1891, 374, 475; 1893, 477
Saratoga Springs, New York State Fair at, 1847, 304
Scaleby Castle, Abyssinian cattle at, 140; shorthorn herd formed at, 148; Mr. Fawcett's sales at, 1875, 429; 1878, 439; dispersion sale at, 1894, 376
Scampston, 32
Scarborough, Yorkshire show at, 1847, 304
Scorton, 274
Scots Highlanders, their quick perception, 68

Scott, Sir Walter, his account of the firing of the beacons, 63
Scratton, Mr. D. R., 376
Seaton, Mr., 66 n., 265
Seaton Delaval, 220; North, 50
Sedgefield, 240, 319
Sedgwick, Mr., of Stone Gap, 249
Septimius Severus, 24
'Shades,' Bakewell's, 41
Shafto, Mr., of Whitworth, his Devons and Kyloes, 110 n.
Sharpley, Mr. H., his sale at Limber, 1894, 376
Sharter, Mr., 33
Sheldon, Mr. J. O., of Geneva, Illinois, 355; his sale at Windsor, 1867, 355-7, 409; disposal of his herd, 365; Mr. H. J., of Brailes, 366 n.; his sales, 1869, 411; 1871, 415; 1873, 419; 1875, 432; 1879, 452; 1888 (Crouch Farm), 372, 472
Sherlowe, Messrs. Evans's sale at, 1894, 375
Sherwood, Col., of Auburn, U.S.A., 303, 304, 308, 322
Shipman, Mr., 315
Sholebroke Lodge, 350-2; Captain Oliver's sales at, 1876, 367; 1877, 366, 438; (private), 1881, 369; 1884, 370, 461
Shorthand, introduction of, 206
Shorthorn Society of Great Britain and Ireland, its prize-scheme, 377; its stereotyped policy, 384
Shorthorns, mediaeval, 23; ancient, 33; on the Tees, 37 n., 47 n.; on Tyneside, 13, 68; George Culley's opinion of, 41; a history of, necessary, 197; Bates's intention to write one, 282, 324; duplex qualification of, 379; Scottish, 381
Siddington, Mr. Bowly's sales at, 1869, 410; 1872, 415; 1875, 426; 1877, 439; 1881, 369
Simson, Col., 53; General, 59, 126, 127, 129; his sale at Pitcorthie, 1818, 151
Simunburn, 172
Sinclair, Sir John, 72, 79; fails to acknowledge Bates's Address, 84; his 'political gratitude,' 119; applies to Bates for the results of experiments, 138

Singleton, Sir H. H. Vivian's sale at, 1891, 475
Sittyton, 381
Sitwell, Mr., 71
Skelmersdale, Lord, 365; his sales at Lathom, 1875, 430; 1879, 368, 450; (Earl of Lathom), 1884, 464
Skerningham, 38
Skipton Bridge, 248
Skipworth, Philip, letters from, 196, 197; at Kirklevington, 199
Slye, Mr. W. W., his sales at Beaumont Grange, 1870, 411; 1872, 416; 1876, 434
Smeaton, near Prestonkirk, 91, 136, 146
Smith, Jacob, of Givendale, 48; Thos. ('Handy'), 319, 397; his story, 399; Wm., 267, 273, 296; Mr., of Haughton Castle, 108
Smithfield Club, origin of, 79; overfed animals at, 281; reform of, 377
Smithson, family of, 28; Sir Hugh, 43
Snowdon, George, 36, 387
Sockburn, 44
Sockburn Short-horns, Origin and Pedigrees of the, 176
Somerset, red cattle in, 26
Somerville, Lord, 77; his dinner, 1808, 79
Southrop, Mr. W. J. Edmonds's sale at, 1894, 477
Spedding, John, of Mirehouse, 86, 123; letters to, 86, 89, 92
Spencer, (3rd) Earl (Lord Althorp), gains champion prize at York, 1838, 263; opposes reaping-machines, 264; at Liverpool, 276; (5th) Earl, 352
Spennithorne, 153
Spraggon, Mr., 404
Springfield, 337
Squires, Mr., 404
Stagshawbank Fair, 1808, 83; Close House, 66
Stamfordham, Bates christened at, 8; milch cow, 8
Standard-bearer, Royal, in Wales, office, 7 n.
Stanhope, Banks, 285, 286, 289; his sale at Revesby Abbey, 1848, 316
Stanwick, herd at, 28, 43, 44

Stapylton of Myton, his sale, 1824, 201
Starkie, Mrs., her sale at Ashton Hall, 1883, 461
Steel pens, introduction of, 98, 99
Stephenson, Mr., of Ketton, 35, 36, 236; John, of Wolviston White House, 227, 230, 231, 236, 237, 244, 316, 323; Thos., 231, 266, 267 n.
Stevens, Ambrose, of Batavia, U.S.A., 303, 307, 311, 314; letters from, 315-7, 320-3
Stewart, Lord, 186
Stick-a-Bitch, 21
Stockton magistrates, 195
Stoke Park, Mr. E. J. Coleman's sale at, 1876, 433
Stokes, Mr. Chas., of Kingston, Notts, 312
Stokesley magistrates, 195
Stone Cross, Mr. A. Brogden's sale at, 1878, 446
Stone Gap, near Skipton, 249
Storer, Rev. John, 29 n., 338, 353, 361-3
Strafford, Mr. Henry, 345, 346, 355, 356
Stratton, Mr. Richard, letter to, 275
Streatlam Grange, 237, 318
Studley Royal, 29, 32
Suffolk dun polls, 27; alleged to have come from Poland, 96
Swaffield, Mr. Benjamin, of Chatsworth, 311
Swalwell, Thos., 231
Swarcliffe, 191
Swansea, Lord: *see* Vivian, Sir H. H.
Swinburne, Sir John, 92

TANQUERAY, Mr., 339; his sale at Hendon, 1855, 342, 406
Taylor, George, of St. Helen's, Auckland, his Devons, 110 n.; Mr. Darcy, his sale, 1895, 376; Mr., of Wetherby, 358
Tees, early shorthorns on the, 28, 37 n., 47, 96
Tempest, Mr., 254; Sir Chas., 311 n.
Temple, of Hilton Castle, 86
Tetley, near Crowle, 281
Thomas, Mr., of Mount Pleasant, near Barnard Castle, 336 n.
Thompson, Mr. Robert, of Stamford, 21, 74; of Chillingham Barns, 114; Mr. P. B., 263; Tommy, cowman at Halton, 136; Robert, cowman at Kirklevington, 357
Thorne, Samuel, U.S.A., 338, 339, 343, 344; E., 356
Thornton, Gabriel, 46
Thorpe Thewles, 238
Threshing-machines, 62
Tindale Ward, Agricultural Society of: *see* Tyneside Agricultural Society
Tinsley, near Sheffield, 312 n.
Titley, Mr. T. A., his sale at Dringhouses, 1880, 470
Topham, James, of Keal Hall, 273, 298, 405; letter from, 302
Torr, Wm., of Aylesby, 312; of Riby, 316, 331
Tortworth, 331; Earl Ducie's sale at, 1853, 339; Lord Moreton's sale at, 1890, 374
Towneley, Col., 354; his sale at Towneley, 1873, 418
Tredegar, 176
Trevelyan, Wm. Blackett, of Netherwitton, 99; letter from, 100; letter to, 106
Trotter, Mr., of Bishop's Middleham, 302; his sale, 1852, 339
Turner's Hill, sale of Mr. Downing's stock from, 1872, 360
Turnips, storing, 87, 101
Turveylaws, near Wooler, 49, 298
Tyneside, shorthorns on, 13; no vestige of them left, 256
Tyneside Agricultural Society, its formation, 69; factions in, 84; slow progress of, 87, 128, 132; decides Bates is to have no prizes, 135; its disruption, 252

UNDERLEY, 365; Lord Bective's sales at, 1874, 423; 1880, 368, 453; 1885, 371, 467; 1887, 372, 470; 1892, 374, 476; 1894, 376, 477
Unsworth, Mr., 402

VAIL, George, of Troy, U.S.A., 330, 401 n., 405; letters to, 284, 304; letters from, 304, 306, 320, 321
Van Mildert, Wm., bishop of Durham, 211, 212
Vane, Rev. J., of Wrington, Somerset, 402 n.

Vane Tempest, Sir Henry, 160, 235, 240, 319; his sale at Wynyard, 1813, 231-4

Vansittart, Rt. Hon. Nicholas, 141-144; letter to, 141; Henry, of Kirkleatham, 141 n., 240

Vaughan, Wm., letter to, 323

Virgil on *dams*, 34

Vivian, Sir H. Hussey, his sales, 1886, 372; 1891, 379, 475; (Lord Swansea), 1893, 375

WAISTELL, Robert, 42, 316, 319 n., 320; his tales, 395-8

Wakefield, 246, 247; Yorkshire Show at, 1846, 281, 282

Wakeman, Mr., U.S.A., 305

Walcott and Campbell, Messrs., U.S.A., 365, 415

Waldy, Mr., sells Kirklevington, 103; led all manure on to grass, 104; young, buys Acklam Red Rose, 150, 169

Wales, H.R.H. the Prince of, sales at Sandringham, 1886, 468; 1889, 473; 1891, 374, 475; 1893, 477

Walker, Mr., near Leyburn, 36; J., at Ketton, 38 n.; Dr., sceptical as to wild cattle, 96; Mr., butcher at South Shields, 214; James, Newberries, Herts, 311

Walkeringham, Mr. Watson's herd at, 316

Wallington, Sir W. C. Trevelyan's sale at, 1879, 448

Walton, Joseph, of Stanhope, 112-4; letter to, 111

Wardon, 82

Wark Eals, 18, 172

Warlaby, sale at, 1895, 377

Warwick, Mr. Holford's sale at, 1892, 476

Wastell, Wm., visits Aydon, 9; spelling of his name, 9 n.; his heifer, 22; names Tripes, 34, 36; sells Jolly's bull, 47

Waterfall, near Guisbrough, Col. Chaloner's sale at, 201 n.

Waterloo, battle of, 139

Watson, Mr., of Walkeringham, his herd, 316

Webb, Mr., 311 n.; Mr. Jonas, of Babraham, 345

Webster, Daniel, his oration at Oxford, 268

Weeting Hall, Mr. Angerstein's sale at, 1876, 366, 436

Weight, loss of live, in cattle by travelling, 286

Welham, Major Bower's sale at, 1835, 251

Wentworth, Mr. Godfrey, 405

West Highland cattle, 60, 88; Bakewell and Marshall's opinion of, 96; compared with mixed breed, 124

Wetherby Grange, 'the Druid's' visits to, 1859, 342; 1868, 357

Wetherell, Mr., his sale at Holme House, 1828, 227; of the Isle, 232; tries to buy bulls from Bates, 257, 387, 388

Wharfdale, Bates's opinion of, 342

Whitaker, Jonas, 155; aided by Bates, 156; they visit Elemore, 156; his opinion of Americans, 158, 165; letters from, 169, 170, 183, 185, 190, 192, 194, 197-202; letter to, 228; Bates seeks a prize-winner for, 181, 191, 213, 245; Bates's only favour from him, 245; his abuse of Duchess 33rd retracted, 249; on Rose of Sharon and Duchess 19th, 250; controls the American trade, 251; buys Lady Colling, 252; his violence, 288; his son, 254

White cows from Flanders, 27

Whitehall, Mr. George Moore's sale at, 1875, 430

Whittington, 98

Whittlebury, Sir Robert Loder's sale at, 1889, 372, 472

Wicken Park, Lord Penrhyn's sales at, 1869, 411; 1873, 418; 1875, 426; 1878, 439; 1880, 368, 454; 1887, 372, 469

Willis's Rooms, sale at, 1865, 346, 347

Wilson, Rev. George, vicar of Corbridge, letter from, 97; Mr., of Shotley, 402; his sale, 1878, 444; Mr., of Cumledge, 401; T., animal painter, 80 n., 119; Professor, 208

Windsor, Mr. J. O. Sheldon's sales at, 1867, 355-7, 409

Winship, Mr., 99

Winterfold, Messrs. Harward and Downing's sales at, 1871, 413; 1872, 360-3, 416

Wiseton, Bates's visit to, 164, 179; sales at, 202; 1846, 335; 1848, 337

Witham, Mr., of Lartington, his sales, 149

Wittenhall, Mr., Canada, 307

Witton Castle, 6, 31

Witton-le-Wear, Bates at school at, 10, 11

Wolviston White House, 227, 229, 236, 244, 317; the Sandy Lane at, 231

Wood, Mr. John, 213, 235, 319, 393; Richard, of the Close, 231

Woodside, Mr. J. H. Blundell's sale at, 1878, 442

Wooler, 181

Wordsworth, Wm., 177

Workington meetings, 87, 92

Worms, Mr. Hayem, 402

Worsall, Mr. Jolly's sale at, 192

Worton, near Devizes, 303

Wright, Mr. John, 273

Wrightson, Mr., of Cusworth, 176

Wynyard, dinner at, 67; agricultural meeting at, 160, 186; the herd, 224, 229, 240, 319; Sir H. Vane Tempest's sale at, 1813, 231; Lady Antrim's sale at, 1818, 236

YANWATH, near Penrith, 302

Yarm, 199, 274; fair, John Bates at, 6, 31; Cherry bought at, 30, 246; magistrates, 215, 216; Orlando's produce round, 176

York, Septimius Severus at, 24; Yorkshire Shows at, 1838, 260-4; 1842, 279-81; Royal Show at, 1848, 308-15

Yorkshire cattle, originally black, 24, 26; Agricultural Society, its Shows at York, 1838, 260-4; at Northallerton, 1839, 269; at Hull, 1841, 278; at York, 1842, 279; at Doncaster, 1843, 284; at Richmond, 1844, 285; at Beverley, 1845, 296 n.; at Wakefield, 1846, 281, 282; Scarborough, 1847, 304

Young, Arthur, his criticism on Bakewell's cattle, 41

SHORTHORN INDEX

ACKLAM RED ROSE, 150, 161, 164
Acomb, 336
  2nd, 448
Adelaide, 243
Airdrie Duchess 3rd, 454
  7th, 459
  9th, 457
  10th, 457
Airdrie Geneva, 427
Albion (by Wynyard), 234
Alcock's (Ralph) bull, 48, 389
Alexandra, 433
Alexina, 141 n., 233
Alfrede, 156, 183, 184, 190
Alonzo, 185
Alpha, 153
American Cow, 150
Anabella, 401 n.
Anecdote, 273, 401 n., 402 n.
Angelina, 231-3, 235-7, 318, 319,
    392, 394, 396
  2nd, 237, 318, 391, 392
Anna Boleyne, 35, 252
Anna-by-Lawnsleeves, 225, 236, 237,
    318, 391, 395, 396
Annette 2nd, 418
Antoinette, 415
Antonia, 415
Archduchess of Oxford 2nd, 469
  3rd, 469
Archduke, 342
Arethusa, 419
  2nd, 419
Ariel Countess, 426
  2nd, 439
Ariel Marchioness, 425
Ariel Viscountess, 448
Artless, 234
Asia, 407
Ayah, 407

BARBARA, 407
Barden, 414
Barforth, 36, 37
Barker's (Richard), bull, 51
  (Wm.), bull, 34
Barming Barrington 2nd, 474
  Duke, 475
Barming Cherry Duchess, 474
Barming Foggathorpe, 475
Barming Waterloo Emperor, 475
Barmpton, 156
Barningham, 191
Baron (58), 237, 239
Baron Barrington 4th, 422
  5th, 422, 430
  6th, 423
Baron Foggathorpe, 406 n.
Baron Graham, 447
Baron Lightburne, 418
Baron Oxford, 354, 409, 418
  2nd, 354, 409
  3rd (25579), 410
  3rd (42737), 455
  4th, 359
  5th, 416, 422
  16th, 461
  17th, 461
Baron Tregunter, 423
Baron Wastwater, 434
Baron Winsome 3rd, 422
  5th, 446
  22nd, 474
Baroness, 69, 112, 115
Baroness Fawsley 3rd, 428
  6th, 428
Baroness Oxford 3rd, 422
  4th, 460
  5th, 445, 464
  9th, 460
  12th, 460

Baroness Oxford 19th, 472
   23rd, 474
   24th, 474
   27th, 474
Barrington, 407
Barrington Beauty, 375
Barrington Prince, 473
Barringtonia, 413
Beaming Eyes 2nd, 459
   3rd, 459
Bearl, 403 n.
Beau of Oxford 2nd, 427
Beauty of Oxford, 344
Bella, 99
Belle, 402 n.
   2nd, 401 n.
Belle of Oxford 5th, 427
   6th, 439
   7th, 440
   8th, 454
Belle of Worcester, 441
   2nd, 465
   3rd, 465
Belleville, 296, 297, 339
Belvedere, 225, 229-31, 237, 242, 244-7, 255, 256, 338, 363, 391, 392
Ben, 53, 54, 59, 151
Benson family, 99
Bertha Waterloo, 451
Bertram, 228, 229
Beverley, 334
Birthday, 282 n., 285 n.
Blackwell Ox, 44
Blanche, 242, 244, 253
   (Lord Feversham's), 323
   2nd, 403 n.
   3rd, 405
   3rd (Holker), 359
   4th, 405 n.
   4th (Mr. Salt's), 433
   5th (Mr. Salt's), 433
   6th, 405
   10th, 436
   15th, 445
Blanche Winsome, 458
Blossom, 406
Blucher, 244, 363
Blush, 407
Blythesome Eyes, 429, 450
Bolingbroke (Lord), 48, 52-5, 203
Bolingbroke, grandson of, 55, 156
Bonny, 157
Bonny Lad, 323

Bracelet, 278, 279
Bride of Oxford, 344
Bridecake, 346
Bridesmaid, 413
Bright Eyes, 35, 387, 388
   (Old), 35, 155, 156, 229
   (of 1817), 191, 203
   (Wetherby), 357
   4th, 432
   5th, 413, 422
   6th, 437
   9th, 432
Briton, 224 n.
Broken Horn, 53, 193, 194
Brokenleg : see Duchess 34th
Brown's (James) Old Red Bull, 34, 155
   White Bull, 52, 168
Brown Cow, 224
   2nd, 224 n., 401 n.
Buchan Hero, 278
Butterfly Princess 24th, 429

CALISTA, 233
Cambridge Duke 3rd, 409
   26th, 375
Cambridge Lady 2nd, 436
Cambridge Premium Duke 2nd, 375, 476
   Rose, 273
   2nd, 256
Cambridge Rose 5th, 332
   6th, 333, 339, 344, 345, 353, 407
   7th, 334
   8th, 468
   10th, 465
Captain Shafto, 296, 297 n., 304
Carcase, 261
Careless, 234
Carham, 151
Carlisle, 442
Carolina 4th, 436
   5th, 414
   7th, 423
   Craggs, 447
Carrie Craggs, 426
Carry 2nd, 422
Cecil, 168
Cecilia, 175
Charger, 346
Charming Daisy, 457
   Duchess 6th, 452
      Echter, 443
Cherry, 38, 49

Cherry (Old), 154
  Countess, 413
  Duchess 1st, 437
    20th, 441
    22nd, 427
    24th, 426
    25th, 427
    28th, 455
    30th, 455
    of Brailes 3rd, 452, 460
      Elmhurst 2nd, 456
      4th, 456
      Hillhurst, 441
      Oneida, 441
  Grand Duchess 4th, 438
    8th, 439
    11th, 451, 463
  Queen, 440
    2nd, 461
Chevalier, 334
Chevy Chase, 405
Chieftain (135), 64 n., 68, 107
  (10048), 334
Christmas Gwynne, 436
  Rose 3rd, 449
Clear Star, 360, 363
Clementi, 273
Cleopatra, 441
Cleveland (146), 169, 179, 182-4
Cleveland Lad, 272, 273, 276-8, 283, 298, 301, 323, 403
  2nd, 284, 285, 298
Colling's (Robert) White Bull, 66, 265, 388
Colonel (152), 165
Colonel's dam, 53
Comet, 48, 59, 120, 122, 138, 139, 229, 271, 365
Comet Halley, 252
Conishead Fuchsia, 437
  Lally 3rd, 451
  Wild Eyes, 438
    2nd, 451, 453
Coral Waterloo, 470
Countess (Middlesbro'), 241
  6th, 448
  14th, 463
  of Barrington, 422
    2nd, 420
    4th, 359
    6th, 422
    7th, 445, 449
    9th, 445
    10th, 460

Countess of Barrington 13th, 460
    14th, 460
  Clarence, 442
    2nd, 443
  Kirklevington, 448
    2nd, 448, 451, 453
  Oxford, 356, 409
    (Mr. Lovatt's), 453, 454
    2nd, 420
    3rd, 420
  Waterloo 4th, 473
    5th, 473
  Worcester (Mr. Larking's), 440
    (Mr. Lloyd's), 458
Craggs, 402
Cramer, 286-8
Crofter, 402 n.
Crooked Tail Bull, 47

DAINTY 2nd, 445
Daisy, 46, 49
  (Old), 150, 229
Daisy (*H. B.* i. 269), 169
Daisy Bull, 59, 63, 64 n., 118, 156, 228
Daisy (Young), 175
Dalton Bull (Duke), 40, 387
Dan O'Connell, 269
Dandy, 190
Dandy (951), 203
Darlington, 336, 404
  5th, 407
  8th, 407
  12th, 411, 415
  13th, 412
  15th, 411, 412
  17th, 411, 412, 423
  18th, 411, 412
  19th, 423
  Belle 3rd, 464
Deception, 311, 356 n.
Deepdale, 424
  3rd, 456
Dentsdale, 423
  2nd, 424
  8th, 470
Desdemona, 338
Diadem, 407
Diana, 149
Dolly Gwynne, 433
Dora, 412, 415
Doralice, 412
Dowager 3rd, 457, 468
  Duchess 4th, 427, 449

Dowager Duchess 5th, 427, 449
   of Barrington, 476
Driffield Cow, 160
Duchess (Appleby's), 34, 43, 49
   by Foljambe 21, 114, 118
   by Favourite, 54
   by Daisy Bull, 64, 66-70, 107-9, 114, 147, 192
     1st (Young or Younger), 108, 109, 114, 117, 147 n., 149, 192, 196, 283
     2nd, 192
     3rd, 155, 158, 160, 164, 179, 192
     4th, 192
     5th, 192
     6th, 148, 179, 192
     7th, 179, 192
     8th, 192
     9th, 192
     19th, 250
     30th, 271
     32nd, 255
     33rd, 248, 249, 253
     34th (Brokenleg), 248, 253, 255, 271, 280, 283, 310, 312
     38th, 253
     41st, 255, 261, 263
     42nd, 261, 262, 266-8, 272, 278
     43rd, 261, 262, 263, 266-8, 272, 278
     51st, 332
     54th, 332, 342
     55th, 331, 332, 340, 342
     56th, 332
     59th, 310, 312, 331, 333, 336, 339, 340, 343
     61st, 333, 338
     62nd, 333
     64th, 334, 339, 343
     66th, 340, 343, 344, 365
     67th, 337, 340, 342, 343
     68th, 340, 343
     69th, 340, 342, 343
     70th, 340, 343
     71st, 344
     72nd, 342, 343
     73rd, 343
     75th, 343
     77th, 343, 357, 358
     78th, 343, 357
     79th, 343, 357
     84th, 358

Duchess 86th, 357
   87th, 357
   88th, 357, 358
   91st, 357, 364
   92nd, 358
   94th, 358
   96th, 358
   97th, 357, 358
   98th, 358
   99th, 358
   100th, 358
   101st, 364
   103rd, 358
   107th, 364
   108th, 364
   114th, 450
   117th, 450, 464
   122nd, 465
   123rd, 465
   124th, 465
   125th, 465
   127th, 464, 477
   131st, 376, 477
   (Darlington), 407
   (Nancy), 272
   8th (Acomb), 444
   20th (Red Rose), 443
   26th (Red Rose), 443
   Carolina, 427, 439
   Craggs 3rd, 462
   Darlington 9th, 463
   Fawsley 9th, 463
   Gwynne, 423, 433, 435, 443
     4th, 423
     5th, 423
     5th, 454
   Lally 3rd, 448
   Wild Eyes, 455
   of Airdrie 8th, 364, 365, 424
     13th, 394, 438
     14th, 364
   Athol, 342
   Barrington, 455
     11th, 472
     24th, 472
   Buckingham, 372, 472
   Clarence 3rd, 432
     7th, 435
     10th, 432
     12th, 432
   Darlington 13th, 372
   Fordham, 344
   Geneva 4th, 366
     6th, 365

Duchess of Geneva 7th, 355, 409
    8th, 365, 420
    10th, 365, 420
    12th, 365, 418
  Glo'ster, 425
  Hillhurst 3rd, 366, 437
    5th, 366, 437
  Hindlip 3rd, 464
    4th, 465, 470
    7th, 465
  Holker 3rd, 376, 474, 477
    4th, 477
  Killhow, 372, 469
    2nd, 472
  Lancaster, 374, 475
  Laughton, 23rd, 476
  Leicester, 370, 459, 466
    3rd, 459
    6th, 459
    17th, 376
    18th, 476
    19th, 376, 476, 477
    23rd, 476
  Oneida, 365, 420
    3rd, 420
    4th, 420
    7th, 420
    8th, 421
    9th, 421
    10th, 421
  Ormskirk, 368, 451
    6th, 464
    7th, 464
  Orpines, 470
  Oxford, 458
    2nd, 458
  Rowfant, 466
  Surrey, 373, 473
    3rd, 374, 476
    4th, 473
    5th, 476
    6th, 476
  Thorndale, 434
    2nd, 434
    3rd, 435
    12th, 420
    13th, 420
  Underley 3rd, 453
  Vittoria 2nd, 459
  Wappenham, 463
  Waterloo, 442
    2nd, 442
  Wellington 6th, 452
    7th, 458

Duchess of Whittlebury 6th, 472
    9th, 472
    10th, 472
    11th, 472
    13th, 373, 472
    14th, 472
    15th, 472
    17th, 472
    18th, 472
    19th, 472
    20th, 472
    22nd, 475, 477
  York 5th, 471
    6th, 471
Duke (224), 66, 67
  (226), 150, 158, 160, 161, 164, 166, 168 n., 179
  2nd, 179
  Lally, 436
  of Airdrie 27th, 447
    Athol, 335
      2nd, 342
    Barrington 4th (30924), 420
      (39712), 446
      6th (33576), 432
      (39714), 446
      7th (39715), 446
      9th (44650) 452
    Brunswick, 465
    Cambridge (3637), 273, 275 n., 278
      (12742), 342, 354, 407
      2nd (3638), 402 n.
      (12743), 337
      3rd (5941), 298-302, 321, 322, 403 n., 406
    Carolina 2nd, 427
    Charmingland, 54th, 472
    Chatsworth 3rd, 375, 474
    Claro 3rd, 358, 360, 417, 428
    Cleveland, 250
    Collingham 2nd, 358, 410
    Connaught, 366, 367, 370, 429
    Cornwall 2nd, 450
      4th, 463
    Cumberland, 465
    Dentsdale 2nd, 424
      3rd, 424
    Fussbox, 412, 415
    Geneva, 349
      3rd, 355, 409
      8th, 360, 361, 418, 428
    Glo'ster, 340, 344
      3rd, 419, 424, 434, 438, 452

Duke of Glo'ster 5th, 455
    6th, 452
    7th, 438
  Hillhurst, 453
    2nd, 437
    3rd, 364, 366, 429, 441
  Holker, 7th, 460
  Kirklevington (25982), 410
    (33683), 424
  Leicester, 368, 455
    3rd, 370, 459
    4th, 459
    6th, 459
    11th, 475
    13th, 475
    18th, 476
  Lincoln, 286
  Norfolk, 405 n.
  Northumberland, 253-6, 261, 266-8, 271, 281, 283, 403
    2nd, 274, 401 n., 402-4
    3rd, 275 n.
    4th, 275 n., 404 n.
  Oneida, 364
    2nd, 365, 420
    4th, 421
    7th, 421
  Ormskirk 5th, 469
  Oxford 2nd, 304, 309, 311, 312, 334, 356 n.
    3rd, 309, 311, 334, 356 n.
    5th, 340, 341
    6th, 342, 343, 407
    17th, 410
    18th, 358, 420, 434
    19th, 414
    20th, 414, 445
    21st, 414
    22nd, 414
    23rd, 414
    24th, 364
    26th, 423, 434
    28th, 423
    29th, 423
    30th, 423
    31st, 444
    32nd, 442
    39th, 456
    44th, 445, 464
    45th, 446
    46th, 446
    60th, 460
    61st, 460
    63rd, 460
Duke of Oxford 65th, 460
    90th, 474
  Richmond, 334, 405
  Rosedale 8th, 449
    9th, 449
    12th, 373, 472
    19th, 467
    20th, 467
  Rothesay, 366, 436
  Saxony, 465
  Siddington 2nd, 476
  Somerset (9048), 303 n.
    (49485), 465
  Surrey 2nd, 373, 473
  Thorndale 3rd, 354
    4th, 345, 349, 358
    12th, 356, 409
  Tosca, 424
  Tregunter, 358, 413
    2nd, 358, 360, 415
    5th, 369
  Vittoria 10th, 471
    13th, 475
  Waterloo 3rd, 410
    11th, 473
  Wellington (3654), 272, 274, 305-7, 321
    4th, 418
    9th, 427
  Wetherby 5th, 445
  Wharfdale 3rd, 360
    5th, 358
  Whittlebury 2nd, 370
  Woodford, 447
  York (1941), 249, 255
    2nd, 404
    3rd, 315, 316, 334, 339
    4th, 331, 334, 340, 341
    5th, 335, 339
    6th, 335
    7th, 359
    8th, 415, 430
    9th, 372, 471
    10th, 471
Dunse, 401 n.
Durham ox: *see* Ketton ox
Dustie, 359

EARL OF BEVERLEY 2nd, 301 n., 405
  Darlington, 248
    2nd, 400
    3rd, 400
  Durham (9059), 402 n.
  Oxford, 428

Earl of Oxford (41487), 452
    3rd, 471
    10th, 421
  Southrop 116th, 375, 477
  Strafford, 418
Earl (Percy): see The Earl
Ebor, 334
Edward, 402 n.
Elmhurst Princess, 437, 443
Elvira, 35, 148, 232
Empress, 166
  of Oxford, 424
Enchanter, 182-4, 190, 191, 194, 197-9
Euclid, 302, 334, 405, 406

FAIR GWYNNE, 454
Fair Kirklevington, 444
Fairfax, 288
Fairy, 150, 182
Faith Gwynne, 451
Fame Gwynne, 435
Fancy Gwynne 2nd, 435
Fantail 5th, 414, 436
  6th, 430
  8th, 436
  9th, 436
  10th, 436
  11th, 436
Fantail's Duchess, 419
  2nd, 425, 433
Favourite Cow: see Lady Maynard
Favourite, 48, 54, 66, 203
Firby, 165
Fitz-Duke, 182 n.
Flavia Gwynne, 433
Fletcher Cow, 240, 336
Fleur-de-lys Gwynne, 442
Flighty Gwynne, 430
Flippant Gwynne, 442, 456
Flora (Emmerson's), 220 n.
  Gwynne, 446
Florence Worcester, 451
Flossy Gwynne, 433
  2nd, 433
  3rd, 433
Fluffy Gwynne, 430
Foggathorpe, 261, 265, 266, 284, 400 n.
  2nd, 285 n., 332
  4th, 332
  6th, 334
Foljambe, 35, 51, 52, 386
Fordham Duke of Oxford 4th, 447

Fortune, 68
  Gwynne, 451
Fragrant Gwynne, 442
Frederick, 213
Frosty Gwynne, 451
Fuchsia 9th, 427
  10th, 440
  12th, 428, 437
  13th, 429, 440
  14th, 429
  15th, 450
  of Hillhurst, 442
Fuchsia's Duchess, 440
  2nd, 440
  4th, 441
Furioso, 175, 176
Fusilier, 441

GAMBIER, 229, 246
Gaudy (Mason's), 191
  (Parrington's), 241
Gazelle 8th, 446
  25th, 441
  26th, 440
  29th, 440, 453
  32nd, 441
  37th, 439
  of Chamonix, 441
  Isis, 441
  Oxford, 442
General Sale, 285 n.
  Washington, 402 n.
Geneva's Kirklevington Duchess, 437
  Minstrel, 419, 433
  Rose of Sharon, 443
George (273), 157
  (3884), 275 n., 402 n.
Georgiana, 336
Georgie Clarence, 426
  Hillhurst, 426
  2nd, 426
  Sandgrove, 475
Glo'ster's Oxford, 344
Golden Pippin, 155, 169, 183, 190
Graceful Duchess, 430
Grand Duchess 1st, 337
  2nd, 337, 338
  5th, 347, 350, 408
  7th, 347, 348
  8th, 347, 358, 408
  9th, 347, 408
  10th, 347, 348
  11th, 347, 408
  12th, 347, 348

Grand Duchess 13th, 347, 348
  14th, 347, 348
  15th, 347
  17th, 347, 350-2, 370, 408
  18th, 347, 351, 408
  19th, 352, 408
  20th, 352, 408
  21st, 352, 408
  23rd, 366, 438
  28th, 369
  29th, 366, 438
  32nd, 461
  35th, 369
  37th, 461
  39th, 461, 466
  40th, 461
  41st, 462
  42nd, 462
  43rd, 462
  44th, 462
  46th, 462
  47th, 462
  48th, 462
  49th, 462
  51st, 462
  52nd, 462
  53rd, 462
  58th, 375, 476
  59th, 372, 470
  60th, 470
  66th, 376
  (Wild Eyes), 3rd, 411
    5th, 411
    6th, 414
  Carolina, 414
    2nd, 436
    3rd, 436
  Surmise, 414
  Acomb, 4th, 463
  of Athol, 416
    Babingley, 477
    Barrington, 415
      2nd, 432
      3rd, 432
    Barringtonia 4th, 439, 448
    Darlington 2nd, 463
    Geneva 4th, 421
      7th, 463
    Holker, 474
      2nd, 474
    Kirklevington, 447
    Oxford 5th, 359
      6th, 422
      7th, 359, 413

Grand Duchess of Oxford 11th, 422, 430
  12th, 359, 422, 446
  16th, 414
  18th, 414
  19th, 445
  21st, 445
  22nd, 445
  25th, 422
  28th, 422
  33rd, 459, 467
  39th, 460
  40th, 369, 456
  48th, 460
  49th, 460
  59th, 460
  60th, 460
  63rd, 460
  67th, 460
  69th, 460
  79th, 473
  80th, 474
  82nd, 474
  84th, 474
  89th, 474
  90th, 474
  102nd, 474
  107th, 474
  (Lord Braybrooke's), 5th, 449
    8th, 449
Grand Duke, 332, 334, 337, 338, 343
  2nd, 343
  3rd, 343
  4th, 352, 354, 408
  6th, 347
  9th, 347, 348
  10th, 347, 359
  13th, 347
  15th, 347, 348
  16th, 349, 353, 408
  17th, 353, 408
  25th, 366, 439
  29th, 366, 439
  31st, 366, 439
  37th, 466
  41st, 462
  44th, 462
  46th, 462
  47th, 464
  48th, 469
  51st, 374, 471, 475
  52nd, 474
  of Connaught, 471
    Geneva, 433

Grand Duke of Geneva 3rd, 463
    Kent 2nd, 424
        3rd, 471
    Lightburne 3rd, 418
    Waterloo, 432
    Weston 3rd, 352
Grand Princess of Lightburne 3rd, 446
Grandson of Bolingbroke : *see* Bolingbroke
Grey Bull, 65, 66, 201
Gwynne Duchess, 448
    Princess 8th, 448

HALTON (Canada), 307
Harefield Darlington, 437
    Gwynne, 437
Harlsonio, 285, 378
Harmless, 406
Harold, 170, 184
Hart cow, 336
Haughton, 35, 38, 51
Havering Waterloo 3rd, 455
Hawser Trunnion, 244
Hazel Eyes, 429
Hecatomb, 261-3, 378
Helen, 234
Hermit, 156-8, 225, 249
Hero, 273, 378
Heydon Rose, 345
    2nd, 449
    3rd, 449
    5th, 449
Hill's Red Bull, 51, 386
    White Bull, 44
Hilpa, 305, 306
His Grace, 162-7, 169, 170, 178, 182
Hollon, Little, 44
Honer, Young, 156, 183, 184, 190
Honey 62nd, 447
    64th, 448
Horsted Rose 4th, 375, 475
Hubback, 37, 46, 49-51, 59, 66, 194, 203, 228, 378, 387
    2nd, 195, 218, 227, 228, 245, 255, 391
    son of 2nd, 245
Hudson Cow, 336
Hudson (9228), 311, 356 n.

IMPERIAL OXFORD, 346, 347, 352
Ivanhoe, 203

JENNY (J. Booth's), 269
    (Parrington's), 241

Johanna, 53, 55, 168
John, 402 n.
Jolly's Bull, 47-9, 191, 298, 389
Julia, 243
Juliet, 244

KATE, 175 n., 177 (Mr. Pilley's Cow), 180, 182-4, 190, 194
Ketton Ox, 21, 54, 59, 136
Ketton (709), 68, 89, 129, 139, 140, 154, 162, 179, 191, 195
    2nd, 148, 154, 155
    3rd, 154, 169, 170, 178
    (346), 168, 169
King of Oxford, 468
Kirklevington, 336 n.
    by Short Tail, 401 n.
    12th, 410
    16th, 413, 423
    17th, 417
    18th, 425
    19th, 413
    20th, 428
    21st, 417
    22nd, 413
    24th, 413
    25th, 417, 431, 443
    26th, 431
    27th, 432
    Countess, 458
    Duchess 2nd, 425
        6th, 425
        7th, 431
        9th, 440
        12th, 425
        15th, 430, 446
        16th, 448
        21st, 451
        25th, 458
        26th, 458
    Empress 2nd, 447
    Lady (Green's), 458
        2nd, 458, 468
        3rd, 458, 466
        3rd (Allen's), 443
        7th (Allen's) 444
        10th (Leney's), 463
    Princess (Larking's), 440, 457
        3rd, 440
        4th, 440
        5th, 440, 458
        6th, 441, 453
    Queen 5th, 465
Knight of Oxford, 6th, 459

Knight of Oxford 8th, 463
  12th, 465

LADY, 48, 55
  (*H. B.* i. 359), 201
  Angelina, 438
    2nd, 443
    3rd, 452
  Ann 2nd, 453
  Ashton Wild Eyes 4th, 461
    5th, 461
    7th, 461
  Barrington, 240, 254
    3rd, 305-7, 405
    4th, 404 n.
    8th, 407, 413
    9th, 427, 439
    10th, 440
    11th, 431
  Bates, 407 *bis.*
    7th, 413
  Bates 4th (Fletcher), 420
    5th, 420
    6th, 420
    7th, 421
    8th, 420
  Bates (Hart), 436
    2nd, 421
  Bickerstaffe, 426
  Blanche, 407
    2nd, 468
    Rose, 468
  Blithfield 4th, 468
  Bright Eyes 2nd, 444
    3rd, 416, 445
    3rd (Holford's), 459
  Clarence Bates, 421
  Colling, 252
  Cumberland, 456
    2nd, 457
  Edith Bates, 453
  Elizabeth (Princess), 425
  Ellen Barrington, 411, 445
  Emily Darlington, 419
  Fuchsia, 427
    2nd, 427
    3rd, 427
  Geneva Waterloo, 430, 442
  Gracious, 447
  Hillhurst Bates, 442
  Hudson's Duchess 4th, 434
  Huntsland Barrington, 467
  Laura Barrington, 414, 423
  Mary, 416

Lady Maynard, 20, 46-8, 132, 389
  Mild Eyes, 431, 443, 448
  Montagu 3rd, 456
  of Oxford 8th, 356, 410
    13th, 438
    15th, 425
    16th, 425
  Oxford 4th, 354
    5th, 354, 358, 359, 409
    6th, 354
    8th, 468
    9th, 464
    10th, 468
    12th, 468
    Barrington, 453
    Bates (Slye's), 421
      (Lovatt's), 456
    Waterloo 5th, 372
  Rosedale Barrington, 467
    2nd, 467
    3rd, 467
  Sale of Putney, 419
  Sarah, 213
  Secret, 426
  Surmise, 437, 468
    7th, 468
  Sussex, 467
    2nd, 467
  Thorndale Bates, 434, 441
    2nd, 411, 416
    Glo'ster Bates, 439
  Tregunter 2nd, 441
    Bates, 414
  Turncroft Bates 8th, 461
  Underley Barrington 2nd, 467
  Walton 2nd, 434
  Warner, 457
  Waterloo, 417
    14th, 419
    15th, 419
    16th, 413
    18th, 431
    22nd, 425
    23rd, 436
    24th, 436
    25th, 419
    26th, 419
    27th, 419
    28th, 425, 443
    29th, 425
    30th, 448
  Surmise, 422
  Wellesley (Princess), 425, 453
    2nd, 438

Lady Wellesley 3rd, 452
    4th, 452
    Wild Eyes, 417, 431
      6th, 447
      7th, 432
      9th, 447
      10th, 447
      11th, 457
      12th, 447
      14th, 447
      Surmise, 442
    Worcester, 364
      3rd, 417
      4th, 420
      5th, 417, 428
      5th (Campbell's), 420
      6th, 417, 440
      7th, 417
      9th, 360, 417, 428
      10th, 417
      11th, 429
      12th, 429
      13th, 429
      14th, 429
      15th, 429
      17th, 450
      18th, 450
      20th, 450, 464
      21st, 450
      23rd, 453, 467
      26th, 374, 475
      Wild Eyes, 437
    York and Underley Bates, 448
Ladybird 6th, 413, 416
Laird, 68, 79, 89, 92, 119 n., 120, 122, 201
Lakeland's Bull, 34
Lally 7th, 417
    8th, 417
    9th, 417
    10th, 417
    13th, 417
    14th, 417
    15th, 360, 417
    16th, 417
    18th, 419
    19th, 432
    Barrington 2nd, 451
    Duchess 2nd, 431
    of Ellington, 456
      Littlehurst, 456
      Lymage, 456
Lame Bull (357), 52, 154, 168, 184
Lancaster, 155

Lawnsleeves, 157, 225, 235, 319, 394, 395
Leonidas, 337, 348
Leven, 170
Lightburne Winsome, 431
Lightburne's Duke of Oxford, 446
    2nd, 446
Lily, 48, 68
    (*H. B.* i. 379), 196
    (Leney's), 421
    Gwynne, 431
Locomotive, 262, 272-4
    2nd, 403 n.
Lord Barrington, 415
    Bolingbroke : *see* Bolingbroke
    Claro 2nd, 418
    Collingham, 412
    George (10439), 342
      Bentinck, 334, 405 n.
    Hardinge, 302
    Lunesdale Bates, 424
    of Garsdale, 424
      Oxford 5th, 415, 433
        6th, 421
      the Forth 2nd, 435
        Tweed, 454
    Oxford, 359
      2nd, 355
      7th, 433
      13th, 452
    Rosebery, 465
    Wild Eyes, 438
      5th, 411
Lovely Eyes, 463
Lupin, 242-4, 363

MAID OF KENT, 244
    Oxford, 344
      2nd, 420
      3rd, 420
      4th, 356, 409
      5th, 356, 360, 409, 417
      6th, 356, 409, 418
      7th, 356, 410, 427
      10th, 438
      11th, 438
      12th, 421
      (Lord Bective's), 453
Major (401), 201
Malibran, 262
Mantalini, 44
Maplewell 2nd, 455
    4th, 455
    Darling, 455

Marcella, 44
Marchioness, 175
  2nd, 416
  3rd, 416, 423
  5th, 416
  12th, 453
  13th, 453
  of Barrington, 437
    Bickerstaffe, 425
    Kirklevington 7th, 375, 476
    Oxford, 416
      2nd, 416
      3rd, 369, 429, 455
      2nd (Vivian's), 374, 475
      3rd (Taylor's), 376
    Worcester, 440
Marian, 243
Marius, 305
Marlborough, 265
Marmaduke, 353, 407
Marquis (407), 107
  3rd, 416, 424
  4th, 424
  6th, 424
  10th, 454
  of Hillhurst, 442
    Oxford, 429, 442
      2nd, 450
    Worcester 2nd, 429
      8th, 450
Marske, 35, 149, 155, 156, 159
  Young, 197, 199, 202
Mary, 59
Master Wild Eyes, 432
Masterman's Bull, 9, 37, 46, 387
Matchem, 224, 225, 362
  Cow, 219, 220, 223-6, 248, 250, 251
    Young : see Oxford Premium Cow
Matchless (Chaloner's), 201 n.
  (Wynyard), 234
Matilda, 229
Maud Waterloo, 450
Meg, 23
Meteor (11811), 305-7, 321
  Young, 305
Meteora, 156, 183, 184, 190, 194
Mild Eyes, 357
  4th, 437
Minstrel 5th, 435
  7th, 447
  8th, 447
  Gwynne, 451
  Maid, 435
  Princess, 446

Miranda, 191
Miss Fairfax, 288
  Jewsbury, 244
  Lax, 48, 52, 390
Mrs. Motte, 315
Monarch, Young, 221 n., 223
Morning Star, 359
Moss Rose, 191
  (Stephenson's), 237
  (Combe's), 345, 353, 359, 407, 408, 410, 411
  2nd, 408, 411
Musical (Angerstein's) 16th, 436
  18th, 436
  19th, 436
  2nd (Hetherington's), 445
  18th (Cheney's), 447

NECKLACE (Eryholme), 47, 48, 390
  (by Priam), 278-82
Nell Gwynne, 233, 394
Nettle, 336 n.
Noble, 234
Nonpareil (Barmpton), 149
  (Norfolk's dam), 245, 253, 285
Nonsuch Bull, 401 n.
Norfolk, 245, 247, 253, 270, 285, 286
Norman Willy, 175, 176
North Star, 151
Northallerton Cow, 336
Nosegay, 336 n.

O'CALLAGHAN's Son of Bolingbroke, 55
Olive Leaf, 406
Omega, 262, 400 n.
Oneida Gwynne, 448
Orlando, 175, 176
Orpines Grand Duchess of Oxford, 470
Oxford (4636), 267
Oxford 1st (Premium Cow), 225, 250, 251, 261, 262, 266-9, 272, 278, 279
  2nd, 304, 309, 331, 332
  4th, 310, 312, 332
  5th, 332, 344
  6th, 309, 333, 340, 344
  7th, 310, 312
  9th, 333
  10th, 333
  11th, 334, 340, 407
  12th, 334
  13th, 334, 344
  14th, 334

Oxford 15th, 340, 341
  16th, 340, 407
  17th, 344
  20th, 344
  26th, 465
  29th, 465
  30th, 465
  Augusta, 456
  Beau, 412
    7th, 448
  Belle 5th, 448
  de Vere 5th, 462
    9th, 479
    10th, 471
  Duchess, 416
    2nd, 428
  Duchess of Killhow 3rd, 469
    4th, 457
    7th, 469
    8th, 469
    9th, 469
  Duke of Calthwaite 1st, 469
    2nd, 470
    of Killhow 2nd, 454
      4th, 457
      6th, 463
      7th, 463
  Grand Duchess 2nd, 473
    3rd, 473
  Helen, 471
  Helene, 475
  Ida, 426
  Kate, 475
  Mary, 369
  Rose, 359
    2nd, 359
    5th, 445, 449
    10th, 445
    12th, 445, 450
Oxford's Baronet, 432
  Waterloo 4th, 418

PALM-FLOWER, 164
Papaver, 426
  2nd, 426
Parentia, 169 n.
Paroquet, 233
Parrington, 334
Patriot, 191
Peach Blossom 8th, 425
  10th, 429
  13th, 439
  14th, 439
Pearlie Gwynne, 443

Pedigree, 141 n.
Peerless (Wiseton), 168
  (Wynyard), 234
Peg Woffington, 141 n., 233
Penelope, 169
Percy (7326), 404
  (9472), 303 n., 404 n.
Philippa, 202
  2nd, 202 n., 218, 219
Phœnix, 48, 52, 54, 132
  (*H. B.* i., 435), 201
Phœnomenon, 395
Piercing Eyes, 444
Pilot (Wynyard), 232, 234, 395
Polly Gwynne, 10th, 435
  13th, 454
  14th, 455
Premium, 164, 168
Priestess, Young, 156, 169 n.
Prince : *see* St. Alban's
Prince Albert (4781), 400 n.
  of Brailes, 452
  Northumberland, 269
  Waterloo 5th, 471
Princess, 35, 231, 232, 234, 235, 366,
    388, 389
  (Parrington's), 241
  4th, by Baron Oxford 3rd, 431
    by Royal Cambridge, 446
  6th, 424
  8th, 453
  Airdrie, 456
  Alexandra 2nd, 438
  Gwynne (by 5th Duke of Oxford),
    433
    (by Royal Cambridge, 434)
    2nd, 423
  of Blythe, 451
    4th, 464
    5th, 464
  Geneva, 425
    2nd, 425
  Lightburne, 431, 434
    2nd, 431
    3rd, 446
  the Valley, 458
  Oneida 2nd, 449
  Sale, 424
    4th, 470
  Victoria 5th, 430
    7th, 415
    10th, 430
    11th, 430
    12th, 430

Priscilla, 169
Punch, 53, 193, 194
Puritan, 345

QUEEN OF OXFORD 3RD, 459
   7th, 473

RADICAL, 168
Red Daisy, 156, 228
   Duke 2nd (8460), 299 n., 301 n., 406 n.
   Princess, 236, 284 n.
Red Rose (Manfield), 34, 53
   by Phenomenon, 233
   by Marmaduke, 353, 408
   Bull, 400
     3rd, 408
     4th, 408
     11th, 248-50, 252 : *see* Rose of Sharon
   13th, 255, 261, 263, 265, 266, 272, 273 : *see* Cambridge Premium Rose
   of Avon, 443
     Balmoral, 429
     Benledi, 450
     Killigrey, 450
     Rodil, 454
     Severn, 437
     Strathearne, 450
     Strathspey, 450
     the Isles, 428
     Virginia, 450
Red Rover, 329, 334
Refiner, 334, 406
Reformer by Pancake, 244
Regent, 149
Remus, 107
Retriever, 335
Rival Gwynne, 435
Roan Duke (8486), 404 n.
Robson's (William) Bull, 35
Romeo's Oxford, 344
Romulus, 107
Rosabella, 167, 168 (Lancaster heifer), 169
Rosalie, 425
Rosalind (by Exmouth), 202
   (by Hector), 203
   4th, 403
Rosanne, 227
   (J. Colling's), 261, 265
Rose of Lightburne 3rd, 437
   Sharon, 255

Rosebud, 170, 284 n.
Rosette, 149, 161, 162, 168, 179
Ross Gwynne, 435
Rowfant Duke of Leicester, 466
   Grand Duchess, 466
   Oxford, 446
   Prince, 473
   Thorndale Rose, 466
Royal Cambridge, 410
   Chester, 359
   Cumberland, 410
   Gwynne, 437
     2nd, 456
Ruby, 190

ST. ALBAN'S, 213, 214
St. John, 68, 69, 107, 133, 151, 191, 204, 288
Sally (*H. B.* i. 487), 169
   dam of Nonpareil, 245, 389
Scots Fusilier, 429
Second Hubback : *see* Hubback 2nd
Secrecy, 444
Secret, 404 n.
Senator, 311, 356
Sequoiah, 456
Shorthorns by Saladin, 240
   4th, 261, 262, 266
Short Tail, 255, 256
Siddington 1st, 410, 416, 440
   4th, 410, 423
   5th, 415
   7th, 411, 416, 423
   7th (Slye's), 434
   9th, 415, 430
   10th, 415
   12th, 435
   13th, 426
   15th, 426, 440
   16th, 426, 440
   Duchess, 434
   Grand Duchess, 435
   Kirklevington, 441
Silence, 407
Silky Gwynne, 433
Sir Glo'ster Barrington, 450
   Lawrence Barrington, 420
   Leoline, 177
   Maurice, 373, 473
   Oliver, 107
   Patrick Spence, 244
   Thomas Fairfax, 270, 285, 286, 288, 289
Sirloin, 163, 167

Skipton Bridge Bull, 248, 251, 252, 284 n.
Smith's (Jacob) Bull, 48
Snowdon's Bull, 36
Snowdrop, 99
Sockburn, Old, 44, 47
    (6509), 404 n.
Sparkles, 150-3, 165, 168
Sparkling Eyes, 429
Spot, 150-3, 165, 168
Standish, 285, 378
Starling, 48, 315, 390
Stephenson's Bull (1480), 242-4, 363
Stick-a-Bitch cow, 21, 54
Strawberry (by Jolly's Bull), 48, 389
Strawberry, Young, 46
Strelly, 284, 378
Studley Bull, 33, 34
Styford, 53, 59, 63-5, 86 n., 165, 179
Sunflower, 243
Surmise, 407
Surplice, 156, 225
Surprise, 407
  by 8th Duke of Geneva, 428
  3rd, 421
  4th, 422
  5th, 428
Symmetry, 284

TEESWATER, 248, 249, 252, 254
Thalia, 168 n.
The Beau, 339, 407
  Beauty, 345, 407
  Briar, 407
  Colonel, 285 n.
  Doctor, 249
  Duke (8676), 324
  Earl (646), 166, 167, 179, 184, 200, 202
  Lord of Hainault, 403 r.
Thorndale Duchess, 421
  2nd, 422
  Duke 3rd, 467
  Grand Duke 2nd, 435
  Premium Rose 5th, 376
    6th, 376
    7th, 376
  Rose, 345, 353, 408
    2nd, 408
    6th, 466
    7th, 368, 449
    9th, 449
    12th, 449
    13th, 449, 464

Thorndale Rose 14th, 449
  22nd, 466
  23rd, 466
  24th, 466
  25th, 466
  26th, 466
  28th, 467
  29th, 467, 470
  30th, 467
Tinsley, 312
Tippoo, 169
Togston (5487), 224 n.
Tragedy, 233
Tregunter Gwynne, 434
Trifle, 141 n., 233
  (Parrington's), 241
Trinket (Wynyard), 232
  (Parrington's), 241
Tripes, 34, 35, 386, 387
Tulip (Wynyard), 233
  (Parrington's), 241
Turcoman 4th, 454
  12th, 470
Turncroft Belle of Oxford, 461
  Duchess of Oxford, 461

UNDERLEY OXFORD, 468
  Princess, 453
Usurer, 336, 358

VAN AMBURG, 400 n.
Velvet Eyes, 426, 436
Venus, 48, 54, 59
Venusta, 443
Violet Eyes, 443
  2nd, 443
Violet Gwynne, 442
Viscount Oxford, 443
  3rd, 368, 455
  Worcester, 441
Viscountess Barrington, 451
  2nd, 451
  Oxford 4th, 459
    5th, 459, 466
    of Ruddington 4th, 476
Visigoth, 424

WALLFLOWER, 243
Walton, 296
Wastell's Bull, 35
  heifer, 22, 238 n.
Water Belle, 437
  Duchess 3rd, 432
    4th, 432
  Flower, 428

Water Girl, 426
  Lass 3rd, 419
  Lily, 429, 443
    2nd, 443, 464
  Witch, 316
Wateringbury Rose 3rd, 463
Waterloo (2816), 231, 237-9, 247, 318, 391-3
  Cow, 237-9, 253
  2nd, 239, 248
  4th, 332
  5th, 286, 322 n.
  9th, 333
  10th, 333
  11th, 333
  12th, 333
  20th, 346
  24th, by Grand Duke 3rd, 409
  24th, by 22nd Duke of Oxford, 435, 441
  26th, by Duke of Geneva, 418
  28th, by 3rd Grand Duke, 410
  30th, by 3rd Duke of Wharfdale, 418
  32nd, by 7th Duke of York, 410
  32nd, by 11th Grand Duke, 426, 444
  33rd, by 11th Grand Duke, 418
  34th, by Wallace, 434
  34th, by Royal Cambridge, 435
  36th, by Grand Duke 11th, 418
  36th, by Earl of Eglinton, 412
  37th, by Oxford Beau, 418, 441
  37th, by Royal Cambridge
  38th, by Grand Duke 11th, 426
  38th, by Earl of Eglinton, 412
  39th, by Earl of Eglinton, 412
  40th, by Grand Duke 20th, 427
  43rd, by Grand Duke of Oxford, 461
  46th, by Beau of Oxford 4th, 454, 467
  47th, by Beau of Oxford 3rd, 455, 468
  59th, by North Pole, 471
  Beau, 471
  Belle, 444, 452
    7th, 465
  Bienvenue, 430
    2nd, 462, 468
    3rd, 468
  Earl Sockburn 3rd, 376
  of Oneida, 441
  Rose, 444
  Victor, 469

Wayward Eyes, 453
Wellington, by Wynyard, 232, 234, 395
  (680), 235, 389
  4th, 409
Wellingtonia, by 3rd Duke of Thorndale, 409, 411
  2nd, by 3rd Duke of Thorndale, 409
  2nd by 12th Duke of Thorndale, 411
  3rd, by Grand Duke of Kent, 428
  4th, by Grand Duke 4th, 409
  4th, by Duke of Oneida 6th, 421
  5th, by Grand Duke 4th, 409
  6th, by Grand Duke 4th, 409
  7th, by Grand Duke 4th, 409
Western Comet, 228, 229, 288
Weston's Gwynne, 437
Wetherby Duchess, 435
Wharfdale Rose, 357
White Bull : see Colling's (Robert) White Bull
  Heifer that Travelled, The, 21
  Rose (Daisy), 200, 201
    (Edwards's), 262
    (Secret), 284, 404
Whitelegs, 244
Wild Boy of the Valley, 452
  Duchess, 413
    4th, 428
  Duchess of Geneva, 419
    2nd, 419
    3rd, 425, 438
    4th, 452
    of Glo'ster, 438
      2nd, 452
    of York, 438
  Duke of Geneva, 432
  Erin, 444
Wild Eyes, 240, 242
  5th, 332
  7th, 332
  8th, 332
  14th, 332
  15th, 332
  16th, 309, 332
  17th, 309, 332
  19th, 332
  21st, 333
  22nd, 333
  23rd, 333
  24th, 333
  24th, by 4th Duke of Oxford, 416

Wild Eyes 25th, 333
  26th, 333
  27th, 333
  28th, 329 n., 333
  28th, by Lord Lally 3rd, 415
  29th, 333
  29th, by Earl of Glo'ster, 417
  30th, 334
  30th, by 7th Duke of York, 417
  32nd, 436
  33rd, 455
  34th, 455
  37th, 437
  42nd, 439
  43rd, 439
  44th, 440
  Duchess, 412, 428
    2nd, 412, 416
  Lad, 418
  Lassie, 417
    3rd, 456
    4th, 457
Wild Eyebright, 429, 444
  Flower Duchess 5th, 418
  Maid, 430
    2nd, 454
  Musical, 441
  Prince 7th, 451
  Princess 2nd, 419
    2nd, by 8th Duke of Geneva, 428
    3rd, 419
    4th, 449
    6th, 463
  Rose, 428
  Winsome 3rd, 444, 458
    4th, 457, 461
    8th, 457
    9th, 457
Wildair by Meteor, 183
  (Parrington's), 241
Wildfire, 415
Wildflower, 243
Willie (11050), 406 n.
Windermere, 410
  2nd, 454
  3rd, 468
Windsor, 151
Winifred, 99
Winsome 2nd, 413, 423, 433
  3rd, 434
  5th, 422
  6th, 427
  7th, 414
  8th, 414

Winsome 9th, 414, 423, 444
  10th, 414
  11th, 414
  12th, 422, 442
  14th, 422
  16th, 422
  17th, 422
  18th, 422
  20th, 445
  25th, 460
  28th, 460
  29th, 460
  29th, by Duke of Rosedale 6th, 466
  Beauty, 451
    4th, 451
    6th, 451
  Care, 458
  Colleen, 444, 458
  Duchess, 434, 453
  Eyes 3rd, 364, 428
  Isis, 444
  Lass, 441
  Oxonia, 444
  Wild Eyes 1st, 427
    2nd, 427, 454
    3rd, 427
    5th, 454
    8th, 463
  Winnie, 444
Winsomedale, 423
  2nd, 424, 440
Wisdom 3rd, 457
  4th, 457
Wizard, 405 n.
Wonderful Ox : see Ketton Ox
Woodbine, 244
Worcester Rose, 465
  2nd, 465
Wynyard, by Phœnomenon, 234
  Young : see Young Wynyard

YARBOROUGH, 154
Yarborough's dam, 154
Yarm, 199
Yellow Rose, 156, 169 n., 183, 190
Yorkshireman, 400 n.,
Young Marske : see Marske, Young
  Matchem Cow : see Matchem Cow, Young
  Monarch : see Monarch, Young
  Star, 151
  Waterloo (2817), 248, 254
  Wynyard (or Young Wellington), 225, 235-7, 362, 392

2 L

HOC SATIS ARMENTIS